Quantum Symmetries: Tensor Categories, TQFTs, and Vertex Algebras

Contemporary Mathematics

813

Centre de Recherches Mathématiques Proceedings

CRM

Quantum Symmetries: Tensor Categories, TQFTs, and Vertex Algebras

Thematic Program
Quantum Symmetries: Tensor Categories, Topological Quantum Field Theories, Vertex Algebras
October 10 – November 4, 2022
Centre de Recherches Mathématiques, Université de Montréal, Québec, Canada

Thomas Creutzig
Julia Plavnik
David Ridout
Editors

AMS American Mathematical Society

2020 *Mathematics Subject Classification.* Primary 17B69, 18M20, 57R56, 81T40.

Library of Congress Cataloging-in-Publication Data

Names: Thematic Program on Quantum Symmetries: Tensor Categories, Topological Quantum Field Theories, Vertex Algebras (2022 : Montréal, Québec) | Creutzig, Thomas, 1980– editor. | Plavnik, Julia, 1985– editor. | Ridout, David, 1976– editor.
Title: Quantum symmetries : tensor categories, TQFTs, and vertex algebras : thematic program on quantum symmetries: tensor categories, topological quantum field theories, vertex algebras, October 10–November 4, 2022, Centre de Recherches Mathématiques, Université de Montréal, Montréal, Québec, Canada | Thomas Creutzig, Julia Plavnik, David Ridout, editors.
Description: Providence, Rhode Island : American Mathematical Society ; Montréal, Québec, Canada : Centre de Recherches Mathematiques, [2025] | Series: Contemporary mathematics, 0271-4132 ; volume 813 | Includes bibliographical references.
Identifiers: LCCN 2024035050 | ISBN 9781470473617 (paperback) | ISBN 9781470478643 (ebook)
Subjects: LCSH: Symmetry (Mathematics) | Quantum theory. | AMS: Nonassociative rings and algebras – Lie algebras and Lie superalgebras – Vertex operators; vertex operator algebras and related structures. | Manifolds and cell complexes – Differential topology – Topological quantum field theories. | Quantum theory – Quantum field theory; related classical field theories – Two-dimensional field theories, conformal field theories, etc..
Classification: LCC QA174.7.S96 T44 2025 | DDC 539.7/25–dc23/eng/20250114
LC record available at https://lccn.loc.gov/2024035050

Contemporary Mathematics ISSN: 0271-4132 (print); 1098-3627 (online)

DOI: https://doi.org/10.1090/conm/813

∞ The paper used in this book is acid-free and falls within the guidelines
established to ensure permanence and durability.
Visit the AMS home page at `https://www.ams.org/`

10 9 8 7 6 5 4 3 2 1   30 29 28 27 26 25

Contents

Preface vii

Affine W-algebras
JETHRO VAN EKEREN 1

OPEs of rank two W-algebras
JUSTINE FASQUEL 47

Fusion rings acting on vertex operator algebras: First steps
ANDREW RIESEN 61

Ground-state degeneracy of twisted sectors of Conway moonshine SCFT
ALISSA FURET and THEO JOHNSON-FREYD 99

Combinatorial principles that enforce a group structure
SIMON LENTNER 117

On near-group centers and super-modular categories
ERIC C. ROWELL, HANNAH SOLOMON, and QING ZHANG 129

Fixed-point-free fusion automorphisms
ANDREW SCHOPIERAY 151

$(3+1)D$ topological orders with only a $\mathbb{Z}_2$-charged particle
THEO JOHNSON-FREYD 175

Grothendieck-Verdier duality in categories of bimodules and weak module functors
JÜRGEN FUCHS, GREGOR SCHAUMANN, CHRISTOPH SCHWEIGERT, and SIMON WOOD 211

Action of subgroups of the mapping class group on Heisenberg homologies
CHRISTIAN BLANCHET, MARTIN PALMER, and AWAIS SHAUKAT 235

Flat principal 2-group bundles and flat string structures
DANIEL BERWICK-EVANS, EMILY CLIFF, LAURA MURRAY, APURVA NAKADE, and EMMA PHILLIPS 257

Categorical center of higher genera
JIN-CHENG GUU 303

Twisted homogeneous racks over the alternating groups
JOSEPH VULAKH 341

Preface

This is the proceedings of a concentration month devoted to "Quantum symmetries" that took place at the Centre de Recherches Mathématiques, Université de Montréal, from October 10 to November 4, 2022. It was one of six such months that constituted the Thematic Semester entitled "Symmetries: Algebras and Physics". Our month featured a one week international conference, three weeks of lecture courses (some introductory and some advanced), problem sessions and additional research seminars. We thank our coorganisers Anne Moreau and Kolya Reshetikhin for their assistance with putting together this program.

Our interpretation of "Quantum symmetries" is that it serves as a common language to connect many important fields of modern mathematics and mathematical physics. This language is that of tensor categories. These mathematical structures were introduced by MacLane and Bénabou in the 1960s and was invigorated in the 1980s by the works of Drinfeld, Jones, Moore and Seiberg, Reshetikhin and Turaev, Verlinde and Witten, among others.

Classically, the symmetries of an object are encoded by a group. This paradigm changed rapidly in both mathematics and physics with the emergence of, and subsequent advances in, fields such as conformal field theory (vertex algebras) and integrability (quantum groups). In their simplest appearances, quantum symmetries arise when considering the space of states of a physical system, typically built from a braided tensor category of modules over some symmetry group or algebra. They moreover imbue the category with beautiful additional features, leading one to the study of what are now called spherical, fusion and modular categories (to name a few). These structures generalise the notion of symmetry far beyond that of a group action.

The particular foci of our concentration month included the theory of abstract tensor categories (*eg.*, fusion, modular and non-semisimple generalisations), as well as applications to Hopf algebras, vertex algebras, invariants of links and 3-manifolds, subfactors, topological quantum field theories and (of course) physics. This theory and its myriad of applications are now absolutely essential parts of the modern mathematician's and mathematical physicist's toolkit. We hope that this collection of reviews and articles will not only be valuable to these researchers, but also inspire them to build even further on the results obtained to date.

We conclude with a brief overview of the articles that appear in this proceedings.

Jethro van Ekeren reviews W-algebras, which are constructed from affine vertex algebras via quantum Hamiltonian reduction associated to a nilpotent element inside the underlying Lie algebra. Depending on the nilpotent element, there are exceptional levels for which the simple W-algebra is strongly rational. In these cases, he also discusses the modular tensor categories formed by their representations.

Justine Fasquel gives explicit formulae for the operator product algebra of W-algebras associated to Lie algebras of rank two. Such formulae are in general hard to obtain and they can (and are) used to find exceptional isomorphisms between vertex algebras.

Andrew Riesen takes the first steps toward an understanding of hypergroup actions on vertex operator algebras. He studies this in the case of strongly rational vertex operator algebras and, in particular, shows that any strongly rational vertex operator subalgebra can be characterized as the fixed points of a hypergroup. These findings are illustrated for the exceptional extensions of the rational affine vertex algebras of $\mathfrak{sl}_2$.

Alissa Furet and Theo Johnson-Freyd consider ground states in twisted sectors of the Conway moonshine holomorphic superconformal field theory. This vertex operator superalgebra has the Conway group, a sporadic group, acting as automorphisms. Accordingly, there is one twisted sector for each conjugacy class and they examine if each sector possesses a single parity or if it has both even and odd states.

Simon Lentner studies a combinatorial problem for finite abelian groups that is motivated by classifying R-matrices for quantum groups.

Modular data is expected to be realized by some category. **Eric C. Rowell, Hannah Solomon and Qing Zhang** accomplish this for some previously unrealized super-modular data by modifying the Drinfeld centers of near-group fusion categories associated with certain abelian groups.

Andrew Schopieray studies fusion ring automorphisms, leaving only the trivial element fixed. He proves that a variety of classical results on fixed-point-free automorphisms of finite groups are true in the generality of fusion rings.

Theo Johnson-Freyd gives proofs of the classification of $(3+1)D$ topological orders with only a $\mathbb{Z}_2$-charged particle, both in the even and odd case.

Jürgen Fuchs, Gregor Schaumann, Christoph Schweigert and Simon Wood discuss under which circumstances the distributors between the two tensor products in Grothendieck–Verdier categories are isomorphisms.

Christian Blanchet, Martin Palmer and Awais Shaukat study the action of subgroups of the mapping class group on Heisenberg homologies.

Daniel Berwick-Evans, Emily Cliff, Laura Murray, Apurva Nakade and Emma Phillips construct, for a weak 2-group $\mathcal{G}$, a bicategory of flat $\mathcal{G}$-bundles over differentiable stacks as a localization of a functor bicategory.

Jin-Cheng Guu is computing the Crane-Yetter theory in (co)dimension 2. Due to the similarity with the Drinfeld center, he calls this the categorical center of higher genera.

Joseph Vulakh makes progress towards a general classification of twisted homogeneous racks of type D by proving that several families of twisted homogeneous racks arising from alternating groups are of type D.

Contemporary Mathematics
Volume **813**, 2025
https://doi.org/10.1090/conm/813/16281

Affine W-algebras

Jethro van Ekeren

ABSTRACT. These are expanded and revised notes for a minicourse entitled "Affine W-algebras", which took place as part of the thematic month "Quantum Symmetries" at the Centre de Recherches Mathématiques in Montréal, Canada in October 2022. The first few sections consist of rapid introductions to vertex algebras, affine Kac-Moody algebras and their integrable and admissible modules, and the homological BRST procedure for quantum Hamiltonian reduction. The affine W-algebras are defined using these ingredients. The remainder of the notes treats the structure and representation theory of the exceptional affine W-algebras, with emphasis on modular tensor categories of their representations.

There are various motivations to study affine W-algebras: they appear in the analysis of completely integrable models, in conformal field theory, representation theory in general, and the geometric Langlands program in particular. In keeping with the theme of the program, our goal will be to understand something of the interesting modular tensor categories that arise as categories of modules over *exceptional* affine W-algebras.

Our story begins with the affine Kac-Moody algebra $\widehat{\mathfrak{g}}$ associated with a finite dimensional simple Lie algebra $\mathfrak{g}$. For each non negative integer k there is a finite set of integrable highest weight $\widehat{\mathfrak{g}}$-modules $L_k(\lambda)$ of level k, whose characters are given explicitly by the Weyl-Kac character formula. This formula exhibits the normalised characters as quotients of theta functions, with the remarkable consequence that they are modular functions [**32**]. The explicit modular transformations were determined by Kac and Peterson [**35**], so that in particular

$$\chi_\lambda(-1/\tau) = \sum_{\lambda'} S_{\lambda,\lambda'} \chi_{\lambda'}(\tau),$$

where S is the matrix

$$S_{\lambda,\lambda'} = \frac{i^{|\Delta_+|}}{|P/(k+h^\vee)Q^\vee|^{1/2}} \sum_{w\in W} \varepsilon(w) e^{-\frac{2\pi i}{k+h^\vee}\left(w(\lambda+\rho),\lambda'+\rho\right)}. \tag{0.1}$$

2020 *Mathematics Subject Classification.* Primary 17B08, 17B67, 17B69, 81R10.
Key words and phrases. Vertex algebras, tensor categories.
I would like to thank the organisers of "Quantum Symmetries" for the opportunity to participate in this stimulating program. For the duration of the program I held a CRM-Simons Scholar position, and I am also very grateful for this support. During the writing I have been supported by grants FAPERJ 201.445/2021, CNPq 310576/2020-2, and Serrapilheira 2023-0001.

Here P and $Q^\vee$ are the weight lattice and coroot lattice of $\mathfrak{g}$, respectively, $h^\vee$ is the dual Coxeter number and ρ the Weyl vector, and sum runs over the Weyl group W of $\mathfrak{g}$. The finite index set for the highest weights λ, λ' here is P_+^k, defined in (2.5) below.

It later emerged that the integrable $\widehat{\mathfrak{g}}$-module $L_k(\mathfrak{g}) = L_k(0)$ is an example of a new kind of algebraic structure known as a vertex algebra [**10**]. In general a vertex algebra is said to be rational if its modules are completely reducible into direct sums of irreducible ones, and if there are finitely many of the latter up to isomorphism. In fact the affine vertex algebra $L_k(\mathfrak{g})$ is rational, and the finitely many irreducible $L_k(\mathfrak{g})$-modules are precisely the $\widehat{\mathfrak{g}}$-modules $L_k(\lambda)$ considered above [**23**].

Theorems of Zhu [**52**] and Huang [**25**] assert, respectively, that if a vertex algebra is rational (and satisfies a few other more technical hypotheses) then appropriately normalised characters of its irreducible modules converge to modular functions, and its category of modules is a modular tensor category whose S-matrix coincides with the modular S-matrix of its characters. Here the tensor product of modules is similar in spirit to, and generalises, the Kazhdan-Lusztig tensor product [**41**] [**26**]. In particular the decomposition of tensor products can be determined through the Verlinde formula [**51**] [**50**], and we see in particular that the formula (0.1) is key to understanding the tensor product structure of the category of $L_k(\mathfrak{g})$-modules.

Affine W-algebras are defined through BRST reduction of affine vertex algebras, more precisely through a special case of the general BRST procedure known as Drinfeld-Sokolov reduction [**14**] [**17**] [**30**]. Varying the level k and the choice of a nonzero nilpotent element $f \in \mathfrak{g}$ which enters the construction, results in a bewildering multitude of algebras. It turns out that the reduction of $L_k(\mathfrak{g})$ for $k \in \mathbb{Z}_+$ vanishes for all nonzero f, but interesting examples are obtained when we pass to so called admissible levels k. There is now a delicate interplay between the dimension of the nilpotent orbit $G \cdot f$, the size of the denominator q of k, and the "size" of the affine W-algebra; for each q there being a distinguished nilpotent orbit $G \cdot f = \mathbb{O}_q$ for which the corresponding BRST reduction of $L_k(\mathfrak{g})$ has a chance of being rational. The reduction is known as an exceptional W-algebra, and the eventual goal of these notes is to state results on rationality and representation theory of exceptional W-algebras.

Characters of irreducible $\widehat{\mathfrak{g}}$-modules $L_k(\lambda)$ of admissible highest weight λ, and their modular transformation properties have been determined by Kac and Wakimoto, generalising (0.1) above [**37**]. The S-matrices of exceptional W-algebras can in turn be computed, and are given by curious variants of (0.1). For example if $\mathfrak{g}$ is a Lie algebra of type D or E, and α_* the root of $\mathfrak{g}$ associated with the trivalent node of its Dynkin diagram, then the following matrix appears as the S-matrix of a modular tensor category of representations of an exceptional W-algebra:

$$S_{\nu,\nu'} = \sum_{w(\alpha^*) \in \Delta_+} \varepsilon(w) \frac{\langle w(\alpha_*), x \rangle}{\langle \alpha_*, x \rangle} e^{-\frac{2\pi i}{q}(y(\nu),\nu')}.$$

Here the sum runs over half of the Weyl group, and $x \in \mathfrak{h}$ is not orthogonal to α_*, but otherwise arbitrary. Details may be found in the text.

We conclude this introduction with a few indications of references. For the basic theory of vertex algebras we recommend the books [**33**] and [**20**], as well as [**36**] (where the construction of the Zhu algebra is described). The introduction to [**13**] also contains a short and clear account of various equivalent definitions. For Kac-Moody algebras we recommend Kac's book [**31**], and for background on category $\mathcal{O}$ methods, such as translation functors, we recommend [**27**]. For BRST reduction in general there is the classic article of Kostant and Sternberg [**45**], and Drinfeld-Sokolov reduction of vertex algebras is described in the book [**20**] for principal nilpotent f. Arakawa's lecture notes [**6**] are a good source, and beyond this there are the primary sources: articles of Kac and Wakimoto and collaborators [**21**, **30**, **37**–**40**] and Arakawa [**2**–**5**].

1. Introduction to vertex algebras

Apart from fixing notation and recording key results and background material, the aim of this section is to dispel the common notion that vertex algebras are difficult to work with and calculate in. To do this we present the λ-bracket calculus as soon as possible, and use it to do a number of example calculations.

We begin by considering the ring

$$H = \mathbb{C}[x_1, x_2, x_3, \ldots]$$

of polynomials in countably many variables. In fact we are not so interested in the ring structure of H, instead we consider H as a vector space, and study the following collection of linear endomorphisms of H: For each integer $n > 0$ we set

$$h_{-n} = x_n, \quad \text{and} \quad h_n = n\frac{\partial}{\partial x_n},$$

and we set $h_0 = 0$. It is easy to confirm that

$$h_m h_n - h_n h_m = m\delta_{m,-n} I_H.$$

Therefore H is a module over the following Lie algebra $\widehat{\mathfrak{a}}$, which is known as the oscillator Lie algebra. As a vector space

$$\widehat{\mathfrak{a}} = \mathbb{C}[t^{\pm 1}] \oplus \mathbb{C}K,$$

and the Lie bracket is defined by

$$[f(t), g(t)] = \operatorname{Res}_t f'(t)g(t)\, K, \quad K \text{ central.}$$

Here the symbol Res_t is the formal residue, which acts by extracting the coefficient of t^{-1}.

The representation of $\widehat{\mathfrak{a}}$ in H is given by $t^m \mapsto h_m$ for all $m \in \mathbb{Z}$ and $K \mapsto I_H$. This $\widehat{\mathfrak{a}}$-module is very well known, and is often referred to as the Fock module, following its appearance in quantum field theory.

The first basic idea of vertex algebras is to manipulate $\{h_n \mid n \in \mathbb{Z}\}$ and similar collections of endomorphisms, using generating functions. In this case we set

$$h(z) = \sum_{n \in \mathbb{Z}} h_n z^{-n-1},$$

noting that at this point the peculiar choice of exponent of z can be thought of as purely conventional. Now we can set about recasting the commutation relations between the h_n in terms of $h(z)$. Indeed

$$\begin{aligned}[h(z), h(w)] &= \sum_{m,n\in\mathbb{Z}} [h_m, h_n] z^{-m-1} w^{-n-1} \\ &= \sum_{m,n\in\mathbb{Z}} m\delta_{m,-n} I_H z^{-m-1} w^{-n-1} \\ &= \sum_{m\in\mathbb{Z}} m I_H z^{-m-1} w^{m-1}.\end{aligned}$$

This last expression is reminiscent of the series expansions

$$i_{z,w}\frac{1}{(z-w)^2} = \sum_{m=0}^{\infty} m z^{-m-1} w^{m-1} \quad \text{and} \quad i_{w,z}\frac{1}{(z-w)^2} = -\sum_{m=0}^{\infty} m z^{m-1} w^{-m-1}$$

convergent, respectively, in the domains $|w| < |z|$ and $|z| < |w|$. Here $i_{z,w}$ and $i_{w,z}$ denote the operations of expansion in positive powers of w, and of z, respectively. These operations can alternatively be characterised as the unique embeddings of fields

$$\mathbb{C}((z))((w)) \leftarrow \mathbb{C}((z,w)) \rightarrow \mathbb{C}((w))((z))$$

compatible with the inclusions of $\mathbb{C}[z^{\pm 1}, w^{\pm 1}]$.

To put the calculation of $[h(z), h(w)]$ into a more general context, it is useful to introduce some notation. The formal delta function is by definition the series

$$\delta(z,w) = \sum_{m\in\mathbb{Z}} z^{-m-1} w^m = i_{z,w}\frac{1}{z-w} - i_{w,z}\frac{1}{z-w}. \tag{1.1}$$

We may then write

$$[h(z), h(w)] = \partial_w \delta(z,w) I_H. \tag{1.2}$$

REMARK 1.1. There are good reasons for applying the name “delta function” to (1.1). By way of justification we do no more than recall the identity

$$\delta(x) = \frac{1}{2\pi i} \lim_{\varepsilon\to 0^+} \left(\frac{1}{x - i\varepsilon} - \frac{1}{x + i\varepsilon} \right)$$

of distributions on $\mathbb{R}$.

Generalising the simple observation that

$$(z-w)\delta(z,w) = 0,$$

we have

$$(z-w)^j \partial_w^{(n)} \delta(z,w) = \begin{cases} \partial_w^{(n-j)} \delta(z,w) & j \le n \\ 0 & j > n, \end{cases}$$

(where $\partial^{(j)}$ denotes $\partial^j/j!$). In particular (1.2) implies

$$(z-w)^2 [h(z), h(w)] = 0.$$

This is abstracted into the following fundamental notion:

DEFINITION 1.2. A formal series in two variables

$$D(z,w) = \sum_{m,n\in\mathbb{Z}} D_{m,n} z^{-m-1} w^{-n-1}$$

is said to be local if there exists a positive integer N such that

$$(z-w)^N D = 0.$$

The following proposition shows that the existence of expansions like (1.2) is a general property of locality.

PROPOSITION 1.3. *The formal series $D(z,w)$ satisfies $(z-w)^N D(z,w) = 0$ if and only if it is expressible in the form*

$$D(z,w) = \sum_{j=0}^{N-1} c_j(w)\partial_w^{(j)}\delta(z,w),$$

for some collection of formal series $c_j(w)$ in one variable.

There is actually a simple formula for the coefficients $c_j(w)$ in Proposition 1.3. If $D(z,w)$ is a local series then

$$c_j(w) = \mathrm{Res}_z (z-w)^j D(z,w).$$

We now say that a pair of series $a(w)$ and $b(w)$ forms a local pair if $[a(z), b(w)]$ is local. In this sense our example of $h(w)$ forms a local pair with itself. In general, a local pair $(a(w), b(w))$ of series yields a finite list of new series $c_j(w)$ via the expansion

$$[a(z), b(w)] = \sum_{j=0}^{N-1} c_j(w)\partial^{(j)}\delta(z,w).$$

The expansion on the right hand side is known as an "operator product expansion", or OPE.

We would like to adopt a perspective in which the fundamental objects are the series themselves, and suppress from our notation the explicit dependence on variables z and w. We thus rewrite the OPE above, purely formally, as

$$[a_\lambda b] = \sum_{j=0}^{N-1} c_j \lambda^{(j)}.$$

For instance if we write $h = h(w)$ and $\mathbf{1} = I_H$ then we have

$$[h_\lambda h] = \lambda \mathbf{1}.$$

To get to the definition of vertex algebra, one further ingredient is required.

DEFINITION 1.4. Let V be a vector space. We say that the formal series

$$a(w) = \sum_{n\in\mathbb{Z}} a_{(n)} w^{-n-1}, \qquad a_{(n)} \in \mathrm{End}(V),$$

is a quantum field on V if, for every $v \in V$, there exists N such that $a_{(n)}v = 0$ for all $n > N$. In other words if $a(z)v \in V((z))$.

It is easy to see that $h(w)$ is a quantum field on H.

The significance of the quantum field property is that it permits the definition of a formal analogue of the normal ordering procedure of quantum field theory. Specifically, for $a(w)$ and $b(w)$ quantum fields on a vector space V, their normally ordered product is by definition

$$: a(w)b(w) := a(w)_+ b(w) + b(w)a(w)_-$$

where

$$a(w)_+ = \sum_{n<0} a_{(n)} w^{-n-1} \quad \text{and} \quad a(w)_- = \sum_{n\geq 0} a_{(n)} w^{-n-1}.$$

PROPOSITION 1.5. *If $a(w)$ and $b(w)$ are two quantum fields on a vector space V, then the normally ordered product $:a(w)b(w):$ is a well defined and is another quantum field on V.*

Happily the quantum field property is also preserved upon passage to OPE coefficients.

PROPOSITION 1.6. *If $a(w)$ and $b(w)$ are two quantum fields on a vector space V which form a local pair, then the series $a(w)_{(j)}b(w)$, for $j \geq 0$, is also a quantum field on V.*

In fact the two constructions are more similar than they look. It is straightforward to verify

$$:a(w)b(w) := \mathrm{Res}_z \left(a(z)b(w)i_{z,w}(z-w)^{-1} - b(w)a(z)i_{w,z}(z-w)^{-1} \right),$$

so that it makes sense to write $:a(w)b(w) := a(w)_{(-1)}b(w)$. In fact the product

$$a(w)_{(n)}b(w) = \mathrm{Res}_z \left(a(z)b(w)i_{z,w}(z-w)^n - b(w)a(z)i_{w,z}(z-w)^n \right)$$

makes sense for any pair of quantum fields $a(w)$ and $b(w)$, and is again a quantum field. On the other hand these n^{th} products, for $n \leq -2$, can easily be recovered from the normally ordered product together with the natural operation of differentiation of a quantum fields:

$$(\partial_w a(w))_{(n)}b(w) = -na(w)_{(n-1)}b(w). \tag{1.3}$$

The relation (1.3) implies the following relation satisfied by the λ-bracket:

$$[(\partial a)_\lambda b] = -\lambda [a_\lambda b]. \tag{1.4}$$

Other relations such as

$$[a_\lambda (\partial b)] = (\partial + \lambda)[a_\lambda b] \quad \text{and} \quad :ab:=:(\partial a)b: + :a(\partial b): \tag{1.5}$$

are also easily established.

We would also like formulas allowing us to compute λ-brackets of normally ordered products of quantum fields, and indeed to manipulate arbitrary combinations of λ-brackets, normally ordered products and derivatives. Vertex algebras provide a context in which we can do this.

DEFINITION 1.7. A vertex algebra consists of a vector space V, an operator $T : V \to V$, a nonzero vector $|0\rangle$ such that $T|0\rangle = 0$, and a set $\mathcal{F}$ of quantum fields on V, such that

- For all $a(w) = \sum_{n \in \mathbb{Z}} a_{(n)} w^{-n-1} \in \mathcal{F}$, we have
$$[T, a(w)] = \partial_w a(w),$$
- All the quantum fields in $\mathcal{F}$ are mutually local,
- V is generated from $|0\rangle$ by $\mathcal{F}$, in the sense that the terms
$$a^1_{(n_1)} \cdots a^k_{(n_k)} |0\rangle$$
together span V, as $a^1(w), \ldots, a^k(w)$ run over $\mathcal{F}$ and $n_1, \ldots, n_k$ run over $\mathbb{Z}$.

The notions of quantum field and mutual locality are incorporated into the definition of vertex algebra. The axiom about generation of V by the coefficients of elements of $\mathcal{F}$ is inessential in the sense that it can be achieved by replacing V by the generated subspace. The two further ingredients in the definition are the vacuum vector $|0\rangle$, and the translation operator T.

It is now possible to set up a "calculus of quantum fields" as outlined above, for the elements of $\mathcal{F}$ and their iterated λ-brackets and normally ordered products. Alongside the formulas (1.4) and (1.5), we can also prove

$$[b_\lambda a] = -[a_{-\lambda-\partial} b]$$

and

$$[a_\lambda :bc:] =: [a_\lambda b]c: + :b[a_\lambda c]: + \int_0^\lambda [[a_\lambda b]_\mu c]\, d\mu.$$

Let us return to our original example H, check that it is indeed a vertex algebra, and do some computations. We equip the vector space H with $|0\rangle = 1$ and $\mathcal{F} = \{h(w)\}$. The operator $T : H \to H$ is determined uniquely by the conditions $T|0\rangle = 0$ and $[T, h(w)] = \partial_w h(w)$, and is well defined since its action can be written down explicitly relative to the basis of H consisting of monomials. It is an easy exercise to check that

$$T = \sum_{j\in\mathbb{Z}_{\geq 1}} h_{-1-j}h_j.$$

So we have a vertex algebra H, which we shall refer to as the Heisenberg vertex algebra.

As an example computation in H, let us consider the quantum field

$$L = \frac{1}{2} :hh:$$

and try to compute $[L_\lambda L]$. Using the rules given above, we confirm that

$$[L_\lambda h] = \partial h + \lambda h,$$

and proceed from there to compute

$$[L_\lambda L] = \partial L + 2\lambda L + \frac{\lambda^3}{12}.$$

Upon decoding the λ-bracket notation, we find

$$[L(z), L(w)] = \partial_w L(w)\delta(z,w) + 2L(w)\partial_w\delta(z,w) + \frac{1}{12}\partial_w^3\delta(z,w), \tag{1.6}$$

and hence, setting $L(z) = \sum_n L_n z^{-n-2}$, we obtain the relations

$$[L_m, L_n] = (m-n)L_{m+n} + \delta_{m,-n}\frac{m^3-m}{12}I_H \tag{1.7}$$

between the endomorphisms $L_n \in \mathrm{End}(H)$. In fact we have just constructed a representation in H of the Virasoro Lie algebra, which is by definition

$$\mathrm{Vir} = \bigoplus_{n\in\mathbb{Z}} \mathbb{C}L_n \oplus \mathbb{C}C$$

with Lie brackets given by C central, and (1.7) with C in place of I_H. In general a Vir-module M in which $C \mapsto cI_M$ is said to be of central charge c, so the Vir-module H is of central charge 1.

We leave it as an exercise to check that the quantum field $B = L + \beta\partial h$ satisfies

$$[B_\lambda B] = \partial B + 2\lambda B + \frac{\lambda^3}{12}(1 - 12\beta^2),$$

so its coefficients endow H with a representation of Vir of central charge $1 - 12\beta^2$.

If we examine the definition of the field $L(w)$ we see that its coefficients L_n are given by the formulas

$$L_n = \frac{1}{2}\sum_{j\in\mathbb{Z}} h_{n-j}h_j, \quad n \neq 0, \quad \text{and} \quad L_0 = \frac{1}{2}h_0^2 + \sum_{j\in\mathbb{Z}_{\geq 1}} h_{-j}h_j.$$

Thus the conversion of H from a $\widehat{\mathfrak{a}}$-module into a Vir-module did not come from anything like a homomorphism between these Lie algebras, we had to step outside the world of Lie algebras to achieve it.

Let us remark at this point, that the coefficient $L_{-1} : H \to H$ of our Virasoro quantum field H happens to coincide with the translation operator $T : H \to H$.

Definition 1.7 is a little peculiar for what purports to be the definition of a class of algebras. But it can be transformed into an equivalent but rather different looking form. We describe the reformulation briefly in the following paragraphs.

Let $(V, |0\rangle, T, \mathcal{F})$ be a vertex algebra. A result known as Dong's lemma guarantees that locality is preserved by n^{th} products. We may thus form the envelope $\overline{\mathcal{F}}$ of $\mathcal{F}$ by repeatedly adding all n^{th} products of existing quantum fields, and pass to the "completed" vertex algebra $(V, |0\rangle, T, \overline{\mathcal{F}})$.

The condition $[T, a(w)] = \partial_w a(w)$ on a quantum field $a(w)$ guarantees that $a(w)|0\rangle \in V[[w]]$. In particular there is a well defined linear map

$$a(w) \mapsto a(w)|0\rangle|_{w=0}, \qquad \overline{\mathcal{F}} \to V.$$

A key result in the basic theory of vertex algebras is that this map is a bijection. We thus denote by $a \mapsto Y(a, w)$ the inverse of this map, then it turns out that

$$Y(|0\rangle, w) = I_V, \quad \text{and} \quad Y(a, w)_{(n)}Y(b, w) = Y(a_{(n)}b, w).$$

Expanding out the latter identity in powers of w yields the following Borcherds identity, valid in any vertex algebra V, for all $a, b, c \in V$ and all $m, k, n \in \mathbb{Z}$;

$$\begin{aligned} &\sum_{j\in\mathbb{Z}_+} \binom{m}{j} (a_{(n+j)}b)_{(m+k-j)}c \\ &\quad = \sum_{j\in\mathbb{Z}_+} (-1)^j \binom{n}{j} \left(a_{(m+n-j)}b_{(k+j)}c - (-1)^n b_{(n+k-j)}a_{(m+j)}c\right). \end{aligned} \tag{1.8}$$

Finally a vertex algebra can equivalently be defined in terms of the Borcherds identity, i.e.,

DEFINITION 1.8. A vertex algebra is a vector space V, with a nonzero vector $|0\rangle \in V$ and a collection of bilinear operations $V \times V \to V$, indexed by $\mathbb{Z}$, denoted $a_{(n)}b$, such that

- for all $a, b \in V$ there exists N such that $a_{(n)}b = 0$ for all $n \geq N$,
- writing $Y(a, w) = \sum_{n\in\mathbb{Z}} a_{(n)}w^{-n-1}$, we have $Y(a, w)|0\rangle \in V[[w]]$ for all $a \in V$,
- $Y(|0\rangle, w) = I_V$ and $Y(a, w)|0\rangle|_{w=0} = a$, and
- for all $a, b, c \in V$ and $n, m, k \in \mathbb{Z}$ the identity (1.8) holds.

As for associative algebras, commutative algebras, Lie algebras, etc., there is an obvious extension of the notion of vertex algebra to that of vertex superalgebra. Vertex superalgebra play an important role in the theory, appearing for example as vertex algebra analogues of Clifford algebras.

We now present a simple example of a vertex superalgebra: the charged free fermions. Let N be a finite dimensional vector space, and consider the sum $S = N \oplus N^*$ together with the canonical symmetric bilinear form given by pairing

$$\langle \varphi, n \rangle = \langle n, \varphi \rangle = \varphi(n), \qquad \text{for } n \in N,\ \varphi \in N^*.$$

We convert S into a purely odd vector superspace, so that $\langle \cdot, \cdot \rangle$ is now a super-skewsymmetric pairing. Consider fields $a = a(z)$, one for each $a \in S$, with λ-bracket defined to be

$$[a_\lambda b] = \langle a, b \rangle .$$

More precisely we set

$$F^{\mathrm{ch}}(N) = U(\widehat{S}) \otimes_{U(\widehat{S}_+)} \mathbb{C}|0\rangle$$

where

$$\widehat{S} = t^{1/2} S[t^{\pm 1}] \oplus \mathbb{C}K, \qquad [at^m, bt^n] = \delta_{m,-n} \langle a, b \rangle K, \qquad K \text{ central},$$

and the action of $\widehat{S}_+ = t^{1/2} S[t] \oplus \mathbb{C}K$ on $\mathbb{C}|0\rangle$ is given by $at^m = 0$ for all $a \in S$ and $K = 1$. It can be confirmed that $F^{\mathrm{ch}}(N)$ is a vertex superalgebra.

Now suppose that σ is a finite dimensional representation of a Lie algebra $\mathfrak{g}$ in our vector space N, let $\{e_1, \ldots, e_r\}$ be a basis of N and $\{\varphi_1, \ldots, \varphi_r\}$ the dual basis of N^*. We form the free fermion vertex algebra $F^{\mathrm{ch}}(N)$ as above, and now for each $x \in \mathfrak{g}$ we define a quantum field

$$F^x = \sum_{i=1}^{r} : (\sigma(x) e_i)\, \varphi_i :$$

on $F^{\mathrm{ch}}(N)$. It can be checked by explicit calculation with the λ-brackets that

$$[F^a_\lambda F^b] = F^{[a,b]} + \lambda\, (a, b)_N,$$

where

$$(a, b)_N = \mathrm{Tr}_N\, \sigma(a) \sigma(b)$$

is the trace form. If we take the collection of fields $F^x(z)$ for $x \in \mathfrak{g}$, and the subspace of $F^{\mathrm{ch}}(N)$ generated from $|0\rangle$ by these fields, then we obtain a vertex algebra, canonically associated to $\mathfrak{g}$ and its representation in N. In Section 2 we will examine a general construction of an "affine" vertex algebra associated with a Lie algebra $\mathfrak{g}$ and an invariant bilinear form on $\mathfrak{g}$.

We now very briefly discuss cohomology of vertex superalgebras. Let V be a vertex superalgebra and let $Q \in V$. It follows from the Borcherds identity that the operator

$$Q_{(0)} : V \to V$$

satisfies

$$[Q_{(0)}, a_{(n)}] = (Q_{(0)} a)_{(n)}, \qquad \text{for all } a \in V,\ n \in \mathbb{Z}.$$

Now, there are natural notions of vertex subalgebra and vertex ideal (and quotient vertex algebra). We observe that $\ker(Q_{(0)}) \subset V$ is a vertex subalgebra. Now suppose that Q is odd and satisfies $Q_{(0)} Q = 0$, then a short computation reveals that $\mathrm{Im}(Q_{(0)}) \subset \ker(Q_{(0)})$ is an ideal. An odd element Q as above thus gives a notion of homology algebra $H(V, Q) = \ker(Q_{(0)}) / \mathrm{Im}(Q_{(0)})$.

Later we will become familiar with the notion of *conformal vertex algebra*, for now we just note that if V is a conformal vertex algebra, then the operator $Q_{(0)} : V \to V$ will preserve conformal weight if $\Delta(Q) = 1$.

2. Affine vertex algebras and the Zhu algebra

Let $\mathfrak{g}$ be a finite dimensional Lie algebra over $\mathbb{C}$, and $\phi : \mathfrak{g} \times \mathfrak{g} \to \mathbb{C}$ an invariant bilinear form. The associated affine Lie algebra is, by definition,

$$\widehat{\mathfrak{g}} = \mathfrak{g}[t, t^{-1}] \oplus \mathbb{C}K$$

with Lie bracket

$$[af, bg] = [a, b]fg + \phi(a, b) \operatorname{Res}_t f'(t)g(t)K, \qquad K \text{ central.} \tag{2.1}$$

The oscillator algebra $\widehat{\mathfrak{a}}$ is recovered as the special case $\mathfrak{g} = \mathbb{C}$ with the bilinear form $\phi(1, 1) = 1$. The Fock space H of $\widehat{\mathfrak{a}}$ is generalised by the construction of the vacuum $\widehat{\mathfrak{g}}$-module as follows. Set

$$V^\phi(\mathfrak{g}) = U(\widehat{\mathfrak{g}}) \otimes_{U(\mathfrak{g}[t]+\mathbb{C}K)} \mathbb{C}|0\rangle.$$

Here $\mathbb{C}|0\rangle$ is a one dimensional vector space furnished with an action of the subalgebra $\mathfrak{g}[t] + \mathbb{C}K \subset \widehat{\mathfrak{g}}$ by $K \mapsto 1$ and $\mathfrak{g}[t] \mapsto 0$. The vacuum module $V^\phi(\mathfrak{g})$ is now equipped with quantum fields $x(z) = \sum_{n\in\mathbb{Z}} xt^n \, z^{-n-1}$, one for each $x \in \mathfrak{g}$, and these generate a vertex algebra structure.

For $x \in \mathfrak{g}$ we write $x_n \in \operatorname{End}(V^\phi(\mathfrak{g}))$ as a shorthand for the action of xt^n, and we write $x_{-1}|0\rangle \in V^\phi(\mathfrak{g})$ simply as x. Comparing these conventions with those of Section 1 we see that $x_{(n)} = x_n$ for all $n \in \mathbb{Z}$ for $x \in \mathfrak{g}$. See also the remarks following Definition 2.1 below.

The construction of fields $F^x(z)$ in the free fermions vertex superalgebra $F^{\text{ch}}(N)$ of a $\mathfrak{g}$-module N, may now be interpreted as the construction of a homomorphism of vertex algebras

$$V^\phi(\mathfrak{g}) \to F^{\text{ch}}(N),$$

where ϕ is the trace form of N.

The construction of $V^\phi(\mathfrak{g})$ is most commonly carried out in the following special case. Let $\mathfrak{g}$ be a finite dimensional simple Lie algebra, $(\cdot, \cdot) : \mathfrak{g} \times \mathfrak{g} \to \mathbb{C}$ the normalised bilinear form (so that $(\theta, \theta) = 2$ for θ the highest root of $\mathfrak{g}$), and $\phi(x, y) = k \cdot (x, y)$. We then write $V^k(\mathfrak{g})$ without ambiguity.

Let $\{a^1, \ldots, a^m\}$ be a basis of $\mathfrak{g}$. By the PBW theorem, a basis of $V^k(\mathfrak{g})$ is provided by the monomials

$$a^{i_1}_{(-n_1)} a^{i_2}_{(-n_2)} \cdots a^{i_s}_{(-n_s)} |0\rangle \tag{2.2}$$

where $n_1 \geq n_2 \geq \ldots \geq n_s \geq 1$ and $1 \leq i_k \leq m$ for each k, and if $n_k = n_{k+1}$ then $i_k \geq i_{k+1}$. A visual impression of $V^k(\mathfrak{sl}_2)$ appears in Figure 1.

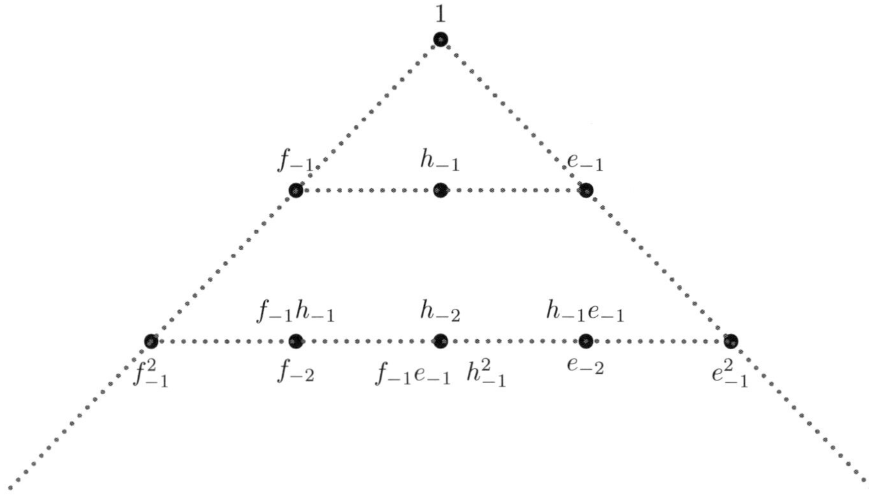

FIGURE 1. Sketch of the vertex algebra $V^k(\mathfrak{sl}_2)$

As usual we let $\mathbf{1}$ denote the quantum field I_V on $V = V^k(\mathfrak{g})$. To verify mutual locality of the fields $a(w)$ introduced already, it suffices to convert the commutation relation (2.1) into a relation between fields $a(z)$ and $b(w)$, which is then seen to be equivalent to the following λ-bracket relation

$$[a_\lambda b] = [a, b] + k(a, b)\lambda\mathbf{1}.$$

We continue to assume $\mathfrak{g}$ is simple. Let $h^\vee$ denote the dual Coxeter number of $\mathfrak{g}$, and let $\{a_i\}$ and $\{b_i\}$ be bases of $\mathfrak{g}$ dual with respect to $(,)$. If $k \neq -h^\vee$ then the Sugawara construction yields a vector

$$L = \frac{1}{2(k+h^\vee)} \sum_i : a_i b_i :$$

which satisfies Virasoro algebra relations, i.e.,

$$[L_\lambda L] = \partial L + 2\lambda L + \frac{\lambda^3}{12} c\, \mathbf{1},$$

where in this case the central charge is given by

$$c = \frac{k \dim(\mathfrak{g})}{k + h^\vee}.$$

Virasoro vectors in vertex algebras turn out to be important in general for the development of the geometric side of the theory.

DEFINITION 2.1. We say the vertex algebra V is *conformal* if it contains a vector L, the modes of whose associated quantum field

$$Y(L, w) = L(w) = \sum_{n\in\mathbb{Z}} L_n w^{-n-2}$$

satisfy the relations

$$[L_m, L_n] = (m-n)L_{m+n} + \delta_{m,-n}\frac{m^3 - m}{12} c\, I_V,$$

such that $T = L_{-1}$, and such that V is decomposed as $V = \bigoplus_{n\in\mathbb{Z}_+} V_n$ into finite dimensional eigenspaces of L_0. The eigenvalue n of an element a of V_n is called the conformal weight of a and denoted $\Delta(a)$.

As a straightforward consequence of the axioms of a vertex algebra, we find that

$$a_{(n)} : V_\alpha \to V_{\alpha+(\Delta-n-1)}$$

for $a \in V_\Delta$. We thus write $a_n = a_{(n+\Delta-1)}$, so that

$$a_n : V_\alpha \to V_{\alpha-n}.$$

The vertex algebra $V^k(\mathfrak{g})$, with its Sugawara vector L, is conformal, and the conformal weight grading is the vertical grading implicit in Figure 1 above. More precisely the conformal weight of the vector (2.2) is $n_1 + \cdots + n_s$.

The Borcherds identity can be rewritten in terms of the conformal grading on modes, and becomes

$$\begin{aligned}\sum_{j\in\mathbb{Z}_+} \binom{m+\Delta(a)-1}{j} (a_{(n+j)}b)_{n+m+k}c \\ = \sum_{j\in\mathbb{Z}_+} (-1)^j \binom{n}{j} \left(a_{m+n-j}b_{k+j}c - (-1)^n b_{n+k-j}a_{m+j}c\right).\end{aligned} \tag{2.3}$$

Now we discuss modules over vertex algebras. To avoid grappling with the several technically inequivalent definitions of V-module that arise in the non conformal case, we suppose from the outset that V is conformal.

DEFINITION 2.2. Let V be a conformal vertex algebra. A (positive energy) V-module is a direct sum of graded vector spaces of the form

$$M = \bigoplus_{\alpha-\alpha_0\in\mathbb{Z}_{\geq 0}} M_\alpha,$$

equipped with an action

$$a \mapsto Y^M(a,z) = \sum_{n\in\mathbb{Z}} a_n^M z^{-n-\Delta(a)}$$

satisfying (2.3) for all $a, b \in V$ and $c \in M$.

Let us think about the restriction of (2.3) to the lowest graded piece $c \in M_{\text{low}} = M_{\alpha_0}$. In this piece we have $a_n^M = 0$ for all $n > 0$. Set $m = 1$, $k = 0$ and $n = -1$ in Borcherds identity. It becomes

$$\sum_{j\in\mathbb{Z}_+} \binom{\Delta(a)}{j} (a_{(j-1)}b)_0^M c = a_0^M b_0^M c.$$

This motivates us to define

$$a * b = \sum_{j\in\mathbb{Z}_+} \binom{\Delta(a)}{j} a_{(j-1)}b$$

in V. Similarly we discover that, if

$$a \circ b = \sum_{j\in\mathbb{Z}_+} \binom{\Delta(a)}{j} a_{(j-2)}b$$

then $(a \circ b)_0^M = 0$ in the restriction to M_{low}. This motivates us to define the quotient

$$\text{Zhu}(V) = \frac{V}{V \circ V}, \qquad \text{with the induced product } a * b.$$

A number of miracles now occur. The subspace $V \circ V$ is really an ideal with respect to the product $*$. The quotient is an associative algebra with unit $|0\rangle$. Furthermore the assignment $M \mapsto M_{\text{low}}$, while not so strong as an equivalence of categories, is bijective between the set of irreducible positive energy V-modules, and the set of irreducible left Zhu(V)-modules [**52**]. For N an irreducible left Zhu(V)-module we denote by $\mathbb{L}(N)$ the unique irreducible positive energy V-module for which $\mathbb{L}(N)_{\text{low}} \cong N$.

The Zhu algebra of $V^k(\mathfrak{g})$ is isomorphic to the universal enveloping algebra $U(\mathfrak{g})$ (no matter what the value of k). Roughly speaking, this is because

$$a \circ v = a_{(-2)}v + a_{(-1)}v,$$

so the spanning monomials (2.2) of $V^k(\mathfrak{g})$ all reduce to monomials of the form

$$a^{i_1}_{(-1)} a^{i_2}_{(-1)} \cdots a^{i_s}_{(-1)} |0\rangle, \tag{2.4}$$

which are then identified with corresponding monomials

$$a^{i_1} a^{i_2} \cdots a^{i_s} \in U(\mathfrak{g}).$$

Actually things are not quite so simple: to achieve an isomorphism $\text{Zhu}(V^k(\mathfrak{g})) \to U(\mathfrak{g})$ not just of vector spaces, but of algebras, the monomial (2.4) must be identified with $a^{i_s} a^{i_{s-1}} \cdots a^{i_1}$. It is a good exercise to verify this, using (2.3) with $m = 0$, $k = 1$. See [**23**].

DEFINITION 2.3. A $\widehat{\mathfrak{g}}$-module M is said to be smooth if for all $a \in \mathfrak{g}$ and $x \in M$, there exists $N \in \mathbb{Z}_+$ such that $at^n \cdot x = 0$ for all $n \geq N$.

If we wish to try and build a $V^k(\mathfrak{g})$-module structure on a $\widehat{\mathfrak{g}}$-module M, by setting $a(z) = \sum_{n \in \mathbb{Z}} at^n z^{-n-1}$, then clearly it is a prerequisite that M be smooth. On the other hand it is sufficient. See [**36**] (also [**46**]) for the following theorem and its proof.

THEOREM 2.4. *There is an equivalence between the category of $V^k(\mathfrak{g})$-modules and the category of smooth $\widehat{\mathfrak{g}}$-modules of level k.*

If the level k is irrational then $V^k(\mathfrak{g})$ is simple, and that's about all there is to the story. If, on the other hand, k is rational, then it is possible to pass to a simple quotient vertex algebra

$$L_k(\mathfrak{g}) = V^k(\mathfrak{g})/N_k.$$

Here $N_k \subset V^k(\mathfrak{g})$ is the maximal proper vertex algebra ideal which, in this case, is the same thing as the maximal proper $\widehat{\mathfrak{g}}$-submodule. It is a very interesting question to ask which $V^k(\mathfrak{g})$-modules descend to $L_k(\mathfrak{g})$-modules. If we content ourselves with the irreducible positive energy modules, then it is enough to work with the Zhu algebra, and in general we have

$$\text{Zhu}(L_k(\mathfrak{g})) = \text{Zhu}(V^k(\mathfrak{g})/N_k) = U/I_k$$

where I_k denotes the two-sided ideal in Zhu(V) generated by the image of N_k.

For instance in $V^1(\mathfrak{sl}_2)$ the maximal proper ideal is generated by $e^2_{(-1)}|0\rangle$. Writing $U = U(\mathfrak{sl}_2)$, we obtain

$$\text{Zhu}(L_1(\mathfrak{sl}_2)) \cong U/(e^2).$$

It is now not hard to compute $U/(e^2) \cong \mathbb{C} \oplus \mathrm{Mat}_{2\times 2}(\mathbb{C})$ as associative $\mathbb{C}$-algebras. Indeed $e^2 \in I$ implies

$$[f, e^2] = -(he + eh) = -2eh - 2e \in I,$$

and after applying $\mathrm{ad}(f)$ again we discover that $2ef - (h^2 + h) \in I$, etc. Continuing in this way we eventually conclude that all elements of U can be reduced modulo I to combinations of the 5 basis elements indicated in Figure 2. We can also compute the products of these elements explicitly.

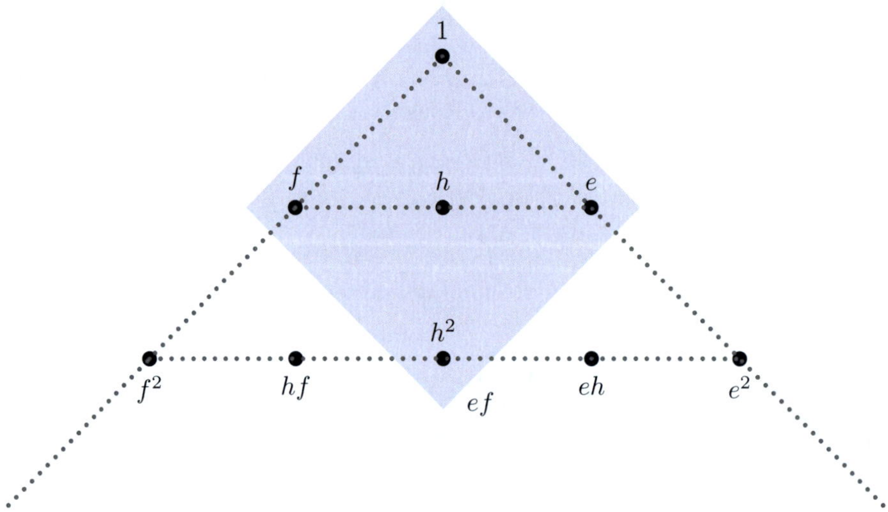

FIGURE 2. Sketch of U with a basis of $U/(e^2)$ highlighted.

The algebra $U/(e^2)$ has two irreducible modules; the one dimensional trivial $\mathfrak{sl}_2$-module $L(0)$, and the two dimensional irreducible $\mathfrak{sl}_2$-module $L(\varpi)$ of highest weight ϖ. Here ϖ is the fundamental weight dual to the single simple root α of $\mathfrak{sl}_2$.

There are thus two irreducible positive energy $L_1(\mathfrak{sl}_2)$-modules. Since $L_1(\mathfrak{sl}_2)$ is simple, it is among these, and indeed

$$\mathbb{L}(L(0)) = L_1(\mathfrak{sl}_2).$$

Continuing along these lines, the following important theorem is established.

THEOREM 2.5 ([**23**, Theorem 3.1.3]). *Let $k \in \mathbb{Z}_+$. The complete set of irreducible positive energy $L_k(\mathfrak{g})$-modules is given by $\mathbb{L}(L(\lambda))$ as λ ranges over*

$$P_+^k = \{\lambda \in P_+ \mid \langle \lambda, \theta^\vee \rangle \le k\}. \tag{2.5}$$

The modules $\mathbb{L}(L(\lambda))$ are irreducible $\widehat{\mathfrak{g}}$-modules, and it is easy to see they are highest weight $\widehat{\mathfrak{g}}$-modules in fact. We discuss the theory of highest weight modules in greater depth in the next section.

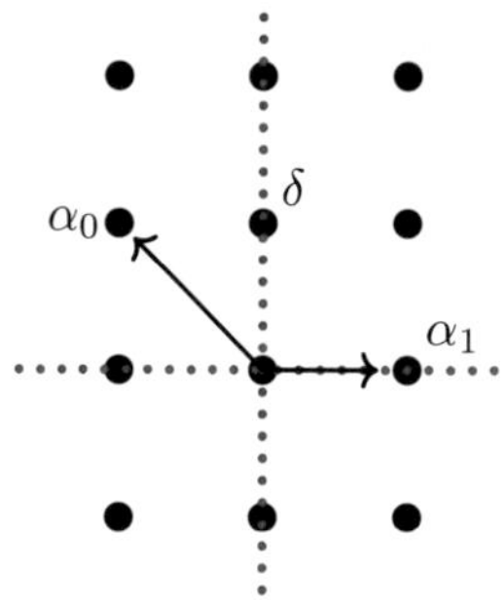

FIGURE 3. The root system of $\widehat{\mathfrak{sl}}_2$

3. Kac-Moody algebras and admissible weights

Let $\mathfrak{g}$ be a finite dimensional simple Lie algebra of rank ℓ. Strictly speaking the affine Lie algebra $\widehat{\mathfrak{g}}$ is not quite a Kac-Moody algebra, while its extension

$$\widetilde{\mathfrak{g}} = \widehat{\mathfrak{g}} \oplus \mathbb{C}d, \qquad [d, at^n] = nat^n, \quad [d, K] = 0,$$

is. The Cartan subalgebra of $\widetilde{\mathfrak{g}}$ is $\mathfrak{h} \oplus \mathbb{C}K \oplus \mathbb{C}d$ and, roughly speaking, the role of the extra vector d is to enlarge the dual $\widehat{\mathfrak{h}}^*$ enough for it to contain the root system of $\widehat{\mathfrak{g}}$. Setting δ and Λ_0 as dual vectors to d and K, respectively, the standard invariant bilinear form $(,)$ on $\mathfrak{h}^*$ extends to $\widetilde{\mathfrak{h}}^*$ with

$$(\delta, \delta) = (\Lambda, \Lambda) = 0, \quad (\delta, \Lambda_0) = 1.$$

The root system of $\widehat{\mathfrak{g}}$ (or rather, of $\widetilde{\mathfrak{g}}$) is

$$\widehat{\Delta} = \{n\delta + \alpha \mid n \in \mathbb{Z}, \ \alpha \in \Delta\} \cup \{n\delta \mid n \in \mathbb{Z}\},$$

with the former and latter sets of roots being referred to as *real* and *imaginary*, respectively. Positive roots of $\widetilde{\mathfrak{g}}$ are those of the form $n\delta + \alpha$ where either $n > 0$ or else $n = 0$ and $\alpha \in \Delta_+$. Simple roots are defined in the usual way, and the set of simple roots turns out to consist of the simple roots $\alpha_1, \ldots, \alpha_\ell$ of $\mathfrak{g}$, together with $\alpha_0 = -\theta + \delta$, where $\theta \in \Delta_+$ denotes the highest root of the finite root system.

The norm of the root $\alpha + n\delta$ is the same as that of $\alpha \in \Delta$, and the set of real coroots is

$$\widehat{\Delta}^{\vee,\mathrm{re}} = \{\nu^{-1}(\alpha)/(\alpha, \alpha) \mid \alpha \in \widehat{\Delta} \text{ real}\}.$$

In Figure 3 part of the root system of $\widehat{\mathfrak{sl}}_2$ is displayed. If $\{e, h, f\}$ is the standard basis of $\mathfrak{sl}_2$, then the root spaces associated with α_1, α_0 and δ are spanned by e, ft and ht, respectively, etc.

We shall generally use the notation α for roots of $\mathfrak{g}$ and of $\widehat{\mathfrak{g}}$, relying on context to resolve ambiguity. On the other hand, having fixed a level k, we shall often write $\widehat{\lambda}$ for the affine weight $\widehat{\lambda} = k\Lambda_0 + \lambda$ corresponding to $\lambda \in \mathfrak{h}^*$. For a $\mathfrak{g}$-module M we denote by $V_k(M)$ the corresponding induced $\widehat{\mathfrak{g}}$-module of level k, i.e.,

$$V_k(M) = U(\widehat{\mathfrak{g}}) \otimes_{U(\widehat{\mathfrak{g}}_+)} M,$$

where $\widehat{\mathfrak{g}}_+ = \mathfrak{g}[t] \oplus \mathbb{C}K$, and its action on M is given by extending that of the subalgebra $\mathfrak{g}$ by setting $at^{>0} = 0$ and $K = k$. If M is a highest weight module of highest weight $\lambda \in \mathfrak{h}^*$ then $V_k(M)$ lies between the corresponding affine Verma module $M(\widehat{\lambda})$ and its irreducible quotient $L(\widehat{\lambda})$,

$$M(\widehat{\lambda}) \twoheadrightarrow V_k(M) \twoheadrightarrow L(\widehat{\lambda}).$$

In particular if $M = M(\lambda)$ is itself a Verma module, then $V_k(M) \cong M(\widehat{\lambda})$. At the other extreme, if $M = L(\lambda)$ is irreducible, then $V_k(M) \cong L(\widehat{\lambda})$ if the level k is irrational, but in general the surjection can be nontrivial. If $M = L(\lambda)$ is finite dimensional irreducible, then $V_k(M)$ is known as a Weyl module. We shall denote it by $V_k(\lambda)$. Thus for example $V^k(\mathfrak{g}) = V_k(0)$ as a $\widehat{\mathfrak{g}}$-module. Clearly $V_k(M)_{\text{low}} = M$ in general.

Much as in the finite case, an affine weight λ is said to be dominant integral if

$$\langle \lambda + \widehat{\rho}, \alpha \rangle \in \mathbb{Z}_{>0} \quad \text{for all } \alpha^\vee \in \widehat{\Delta}_+^{\vee,\text{re}}.$$

Here the Weyl vector $\widehat{\rho} \in \widehat{\mathfrak{h}}^*$ is characterised by the property $\widehat{\rho}(\alpha_i^\vee) = 1$ for $i = 0, 1, \ldots, \ell$. Unlike the finite case, in the affine case this does not characterise $\widehat{\rho}$ uniquely, so we make the choice $\widehat{\rho} = h^\vee \Lambda_0 + \rho$ for definiteness.

An interesting generalisation of dominant integral in the affine case was identified by Kac and Wakimoto in [**37**]. The integral coroot system of an affine weight λ is by definition

$$\widehat{\Delta}^{\vee,\text{re}}(\lambda) = \{\alpha^\vee \in \widehat{\Delta}^{\vee,\text{re}} \mid \langle \lambda + \widehat{\rho}, \alpha^\vee \rangle \in \mathbb{Z}\}.$$

Now λ is said to be *admissible* if, in essence, it is dominant integral with respect to its integral coroot system. More precisely, if

- the rank of $\widehat{\Delta}^{\vee,\text{re}}(\lambda)$ equals that of $\widehat{\Delta}^\vee$, and
- $\langle \lambda + \widehat{\rho}, \alpha^\vee \rangle \in \mathbb{Z}_{>0}$ for all $\alpha^\vee \in \widehat{\Delta}^{\vee,\text{re}}(\lambda) \cap \widehat{\Delta}_+$.

As a simple example, we examine for which k the vacuum weight $k\Lambda_0$ is admissible. The simple coroots in $\widehat{\Delta}^{\vee,\text{re}}(k\Lambda_0)$ must be of the form

$$\Pi^\vee(\lambda) = \{\alpha^\vee \in \Delta^{\vee,\text{re}}(\lambda) \cap \widehat{\Delta}_+ \mid \alpha \text{ indecomposable in } \Delta^{\vee,\text{re}}(\lambda) \cap \widehat{\Delta}_+\}$$

must be

$$\Pi^\vee(k\Lambda_0) = \Pi_q^\vee = \{\alpha_1^\vee, \ldots, \alpha_\ell^\vee, -\theta^\vee + qK\}.$$

And in turn we must have

$$k = -h^\vee + p/q, \qquad p \geq h^\vee.$$

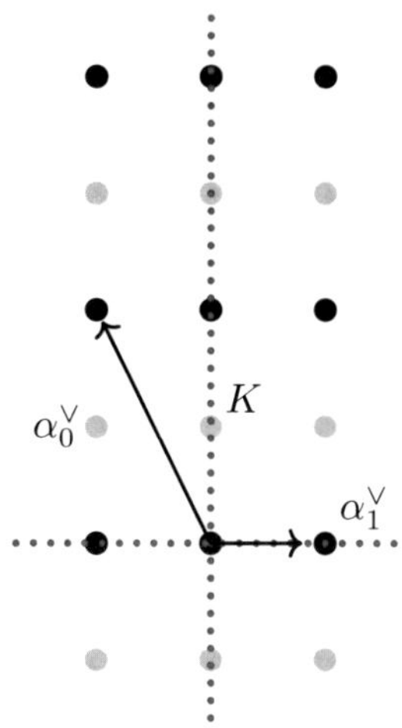

FIGURE 4. The coroot subsystem $\widehat{\Delta}^\vee(k\Lambda_0)$ for $\widehat{\mathfrak{sl}}_2$ and k admissible with denominator $q = 2$

REMARK 3.1. If $\mathfrak{g}$ is simply laced, then the intersection of $\widehat{\Delta}^{\vee,\mathrm{re}}$ with the set $sK + \mathfrak{h}$ is, depending on the value of s, either empty or of the form $\Delta^\vee$, where $\Delta^\vee$ is the coroot system of $\mathfrak{g}$. This is no longer the case when $\mathfrak{g}$ is not simply laced; in this case the intersection is either $\Delta^\vee$ or its proper subset $\Delta^\vee_{\mathrm{short}}$ consisting of short coroots. There is therefore a marked difference in the structure of $\Pi_q^\vee$ between the cases $(q, r^\vee) = 1$ and $(q, r^\vee) \neq 1$ when $\mathfrak{g}$ is non simply laced with lacing number $r^\vee$. This issue is present throughout the whole story, but we ignore it for the sake of simplicity.

We now state an important generalisation, due to Arakawa, of Theorem 2.5

THEOREM 3.2 (Arakawa, [**5**]). *Let k be an admissible level with denominator q. Recall $V^k(\mathfrak{g})$ the universal affine vertex algebra of level k and $L_k(\mathfrak{g})$ its simple quotient. The $V^k(\mathfrak{g})$-module $L_k(\lambda)$ descends to a $L_k(\mathfrak{g})$-module if and only if λ is admissible with $\Pi^\vee(\lambda)$ conjugate to $\Pi_q^\vee$.*

The proof of this theorem uses a number of advanced techniques. The "only if" part is proved by reduction to the case of $\mathfrak{g} = \mathfrak{sl}_2$ via *semi-infinite reduction.* The case of $\mathfrak{sl}_2$, in turn, had been settled earlier by Adamovic and Milas using the Zhu algebra [**1**].

The "if" part is proved first for G-integrable admissible weights, and the case of general admissible weight is reduced to the G-integrable case using the theory of characteristic varieties. For $\mathfrak{g}$ a finite dimensional semisimple Lie algebra, and a two-sided ideal $I \subset U(\mathfrak{g})$, the *characteristic variety* $V(I) \subset \mathfrak{h}^*$ is introduced as the spectrum of the image of I under the projection to the second factor in

$$U(\mathfrak{g}) \to (\mathfrak{n}_- U(\mathfrak{g}) + U(\mathfrak{g})\mathfrak{n}_+) \oplus U(\mathfrak{h}).$$

If $\lambda \in V(I)$ then $IL(\lambda) = 0$. For reflections r_i in simple roots, it is a classical fact due to Joseph [**29**] that if $\lambda \in V(I)$ then $r_i \circ \lambda \in V(I)$ too. A generalisation to affine case allows the reduction from general admissible weights to G-integrable admissible weights to be carried out.

The case of G-integrable $L_k(\lambda)$ is handled using a theory of "Kazhdan-Lusztig translation functors" developed by Feigin and Malikov. Although it takes us a little far afield to describe this theory, and we make no pretense at completeness, we sketch the ideas in Section 5 below.

4. Characters formulas for irreducible $\widehat{\mathfrak{g}}$-modules

Let $k \in \mathbb{Z}_+$ and recall the set

$$P_+^k = \{\lambda \mid \langle \lambda, \theta^\vee \rangle \leq k\}$$

of dominant integral weights of level k. The corresponding irreducible highest weight $\widehat{\mathfrak{g}}$-modules $L_k(\lambda)$ are examples of *integrable* $\widehat{\mathfrak{g}}$-modules, and their formal characters are given by the Weyl-Kac character formula:

$$\chi_{L(\lambda)} = \sum_{w \in \widehat{W}} \varepsilon(w) \chi_{M(w \circ \lambda)},$$

where $\widehat{W} = W \ltimes t_{Q^\vee}$ is the affine Weyl group, and

$$\chi_{M(\lambda)} = \frac{e^\lambda}{\prod_{\alpha \in \Delta_+} (1 - e^{-\alpha})}$$

is the character of the Verma module $M(\lambda)$. Here the notation $w \circ \lambda$ represents the "dot action" of W on the weight $\lambda \in \widehat{\mathfrak{h}}^*$, defined to be

$$w \circ \lambda = w(\lambda + \widehat{\rho}) - \widehat{\rho}.$$

As we have indicated in the previous paragraph, the structure of the Weyl group of $\widetilde{\mathfrak{g}}$ is that of a semidirect product of the finite Weyl group W with an abelian group of "translations" t_α indexed by the coroot lattice $\alpha \in Q^\vee$. We make this more explicit now. The actions of t_α on $\mathfrak{h}^*$ and $\mathfrak{h}$, respectively, are:

$$\begin{aligned} t_\alpha(\lambda) &= \lambda + \lambda(K)\alpha - \left[(\alpha, \lambda) + \frac{|\alpha|^2}{2} \lambda(K) \right] \delta \\ t_\alpha(x) &= x - \alpha(x) K \bmod \mathbb{C}d. \end{aligned} \tag{4.1}$$

To illustrate we let $\mathfrak{g} = \mathfrak{sl}_2$ and we consider the integrable $\widehat{\mathfrak{g}}$-module $L_1(\mathfrak{g}) = L(\Lambda_0)$ of level $k = 1$. We have $h^\vee = 2$ and so $\widehat{\rho} = 2\Lambda_0 + \alpha/2$, where α is the unique simple root of $\mathfrak{sl}_2$. A few elements of $\widehat{W} \circ \Lambda_0$ are shown in Figure 5.

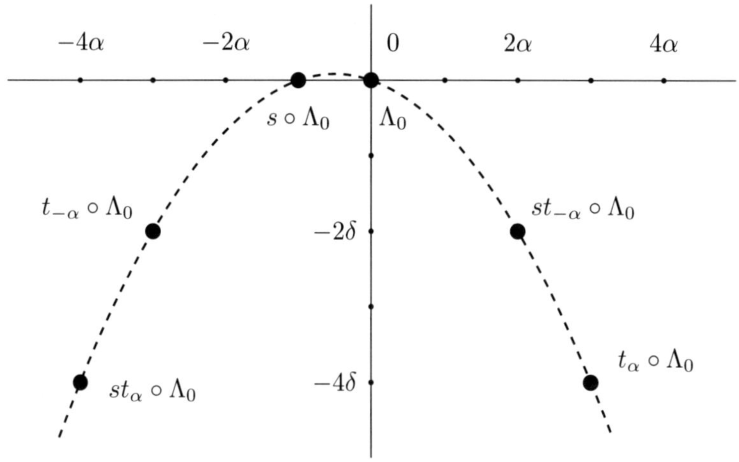

FIGURE 5. Dot orbit of Λ_0 under the affine Weyl group

Putting

$$q = e^{-\delta}, \quad \text{and} \quad y = e^{\alpha/2},$$

we easily obtain from the Weyl-Kac character formula

$$\chi_{L(\Lambda_0)} = \frac{\sum_{n\in\mathbb{Z}} y^{3n}q^{3n^2+n} - \sum_{n\in\mathbb{Z}} y^{3n-1}q^{3n^2-n}}{\prod_{n=1}^{\infty}(1-y^{-1}q^{n-1})(1-q^n)(1-yq^n)}. \tag{4.2}$$

Notice that the vertex algebra $L_1(\mathfrak{sl}_2)$, as an $\widehat{\mathfrak{sl}}_2$-module, is isomorphic to $L(\Lambda_0)$. It is not difficult to confirm that under this identification $d = -L_0$. So the graded dimension of the vertex algebra relative to its conformal $\mathbb{Z}_+$-grading, is just the $y = 1$ specialisation of (4.2).

We now make a brief tangential remark. If Q is an even integral lattice, then there is the well known construction of a lattice vertex algebra V_Q [**33**]. This vertex algebra is conformal, and from its construction the graded dimension χ_{V_Q} is easy to write down, it is

$$\prod_{n=1}^{\infty}(1-q^n)^{-1} \cdot \sum_{m\in\mathbb{Z}} e^{\alpha} q^{(\alpha,\alpha)/2}.$$

The Frenkel-Kac-Segal theorem asserts that if $\mathfrak{g}$ is a simply laced simple Lie algebra with root lattice Q, then

$$L_1(\mathfrak{g}) \cong V_Q.$$

So as a consequence of the isomorphism in the case of $\mathfrak{g} = \mathfrak{sl}_2$, we ought to recover the identity

$$\prod_{n=1}^{\infty}(1-y^{-1}q^{n-1})(1-yq^n) \cdot \sum_{m\in\mathbb{Z}} y^m q^{m^2} = \sum_{n\in\mathbb{Z}} y^{3n}q^{3n^2+n} - \sum_{n\in\mathbb{Z}} y^{3n-1}q^{3n^2-n}.$$

This is in fact a classical identity.

One of the key features of the admissible weights λ is that the corresponding highest weight modules $L(\lambda)$, although not integrable, have characters given by formulas essentially similar to the Weyl-Kac character formula. Indeed let λ be an admissible weight for $\widehat{\mathfrak{g}}$. In [**37**] Kac and Wakimoto proved that

$$\chi_{L(\lambda)} = \sum_{w \in \widehat{W}(\lambda)} \varepsilon(w) \chi_{M(w \circ \lambda)}, \tag{4.3}$$

where now $\widehat{W}(\lambda) \subset \widehat{W}$ is the subgroup generated by reflections in the coroots that make up $\Pi^\vee(\lambda)$.

In order to sketch the proof, we require the notions of blocks and linkage. A good introductory reference, for the case of finite dimensional $\mathfrak{g}$, is [**27**]. See the article [**34**] and books [**31**] and [**48**] for the Kac-Moody case.

The block decomposition is a fundamental tool for organising our understanding of category $\mathcal{O}$ of a semisimple (or Kac-Moody) Lie algebra. Basic questions in the theory are to determine (1) for which weights μ the irreducible module $L(\mu)$ may occur in a composition series of the Verma module $M(\lambda)$, and (2) with what multiplicities. A definitive answer to the second question came with the Kazhdan-Lusztig conjecture and its solution, and extensions thereof. As for the first question, an important necessary condition (besides the obvious $\lambda - \mu \in Q_+$) is that there exist $w \in W$ such that

$$\mu = w(\lambda + \rho) - \rho.$$

We write $w \circ \lambda$ for the term on the right hand side, and say that weights satisfying such a relation are "linked". In fact if $\lambda \notin P$ then not every $w \circ \lambda$ differs from λ by an element of Q, so it is possible to make an obvious refinement to the notion of linkage: we define $W(\lambda) \subset W$ the subgroup generated by reflections in those coroots α such that $\langle \lambda + \rho, \alpha^\vee \rangle \in \mathbb{Z}$ (for an admissible weight λ of $\widehat{\mathfrak{g}}$, this group is precisely the group $\widehat{W}(\lambda)$ appearing in the character formula (4.3)), and linkage is defined as the equivalence relation generated by $\mu \in W(\lambda) \circ \lambda$.

The objects of $\mathcal{O}$ possessing a composition series all of whose irreducible factors have highest weight in a fixed "linkage class", forms an abelian subcategory of $\mathcal{O}$. There are no nonzero morphisms between objects of these subcategories, which are known as "blocks", and $\mathcal{O}$ decomposes into a direct sum of blocks.

It can be convenient to translate results from one block to another by means of so called translation functors. Let $\mathcal{O}_\lambda$ and $\mathcal{O}_\mu$ denote blocks associated with weights λ and μ. If $\lambda - \mu \in P$ then the translation functor $T_\lambda^\mu : \mathcal{O}_\lambda \mapsto \mathcal{O}_\mu$ is defined by

$$T_\lambda^\mu(M) = \mathrm{pr}_\mu \left(M \otimes L(\nu) \right),$$

where ν is the unique representative of $P_+ \cap W(\lambda - \mu)$. Here pr_μ means projection from $\mathcal{O}$ to the summand $\mathcal{O}_\mu$. The module $L(\nu)$ is integrable, which turns out to be important to guarantee exactness of the translation functor.

If μ is in the same Weyl chamber as λ, or a facet of it (in the sense that for each $\alpha^\vee \in \Delta^\vee(\lambda)$ we have $\langle \lambda + \rho, \alpha^\vee \rangle > 0$ (resp. < 0) implies $\langle \mu + \rho, \alpha^\vee \rangle \geq 0$ (resp. ≤ 0)), then

$$T_\lambda^\mu M(w \circ \lambda) = M(w \circ \mu).$$

The proof of this fact relies on the following standard lemma (see [**27**, Lemma 7.5]).

Lemma 4.1. *Let $\lambda, \mu \in P_+$, and let ν be the unique element of the intersection of $W(\mu - \lambda)$ with P_+. Then the only weight $\lambda + \nu'$ (where ν' is a weight of $L(\nu)$) that is linked to μ, is $\mu = \lambda + \nu$ itself.*

Returning now to the character of admissible $\widehat{\mathfrak{g}}$-modules. By the linkage principle

$$\chi_{L(\lambda)} = \sum_{w\in\widehat{W}(\lambda)} m(\lambda, w)\, \chi_{M(w\circ\lambda)}$$

for some set of integer coefficients $m(\lambda, w)$. Let $s = r_{\alpha_i}$ be a simple reflection in $\widehat{W}(\lambda)$, and choose μ so that $\widehat{W}(\mu - \lambda)$ meets P_+, and so that $\langle \mu + \rho, \alpha^\vee\rangle \in \mathbb{Z}_{\geq 0}$ for all $\alpha^\vee \in \Pi^\vee(\lambda)$, with $\langle \mu + \rho, \alpha^\vee\rangle = 0$ for exactly the one coroot $\alpha^\vee = \alpha_i^\vee$. We apply a translation functor to the surjection

$$M(\lambda)/M(s \circ \lambda) \to L(\lambda).$$

Since $s \circ \mu = \mu$, we get $T_\lambda^\mu L(\lambda) = 0$ (an exact functor sends surjections to surjections), hence

$$\sum_{w\in\widehat{W}(\lambda)} m(\lambda, w)\chi_{M(w\circ\mu)} = 0,$$

and so $m(\lambda, w) + m(\lambda, s\circ w) = 0$. This goes for each $s = r_{\alpha_i}$, and so $m(\lambda, w) = \varepsilon(w)$ for all $w \in \widehat{W}(\lambda)$, thus proving the formula.

5. The Kazhdan-Lusztig tensor product

The tensor product, as vector spaces, of a $\widehat{\mathfrak{g}}$-module of level k with one of level ℓ is again a $\widehat{\mathfrak{g}}$-module in the usual way, and has level $k + \ell$. On the other hand Kazhdan and Lusztig have defined a tensor product for modules of a single level k, which we describe briefly in this section. The main reference is the original series of four articles [**41**–**44**].

Above we have discussed G-integrable modules already, it is convenient to introduce notation for this now. Let KL_k denote the category of G-integrable positive energy $\widehat{\mathfrak{g}}$-modules of level k. This category contains the Weyl modules $V_k(\lambda)$ for $\lambda \in P_+$ in particular.

Let $\Gamma(\mathbb{C}P^1\backslash\{0, 1, \infty\}, \mathcal{O})$ denote the ring of rational functions on the complex projective space $\mathbb{C}P^1$ of dimension one, with poles only at 0, 1 and ∞. Let t denote the standard coordinate on $\mathbb{C} = \mathbb{C}P^1\backslash\{\infty\}$. Following Kazhdan and Lusztig, we consider the Lie algebras

$$\Gamma = \mathfrak{g} \otimes \Gamma(\mathbb{C}P^1\backslash\{0, 1, \infty\}, \mathcal{O})$$
$$\text{and} \quad \widehat{\Gamma} = \Gamma \oplus \mathbb{C}K$$

with Lie brackets

$$[x\, f, y\, g] = [x, y]\, fg$$
$$\text{and} \quad [x\, f, y\, g] = [x, y]\, fg + \mathrm{Res}_{t=0}\, f'(t)g(t)K.$$

There is now a homomorphism $\widehat{\Gamma} \to \widehat{\mathfrak{g}}$ induced by Laurent series expansion of rational functions

$$i_0 : \Gamma(\mathbb{C}P^1\backslash\{0, 1, \infty\}, \mathcal{O}) \to \mathbb{C}((t))$$

at 0 in powers of t.

Let $A, B \in \mathrm{KL}_k$. We turn the vector space tensor product

$$W = A \otimes B$$

into a $\widehat{\Gamma}$-module, by equipping it with the action defined by

$$xf(a \otimes b) = (x\, i_1 f)a \otimes b + a \otimes (x\, i_\infty f)b, \qquad K = -k \,\mathrm{Id}\,.$$

The choice of action of $K = -k\,\mathrm{Id}$ here was not made arbitrarily, it is the only choice consistent with the action of terms xf as defined. This is a consequence of the residue theorem. We denote by Z the algebraic dual of W. This carries an action of $\widehat{\Gamma}$ of level k.

Now Z is almost the module we want; a module of level k constructed from a pair of modules A and B of level k. However, as an algebraic dual of an infinite dimensional vector space, Z has larger cardinality than A and B, whereas we ultimately want to construct a tensor product $\dot{\otimes} : \mathrm{KL}_k \times \mathrm{KL}_k \to \mathrm{KL}_k$.

Inside Z we consider subspaces

$$Z^N = \{\varphi \in Z \mid \varphi(\Gamma_0^N W) = 0\},$$

where $\Gamma_0 \subset \widehat{\Gamma}$ denotes the subspace of $\mathfrak{g}$-valued functions that vanish at 0. These form an increasing sequence, and we denote by Z^∞ their union. Let $f(t) \in \mathbb{C}((t))$ and $x \in \mathfrak{g}$. How do we define the action of $x\,f$ on $\varphi \in Z^\infty$? Noting that by definition $\varphi \in Z^N$ for some $N \in \mathbb{Z}_+$, the answer is to approximate $f(t)$ by a rational function $\widetilde{f}(t) \in \Gamma(\mathbb{C}P^1 \backslash \{0, 1, \infty\}, \mathcal{O})$, and define $x\,f \cdot \varphi$ by

$$[x\,f \cdot \varphi](a \otimes b) = \varphi(x\,\widetilde{f}(a \otimes b)).$$

If the chosen approximation is accurate to order at least t^N, then φ ought to annihilate the difference between $x\,f$ and $x\,\widetilde{f}$, and so this definition of the action of $\widehat{\mathfrak{g}}$ is reasonable.

THEOREM 5.1 (KL1, Section 7). *If $A, B \in \mathrm{KL}_k$ then Z^∞ is again a $\widehat{\mathfrak{g}}$-module in KL_k.*

If we are given three objects A, B and C of KL_k then we can consider the action of Γ on $A \otimes B \otimes C$ given by

$$x\,f(a \otimes b \otimes c) = (x\,i_1 f)a \otimes b \otimes c + a \otimes (x\,i_\infty f)b \otimes c + a \otimes b \otimes (x\,i_0 f)c.$$

This is well-defined because of the residue theorem. Then we consider the vector space

$$\mathcal{C}(A, B, C) = \left(\frac{A \otimes B \otimes C}{\Gamma \cdot (A \otimes B \otimes C)} \right)^*, \tag{5.1}$$

which is finite dimensional under favourable circumstances, and is known as the space of conformal blocks. From A and B we construct Z^∞ as above, and then it turns out that

$$\mathrm{Hom}_{\widehat{\mathfrak{g}}}(C, Z^\infty) \cong \mathcal{C}(A, B, C).$$

Reflecting on the tensor-hom adjunction, this relation shows that Z^∞ behaves like the dual of a tensor product. This is not surprising since we have defined Z^∞ inside an algebraic dual. The final step in defining $A \dot{\otimes} B$ is to take a graded (or "contragredient") dual $D(-)$, twisting the action of $\widehat{\mathfrak{g}}$ in the process. We do not enter into details here, merely remarking that the procedure is much the same as that employed to define duals in category $\mathcal{O}$ of a finite dimensional semisimple Lie algebra [**27**]. The Kazhdan-Lusztig tensor product is thus the contragredient dual of Z^∞:

$$A \dot{\otimes} B = D(Z^\infty).$$

The construction of $A\dot{\otimes}B$ can be carried out with A and B located, not at ∞ and 1 respectively, but at any pair of points (p_1, p_2) distinct from 0 and from each other. An important part of the theory, which it would be too large a detour to describe in detail in these notes, is the construction of a connection on the vector bundle $\mathcal{C}$ over the configuration space of such pairs (p_1, p_2), whose fibres are the corresponding spaces of conformal blocks. Using this connection, one can let the points move and, importantly, loop around one another and exchange positions. In this way for example, braiding isomorphisms

$$A\dot{\otimes}B \to B\dot{\otimes}A$$

can be defined. Associativity isomorphisms are also defined, and correspond to configurations of three points (distinct from 0) passing to degenerate configurations, in which two points "collide" in distinct ways. In the end a braided monoidal category is obtained. For a complete definition, and theory, of (braided) monoidal category, see [**15**]. For an account emphasising links with vertex algebras, see [**9**].

DEFINITION 5.2. A monoidal category is a category $\mathcal{C}$ with a bifunctor $\otimes$: $\mathcal{C} \times \mathcal{C} \to \mathcal{C}$ and natural isomorphism

$$a_{X,Y,Z} : (X \otimes Y) \otimes Z \to X \otimes (Y \otimes Z),$$

a unit $\mathbf{1} \in \mathcal{C}$ and unit natural isomorphism, satisfying certain identities known as the pentagon and triangle identities. A braiding on a monoidal category is a further natural isomorphism

$$c_{X,Y} : X \otimes Y \to Y \otimes X,$$

compatible with the associativity and unit isomorphisms through identities known as the hexagon identities.

The Kazhdan-Lusztig category is easier to describe for irrational level k than for rational level. In this case the Weyl modules $V_k(\lambda)$, for $\lambda \in P_+$, are irreducible, and the Kazhdan-Lusztig tensor product is compatible with induction in the sense that

$$V_k(\lambda)\dot{\otimes}V_k(\mu) \cong \bigoplus_{\nu \in P_+} V_k(\nu)^{\oplus m_\nu}, \quad \text{where} \quad L(\lambda) \otimes L(\mu) \cong \bigoplus_{\nu \in P_+} L(\nu)^{\oplus m_\nu}.$$

The situation at rational values of the level k is more complicated, but what can be said in general is that

$$V_k(\lambda)\dot{\otimes}V_k(\mu)$$

possesses a filtration, whose composition factors are the Weyl modules $V_k(\nu)$ for $\nu \in P_+$ with the same multiplicities m_ν as for irrational level k [**44**] [**22**, Theorem 3.1.2].

The category KL_k is a subcategory of $\mathcal{O}$ and as such splits into blocks, corresponding to the equivalence relation

$$\mu = w \circ \lambda, \qquad w \in \widehat{W}_k.$$

Here the group $\widehat{W}_k = W \ltimes pQ^\vee = \widehat{W}(k\Lambda_0)$ appears naturally since objects of KL_k are by definition G-integrable, and by (4.1) translation $t_\alpha(\lambda)$ may satisfy $t_\alpha(\lambda) - \lambda \in Q$ only if $t_\alpha(\lambda) - \lambda \in pQ$.

The geometry of weights under the action of W_k is substantially similar to the case of $\mathfrak{g}$ semisimple, and linkage and translation functors can be defined in this context. For $\lambda, \mu \in P_+^k$ regular, let ν be the unique weight in $P_+^k \cap W(\mu - \lambda)$, then for $M \in \mathcal{O}_\lambda$ one defines

$$T_\lambda^\mu(M) = \mathrm{pr}_\mu(M \dot{\otimes} V_k(\nu)).$$

Notice that the tensor product is taken with the Weyl module $V_k(\nu)$ rather than the irreducible quotient $L(\lambda)$. It can be proved that T_λ^μ sends Weyl modules to Weyl modules, and more specifically that

$$T_\lambda^\mu(V_k(w \circ \lambda)) \cong V_k(w \circ \mu),$$

whenever $w \circ \lambda, w \circ \mu \in P_+$. This is a consequence of a version of Lemma 4.1 for $\widehat{W}_k$, which is contained in Jantzen's book [**28**, Lemma 7.7], see also [**22**, Lemma 2.1.1]. Next we would like to prove that

$$T_\lambda^\mu(L_k(w \circ \lambda)) \cong L_k(w \circ \mu). \tag{5.2}$$

Conceptually this is done using the notions of duality and rigidity in monoidal categories.

Definition 5.3. For objects X and Y in a monoidal category, a duality datum (exhibiting X as a left dual of Y and Y as a right dual of X) is a pair of morphisms

$$\mathrm{ev} : Y \otimes X \to \mathbf{1}, \qquad \mathrm{coev} : \mathbf{1} \to X \otimes Y$$

such that the compositions

$$X \to \mathbf{1} \otimes X \to X \otimes Y \otimes X \to X \otimes \mathbf{1} \to X$$

and

$$Y \to Y \otimes \mathbf{1} \to Y \otimes X \otimes Y \to \mathbf{1} \otimes Y \to Y$$

are the identities.

These axioms capture the essential properties of the notion of dual of a finite dimensional vector space over a field $\mathbb{F}$ (and of dual module for modules of a group or Lie algebra, etc.). The evaluation and coevaluation morphisms for such a vector space V and its dual V^* are

$$\mathrm{ev} : V^* \otimes V \to \mathbb{F}, \qquad \varphi \otimes v \mapsto \varphi(v)$$
$$\text{and} \quad \mathrm{coev} : \mathbb{F} \to V \otimes V^*, \qquad 1 \mapsto \sum_i v_i \otimes \varphi_i,$$

where $\{v_i\}$ is any basis of V and $\{\varphi_i\}$ is its dual basis of V^*.

The following is a standard result in the theory.

Proposition 5.4. *If (X, X^*) is a duality datum in an abelian monoidal category, i.e., if there are evaluation and coevaluation maps making X^* a left dual of X, then*

$$\mathrm{Hom}(U, X \otimes V) \cong \mathrm{Hom}(X^* \otimes U, V)$$

and $X \otimes (-)$ is an exact functor.

A monoidal category $\mathcal{C}$ is said to be *rigid* if there is a contravariant functor $X \mapsto X^*$ and natural duality data relating X and X^* for all $X \in \mathcal{C}$.

The category $(\mathrm{KL}_k, \dot{\otimes})$ is rigid if $k \notin \mathbb{Q}$, while for $k \in \mathbb{Q}_{>0}$ there are in general some objects that do not admit a duality datum (highest weight modules with highest weight that lies on an affine wall for $\widehat{W}_k$). In this case Frenkel-Malikov [**22**] observe that $V_k(\lambda)$ possesses a duality datum for regular weights λ, with dual $V_k(\lambda)^* = V_k(-w_\circ(\lambda))$, where $w_\circ$ is the longest element in the Weyl group of $\mathfrak{g}$.

In general if $V_k(\lambda)$ fits into a duality datum, we have exactness of T_λ^μ, then a standard argument (see [**27**, Proposition 7.8] for the simplest iteration of this argument, and [**22**] for the present context) implies (5.2).

Let $\lambda \in P_+$ and recall that the irreducible $\widehat{\mathfrak{g}}$-module $L_k(\lambda)$ is a quotient of the Weyl module $V_k(\lambda)$. In particular $L_k(\mathfrak{g}) = V_k(0)/I_k$ where I_k is both a $\widehat{\mathfrak{g}}$-submodule and a vertex algebra ideal. Suppose now that $\lambda \in P_+^k$ is a regular dominant integral weight. In order to prove that $L_k(\lambda)$ is a $L_k(\mathfrak{g})$-module, we should like to prove that

$$L_k(\lambda) \cong V_k(\lambda)/I_k \cdot V_k(\lambda). \tag{5.3}$$

Let V be a vertex algebra, $V^k(\mathfrak{g})$ in the case at hand, let M_1, M_2 and N be three V-modules, and let $I \subset V$ be an ideal. The following is an easy but key observation:

$$\mathcal{C}(M_1/IM_1, M_2, N) = \mathcal{C}(M_1, M_2/IM_2, N).$$

Indeed let us write Γ for global sections as in (5.1). Then the left and right hand sides are identified, respectively, with the subsets

$$\begin{aligned} &\{\varphi : M_1 \otimes M_2 \otimes N \to \mathbb{C} \mid \varphi(i_{(n)}M_1, M_2, N) = 0 \text{ for all } i \in I \text{ and } n \in \mathbb{Z}\} \\ \text{and } \ &\{\varphi : M_1 \otimes M_2 \otimes N \to \mathbb{C} \mid \varphi(M_1, i_{(n)}M_2, N) = 0 \text{ for all } i \in I \text{ and } n \in \mathbb{Z}\} \end{aligned}$$

of

$$\mathcal{C}(M_1, M_2, N) = \{\varphi : M_1 \otimes M_2 \otimes N \to \mathbb{C} \mid \varphi(\Gamma \cdot (M_1 \otimes M_2 \otimes N)) = 0\}.$$

All we have to do is convince ourselves that these two subsets coincide.

From (5.2) we deduce $L_k(\lambda) \cong T_0^\lambda(L_k(\mathfrak{g}))$. We combine this with the other ingredients above in the following simple calculation:

$$\begin{aligned} \operatorname{Hom}_{\widehat{\mathfrak{g}}}(L_k(\lambda), D(N)) &\cong \operatorname{Hom}_{\widehat{\mathfrak{g}}}(T_0^\lambda(V_k(0)/I_k), D(N)) \\ &\cong \mathcal{C}(V_k(0)/I_k, V_k(\lambda), N) \\ &= \mathcal{C}(V_k(0), V_k(\lambda)/I_k \cdot V_k(\lambda), N) \\ &\cong \operatorname{Hom}_{\widehat{\mathfrak{g}}}(V_k(\lambda)/I_k \cdot V_k(\lambda), D(N)). \end{aligned}$$

So by the representing property of the tensor product, we have (5.3) as desired.

6. The BRST complex and its quantisation

Hamiltonian reduction of symplectic varieties admitting Lie group symmetry is an important basic construction in mathematical physics, admitting many refinements and variations. Firstly the notion of Marsden-Weinstein quotient of a symplectic manifold by a good Hamiltonian group action is given a homological interpretation in the form of the BRST complex [**45**]. This homological construction opens the door both to quantisation, and to generalisation to the infinite dimensional context, where it yields a Hamiltonian reduction functor for vertex algebras. Via this construction the *affine W-algebras* are defined and studied.

Let M be a smooth Poisson manifold equipped with a Hamiltonian action of a Lie group N. Let $\mathfrak{n}$ be the Lie algebra of N. Then $\mathcal{O} = C^\infty(M)$ is a Poisson algebra, and the Hamiltonian action basically amounts to a homomorphism of Lie algebras

$$\alpha : \mathfrak{n} \to \mathcal{O}.$$

The moment map $\mu : M \to \mathfrak{n}^*$ takes a point $p \in M$ to the linear functional on $\mathfrak{n}$ that sends $x \in \mathfrak{n}$ to the number obtained by evaluating $\alpha(x) \in \mathcal{O}$ at p.

The algebra of functions on $\mu^{-1}(0)$ is the quotient $\mathcal{O}/(\alpha(\mathfrak{n}))$, where $(\ldots)$ denotes "ideal generated" in the commutative algebra $\mathcal{O}$. Under favourable hypotheses on the structure of the action, this quotient algebra is resolved by the Koszul complex:

$$\mathcal{O} \otimes \wedge^\bullet \mathfrak{n},$$

the differential being given by

$$\begin{aligned} f \otimes \phi &\mapsto \alpha(\phi) f, \\ f \otimes \phi_1 \wedge \phi_2 &\mapsto \alpha(\phi_1) f \otimes \phi_2 - \alpha(\phi_2) f \otimes \phi_1, \end{aligned}$$

etc.

So far we have not used either the Lie algebra structure of $\mathfrak{n}$ nor the Poisson bracket in $\mathcal{O}$. We do use these structures, however, to furnish the whole Koszul complex with the structure of a complex of $\mathfrak{n}$-modules. Finally the subalgebra of $\mathcal{O}/(\alpha(\mathfrak{n}))$ consisting of N-invariant functions on $\mu^{-1}(0)$ is resolved by the Chevalley-Eilenberg complex of the $\mathfrak{n}$-module $\mathcal{O}\otimes$, namely $(\mathcal{O} \otimes \wedge^\bullet \mathfrak{n}) \otimes \wedge^\bullet \mathfrak{n}^*$. The map $\mathcal{O} \to \mathcal{O} \otimes \mathfrak{n}^*$ is the natural one obtained from the action of $\mathfrak{n}$ through tensor-hom adjunction:

$$\mathrm{Hom}(\mathfrak{n} \otimes \mathcal{O}, \mathcal{O}) \cong \mathrm{Hom}(\mathcal{O}, \mathcal{O} \otimes \mathfrak{n}^*).$$

More explicitly, if $\{\phi_1, \ldots, \phi_n\}$ is a basis of $\mathfrak{n}$ and $\{\phi_1^*, \ldots, \phi_n^*\}$ is the corresponding dual basis of $\mathfrak{n}^*$, then

$$f \mapsto \sum_{i=1}^n \{\alpha(\phi_i), f\} \otimes \phi_i^*.$$

The other vertical morphisms are constructed in a similar way.

$$\begin{array}{ccccccc} & & \uparrow & & \uparrow & & \uparrow \\ \longrightarrow & & (\mathcal{O} \otimes \wedge^2 \mathfrak{n}) \otimes \mathfrak{n}^* & \longrightarrow & (\mathcal{O} \otimes \mathfrak{n}) \otimes \mathfrak{n}^* & \longrightarrow & \mathcal{O} \otimes \mathfrak{n}^* \\ & & \uparrow & & \uparrow & & \uparrow \\ \longrightarrow & & \mathcal{O} \otimes \wedge^2 \mathfrak{n} & \longrightarrow & \mathcal{O} \otimes \mathfrak{n} & \longrightarrow & \mathcal{O} \end{array}$$

So we have constructed a bicomplex, from which we may pass to the total complex $C(\mathfrak{n}, \mathcal{O})$ with total differential d_{Tot}. Under favourable circumstances, the total complex has cohomology concentrated in degree 0, recovering the ring of N-invariant functions on $\mu^{-1}(0)$.

Now the quantisation proposal: Suppose we knew how to quantise the Poisson manifold M, that is, suppose we had a filtered associative algebra A whose associated graded (i.e., classical limit) recovers $\mathcal{O} = \mathrm{gr}(A)$. Suppose further that we knew how to quantise M equivariantly, that is, suppose A comes with a compatible homomorphism $\alpha : U(\mathfrak{n}) \to A$. Then a natural quantisation can be found for the BRST complex, and thus for $\mu^{-1}(0)/N$.

Indeed, let $\mathfrak{a}$ be a vector space and $\langle \cdot, \cdot \rangle : \mathfrak{a} \times \mathfrak{a} \to \mathbb{C}$ a symmetric bilinear form. The corresponding Clifford algebra is the quotient

$$\mathcal{C}\ell(\mathfrak{a}) = \frac{T(\mathfrak{a})}{(ab + ba - \langle a, b \rangle)}$$

of the tensor algebra $T(\mathfrak{a})$. We may also consider $\mathfrak{a}$ as a purely odd vector superspace, and $T(\mathfrak{a})$ as an associative superalgebra. The canonical grading on $T(\mathfrak{a})$ by word length induces a filtration on $\mathcal{C}\ell(\mathfrak{a})$, and the associated graded is isomorphic to the exterior algebra $\wedge(\mathfrak{a})$. Let A be a filtered associative algebra, we can form the filtered associative superalgebra

$$A \otimes \mathcal{C}\ell(\mathfrak{a}).$$

Now we suppose, in the context of the Hamiltonian action of N on M as above, that our associative algebra A is a quantisation of M in the sense that the associated graded $\mathrm{gr}(A)$ recovers the Poisson algebra $\mathcal{O}$. If we take $\mathfrak{a} = \mathfrak{n} \oplus \mathfrak{n}^*$, where $\mathfrak{n}$ is our Lie algebra from above, then $\mathfrak{a}$ can be given the tautological symmetric bilinear form $\langle \cdot, \cdot \rangle$ of dual pairing between $\mathfrak{n}$ and $\mathfrak{n}^*$. Also $\mathfrak{a}$ acquires a semidirect product Lie algebra structure, with Lie brackets as follows given by the adjoint and coadjoint actions

$$\begin{aligned} \mathfrak{n} \times \mathfrak{n} &\to \mathfrak{n} \\ \mathfrak{n} \times \mathfrak{n}^* &\to \mathfrak{n}^* \\ \mathfrak{n}^* \times \mathfrak{n}^* &\to 0. \end{aligned}$$

The form $\langle \cdot, \cdot \rangle$ is invariant for this Lie bracket. The associated graded of the filtered superalgebra $A \otimes \mathcal{C}\ell(\mathfrak{a})$ now recovers

$$\mathcal{O} \otimes \wedge^\bullet(\mathfrak{n} \oplus \mathfrak{n}^*).$$

Somewhat remarkably, there exists in $A \otimes \mathcal{C}\ell(\mathfrak{a})$ a cubic element

$$Q = \sum_i \alpha(\phi_i) \otimes \phi_i^* - 1 \otimes \frac{1}{2} \sum_{i,j} \phi_i^* \phi_j^* [\phi_i, \phi_j]$$

(the summation indices run over a basis of $\mathfrak{n}$) with the property that $Q^2 = 0$ (which implies that $\mathrm{ad}(Q)^2 = 0$ and hence $\mathrm{ad}(Q)$ defines a differential in $A \otimes \mathcal{C}\ell(\mathfrak{a})$), and in the associated graded we recover the total differential as

$$d_{\mathrm{Tot}} = \{Q, -\}.$$

The proposal, therefore, is to regard

$$H^\bullet(A \otimes \mathcal{C}\ell(\mathfrak{a}), \mathrm{ad}(Q))$$

as a quantisation of $\mu^{-1}(0)/N$.

For simplicity we have described the reduction of $\mu^{-1}(0)$ and its quantisation. All of this can be generalised to $\mu^{-1}(p)$ for suitable $p \in M$ by modifying the differential Q.

7. Slodowy slices and Drinfeld-Sokolov reduction

We wish to apply the machinery described above to a particular class of examples of Hamiltonian reductions, known as Slodowy slices. In this particular case, the relevant structures all have vertex algebraic analogues, and so a parallel story can be developed for vertex algebras. In this context the construction is known as quantised Drinfeld-Sokolov reduction.

Let $\mathfrak{g}$ be a simple Lie algebra with invariant bilinear form $(\cdot, \cdot)$, and let $f \in \mathfrak{g}$ be a nilpotent element. The Jacobson-Morozov theorem [**11**] asserts that f can be included into an $\mathfrak{sl}_2$-triple $\{e,h,f\} \subset \mathfrak{g}$. The endomorphism $\operatorname{ad}(h)$ induces a decomposition of $\mathfrak{g}$ into eigenspaces

$$\mathfrak{g} = \bigoplus_{j\in\mathbb{Z}} \mathfrak{g}_j, \qquad \operatorname{ad}(h)|_{\mathfrak{g}_j} = j. \tag{7.1}$$

At the same time, $\mathfrak{g}$ decomposes into finite dimensional irreducible $\mathfrak{a}$-modules, where $\mathfrak{a} \subset \mathfrak{g}$ denotes the subalgebra isomorphic to $\mathfrak{sl}_2$ spanned by e, h and f. If $\mathfrak{g}_j$ vanishes for odd j then we say the nilpotent element f is even. Although it is not essential, in these notes we restrict our discussion to even nilpotent elements.

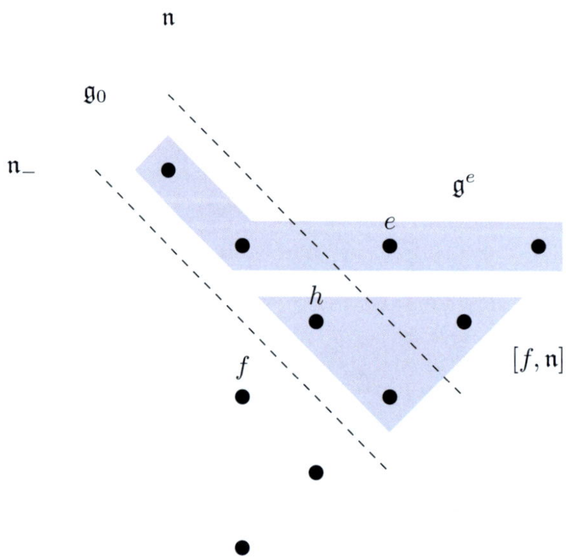

FIGURE 6. Decomposition of $\mathfrak{g}$ with $\mathfrak{g}^e$ and $[f,\mathfrak{n}]$ highlighted.

Let us consider the subalgebras $\mathfrak{n} = \mathfrak{g}_{>0}$ and $\mathfrak{g}^e = \{x \in \mathfrak{g} \mid [e,x] = 0\}$, and the subspace $\mathfrak{n}^\perp = \{x \in \mathfrak{g} \mid (x,\mathfrak{n}) = 0\}$ (which in this case is also a subalgebra, in fact). It is not difficult to see that

$$[\mathfrak{n}, f] + \mathfrak{g}^e = \mathfrak{g}_{\geq 0} = \mathfrak{n}^\perp. \tag{7.2}$$

This equality has been upgraded by Gan and Ginzburg [**24**] to an isomorphism of algebraic (Poisson) varieties

$$N \times (f + \mathfrak{g}^e) \cong f + \mathfrak{n}^\perp,$$

where N is the unipotent Lie group with Lie algebra $\mathfrak{n}$ and the isomorphism is given by the action map. The isomorphism (7.2) is recovered as the infinitesimal expression. This motivates us to interpret the following natural maps

$$\alpha : \mathfrak{n} \to \mathfrak{g} \subset S(\mathfrak{g}) = \mathbb{C}[\mathfrak{g}^*], \qquad \mu : \mathfrak{g}^* \to \mathfrak{n}^*,$$

as Hamiltonian action and momentum map. The affine variety

$$S = S_f = f + \mathfrak{g}^e$$

is known as the *Slodowy slice* associated with f. The theorem of Gan and Ginzburg, generalising the case of principal nilpotent due to Kostant, is as follows.

THEOREM 7.1 ([**24**]). *Let $\chi \in \mathfrak{g}^*$ identified with $f \in \mathfrak{g}$ under the invariant bilinear form $(\cdot,\cdot)$. Then the Hamiltonian reduction*

$$\mu^{-1}(\chi)/N$$

is an affine Poisson variety, naturally isomorphic to the Slodowy slice

$$S = f + \mathfrak{g}^e.$$

In the case of the Poisson varieties $\mathfrak{g}^*$ and S_f, enough of the notions and objects have clear infinite dimensional analogues to permit the rendition of the BRST construction to the context of vertex algebras.

We take the universal affine vertex algebra $V = V^k(\mathfrak{g})$ and the charged free fermions vertex superalgebra $F = F^{\mathrm{ch}}(\mathfrak{n})$ based on the subspace $\mathfrak{n} \subset \mathfrak{g}$. Recall that F is generated by odd fields ϕ_i and ϕ_i^* indexed by dual bases of $\mathfrak{n}$ and $\mathfrak{n}^*$, with

$$[\phi_i{}_\lambda \phi_j^*] = \delta_{ij}.$$

We overload notation by writing $\alpha(\phi)$ for the field in V corresponding to $\alpha(\phi) \in \mathfrak{g}$. Let $C = V \otimes F$ and put

$$Q^{\mathrm{st}} = \sum_i \alpha(\phi_i) \otimes \phi_i^* - \mathbf{1} \otimes \frac{1}{2} \sum_{i,j} :[\phi_i, \phi_j]\, \phi_i^* \phi_j^* : .$$

Using the λ-bracket calculus, we may compute explicitly

$$[Q^{\mathrm{st}}_\lambda Q^{\mathrm{st}}] = 0,$$

and so $d_{\mathrm{st}} = Q^{\mathrm{st}}_{(0)}$, which satisfies $d_{\mathrm{st}}^2 = 0$, will be a differential.

The definition of *semi-infinite homology* has been introduced by B. Feigin [**18**], and the complex (C, d_{st}) coincides with the complex of semi-infinite homology of the $\mathfrak{n}[t,t^{-1}]$-module V [**20**, p. 246]. For this reason its homology is often met in the literature under the following notation

$$H_i(C, d_{\mathrm{st}}) = H^{\frac{\infty}{2}+i}(\mathfrak{n}[t,t^{-1}], V).$$

Once again, this homology corresponds to a quantisation of a Hamiltonian reduction of $\mu^{-1}(0)$. The modification from $\mu^{-1}(0)$ to $\mu^{-1}(\chi)$ entails replacement of Q^{st} with

$$Q^{\mathrm{st}} + p, \quad \text{where} \quad p = \sum_i (f, \phi_i)\mathbf{1} \otimes \phi_i^*.$$

It may be verified, again by direct calculation with λ-brackets, that $d = (Q^{\mathrm{st}} + p)_{(0)}$ satisfies $d^2 = 0$ [**30**, p. 316]. Finally the homology vertex algebra

$$W = H(C, d)$$

is the affine W-algebra.

8. Structure, modules and characters of W-algebras

We have just defined a vertex algebra, really a class of vertex algebras, $W = H(C, d)$. But we have little idea what these algebras “look like”. Vertex algebras are infinite dimensional, but a grading (for example a conformal grading) with finite dimensional components allows us to write down a character, or graded dimension, of V. Such characters are a useful tool in the theory.

We have already seen that $V = V^k(\mathfrak{sl}_2)$ has a conformal structure, and thus a grading (in this case a $\mathbb{Z}_+$-grading) by eigenvalues of L_0. Counting the monomials (2.2) with conformal weight yields the character

$$\chi_V(q) = \mathrm{Tr}_V\, q^{L_0} = \prod_{n=1}^{\infty} \frac{1}{(1-q^n)^3}.$$

The vertex superalgebra $F = F^{\mathrm{ch}}(\mathbb{C}\phi)$ is also conformal, with conformal vector

$$L^{\mathrm{ch}} =: (T\phi)\phi^* := - : \phi^*(T\phi) : .$$

Under this conformal structure we have $\Delta(\phi) = 0$ and $\Delta(\phi^*) = 1$. For a vector superspace $U = U_{\bar{0}} \oplus U_{\bar{1}}$ the super dimension $\dim(U_{\bar{0}}) - \dim(U_{\bar{1}})$ is more natural than the dimension. The supercharacter of F is

$$\chi_F(q) = \prod_{n=1}^{\infty} (1-q^{n-1})(1-q^n) = 0,$$

in this case the even and odd components of F have matching characters, which cancel out. An additional $\mathbb{Z}$-grading by "charge" can be introduced in F so that $F = \bigoplus_{m\in\mathbb{Z}} F^m$, in which ϕ^* has charge $+1$ and ϕ has charge -1. In fact, a λ-bracket computation reveals that the charge grading coincides with the grading by eigenvalues of ξ_0, where $\xi =: \phi^*\phi :$. We now have a two-variable character

$$\chi_F(q,y) = \mathrm{Tr}_F\, y^{\xi_0} q^{L_0} = \prod_{n=1}^{\infty} (1-y^{-1}q^{n-1})(1-yq^n). \tag{8.1}$$

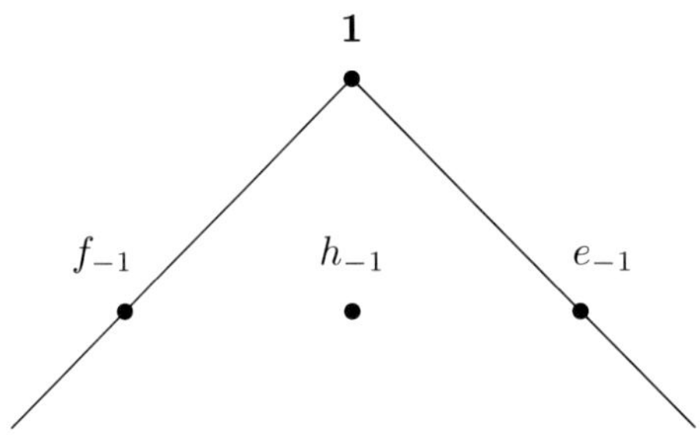

FIGURE 7. The vertex algebra $V^k(\mathfrak{sl}_2)$.

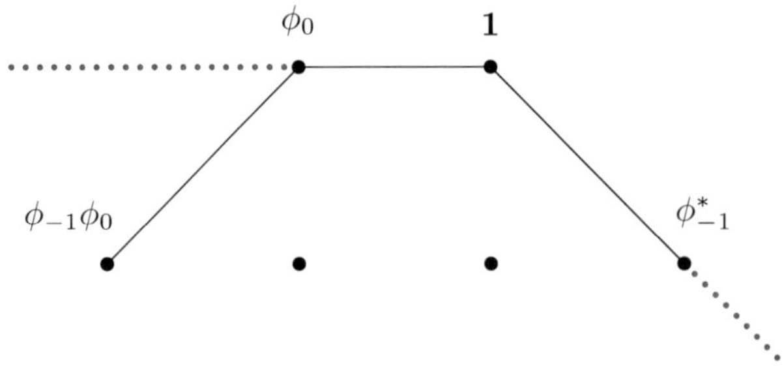

FIGURE 8. The rank 1 charged free fermions vertex algebra F^{ch}.

We have seen in the previous section that the vertex algebra BRST complex associated with a nilpotent element is given by a tensor product of the form $C = V \otimes F$ where $V = V^k(\mathfrak{g})$ and $F = F^{\text{ch}}(\mathfrak{n})$. In fact the homological $\mathbb{Z}$-grading on C is precisely that inherited from the charge grading on F.

If the differential d in the complex C is compatible with the conformal grading, i.e., if $[L_0, d] = 0$, then we have

$$\sum_{n\in\mathbb{Z}} \chi_{H^n(C(M))}(q) = \sum_{n\in\mathbb{Z}} \chi_{C^n(M)}(q) \tag{8.2}$$

by the Euler-Poincaré principle (we work with superdimensions here, so components in odd homological degree are odd vector superspaces, and have negative contribution to the superdimension). Later we will see that under favourable circumstances

$$H^{\neq 0}(C) = 0,$$

which implies that the character of the BRST reduction can be computed explicitly from knowledge of the character of $V^k(\mathfrak{g})$. Similar remarks apply in the case of nontrivial $V^k(\mathfrak{g})$-modules and their BRST reductions.

It turns out that the zero mode of the most obvious conformal vector on C, namely $L \otimes \mathbf{1} + \mathbf{1} \otimes L^{\text{ch}}$, does *not* commute with d, but an easy modification fixes the problem: If we set

$$L^C = \left(L + \frac{1}{2}Th\right) \otimes \mathbf{1} + \mathbf{1} \otimes L^{\text{ch}} \tag{8.3}$$

then $[L_0^C, d] = 0$. A similar manoeuvre can be performed in the general case. It transpires that in order to obtain a conformal structure on C compatible with the differential $d = (Q^{\text{st}} + p)_{(0)}$, we should take

$$L^{\text{ch}} = \sum_i \left(\text{ht}(\phi_i) : (T\phi_i)\phi_i^* : + (\text{ht}(\phi_i) - 1) : \phi_i(T\phi_i^*) :\right),$$

where $\{\phi_i\}$ is a basis of $\mathfrak{n}$ homogeneous with respect to the decomposition (7.1), and we have written $\text{ht}(\phi) = j$ for $\phi \in \mathfrak{g}_j$. This is forced on us by the requirement that $\Delta(p) = 1$. And we should take L^C as in formula (8.3), where now $h \in \mathfrak{g}$ is the semisimple element of chosen $\mathfrak{sl}_2$-triple for f.

Returning to the case of $\mathfrak{sl}_2$, the conformal vector (8.3) gives $e, h, f \in V$ conformal weights 0, 1 and 2 respectively. This is bad news for our computation of the character of $H^0(C)$ via (8.2), because now the character of V has become ill-defined; the infinite set of vectors e_{-1}^N all have the same conformal weight of 0.

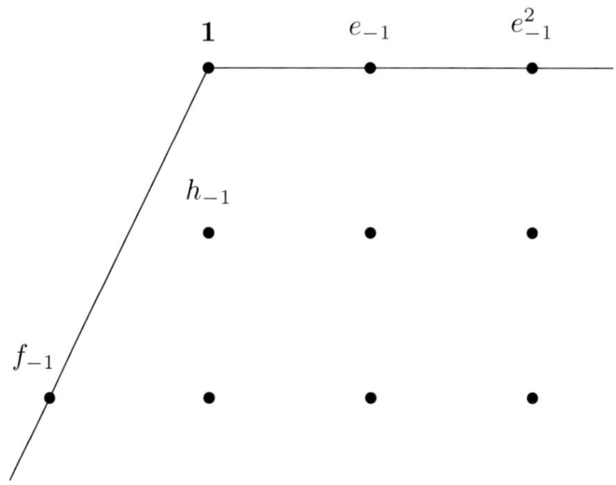

FIGURE 9. The vertex algebra $V^k(\mathfrak{sl}_2)$ with modified conformal grading.

We can obtain a well-defined character by introducing a second gradation, in which e, h and f have degrees -1, 0 and 1, respectively. In other words, gradation by eigenvalues of $-x_0$ where $x = h/2$. The character of V becomes

$$\chi_V(q) = \operatorname{Tr}_V y^{-x_0} q^{(L+Tx)_0} = \prod_{n=1}^{\infty} \frac{1}{(1-y^{-1}q^{n-1})(1-q^n)(1-yq^{n+1})}. \tag{8.4}$$

Upon passage to homology, the infinite dimensional components of V will be reduced to finite dimension, and $H^0(C)$ will turn out to be conformal with finite dimensional graded pieces (see Theorem 8.1 below). At the level of characters, this fact corresponds to a cancellation between the factor $1 - y^{-1}$ in the expression (8.1) for the character of F (which tends to zero as y tends to 1), with the factor $(1 - y^{-1})^{-1}$ in the expression (8.4) for the character of V (which tends to infinity as y tends to 1). Let us then set

$$\widetilde{x} = x \otimes \mathbf{1} + \mathbf{1} \otimes :\phi\phi^*: \in V \otimes F$$

so that

$$\begin{aligned}\chi_C(q, y) = \operatorname{Tr}_C y^{-\widetilde{x}_0} q^{L_0^C} &= \prod_{n=1}^{\infty} \frac{(1-y^{-1}q^n)(1-yq^{n-1})}{(1-y^{-1}q^{n+1})(1-q^n)(1-yq^{n-1})} \\ &= (1-yq) \prod_{n=1}^{\infty} \frac{1}{1-q^n}.\end{aligned}$$

Finally we can pass to the $y \to 1$ limit and obtain the graded dimension of the BRST reduction

$$\chi_{H^0(C)}(q) = \prod_{n=2}^{\infty} \frac{1}{1-q^n}.$$

This is precisely the character of the universal Virasoro vertex algebra, and in fact

$$H^0(C) \cong \operatorname{Vir}^c$$

at central charge

$$c = -\frac{6k^2 + 11k + 4}{k+2}.$$

(Again, see Theorem 8.1 below.) There is one problem in the story outlined so far, namely that the zero mode $\widetilde{x}_0$ does not commute with d. It turns out that $\widetilde{x}_0$ does commute with $Q^{\text{st}}_{(0)}$, and the required character formula for $H^0(C)$ can be obtained by splitting d as $d = Q^{\text{st}}_{(0)} + p_{(0)}$ and considering the corresponding spectral sequence. Already after passage to the $Q^{\text{st}}_{(0)}$-homology the L_0^C graded pieces become finite dimensional, after which point the limit $y \to 1$ can be taken legitimately.

Notice that $:\phi\phi^*:$ above is precisely F^x, where we are thinking of $\mathfrak{n} = \mathbb{C}\phi$ as a $\mathfrak{h}$-module.

In general we consider

$$F^h = \sum_i :(\mathrm{ad}(h)\phi_i)\phi_i^*: \quad \text{and} \quad \widetilde{h} = h \otimes \mathbf{1} + \mathbf{1} \otimes F^h,$$

and the arguments go through as above.

The following theorem was proved by Feigin and Frenkel [**17**] for f a principal nilpotent element, and in the general case by Kac and Wakimoto [**40**].

THEOREM 8.1 ([**40**, Theorem 4.1]). *The BRST homology $H^i(C,d)$ vanishes for $i \neq 0$, and for $i = 0$ is a vertex algebra with a set of strong generators parametrised by a basis of $\mathfrak{g}^f = \{x \in \mathfrak{g} \mid [f,x] = 0\}$. We denote it*

$$W^k(\mathfrak{g}, f).$$

Furthermore, if $\{a_1, \dots, a_r\}$ is a basis of $\mathfrak{g}^f$ homogeneous with respect to the grading $\mathfrak{g} = \bigoplus_{j\in\mathbb{Z}} \mathfrak{g}_j$ and $a_i \in \mathfrak{g}_{-j}$ then the corresponding generator of $W^k(\mathfrak{g}, f)$ has conformal weight $j+1$.

Here is a sketch of the proof, following the approach originated by de Boer and Tjin [**12**] in the case of f principal and $\mathfrak{g} = \mathfrak{sl}_n$, and extended to other cases in [**20**] and [**40**]. One introduces subcomplexes

$$C_0 = \left\langle \widetilde{\mathfrak{h} + \mathfrak{n}_-}, \phi^* \right\rangle \qquad \text{and} \qquad C' = \langle \widetilde{\mathfrak{n}}, \phi \rangle$$

of $C = V^k(\mathfrak{g}) \otimes F(\mathfrak{n})$, where $\widetilde{\mathfrak{n}}$ stands for $\{\widetilde{x} \mid x \in \mathfrak{n}\}$, etc.[1] It turns out that $C = C_0 \otimes C'$ as complexes. It is easy to show that $H(C')$ is $\mathbb{C}$ concentrated in degree 0. So we have

$$H(C) = H(C_0).$$

To continue, we split $d = Q^{\text{st}}_{(0)} + p_{(0)}$, compute cohomology first with respect to $p_{(0)}$, and then with respect to $Q^{\text{st}}_{(0)}$. A further factorisation occurs, with $(C_0, p_{(0)}) = V^\phi(\mathfrak{g}^f) \otimes \widetilde{C}_0$ as complexes, where the cohomology of $\widetilde{C}_0$ is $\mathbb{C}$ concentrated in degree 0. The spectral sequence collapses on the next page, and we obtain

$$H^0(C) = V^\phi(\mathfrak{g}^f)$$

as (graded) vector spaces, as required.

Determining the vertex algebra structure of $W^k(\mathfrak{g}, f)$ explicitly requires lifting the elements of $V(\mathfrak{g}^f)$ to d-cocycles, which turns out to be very complicated in

[1] In fact $\mathfrak{n}$ is not a $\mathfrak{n}_-$-module, so the meaning of F^x for $x \in \mathfrak{n}_-$ requires explanation. A basis of $\mathfrak{n}$ is extended to a basis of $\mathfrak{g}$, the action of $\mathfrak{n}_-$ is written relative to this basis, and components of the action of $\mathrm{ad}(x)$ in basis vectors outside $\mathfrak{n}$ are discarded.

general. Nevertheless, we have the character of $W^k(\mathfrak{g}, f)$; it is

$$\prod_{j \in \mathbb{Z}_+} \prod_{n=1}^{\infty} \frac{1}{1 - q^{n+\dim(\mathfrak{g}^f_{-j})+1}}$$

The algebra of functions on the Slodowy slice S_f is the BRST reduction of the Poisson algebra $S(\mathfrak{g})$ by a Hamiltonian action. Replacing $S(\mathfrak{g})$ by its obvious quantisation $U(\mathfrak{g})$ and carrying out the BRST recipe, yields a class of associative algebras $U(\mathfrak{g}, f)$ known as finite W-algebras. One might ask for the relation between $W^k(\mathfrak{g}, f)$ and $U(\mathfrak{g}, f)$. The answer turns out to be as simple as one could hope for:

THEOREM 8.2 ([**13**]). *The Zhu algebra of $W^k(\mathfrak{g}, f)$ is the finite W-algebra $U(\mathfrak{g}, f)$.*

The case of principal nilpotent element f had been proved earlier by Arakawa [**2**], in this case the finite W-algebra coincides with the centre $Z(\mathfrak{g})$ of the associative algebra $U(\mathfrak{g})$.

The argument, in a nutshell, is this. The complex $C = V^k(\mathfrak{g}) \otimes F^{\text{ch}}(\mathfrak{n})$ is itself a vertex superalgebra, and so has a Zhu algebra. This associative algebra inherits a differential, and there is a natural map

$$\text{Zhu}(W^k(\mathfrak{g}, f)) \to H^0(\text{Zhu}(C^\bullet), d).$$

One directly verifies that the Zhu algebra complex is nothing other than the finite BRST complex which computes $U(\mathfrak{g}, f)$ by the definition of the latter.

There is a complication, however. The preceding statements are true of the vertex algebra $V^k(\mathfrak{g}) \otimes F^{\text{ch}}(\mathfrak{n})$ in which $V^k(\mathfrak{g})$ is equipped with its standard conformal vector. But we were forced to modify this vector

$$L \to L + Tx$$

to ensure compatibility with d. This change alters the conformal weights and conformal gradings of the quantum fields in a systematic way, specifically

$$a_n \mapsto a_{n-j} \quad \text{for } a \in \mathfrak{g}_j,$$

and thus alters the Zhu algebra. The effect of this modification is ultimately that the Zhu algebra is still isomorphic to $U(\mathfrak{g}, f)$, but the bijection between irreducible positive energy $W^k(\mathfrak{g}, f)$-modules and irreducible $U(\mathfrak{g}, f)$-modules is more convoluted than one might expect.

We continue in the principal case for simplicity. Irreducible $Z(\mathfrak{g})$-modules are one dimensional and are known as central characters. For example any Verma $\mathfrak{g}$-module $M(\lambda) = U(\mathfrak{g})v_\lambda$ induces a central character

$$\gamma_\lambda : Z(\mathfrak{g}) \to \text{End}(\mathbb{C})$$

defined by $zv_\lambda = \gamma_\lambda(z)v_\lambda$ for all $z \in Z(\mathfrak{g})$. Chasing through the identifications sketched above, one discovers that the $Z(\mathfrak{g})$-module corresponding to the principal affine W-algebra itself is

$$W^k(\mathfrak{g}, f_{\text{prin}})_{\text{low}} \cong \gamma_{-(k+h^\vee)x}.$$

We now describe a variant of the BRST reduction functor, which has superior formal properties for applications in representation theory. We build a charged free fermion vertex superalgebra based on $\mathfrak{n}_-$ now, that is $F_- = F^{\text{ch}}(\mathfrak{n}_-)$. We equip F_- with a conformal structure such that $\Delta(\mathfrak{n}_-) = 0$ and $\Delta(\mathfrak{n}_-^*) = 1$.

Let M be a $V^k(\mathfrak{g})$-module. We define $H_-(M)$ to be the cohomology of the complex

$$C_-(M) = M \otimes F_-, \quad d = (Q^- + p_-)_{(0)}$$

where

$$\begin{aligned} Q^- &= \sum_{i \in \Delta_-} \alpha(\phi_i) \otimes \phi_i^* - \mathbf{1} \otimes \frac{1}{2} \sum_{i,j \in \Delta_-} :[\phi_i, \phi_j]\, \phi_i^* \phi_j^* : \\ p_- &= \sum_i (e, \phi_i) \mathbf{1} \otimes \phi_i^*. \end{aligned} \tag{8.5}$$

The vector space $H_-(M)$ carries a natural $W^k(\mathfrak{g}, f)$-module structure. The explicit description of this action involves $w_\circ$ the longest element in the Weyl group of $\mathfrak{g}$ and a few other ingredients. We refer the reader to [**2**, p. 291] for details.

Arakawa proved the following vanishing theorem for $H_-(-)$. Recall the decomposition $\mathfrak{g} = \bigoplus_{j \in \mathbb{Z}} \mathfrak{g}_j$ associated with f. Recall also that the category of $V^k(\mathfrak{g})$-modules is equivalent to the category of smooth $\widehat{\mathfrak{g}}$-modules of level k. Let $\mathcal{O}_k$ denote category of $\widehat{\mathfrak{g}}$-modules M of level k, admitting a weight space decomposition $M = \bigoplus_{\mu \in \widehat{\mathfrak{h}}^*} M_\mu$ with finite dimensional weight spaces, such that $\text{Supp}(M) \subset \cup_i (\mu_i - Q_+)$, where i runs over some finite set. Further let $\mathcal{O}_k^{(0)}$ be the full subcategory of $\mathcal{O}_k$ consisting of $\widehat{\mathfrak{g}}$-modules integrable as modules over the subalgebra $\mathfrak{g}_0 \subset \mathfrak{g} \subset \widehat{\mathfrak{g}}$. If f is principal then $\mathcal{O}_k^{(0)} = \mathcal{O}_k$ because $\mathfrak{g}_0 = \mathfrak{h}$.

THEOREM 8.3 ([**2**, Theorem 7.6.1] [**3**, Theorem 5.4.1]). *Let $\mathfrak{g}$ be a simple Lie algebra, $f \in \mathfrak{g}$ nilpotent, and other constructions as described above.*

- *If f is principal, and $M \in \mathcal{O}_k$, then $H_-^i(M) = 0$ for all $i \neq 0$,*
- *If f is even, and $M \in \mathcal{O}_k^{(0)}$, then $H_-^i(M) = 0$ for all $i \neq 0$,*
- *More generally if f possesses a good even grading, and $M \in \mathcal{O}_k^{(0)}$, then $H_-^i(M) = 0$ for all $i \neq 0$ (though the $H_-^i(M)$ are not $W^k(\mathfrak{g}, f)$-modules, rather they are Ramond twisted $W^k(\mathfrak{g}, f)$-modules).*

Next comes the following theorem, which establishes the structure of $H_-(-)$ reduction of Verma $\widehat{\mathfrak{g}}$-modules. To state the theorem, we need to elaborate a little on the relation between irreducible positive energy V-modules and irreducible $\text{Zhu}(V)$-modules. It is possible to construct a Lie algebra $\text{Lie}(V)$ canonically from V, whose elements are symbols a_n as a ranges over V and n ranges over $\mathbb{Z}$, with the property that any V-module becomes a $\text{Lie}(V)$-module in the obvious way. This Lie algebra is $\mathbb{Z}$-graded and, it turns out, there is a homomorphism $\text{Lie}(V)_0 \to \text{Zhu}(V)$, where $\text{Zhu}(V)$ is equipped with the commutator bracket. Starting from a $\text{Zhu}(V)$-module N, which we can convert into a $\text{Lie}(V)_0$-module, we may impose an action of $\text{Lie}(V)_{>0}$ in N by 0, and then induce to obtain a $U(\text{Lie}(V))$-module

$$\widetilde{M} = U(\text{Lie}(V)) \otimes_{U(\text{Lie}(V))_{\geq 0}} N.$$

A quotient of $\widetilde{M}$ by a subspace $(B) \cdot \widetilde{M}$ (which corresponds to the imposition of Borcherds identity), results in a V-module

$$\mathbb{M}(N) = \widetilde{M}/(B) \cdot \widetilde{M}$$

which is positive energy with $\mathbb{M}(N)_{\text{low}}$. The irreducible module $\mathbb{L}(N)$, already introduced in Section 2, is the quotient of $\mathbb{M}(N)$ by its maximal proper V-submodule.

The relationship between $\mathbb{M}(N)$ and $\mathbb{L}(N)$ is reminiscent of the relationship between Weyl $\widehat{\mathfrak{g}}$-modules and irreducible highest weight $\widehat{\mathfrak{g}}$-modules.

THEOREM 8.4 ([**2**, Theorem 7.5.1]). *The reduction of Verma modules*

$$H^0_-(M_k(\lambda)) \cong \mathbb{M}(\gamma_{\lambda^*})$$

Here $\lambda^* = -w_\circ(\lambda)$ *is the dual weight.*

To prove this theorem, one first checks that the central character of the lowest piece is as stated. Note that λ is "twisted" to λ^*, this is due to the structure of the action of $W^k(\mathfrak{g}, f)$ on $H_-(M)$ as commented after equation (8.5). By an analysis similar to the proof of Theorem 8.1, it is shown that $H^0_-(M(\lambda))$ has a PBW basis. This implies that the canonical map

$$\mathbb{M}(\gamma_{\lambda^*}) \to H^0_-(M_k(\lambda))$$

must be an isomorphism.

Finally, using the previous theorem and exactness of $H^0_-(-)$, Arakawa proves the following theorem.

THEOREM 8.5 ([**2**, Theorem 7.6.3]). *For* λ *antidominant, the reduction of* $L_k(\lambda)$ *is*

$$H^0_-(L_k(\lambda)) \cong \mathbb{L}(\gamma_{\lambda^*})$$

Here $\lambda^* = -w_0\lambda$ *is the dual weight.*

This result really is key since with it, knowledge of the characters of $V^k(\mathfrak{g})$-modules can be used to compute the characters of the corresponding $W^k(\mathfrak{g}, f)$-modules.

We sketch the proof. By exactness of $H^0_-(-)$ we have a surjection

$$\mathbb{M}(\gamma_{\lambda^*}) = H^0_-(M_k(\lambda)) \to H^0_-(L_k(\lambda)),$$

and an injection

$$H^0_-(L_k(\lambda)) \to H^0_-(D(M_k(\lambda))) = D(\mathbb{M}(\gamma_\lambda)).$$

A standard argument (see [**27**, Theorem 3.3] for the prototype case of dual Verma modules over finite dimensional simple Lie algebras) says that image of $\mathbb{M}(\gamma_{\lambda^*})$ in $D(\mathbb{M}(\gamma_\lambda))$ is the unique simple highest weight submodule of the latter, isomorphic to $\mathbb{L}(\gamma_{\lambda^*})$.

9. Exceptional W-algebras

The universal affine vertex algebra $V^k(\mathfrak{g})$ is simple for generic values of the level k, but at special levels $V^k(\mathfrak{g})$ has a nontrivial simple quotient $L_k(\mathfrak{g})$ whose representation theory is quite interesting. In Theorem 2.5 for instance we saw that $L_k(\mathfrak{g})$ has finitely many irreducible positive energy modules when $k \in \mathbb{Z}_+$. Furthermore in this case $L_k(\mathfrak{g})$ is an example of a *rational* vertex algebra, which loosely means that all of its modules are completely reducible into direct sums of irreducible modules, and that it has finitely many irreducible modules. For admissible level k the vertex algebra $L_k(\mathfrak{g})$ is not rational, but we have seen in Theorem 3.2 that some parallels with the positive integral case persist.

On the other hand we have examined the BRST procedure, and obtained a class of vertex algebras $W^k(\mathfrak{g}, f)$, the universal affine W-algebras. Once again these vertex algebras are simple for generic level k, but have interesting simple quotients at special levels. A basic question is: for which combinations of $\mathfrak{g}$, f and k is the simple quotient $W_k(\mathfrak{g}, f)$ of $W^k(\mathfrak{g}, f)$ a rational vertex algebra?

A natural class of candidates is given by the *exceptional W-algebras*, which we define below. In this section we describe some results from our article [8] on the structure and representation theory of exceptional W-algebras, including rationality in many cases.

In Theorem 2.5 we saw the classification of irreducible positive energy $L_k(\mathfrak{g})$-modules, for $k \in \mathbb{Z}_+$. By the correspondence with irreducible modules over the Zhu algebra, it follows that

$$\mathrm{Zhu}(L_k(\mathfrak{g})) \cong \bigoplus_{\lambda \in P_+^k} \mathrm{End}(L(\lambda)).$$

In [8] we generalise this result from positive integral to other admissible levels. To state the result, we first fix some notation. Let $\mathfrak{g}$ be a finite dimensional simple Lie algebra, and $U = U(\mathfrak{g})$. For a U-module M we consider the two-sided ideal

$$\mathrm{Ann}_U(M) = \{u \in U \mid uM = 0\},$$

and for $\lambda \in \mathfrak{h}^*$ we write J_λ for $\mathrm{Ann}_U(L(\lambda))$. Recall that a weight λ is said to be dominant if $\langle \lambda + \rho, \alpha^\vee \rangle \notin \mathbb{Z}_{\leq 0}$ for all $\alpha \in \Delta_+$. A result of Joseph asserts that if λ and $w \circ \lambda$ are dominant then $J_{w\circ\lambda} = J_\lambda$. In particular admissible weights are dominant, so if we write Pr_k for the set of admissible weights of admissible level k, then J_λ depends on $\lambda \in \mathrm{Pr}_k$ only through its image in

$$[\mathrm{Pr}_k] = \mathrm{Pr}_k / \sim,$$

where $\sim$ denotes the equivalence relation on $\mathfrak{h}^*$ induced by the $W \circ (-)$ action.

THEOREM 9.1 ([8]). *Let k be an admissible level for $\mathfrak{g}$, then*

$$\mathrm{Zhu}(L_k(\mathfrak{g})) \cong \prod_{\lambda \in [\mathrm{Pr}_k]} U/J_\lambda.$$

The Zhu algebra of $L_k(\mathfrak{g})$ (for any level k) is isomorphic to the quotient U/I_k for some ideal I_k which depends on k. We consider the tensor product of $\mathrm{Zhu}(L_k(\mathfrak{g}))$, over the centre $Z = Z(U)$, with a central character. The basic observation in the proof of the theorem is that

$$\mathrm{Zhu}(L_k(\mathfrak{g})) \otimes_Z \mathbb{C}\gamma_\lambda = U/J$$

for some two sided ideal J, and since $M(\lambda)/JM(\lambda)$ would be a highest weight $\mathrm{Zhu}(L_k(\mathfrak{g}))$-module, it can only be nonzero if $\lambda \in \mathrm{Pr}_k$ and furthermore we must have $M(\lambda)/JM(\lambda) = L(\lambda)$, and hence $J = J_\lambda$.

Recall that $\mathrm{Zhu}(W^k(\mathfrak{g}), f) \cong U(\mathfrak{g}, f)$. Since $L_k(\mathfrak{g})$ is a quotient of $V^k(\mathfrak{g})$, the Zhu algebra of $H^0(L_k(\mathfrak{g}))$ is a certain quotient of $U(\mathfrak{g}, f)$, which we would now like to determine via BRST reduction, using the description obtained above of the Zhu algebra of $L_k(\mathfrak{g})$. Here and below let us write

$$W = H^0(L_k(\mathfrak{g})).$$

THEOREM 9.2 ([8]). *Let k be an admissible level for $\mathfrak{g}$, then*

$$\mathrm{Zhu}(W) \cong \prod_{\lambda \in [\mathrm{Pr}_k]} H^0(U/J_\lambda).$$

Note that it can easily happen that $W = 0$. Indeed if $k \in \mathbb{Z}_+$ then $L_k(\mathfrak{g})$ is G-integrable and so $H^0(L_k(\mathfrak{g})) = 0$ can already be deduced from consideration of characters as in Section 8. If $W \neq 0$ then certainly there exists a surjection from W to $W_k(\mathfrak{g}, f)$ the unique simple quotient.

The universal enveloping algebra $U = U(\mathfrak{g})$ is naturally filtered by "word length" and the associated graded algebra $\mathrm{gr}(U)$ is isomorphic to the symmetric algebra $S(\mathfrak{g})$ (in fact this is one formulation of the PBW theorem). Let $\mathfrak{g}$ be finite dimensional and simple, so that $S(\mathfrak{g})$ is identified with the algebra of polynomials on $\mathfrak{g} \cong \mathfrak{g}^*$. An ideal $I \subset U$ induces an ideal $\mathrm{gr}(I) \subset S(\mathfrak{g})$ and the corresponding zero set is known as the associated variety $\mathrm{Var}(I) \subset \mathfrak{g}$. We may also write, more specifically, $\mathrm{Var}(M)$ for $\mathrm{Var}(\mathrm{Ann}_U(M))$.

Let k be an admissible level for $\mathfrak{g}$, then a theorem of Arakawa [**4**, Theorem 5.14] asserts that $\mathrm{Var}(I_k) \subset \mathfrak{g}$ is in fact the closure of a nilpotent orbit, which depends only on the denominator, so we call it $\mathbb{O}_q$.

Now a result of Losev [**47**] asserts that

$$H^0(U/J_\lambda) = \begin{cases} 0 & \mathrm{Var}(J_\lambda) \subsetneq \overline{G \cdot f} \\ \text{finite dimensional} & \mathrm{Var}(J_\lambda) = \overline{G \cdot f} \end{cases}$$

Of course, for all $\lambda \in \mathrm{Pr}_k$ we have

$$\mathrm{Var}(J_\lambda) \subset \mathrm{Var}(I_k).$$

It therefore makes sense to consider the restriction on our choice of f:

HYPOTHESIS 1. *The nilpotent element f is chosen in the open part of* $\mathrm{Var}(I_k)$, *namely in $\mathbb{O}_q$.*

In fact let us elevate this hypothesis to a definition

DEFINITION 9.3. The W-algebra $H^0(L_k(\mathfrak{g})), f)$ will be said to be exceptional if $f \in \mathbb{O}_q$.

(Actually this definition is similar but slightly different to the notion of exceptional introduced in [**39**]). One of the key results of [**4**] is that exceptional W-algebras are C_2-cofinite. In general a vertex algebra V is said to be C_2-cofinite (or "lisse") if $V/V_{(-2)}V$ is finite dimensional. The importance of this condition, which was introduced by Zhu in [**52**], is certainly not immediately clear, but it turns out to imply several other desirable finiteness and convergence properties for objects associated with V. In particular, although C_2-cofiniteness is known not to imply rationality, it suggests it.

Anyway, under Hypothesis 1, we have $\mathrm{Var}(I_k) = \overline{G \cdot f}$, and so $\mathrm{Var}(J_\lambda) \subset \overline{G \cdot f}$ for all $\lambda \in \mathrm{Pr}_k$. Let us call an admissible weight λ "replete" if we have equality $\mathrm{Var}(J_\lambda) = \overline{G \cdot f}$. Intuitively the replete weights λ are those for which we expect a nonzero W-module obtained as Hamiltonian reduction of $L(\lambda)$.

COROLLARY 9.4 ([**8**]). *The Zhu algebra of an exceptional W-algebra is finite dimensional and semisimple.*

REMARK 9.5. McRae has recently proved that a C_2-cofinite vertex algebra V for which $\mathrm{Zhu}(V)$ is finite dimensional and semisimple (and for which a few other more technical hypotheses are satisfied) is rational [**49**]. It follows that exceptional W-algebras are rational.

In order to go further and explicitly describe the irreducible W-modules using Hamiltonian reduction, we have found it convenient to impose some technical conditions. (For some progress in cases outside the scope of these conditions, see [**16**].)

Let us call an admissible weight λ "conservative" for f if

- λ is dominant integral with respect to $\mathfrak{g}_0$ (i.e. $L(\lambda) \in \mathcal{O}_k^{(0)}$), and
- λ is antidominant with respect to all roots of $\Delta_{>0}$ (i.e., $\langle \lambda + \rho, \alpha^\vee \rangle \notin \mathbb{Z}_{>0}$ for all $\alpha \in \Delta_{>0}$).

Then we have

PROPOSITION 9.6 ([**8**]). *If λ is conservative for f, then the associative algebra $H^0(U/J_\lambda)$ has a unique irreducible module E_λ, and*

$$H^0_-(L_k(\lambda)) \cong \mathbb{L}(E_\lambda).$$

We sketch the proof. Results of [**3**], specifically Theorem 5.5.4 guarantee that on the category $\mathcal{O}_k^{(0)}$ of $\mathfrak{g}_0$-integrable $\widehat{\mathfrak{g}}$-modules, the reduction functor $H^0_-(-)$ is exact and sends conservative modules to nonzero modules. Furthermore that $H_-(L_k(\lambda))$ is almost irreducible, meaning that the union of all its proper submodules has trivial intersection with $H_-(L_k(\lambda))_{\text{low}}$. This produces a nonzero $H^0(U/J_\lambda)$-module, which means in particular that this algebra must be nonzero, and so we deduce $\text{Var}(J_\lambda) = \overline{G \cdot f}$. On the other hand, Losev characterised the set of finite dimensional irreducible $H^0(U/J_\lambda)$-modules: they are finite in number and there is a certain group $C(f)$ which acts transitively on this set. Integrability of $L(\lambda)$ under G_0 (the algebraic group corresponding to $\mathfrak{g}_0$) assures us that $C(f)$ acts trivially, and so $H^0_-(L_k(\lambda))_{\text{low}}$ must be the unique irreducible module, hence the isomorphism of the theorem statement.

Proposition 9.6 gives us good control of the Hamiltonian reduction of $L_k(\lambda)$ for conservative λ. We therefore introduce the following:

HYPOTHESIS 2. *The $W \circ (-)$-class of every replete weight has a conservative representative.*

Under this hypothesis, the representation theory of W is sufficiently well controlled by that of $L_k(\mathfrak{g})$, that we may prove the following rationality theorem.

THEOREM 9.7 ([**8**]). *Under the hypothesis above, the exceptional W-algebra is rational.*

We sketch the proof. It is not difficult to argue that the existence of a nontrivial extension

$$0 \to \mathbb{L}(E_\lambda) \to M \to \mathbb{L}(E_{\lambda'}) \to 0,$$

implies $[M_k(\lambda') : L_k(\lambda)] > 0$, which in turn implies linkage of λ and λ' under the admissible level affine Weyl group $W \ltimes pQ^\vee$. But dominance of both λ and λ' then implies that they are actually equal. On the other hand self-extensions of $\mathbb{L}(E_\lambda)$ would be detected by the Zhu algebra, and are thus ruled out since it is semisimple.

10. Characters of exceptional W-algebras

We have seen that application of the Weyl-Kac character formula to integrable highest weight $\widehat{\mathfrak{g}}$-modules yields character formulas in terms of classical theta functions. This phenomenon has its origin in the structure of the affine Weyl group

$\widehat{W} = W \ltimes t_{Q^\vee}$ as a semidirect product of a finite group with a lattice, together with the structure of the action of the lattice $t_{Q^\vee}$ on weights, see formula (4.1). We have also seen that the same goes through for admissible highest weight modules $L(\lambda)$, the group $\widehat{W}$ being replaced by its subgroup $\widehat{W}(\lambda)$ which has substantially the same structure as $\widehat{W}$.

Let Q be an integral lattice of rank ℓ, and $\mathfrak{h} = Q \otimes_{\mathbb{Z}} \mathbb{C}$. By a theta function of Q we mean a function of the general form

$$\theta_{\mu+Q}(\tau, x) = \sum_{\alpha \in \mu+Q} e^{2\pi i(\alpha,x)} e^{\pi i \tau(\alpha,\alpha)}, \tag{10.1}$$

defined on the domain $(\tau, x) \in \mathcal{H} \times \mathfrak{h}$ where $\mathcal{H}$ is the upper half complex plane. A key feature of theta functions is their modular invariance. In particular, for each $\mu + Q \in Q^\vee/Q$ one has

$$\theta_{\mu+Q}(\tau, z)(-1/\tau, x/\tau) = \tau^{\ell/2} e^{\pi i |x|^2/\tau} \sum_{\mu' \in Q^\vee/Q} S_{\mu,\mu'} \theta_{\mu'+Q}(\tau, x),$$

where

$$S_{\mu,\mu'} = \frac{(-i)^{\ell/2}}{|Q^\vee/Q|^{1/2}} e^{-2\pi i(\mu,\mu')}.$$

Evaluation of the Kac-Wakimoto character formula (4.3) on $2\pi i \tau d + x \in \mathfrak{h}^*$ yields an expression

$$\chi_{L(\lambda)}(\tau, x) = B_\lambda(\tau, x)/R(\tau, x),$$

where

(10.2)

$$R(\tau, x) = \prod_{\alpha \in \Delta_+} \Theta(\tau, \alpha(x)), \qquad \Theta(\tau, z) = \prod_{n=1}^{\infty} (1 - e^{-2\pi i z} q^{n-1})(1 - q^n)(1 - e^{2\pi i z} q^n)$$

is the Weyl denominator, and $B_\lambda(\tau, x)$ is a sum of theta functions (10.1) for a rescaled copy of the root lattice Q of $\mathfrak{g}$, the sum being parametrised by the finite Weyl group W.

We now examine two cases: the case of principal nilpotent element f, and the case of subregular nilpotent element f.

Let $k = -h^\vee + p/q$ be an admissible weight for $\mathfrak{g}$. Recall that an exceptional W-algebra $H^0(L_k(\mathfrak{g}))$ is one for which the nilpotent element f lies in the open part of $\mathrm{Var}(I_k) \subset \mathfrak{g}$, which is to say, $f \in \mathbb{O}_q$. It happens that $\mathbb{O}_q$ equals the principal nilpotent orbit for all sufficiently large q, namely $q \geq h^\vee$. So let us assume that $q \geq h^\vee$ throughout this section.

We now describe the admissible weights of level k. In fact the first part of this discussion is valid for any admissible level k, without the assumption $q \geq h^\vee$. It is not hard to see that the admissible weights λ of level k whose integral coroot system is $\Pi^\vee(\lambda) = \Pi^\vee_q$ consists of

$$\lambda = k\Lambda_0 + \nu, \quad \text{for } \nu \in P_+^{p,\mathrm{reg}}.$$

Here $P_+^{p,\mathrm{reg}}$ means the set of weights

$$\{\nu \in P \mid \langle \nu, \alpha_i^\vee \rangle > 0 \text{ for } i = 1, \dots, \ell \text{ and } \langle \nu, \theta^\vee \rangle < p\}$$

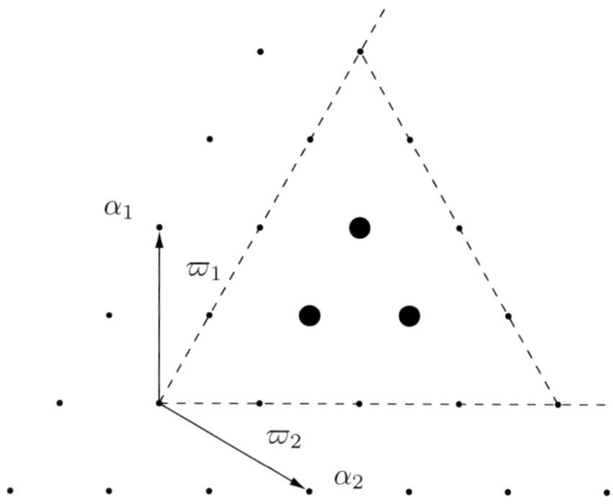

FIGURE 10. The set $P_+^{p,\mathrm{reg}}$ of regular weights of level $p = 4$ for $\mathfrak{g} = \mathfrak{sl}_3$

The remaining admissible weights λ are obtained by classifying the possible sets of simple coroots $\Pi^\vee(\lambda)$, obtained as transformations of $\Pi_q^\vee$

$$\begin{aligned}\Pi^\vee(\lambda) &= yt_{-\eta}(\Pi_q^\vee)\\ \lambda &= yt_{-\eta}(\nu) - \rho,\end{aligned}$$

where $y \in W$ the finite Weyl group and $\eta \in P$ the weight lattice. Roughly speaking, the transformation $t_{-\eta}$, applied to $\Pi_q^\vee$ lifts the simple coroots $\alpha_i^\vee$, $i = 1, \ldots, \ell$ out of $\Delta_+^\vee$ into $\widehat{\Delta}_+^{\vee,\mathrm{re}}$, sending $\alpha_i^\vee \mapsto \alpha_i^\vee + m_i K$ with the coefficients $m_i \in \mathbb{Z}_{\geq 0}$ encoded into $\eta \in P$. Then y sends $\alpha_i^\vee + m_i K \mapsto y(\alpha_i^\vee) + m_i K$. Since the simple coroot $-\theta^\vee + qK$ must remain in $\widehat{\Delta}_+^{\vee,\mathrm{re}}$ at the end of this process, it is required that $\eta \in P_+^q$. In fact one perceives that this condition is not sufficient, and in general the parametrisation of valid pairs (y, η) is quite complicated.

We are analysing the case of f principal nilpotent. And in this case a fortuitous simplification takes place. For such f, an admissible weight λ is replete if it is antidominant, and any replete weight is automatically conservative, because $\mathfrak{g}_0 = \mathfrak{h}$ and integrability with respect to this subalgebra is no condition on λ. We are thus interested in parametrising those $\lambda \in \mathrm{Pr}_k$ which pair non-integrally with each finite simple coroot $\alpha_i^\vee$, $i = 1, \ldots, \ell$. This condition on λ translates directly into the condition $\eta \in P_+^{q,\mathrm{reg}}$. Now since $t_{-\eta}(\Delta_+^\vee) \subset \widehat{\Delta}_+^{\vee,\mathrm{re}} \backslash \Delta_+$ it follows that $yt_{-\eta}(\Pi_q^\vee) \subset \widehat{\Delta}_+^{\vee,\mathrm{re}}$ for any $y \in W$.

We thus obtain (following Kac and Wakimoto [**38**, Lemma 4.2]) a pleasingly symmetrical parametrisation of the relevant admissible weights:

$$(y, \nu, \eta) \in W \times P_+^{p,\mathrm{reg}} \times P_+^{q,\mathrm{reg}}.$$

The BRST reduction sends $L(\lambda)$ and $L(y \circ \lambda)$ to the same irreducible W-algebra module, so the action of the finite Weyl group W is removed from the parametrisation.

The principal nilpotent f can be chosen so that in the decomposition (7.1), the subalgebra $\mathfrak{n} = \mathfrak{g}_{>0}$ coincides with the subalgebra $\mathfrak{n}_+$ of the usual triangular decomposition $\mathfrak{g} = \mathfrak{n}_- \oplus \mathfrak{h} \oplus \mathfrak{n}_+$. A little calculation shows that the character of

$F^{\mathrm{ch}}(\mathfrak{n})$ is "two thirds" of the Weyl denominator R of (10.2). More precisely

$$\chi_{F^{\mathrm{ch}}(\mathfrak{n})}(\tau, x) = \prod_{\alpha \in \Delta_+} \Theta(\tau, \alpha(x)),$$

where

$$\Theta(\tau, z) = \prod_{n=1}^{\infty} (1 - e^{-2\pi i z} q^{n-1})(1 - e^{2\pi i z} q^n).$$

At the level of characters the BRST procedure converts the character $\chi_{L(\lambda)}(\tau, x) = B_\lambda(\tau, x)/R(\tau, x)$ into

$$\chi_{H(L(\lambda))}(\tau) = \lim_{x \to 0} \frac{B_\lambda(\tau, x)}{\prod_{n=1}^{\infty}(1 - q^n)^\ell}.$$

And finally we have the modular S-matrix for the characters

$$\chi_{H(L(\lambda))}(-1/\tau) = \sum_{\lambda'} S_{\lambda,\lambda'} \chi_{H(L(\lambda'))}(\tau)$$

where

$$S_{\lambda,\lambda'} = S_{(\nu,\eta),(\nu',\eta')} = \sum_{y \in W} \varepsilon(y) e^{-2\pi i \frac{p}{q}(y(\eta),\eta')} \sum_{w \in W} \varepsilon(w) e^{-2\pi i \frac{q}{p}(w(\nu),\nu')}. \tag{10.3}$$

We now pass to the case of subregular nilpotent element f. The subregular nilpotent orbit in the finite dimensional simple Lie algebra $\mathfrak{g}$ is uniquely characterised by the property $\dim(\mathfrak{g}_0) = \dim(\mathfrak{h}) + 2$ in the decomposition (7.1). From now on we assume $\mathfrak{g}$ to be simply laced, then it turns out that the subregular nilpotent f can be chosen so that $\mathfrak{g}_0 = \mathfrak{h} + \mathfrak{g}_{\alpha_*} + \mathfrak{g}_{-\alpha_*}$ for a certain root α_*. For types D and E this root α_* corresponds to the trivalent node in the Dynkin diagram. For type A_n the subregular nilpotent f is even only if n is odd, in which case α_* corresponds to the middle node in the Dynkin diagram. The case n even can be treated by introducing a "good even grading" on $\mathfrak{g}$ adapted to f, different than the Dynkin grading. We have avoided discussing this trick in these notes to keep things as simple as possible.

Let $k = -h^\vee + p/q$ be admissible. The denominator(s) q for which $\mathbb{O}_q$ equals the subregular nilpotent orbit were determined in [**4**]. For example for D_n one has $q = 2n - 4$ and $q = 2n - 3$, and for $\mathfrak{g} = E_8$ one has $q = 24, 25, \ldots, 29$. We need to identify the replete weights and conservative weights $\lambda \in \mathrm{Pr}_k$. Let us return to the discussion around $\Pi^\vee(\lambda) = y t_{-\eta}(\Pi_q^\vee)$. In the principal case it was necessary that $t_{-\eta}$ shifted all the simple coroots $\alpha_i^\vee$ out of the finite root system Δ, so that λ would not pair integrally with any finite root. In the present case, λ is replete if $t_{-\eta}(\alpha_i^\vee) = \alpha_i^\vee$ for exactly one $i = 1, \ldots, \ell$ or else $t_{-\eta}(-\theta_i^\vee + qK) = -\theta_i^\vee$. And λ is conservative if furthermore $\langle \lambda, \alpha_* \rangle \in \mathbb{Z}_{\geq 0}$.

Interestingly, the repleteness condition says precisely that η, instead of being regular as it was in the principal case, must lie on exactly one of the walls of the level q affine Weyl chamber:

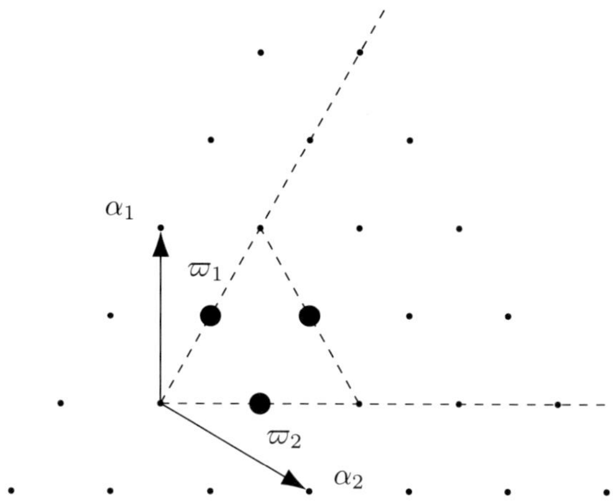

FIGURE 11. The set $P_+^{q,\mathrm{subreg}}$ of "subregular weights" of level $q = 2$ for $\mathfrak{g} = \mathfrak{sl}_3$

The characters $\chi_{L(\lambda)}(\tau, x)$ have the same expression as before. In this case $\mathfrak{n}_+ = \mathfrak{n} \oplus \mathfrak{g}_{\alpha_*}$, so that the cancellation between the character of $F^{\mathrm{ch}}(\mathfrak{n})$ and the Weyl denominator is less perfect than in the principal case. Indeed we obtain

$$\chi_{C(L(\lambda))}(\tau, x) = \frac{1}{\prod_{n=1}^{\infty}(1 - q^n)^{\ell}} \cdot \frac{B_\lambda(\tau, x)}{\Theta(\tau, \alpha_*(x))}. \tag{10.4}$$

At the end of the day we have to take the limit $x \to 0$ in (10.4), but here we have an indeterminate expression: The theta function has a zero along the hyperplane $\alpha_*(x) = 0$, while the numerator has a zero along the hyperplane $\gamma(x) = 0$, where $\gamma \in \Delta_+$ is the unique root relative to which λ is integrable. Of course when λ is conservative these zeros cancel out and the limit is well-defined.

Let us fix a set

$$\{\lambda_1, \lambda_2, \ldots, \lambda_N\}$$

of representatives of the classes of conservative weights, modulo $\sim$. The characters of the modules $M_i = H^0_-(L(\lambda_i))$ are given by the limit of (10.4). We would like to determine the S-matrix

$$\chi_{M_i}(-1/\tau) = \sum_{i'} S_{i,i'} \chi_{M_{i'}}(\tau).$$

In principle this can be done by performing the S-transformation directly upon (10.4) with $\lambda = \lambda_i$, and carefully combining terms $B_\mu(\tau, x)$, for $\mu \notin \{\lambda_1, \ldots, \lambda_N\}$ to obtain cancellations and a well-defined limit. In fact we may bypass much of the difficulty of finding these explicit cancellations by keeping all the indeterminate terms, and simply computing the $x \to 0$ limit using l'Hopital's rule.

The basic approach is to fix $x \in \mathfrak{h}$ appropriately, and take a limit along a ray tx as $t \to 0$. Terms of the form

$$\frac{1 - e^{-2\pi i \gamma(tx)}}{1 - e^{-2\pi i \alpha_*(tx)}} \to \frac{\langle \gamma, x \rangle}{\langle \alpha_*, x \rangle}$$

appear in the sum (10.3), whose derivation is otherwise similar. The upshot is that the S-matrix is

(10.5)
$$S_{(\nu,\eta),(\nu',\eta')} = \sum_{y(\alpha_*)\in\Delta_+} \varepsilon(y)\frac{\langle y(\alpha_*), x\rangle}{\langle \alpha_*, x\rangle} e^{-2\pi i \frac{p}{q}(y(\eta),\eta')} \sum_{w\in W} \varepsilon(w) e^{-2\pi i \frac{q}{p}(w(\nu),\nu')}.$$

The sum

$$\sum_{y(\alpha_*)\in\Delta_+} \varepsilon(y)\frac{\langle y(\alpha_*), x\rangle}{\langle \alpha_*, x\rangle} e^{-2\pi i \frac{p}{q}(y(\eta),\eta')} \tag{10.6}$$

appears here as a sort of degenerate analogue of the affine S-matrix (0.1).

If we choose $p \geq h^\vee$ as small as possible (while respecting that p and our fixed q must be coprime), it often happens that the second sum in (10.5) reduces to a single term, and so the matrix (10.6), like (0.1), is realised as the S-matrix of a rational vertex algebra.

In many cases these rational vertex algebras are easily identified. For example the exceptional subregular W-algebra associated with D_n and level $k = -h^\vee + p/q$ where $p/q = (2n-1)/(2n-3)$, is isomorphic to the Virasoro minimal model $\mathrm{Vir}_{3,n-2}$, and (10.6) recovers the well-known S-matrix in this case [**19**].

Other cases are more intriguing. For example $\mathfrak{g}$ of type E_8 and $p/q = 30/29$ yields a rational exceptional W-algebra whose category of modules has 44 simple objects. The S-matrix (10.6) yields very complicated fusion rules via the Verlinde formula, the largest fusion coefficient being 92 as it happens.

References

[1] Drazen Adamovic and Antun Milas, *Vertex operator algebras associated to modular invariant representations for $A_1^{(1)}$*, Math. Res. Lett. **2** (1995), no. 5, 563–575, DOI 10.4310/MRL.1995.v2.n5.a4. MR1359963

[2] Tomoyuki Arakawa, *Representation theory of $\mathcal{W}$-algebras*, Invent. Math. **169** (2007), no. 2, 219–320, DOI 10.1007/s00222-007-0046-1. MR2318558

[3] Tomoyuki Arakawa, *Representation theory of W-algebras, II*, Exploring new structures and natural constructions in mathematical physics, Adv. Stud. Pure Math., vol. 61, Math. Soc. Japan, Tokyo, 2011, pp. 51–90, DOI 10.2969/aspm/06110051. MR2867144

[4] Tomoyuki Arakawa, *Associated varieties of modules over Kac-Moody algebras and C_2-cofiniteness of W-algebras*, Int. Math. Res. Not. IMRN **22** (2015), 11605–11666, DOI 10.1093/imrn/rnu277. MR3456698

[5] Tomoyuki Arakawa, *Rationality of admissible affine vertex algebras in the category $\mathcal{O}$*, Duke Math. J. **165** (2016), no. 1, 67–93, DOI 10.1215/00127094-3165113. MR3450742

[6] Tomoyuki Arakawa, *Introduction to W-algebras and their representation theory*, Perspectives in Lie theory, Springer INdAM Ser., vol. 19, Springer, Cham, 2017, pp. 179–250. MR3751125

[7] Tomoyuki Arakawa and Jethro van Ekeren, *Modularity of relatively rational vertex algebras and fusion rules of principal affine W-algebras*, Comm. Math. Phys. **370** (2019), no. 1, 205–247, DOI 10.1007/s00220-019-03504-6. MR3982694

[8] Tomoyuki Arakawa and Jethro van Ekeren, *Rationality and fusion rules of exceptional $\mathcal{W}$-algebras*, J. Eur. Math. Soc. (JEMS) **25** (2023), no. 7, 2763–2813, DOI 10.4171/jems/1250. MR4612101

[9] Bojko Bakalov and Alexander Kirillov Jr., *Lectures on tensor categories and modular functors*, University Lecture Series, vol. 21, American Mathematical Society, Providence, RI, 2001, DOI 10.1090/ulect/021. MR1797619

[10] Richard E. Borcherds, *Vertex algebras, Kac-Moody algebras, and the Monster*, Proc. Nat. Acad. Sci. U.S.A. **83** (1986), no. 10, 3068–3071, DOI 10.1073/pnas.83.10.3068. MR843307

[11] David H. Collingwood and William M. McGovern, *Nilpotent orbits in semisimple Lie algebras: An introduction*, Van Nostrand Reinhold Mathematics Series, Van Nostrand Reinhold Co., New York, 1993. MR1251060

[12] Jan de Boer and Tjark Tjin, *The relation between quantum W algebras and Lie algebras*, Comm. Math. Phys. **160** (1994), no. 2, 317–332. MR1262200

[13] Alberto De Sole and Victor G. Kac, *Finite vs affine W-algebras*, Jpn. J. Math. **1** (2006), no. 1, 137–261, DOI 10.1007/s11537-006-0505-2. MR2261064

[14] V. G. Drinfeld and V. V. Sokolov, *Lie algebras and equations of Korteweg-de Vries type* (Russian), Current problems in mathematics, Vol. 24, Itogi Nauki i Tekhniki, Akad. Nauk SSSR, Vsesoyuz. Inst. Nauchn. i Tekhn. Inform., Moscow, 1984, pp. 81–180. MR760998

[15] Pavel Etingof, Shlomo Gelaki, Dmitri Nikshych, and Victor Ostrik, *Tensor categories*, Mathematical Surveys and Monographs, vol. 205, American Mathematical Society, Providence, RI, 2015, DOI 10.1090/surv/205. MR3242743

[16] Justine Fasquel, *Rationality of the exceptional W-algebras $\mathcal{W}_k(\mathfrak{sp}_4, f_{subreg})$ associated with subregular nilpotent elements of $\mathfrak{sp}_4$*, Comm. Math. Phys. **390** (2022), no. 1, 33–65, DOI 10.1007/s00220-021-04294-6. MR4381184

[17] Boris Feigin and Edward Frenkel, *Affine Kac-Moody algebras at the critical level and Gelfand-Dikii algebras*, Infinite analysis, Part A, B (Kyoto, 1991), Adv. Ser. Math. Phys., vol. 16, World Sci. Publ., River Edge, NJ, 1992, pp. 197–215, DOI 10.1142/s0217751x92003781. MR1187549

[18] B. L. Feigin, *Semi-infinite homology of Lie, Kac-Moody and Virasoro algebras* (Russian), Uspekhi Mat. Nauk **39** (1984), no. 2(236), 195–196. MR740035

[19] Philippe Di Francesco, Pierre Mathieu, and David Sénéchal, *Conformal field theory*, Graduate Texts in Contemporary Physics, Springer-Verlag, New York, 1997, DOI 10.1007/978-1-4612-2256-9. MR1424041

[20] Edward Frenkel and David Ben-Zvi, *Vertex algebras and algebraic curves*, 2nd ed., Mathematical Surveys and Monographs, vol. 88, American Mathematical Society, Providence, RI, 2004, DOI 10.1090/surv/088. MR2082709

[21] Edward Frenkel, Victor Kac, and Minoru Wakimoto, *Characters and fusion rules for W-algebras via quantized Drinfel'd-Sokolov reduction*, Comm. Math. Phys. **147** (1992), no. 2, 295–328. MR1174415

[22] F. G. Malikov and I. B. Frenkel, *Annihilating ideals and tilting functors* (Russian, with Russian summary), Funktsional. Anal. i Prilozhen. **33** (1999), no. 2, 31–42, 95, DOI 10.1007/BF02465191; English transl., Funct. Anal. Appl. **33** (1999), no. 2, 106–115. MR1719326

[23] Igor B. Frenkel and Yongchang Zhu, *Vertex operator algebras associated to representations of affine and Virasoro algebras*, Duke Math. J. **66** (1992), no. 1, 123–168, DOI 10.1215/S0012-7094-92-06604-X. MR1159433

[24] Wee Liang Gan and Victor Ginzburg, *Quantization of Slodowy slices*, Int. Math. Res. Not. **5** (2002), 243–255, DOI 10.1155/S107379280210609X. MR1876934

[25] Yi-Zhi Huang, *Rigidity and modularity of vertex tensor categories*, Commun. Contemp. Math. **10** (2008), no. suppl. 1, 871–911, DOI 10.1142/S0219199708003083. MR2468370

[26] Y.-Z. Huang and J. Lepowsky, *A theory of tensor products for module categories for a vertex operator algebra. I, II*, Selecta Math. (N.S.) **1** (1995), no. 4, 699–756, 757–786, DOI 10.1007/BF01587908. MR1383584

[27] James E. Humphreys, *Representations of semisimple Lie algebras in the BGG category $\mathcal{O}$*, Graduate Studies in Mathematics, vol. 94, American Mathematical Society, Providence, RI, 2008, DOI 10.1090/gsm/094. MR2428237

[28] Jens Carsten Jantzen, *Representations of algebraic groups*, 2nd ed., Mathematical Surveys and Monographs, vol. 107, American Mathematical Society, Providence, RI, 2003. MR2015057

[29] A. Joseph, *A characteristic variety for the primitive spectrum of a semisimple Lie algebra*, Non-commutative harmonic analysis (Actes Colloq., Marseille-Luminy, 1976), Lecture Notes in Math., Vol. 587, Springer, Berlin-New York, 1977, pp. 102–118. MR450350

[30] Victor Kac, Shi-Shyr Roan, and Minoru Wakimoto, *Quantum reduction for affine superalgebras*, Comm. Math. Phys. **241** (2003), no. 2-3, 307–342, DOI 10.1007/s00220-003-0926-1. MR2013802

[31] Victor G. Kac, *Infinite-dimensional Lie algebras*, 3rd ed., Cambridge University Press, Cambridge, 1990, DOI 10.1017/CBO9780511626234. MR1104219

[32] V. G. Kac, *Infinite-dimensional Lie algebras, and the Dedekind η-function* (Russian), Funkcional. Anal. i Priložen. **8** (1974), no. 1, 77–78. MR374210

[33] Victor Kac, *Vertex algebras for beginners*, 2nd ed., University Lecture Series, vol. 10, American Mathematical Society, Providence, RI, 1998, DOI 10.1090/ulect/010. MR1651389

[34] V. G. Kac and D. A. Kazhdan, *Structure of representations with highest weight of infinite-dimensional Lie algebras*, Adv. in Math. **34** (1979), no. 1, 97–108, DOI 10.1016/0001-8708(79)90066-5. MR547842

[35] Victor G. Kac and Dale H. Peterson, *Infinite-dimensional Lie algebras, theta functions and modular forms*, Adv. in Math. **53** (1984), no. 2, 125–264, DOI 10.1016/0001-8708(84)90032-X. MR750341

[36] Victor G. Kac, Ashok K. Raina, and Natasha Rozhkovskaya, *Bombay lectures on highest weight representations of infinite dimensional Lie algebras*, 2nd ed., Advanced Series in Mathematical Physics, vol. 29, World Scientific Publishing Co. Pte. Ltd., Hackensack, NJ, 2013, DOI 10.1142/8882. MR3185361

[37] Victor G. Kac and Minoru Wakimoto, *Modular invariant representations of infinite-dimensional Lie algebras and superalgebras*, Proc. Nat. Acad. Sci. U.S.A. **85** (1988), no. 14, 4956–4960, DOI 10.1073/pnas.85.14.4956. MR949675

[38] V. G. Kac and M. Wakimoto, *Branching functions for winding subalgebras and tensor products*, Acta Appl. Math. **21** (1990), no. 1-2, 3–39, DOI 10.1007/BF00053290. MR1085771

[39] Victor G. Kac and Minoru Wakimoto, *On rationality of W-algebras*, Transform. Groups **13** (2008), no. 3-4, 671–713, DOI 10.1007/s00031-008-9028-7. MR2452611

[40] Victor G. Kac and Minoru Wakimoto, *Quantum reduction and representation theory of superconformal algebras*, Adv. Math. **185** (2004), no. 2, 400–458, DOI 10.1016/j.aim.2003.12.005. MR2060475

[41] D. Kazhdan and G. Lusztig, *Tensor structures arising from affine Lie algebras. I, II*, J. Amer. Math. Soc. **6** (1993), no. 4, 905–947, 949–1011, DOI 10.2307/2152745. MR1186962

[42] D. Kazhdan and G. Lusztig, *Tensor structures arising from affine Lie algebras. I, II*, J. Amer. Math. Soc. **6** (1993), no. 4, 905–947, 949–1011, DOI 10.2307/2152745. MR1186962

[43] D. Kazhdan and G. Lusztig, *Tensor structures arising from affine Lie algebras. III*, J. Amer. Math. Soc. **7** (1994), no. 2, 335–381, DOI 10.2307/2152762. MR1239506

[44] D. Kazhdan and G. Lusztig, *Tensor structures arising from affine Lie algebras. IV*, J. Amer. Math. Soc. **7** (1994), no. 2, 383–453, DOI 10.2307/2152763. MR1239507

[45] Bertram Kostant and Shlomo Sternberg, *Symplectic reduction, BRS cohomology, and infinite-dimensional Clifford algebras*, Ann. Physics **176** (1987), no. 1, 49–113, DOI 10.1016/0003-4916(87)90178-3. MR893479

[46] Hai-Sheng Li, *Local systems of vertex operators, vertex superalgebras and modules*, J. Pure Appl. Algebra **109** (1996), no. 2, 143–195, DOI 10.1016/0022-4049(95)00079-8. MR1387738

[47] Ivan Losev, *Finite-dimensional representations of W-algebras*, Duke Math. J. **159** (2011), no. 1, 99–143, DOI 10.1215/00127094-1384800. MR2817650

[48] Robert V. Moody and Arturo Pianzola, *Lie algebras with triangular decompositions*, Canadian Mathematical Society Series of Monographs and Advanced Texts, John Wiley & Sons, Inc., New York, 1995. A Wiley-Interscience Publication. MR1323858

[49] Robert McRae, *On rationality for C_2-cofinite vertex operator algebras*, `arXiv:2108.01898`, 2021.

[50] Gregory Moore and Nathan Seiberg, *Classical and quantum conformal field theory*, Comm. Math. Phys. **123** (1989), no. 2, 177–254. MR1002038

[51] Erik Verlinde, *Fusion rules and modular transformations in 2D conformal field theory*, Nuclear Phys. B **300** (1988), no. 3, 360–376, DOI 10.1016/0550-3213(88)90603-7. MR954762

[52] Yongchang Zhu, *Modular invariance of characters of vertex operator algebras*, J. Amer. Math. Soc. **9** (1996), no. 1, 237–302, DOI 10.1090/S0894-0347-96-00182-8. MR1317233

Instituto de Matemática Pura e Aplicada (IMPA), Rio de Janeiro, Brazil
Email address: `jethro@impa.br`

Contemporary Mathematics
Volume **813**, 2025
https://doi.org/10.1090/conm/813/16282

OPEs of rank two W-algebras

Justine Fasquel

ABSTRACT. In this short note, we provide OPEs for affine W-algebras associated with Lie algebras of rank two and give some direct applications.

1. Introduction

The W-algebra $\mathcal{W}^k(\mathfrak{g}, f)$ is obtained from an Hamiltonian reduction of the affine vertex algebra $V^k(\mathfrak{g})$ [**14**, **17**]. Hence, both structure are closely related. However, if the description of $V^k(\mathfrak{g})$ in term of strong generators and relations (OPEs) is well-known (see for instance [**15**, **16**]), it is no longer the case for the W-algebras. Indeed, the latter do not directly come from Lie algebras and we need the formalism of vertex algebras to describe them. The computation of OPEs is difficult in general and only few examples were computed explicitly [**6**, **10**, **18**, **23**].

In this short note, we provide the OPEs for W-algebras associated with Lie algebras of rank two (i.e. $sl(3)$, $sp(4)$ and G_2). They were computed using the Mathematica package [**21**].

The OPEs consist in a powerful tool to prove important and precise results. For instance, they can be used to prove rationality of W-algebras [**6**, **13**], compute associated varieties [**12**], or find isomorphisms between vertex algebras [**2**, **3**, **19**].

The present paper is organized as follows. In Sect. 2 we recall the structure of affine W-algebras. Sect. 3 presents the OPEs of W-algebras associated to a Lie algebra $\mathfrak{g}$ of rank two. We use these OPEs to emphasize isomorphisms of W-algebras for particular values of the level k. These levels are listed in Sect. 4.

2. Structure of affine W-algebras

We assume the lector familiar with the formalism of vertex algebras and we refer to [**17**] and [**4**, **12**, **18**] for details on the construction of W-algebras.

Let $\mathfrak{g}$ be a complex simple finite dimensional Lie algebra. Consider f a nilpotent element of $\mathfrak{g}$ and k a complex number. The (affine) W-algebra $\mathcal{W}^k(\mathfrak{g}, f)$ is the vertex algebra obtained as the zero-th cohomology of a complex $(\mathscr{C}(\mathfrak{g}, f, k), d_{(0)})$ constructed from the affine vertex algebra $V^k(\mathfrak{g})$. It is a particular case of BRST reduction usually referred as the Drinfeld-Sokolov reduction of $V^k(\mathfrak{g})$. We write

$$\mathcal{W}^k(\mathfrak{g}, f) = H^0_f(V^k(\mathfrak{g})).$$

2020 *Mathematics Subject Classification.* Primary 17B69, 81R10.

The author acknowledges support from the Labex CEMPI (ANR-11-LABX-0007-01). Her research is supported by a University of Melbourne Establishment Grant.

Its simple graded quotient is denoted by $\mathcal{W}_k(\mathfrak{g}, f)$. Provided that k is non-critical, i.e. $k \neq -h^\vee$, $\mathcal{W}^k(\mathfrak{g}, f)$ inherits a structure of conformal vertex algebra from the one of $V^k(\mathfrak{g})$. Denote $L(z)$ the usual conformal field defined in [**17**].

Let us embed f into an $sl(2)$-triple (e, h, f) given by the Jacobson-Morosov theorem. We have

$$[h, e] = 2e, \qquad [h, f] = -2f, \qquad \text{and} \qquad [e, f] = h.$$

The semisimple element $x_0 := h/2$ induces an $\frac{1}{2}\mathbb{Z}$-gradation on $\mathfrak{g}$,

$$\mathfrak{g} = \bigoplus_{j \in \frac{1}{2}\mathbb{Z}} \mathfrak{g}_j, \tag{2.1}$$

where $\mathfrak{g}_j = \{y \in \mathfrak{g} \mid [x_0, y] = jy\}$. To any $a \in \mathfrak{g}_{-j}$, $j \in \frac{1}{2}\mathbb{Z}$, we can associate a field $J^a(z)$ of $\mathscr{C}(\mathfrak{g}, f, k)$. This field has conformal weight $1 + j$ with respect to L. The fields $\{J^a\}_{a \in \mathfrak{g}}$ play an important role in the structure of the W-algebra $\mathcal{W}^k(\mathfrak{g}, f)$. Let $\mathfrak{g}^f$ be the centralizer of f in $\mathfrak{g}$, and set $\mathfrak{g}^f_j := \mathfrak{g}^f \cap \mathfrak{g}_j$ for all $j \in \frac{1}{2}\mathbb{Z}$. By the theory on $sl(2)$, we have

$$\mathfrak{g}^f = \bigoplus_{j \leq 0} \mathfrak{g}^f_j. \tag{2.2}$$

The strong generators of the W-algebra $\mathcal{W}^k(\mathfrak{g}, f)$ are in one to one correspondence with the elements of a basis of $\mathfrak{g}^f$ which respect the grading (2.2).

THEOREM 2.1 ([**18**]). *For each $a \in \mathfrak{g}^f_{-j}$, $j \geq 0$, there exists a field $J^{\{a\}}(z)$ of $\mathcal{W}^k(\mathfrak{g}, f)$ of conformal weight $1+j$ with respect to L such that $J^{\{a\}}(z) - J^a(z)$ is a linear combination of normally order products of the fields $J^b(z)$, where $b \in \mathfrak{g}_{-s}$, $0 \leq s < j$ and their derivatives.*

Let $\{a_i\}_{i \in I}$ be a basis of $\mathfrak{g}^f$ compatible with the graduation (2.2). Then $\mathcal{W}^k(\mathfrak{g}, f)$ is freely strongly generated by the fields $\{J^{\{a_i\}}\}_{i \in I}$.

In practice, to construct $J^{\{a\}}(z)$ from $J^a(z)$, we write a linear combination of fields as in the theorem and find coefficients so that the field $J^{\{a\}}(z)$ is $d_{(0)}$-closed. When f is an even nilpotent element there exists an algorithm to construct the strong generators of the W-algebra $\mathcal{W}^k(\mathfrak{g}, f)$ (see [**12, 18**]). In [**18**], Kac and Wakimoto give formulas to construct strong generators with conformal weights 1, 1/2 and 0, and they compute OPEs between them. This provides in particular an explicit description of all W-algebras associated with minimal nilpotent elements. Indeed, if f belongs to the minimal nilpotent orbit $\mathbb{O}_{\min}$ then the grading (2.1) induced by x is *minimal*:

$$\mathfrak{g} = \mathfrak{g}_{-1} \oplus \mathfrak{g}_{-\frac{1}{2}} \oplus \mathfrak{g}_0 \oplus \mathfrak{g}_{\frac{1}{2}} \oplus \mathfrak{g}_1,$$

with $\mathfrak{g}_{-1} = \mathbb{C}f$ and $\mathfrak{g}_1 = \mathbb{C}e$.

3. OPEs of W-algebras associated with rank two Lie algebras

In this section, we present OPEs of the W-algebra $\mathcal{W}^k(\mathfrak{g}, f)$ where $\mathfrak{g}$ is a simple Lie algebra of rank two and $k \neq -h^\vee$ is a non-critical level. For simplicity, L will always denote the conformal vector. In the cases we consider, we were able to find strong generators which are all – except L – *primary* with respect to the conformal structure induced by L. It means that for any strong generator $J \neq L$ of conformal weight Δ,

$$L(z)J(w) \sim \frac{\Delta}{(z-w)^2}J(w) + \frac{1}{(z-w)}\partial J(w).$$

We take these primary generators in the following. With this choice, the OPEs involving L are uniquely determined. We omit them in the computations.

3.1. W-algebras associated with nilpotent elements of $sl(3)$. The simple Lie algebra $sl(3)$ of type A_2 admits two non-trivial nilpotent orbits: the principal and the minimal one. The corresponding W-algebras are among the first examples of W-algebras which have been described explicitly. They are respectively called the Zamolodchikov vertex algebra [**23**] and the Bershadsky-Polyakov vertex algebra [**8, 20**].

In [**6**], Arakawa gives an explicit description of the Bershadsky-Polyakov vertex algebra which corresponds to the W-algebra associated with a minimum nilpotent element of $sl(3)$. The W-algebra $\mathcal{W}^k(sl(3), f_{\min})$ is strongly generated by the fields $J(z)$, $G^{\pm}(z)$ and $L(z)$ of conformal weights 1, $\frac{3}{2}$ and 2 respectively. These fields satisfy the following relations:

$$\begin{aligned}
J(z)J(w) &\sim \frac{(3+2k)}{3(z-w)^2},\\
J(z)G^{\pm}(w) &\sim \pm\frac{1}{(z-w)}G^{\pm}(w),\\
G^{\pm}(z)G^{\pm}(w) &\sim 0,\\
G^{+}(z)G^{-}(w) &\sim \frac{(1+k)(3+2k)}{(z-w)^3} + \frac{3(1+k)}{(z-w)^2}J(w)\\
&+\frac{1}{(z-w)}\left(-(3+k)L(w) + 3 : J(w)^2 : + \frac{3(1+k)}{2}\partial J(w)\right),
\end{aligned}$$

where

$$c_k = -\frac{(1+3k)(3+2k)}{3+k}.$$

On another side, the W-algebra $\mathcal{W}^k(sl(3))$ is strongly generated by the fields $L(z)$ and $W(z)$ of conformal weights 2 and 3 respectively satisfying the OPE:

$$\begin{aligned}
W(z)W(w) &\sim \frac{w_k c_k}{3(z-w)^6} + \frac{2w_k}{(z-w)^4}L(w) + \frac{w_k}{(z-w)^3}\partial L(w)\\
&+\frac{1}{(z-w)^2}\left(\frac{2(3+k)^3}{3} : L(w)^2 : - \frac{3(3+k)^2(2+k)^2}{4}\partial^2 L(w)\right)\\
&+\frac{1}{(z-w)}\left(\frac{2(3+k)^3}{3} : L(w)\partial L(w) : - \frac{(3+k)^2(18+14k+3k^2)}{18}\partial^3 L(w)\right),
\end{aligned}$$

where

$$c_k = -\frac{2(5+3k)(9+4k)}{3+k},$$

and

$$w_k = -\frac{(3+k)^2(4+3k)(12+5k)}{6}.$$

3.2. W-algebras associated with nilpotent elements of $sp(4)$. There are three non-trivial nilpotent orbits in $sp(4)$ corresponding to the principal, the subregular, the minimal nilpotent orbits.

The minimal W-algebra $\mathcal{W}^k(sp(4), f_{\min})$ is strongly generated by the fields $J(z)$, $F^{\pm}(z)$, $G^{\pm}(z)$ and $L(z)$ of conformal weights 1, 1, $\frac{3}{2}$ and 2 respectively. They satisfy the following OPEs:

$$\begin{aligned}
J(z)J(w) &\sim \frac{(1+2k)}{(z-w)^2},\\
J(z)F^{\pm}(w) &\sim \pm\frac{2}{(z-w)}F^{\pm}(w),\\
F^{\pm}(z)F^{\pm}(w) &\sim 0,\\
F^{+}(z)F^{-}(w) &\sim \frac{(1+2k)}{2(z-w)^2} + \frac{1}{(z-w)}J(w),\\
J(z)G^{\pm}(w) &\sim \pm\frac{1}{(z-w)}G^{\pm}(w),\\
G^{\pm}(z)F^{\pm}(w) &\sim 0,\\
G^{\pm}(z)F^{\mp}(w) &\sim \frac{1}{(z-w)}G^{\mp}(w),\\
G^{\pm}(z)G^{\pm}(w) &\sim \pm\frac{4(2+k)}{(z-w)^2}F^{\pm}(w) \pm \frac{2(2+k)}{(z-w)}\partial F^{\pm}(w),\\
G^{+}(z)G^{-}(w) &\sim \frac{2(1+2k)(2+k)}{(z-w)^3} + \frac{2(2+k)}{(z-w)^2}J(w)\\
&+\frac{1}{(z-w)}\left(-2(3+k)L(w) + 4 :F^{+}(w)F^{-}(w): + :J(w)^2: + k\partial J(w)\right),
\end{aligned}$$

where

$$c_k = -\frac{3(k+1)(2k+1)}{3+k}.$$

The explicit description of the subregular W-algebra $\mathcal{W}^k(sp(4), f_{\text{subreg}})$ firstly appeared in [**13**]. The W-algebra $\mathcal{W}^k(sp(4), f_{\text{subreg}})$ is strongly generated by the fields $J(z)$, $G^{\pm}(z)$ and $L(z)$ of conformal weights 1, 2 and 2 respectively. They satisfy the OPEs:

$$\begin{aligned}
J(z)J(w) &\sim \frac{(2+k)}{(z-w)^2},\\
J(z)G^{\pm}(w) &\sim \pm\frac{1}{(z-w)}G^{\pm}(w),\\
G^{\pm}(z)G^{\pm}(w) &\sim 0,\\
G^{+}(z)G^{-}(w) &\sim -\frac{3(1+k)(2+k)^2}{(z-w)^4} - \frac{3(1+k)(2+k)}{(z-w)^3}J(w)\\
&+\frac{1}{(z-w)^2}\left((2+k)(3+k)L(w) - (3+2k) :J(w)^2: - \frac{3(1+k)(2+k)}{2}\partial J(w)\right)\\
&+\frac{1}{(z-w)}\Bigg((3+k) :L(w)J(w): + \frac{(3+k)(2+k)}{2}\partial L(w) - :J(w)^3:\\
&\qquad -(3+2k) :J(w)\partial J(w): - \frac{(5+4k+k^2)}{2}\partial^2 J(w)\Bigg),
\end{aligned}$$

where

$$c_k = -\frac{2(9+16k+6k^2)}{3+k}.$$

The principal W-algebra $\mathcal{W}^k(sp(4))$ is of type $(2,4)$ strongly generated by the fields $L(z)$ and $W(z)$. The field $W(z)$ satisfies the OPE:

$$\begin{aligned}
W(z)W(w) &\sim \frac{w_0 c_k}{4(z-w)^8} + \frac{2w_0}{(z-w)^6}L(w) + \frac{w_0}{(z-w)^5}\partial L(w) \\
&+\frac{1}{(z-w)^4}\left(2w_2W(w) - w_1\left(21(3+k) :L(w)^2: -\frac{3(162+139k+30k^2)}{2}\partial^2 L(w)\right)\right) \\
&+\frac{1}{(z-w)^3}\left(w_2\partial W(w) - w_1\left(21(3+k) :\partial L(w)L(w): -\frac{(162+139k+30k^2)}{3}\partial^3 L(w)\right)\right) \\
&+\frac{(3+k)}{(z-w)^2}\left(17+12k+2k^2\right)\left(\frac{(684+653k+150k^2)}{2}\partial^2 W(w) - 84(3+k) :L(w)W(w):\right) \\
&+\frac{w_3}{(z-w)^2}\big(4(3+k)^2(65+27k)(69+32k) :L(w)^3: -(3+k)w_4 :\partial^2 L(w)L(w): \\
&\qquad -(3+k)(540+509k+118k^2)(812+701k+150k^2) :\partial L(w)^2: +w_5\partial^4 L(w)\big) \\
&+\frac{(3+k)}{2(z-w)}(17+12k+2k^2)\big((162+139k+30k^2)\partial^3 W(w) \\
&\qquad - 84(3+k)(:L(w)\partial W(w): + :\partial L(w)W(w):)\big) \\
&+\frac{w_3}{(z-w)}\big(6(3+k)^2(65+27k)(69+32k) :\partial L(w)L(w)^2: \\
&\qquad -(3+k)\left(w_6 :\partial^2 L(w)\partial L(w): +w_7 :\partial^3 L(w)L(w):\right) + w_8\partial^5 L(w)\big).
\end{aligned}$$

where

$$c_k = -\frac{2(12+5k)(13+6k)}{3+k},$$

and

$$\begin{aligned}
w_0 &= (3+k)^2(3+2k)(8+3k)(11+5k)(11+6k)(18+7k)(19+8k)(747+674k+150k^2), \\
w_1 &= (3+k)^2(3+2k)(8+3k)(11+5k)(11+6k)(18+7k)(19+8k), \\
w_2 &= 45(3+k)(3+2k)(8+3k)(17+12k+2k^2), \\
w_3 &= \frac{1}{2}(3+k)^2(3+2k)(8+3k), \\
w_4 &= 566676+998131k+658239k^2+192662k^3+21120k^4, \\
w_5 &= 823635+2187341k+2413982k^2+1417736k^3+467512k^4+82100k^5+6000k^6 \\
w_6 &= 3(95769+168158k+110646k^2+32332k^3+3540k^4), \\
w_7 &= 125046+220605k+145649k^2+42666k^3+4680k^4, \\
w_8 &= \frac{3}{20}(823635+2187341k+2413982k^2+1417736k^3+467512k^4+82100k^5+6000k^6).
\end{aligned}$$

3.3. W-algebras associated with nilpotent elements of G_2. Consider the Lie algebra G_2 of rank two. According to the Bala-Carter classification, G_2 has four non-trivial nilpotent orbits: the three canonical nilpotent orbits – $\mathbb{O}_{\min}$,

$\mathbb{O}_{\text{subreg}}$, and $\mathbb{O}_{\text{reg}}$ – and an additional nilpotent orbit of dimension 8, denoted $\tilde{A}_1$ [**9**, Chap. 8].

The W-algebra $\mathcal{W}^k(G_2, f_{\min})$ is strongly generated by the fields $J(z)$, $F^{\pm}(z)$, $G^{\pm}(z)$, $W^{\pm}(z)$, and $L(z)$ of conformal weight 1, 1, $\frac{3}{2}$, $\frac{3}{2}$ and 2 respectively. They satisfy the OPEs:

$$
\begin{aligned}
J(z)J(w) &\sim \frac{2(5+3k)}{(z-w)^2},\\
J(z)F^{\pm}(w) &\sim \pm\frac{2}{(z-w)}F^{\pm}(w),\\
J(z)G^{\pm}(w) &\sim \pm\frac{3}{(z-w)}G^{\pm}(w),\\
J(z)W^{\pm}(w) &\sim \pm\frac{1}{(z-w)}W^{\pm}(w),\\
F^{\pm}(z)F^{\pm}(w) &\sim G^{\pm}(z)G^{\pm}(w) \sim F^{\pm}(z)G^{\pm}(w) \sim 0,\\
F^{+}(z)F^{-}(w) &\sim \frac{(5+3k)}{(z-w)^2} + \frac{1}{(z-w)}J(w),\\
G^{+}(z)G^{-}(w) &\sim \frac{2(4+3k)(5+3k)}{9(z-w)^3} + \frac{(4+3k)}{3(z-w)^2}J(w)\\
&+\frac{1}{(z-w)}\left(-(4+k)L(w) + \frac{2}{3} :F^{+}(w)F^{-}(w): + \frac{1}{3} :J(w)^2: + \frac{(2+3k)}{6}\partial J(w)\right),\\
F^{\pm}(z)G^{\mp}(w) &\sim \frac{1}{(z-w)}W^{\mp}(w),\\
W^{\pm}(z)W^{\pm}(w) &\sim \pm\frac{4(4+3k)}{3(z-w)^2}F^{\pm}(w) \pm \frac{2(4+3k)}{3(z-w)}\partial F^{\pm}(w),\\
W^{+}(z)W^{-}(w) &\sim -\frac{2(4+3k)(5+3k)}{3(z-w)^3} - \frac{(4+3k)}{3(z-w)^2}J(w)\\
&+\frac{1}{(z-w)}\left(3(4+k)L(w) - \frac{10}{3} :F^{+}(w)F^{-}(w): - \frac{1}{3} :J(w)^2: + \frac{(2-k)}{2}\partial J(w)\right),\\
F^{\pm}(z)W^{\pm}(w) &\sim \frac{3}{(z-w)}G^{\pm}(w),\\
F^{\pm}(z)W^{\mp}(w) &\sim \frac{2}{(z-w)}W^{\pm}(w),\\
G^{\pm}(z)W^{\pm}(w) &\sim \pm\frac{2}{3(z-w)} :F^{\pm}(w)^2:,\\
G^{\pm}(z)W^{\mp}(w) &\sim \mp\frac{2(4+3k)}{3(z-w)^2}F^{\pm}(w)\\
&\quad + \frac{1}{(z-w)}\left(-\frac{2}{3} :J(w)F^{\pm}(w): \mp \frac{(2+3k)}{3}\partial F^{\pm}(w)\right),
\end{aligned}
$$

where

$$c_k = -\frac{2k(5+3k)}{4+k}.$$

The W-algebra $\mathcal{W}^k(G_2, f_{\tilde{A}_1})$ is strongly generated by the fields $J(z)$, $F^{\pm}(z)$, $G^{\pm}(z)$, and $L(z)$ of conformal weights 1, 1, $\frac{5}{2}$ and 2 respectively. They satisfy the OPEs:

$$
\begin{aligned}
J(z)J(w) &\sim \frac{(3+2k)}{(z-w)^2},\\
J(z)F^{\pm}(w) &\sim \pm\frac{2}{(z-w)}F^{\pm}(w),\\
J(z)G^{\pm}(w) &\sim \pm\frac{1}{(z-w)}G^{\pm}(w),\\
F^{\pm}(z)F^{\pm}(w) &\sim G^{\pm}(z)F^{\pm}(w) \sim 0,\\
F^{+}(z)F^{-}(w) &\sim \frac{(3+2k)}{2(z-w)^2} + \frac{1}{(z-w)}J(w),\\
F^{\pm}(z)G^{\mp}(w) &\sim \frac{1}{(z-w)}G^{\pm}(w),\\
G^{\pm}(z)G^{\pm}(w) &\sim \mp\frac{2(2+k)(10+3k)(17+6k)}{(z-w)^4}F^{\pm}(w) \mp \frac{(2+k)(10+3k)(17+6k)}{(z-w)^3}\partial F^{\pm}(w)\\
&\quad+\frac{1}{(z-w)^2}\big(\pm 2(4+k)(16+5k) : L(w)F^{\pm}(w) : \mp 16(3+k) : F^{+}(w)F^{\pm}(w)F^{-}(w) :\\
&\qquad \mp 4(3+k) : F^{\pm}(w)J(w)^2 : -(44 \mp 48 + (24 \mp 16)k + 3k^2) : F^{\pm}(w)\partial J(w) :\\
&\qquad + (2+k)(10+3k) : J(w)\partial F^{\pm}(w) : \mp 2(3+k)(38 \mp 4 + 20k + 3k^2)\partial^2 F^{\pm}(w)\big)\\
&\quad+\frac{1}{(z-w)}\big(\pm(4+k)(16+5k) : L(w)\partial F^{\pm}(w) : \mp 16(3+k) : F^{+}(w)F^{-}(w)\partial F^{\pm}(w) :\\
&\qquad \pm(4+k)(16+5k) : \partial L(w)F^{\pm}(w) : \mp 8(3+k) : F^{\pm}(w)^2\partial F^{\mp}(w) :\\
&\qquad \mp 4(3+k) : F^{\pm}(w)J(w)\partial J(w) : \mp 2(3+k) : J(w)^2\partial F^{\pm}(w) :\\
&\qquad + \frac{44+24k+3k^2}{2}(: J(w)\partial^2 F^{\pm}(w) : - : F^{\pm}(w)\partial^2 J(w) :)\\
&\qquad \pm 4(2 \mp 1)(3+k) : \partial J(w)\partial F^{\pm}(w) : \mp \frac{940+736k+195k^2+18k^3}{12}\partial^3 F^{\pm}(w)\bigg),\\
G^{+}(z)G^{-}(w) &\sim \frac{(2+k)(3+2k)(10+3k)(17+6k)}{(z-w)^5} + \frac{(2+k)(10+3k)(17+6k)}{(z-w)^4}J(w)\\
&\quad+\frac{1}{(z-w)^3}\bigg(-(4+k)(3+2k)(16+5k)L(w) + \frac{188+256k+119k^2+18k^3}{2}\partial J(w)\\
&\qquad + (38+34k+7k^2)(4 : F^{+}(w)F^{-}(w) : + : J(w)^2 :)\big)\\
&\quad+\frac{1}{(z-w)^2}\big(-(4+k)(16+5k) : L(w)J(w) : +8(3+k) : F^{+}(w)F^{-}(w)J(w) :\\
&\qquad +(8+36k+11k^2) : F^{+}(w)\partial F^{-}(w) : +(144+100k+17k^2) : F^{-}(w)\partial F^{+}(w) :\\
&\qquad +2(3+k) : J(w)^3 : +(26+30k+7k^2) : J(w)\partial J(w) :\\
&\qquad - \frac{(4+k)(3+2k)(16+5k)}{2}\partial L(w) + (3+k)(42+20k+3k^2)\partial^2 J(w)\bigg)\\
&\quad+\frac{1}{(z-w)}\bigg(\frac{3(4+k)^2}{2} : L(w)^2 : - \frac{3(2+k)(3+k)(4+k)}{2}\partial^2 L(w)\\
&\qquad - \frac{(4+k)(16+5k)}{2} : \partial L(w)J(w) : - \frac{(4+k)(8+5k)}{2} : L(w)\partial J(w) :
\end{aligned}
$$

$$
\begin{aligned}
&+\frac{396+332k+90k^2+9k^3}{12}\partial^3 J(w) - 8(4+k) : L(w)F^+(w)F^-(w) : \\
&-2(4+k) : L(w)J(w)^2 : +\frac{116+72k+15k^2}{2} : \partial F^+(w)\partial F^-(w) : \\
&+\frac{(2+k)(26+15k)}{8} : \partial J(w)^2 : +\frac{96+52k+9k^2}{4} : J(w)\partial^2 J(w) : \\
&+2(27+19k+3k^2) : F^-(w)\partial^2 F^+(w) : +(26+14k+3k^2) : F^+(w)\partial^2 F^-(w) : \\
&+(7+3k) : J(w)^2\partial J(w) : +4(5+k) : J(w)F^-(w)\partial F^+(w) : \\
&+4(1+k) : J(w)F^+(w)\partial F^-(w) : +4(k-1) : F^+(w)F^-(w)\partial J(w) : \\
&+8 : F^+(w)^2F^-(w)^2 : +4 : F^+(w)F^-(w)J(w)^2 : +\frac{1}{2} : J(w)^4 : \Big),
\end{aligned}
$$

where

$$
c_k = -\frac{(92+81k+18k^2)}{4+k}.
$$

The W-algebra $\mathcal{W}^k(G_2, f_{\text{subreg}})$ is strongly generated by the fields $L(z)$, $G^{\pm}(z)$ and $F(z)$ of conformal weights 2, 2 and 3 respectively. They satisfy the OPEs:

$$
\begin{aligned}
G^+(z)F(w) \sim\ & \frac{2(2+k)(16+5k)}{(z-w)^3}G^-(w) + \frac{(2+k)(16+5k)}{2(z-w)^2}\partial G^-(w) \\
&+ \frac{1}{(z-w)}\Big(2 : G^+(w)G^-(w) : -2(4+k) : L(w)G^-(w) : \\
&\quad +2\partial F(w) + \frac{(2+k)^2}{2}\partial^2 G^-(w)\Big), \\
G^-(z)F(w) \sim\ & \frac{2(2+k)(16+5k)}{(z-w)^3}G^+(w) + \frac{(2+k)(16+5k)}{2(z-w)^2}\partial G^+(w) \\
&+ \frac{1}{(z-w)}\Big(- : G^+(w)^2 : -2(4+k) : L(w)G^+(w) : \\
&\quad - : G^-(w)^2 : +\frac{(2+k)^2}{2}\partial^2 G^+(w)\Big), \\
F(z)F(w) \sim\ & -\frac{(2+k)(10+3k)(16+5k)(4+k)c_k}{2(z-w)^6} \\
&- \frac{3(2+k)(4+k)(10+3k)(16+5k)}{(z-w)^4}L(w) \\
&- \frac{3(2+k)(4+k)(10+3k)(16+5k)}{2(z-w)^3}\partial L(w) \\
&+ \frac{1}{(z-w)^2}\Big(-(8+3k) : G^+(w)^2 : +2(4+k)^2(10+3k) : L(w)^2 : \\
&+(8+3k) : G^-(w)^2 : -\frac{3(2+k)(4+k)(8+3k)(10+3k)}{4}\partial^2 L(w)\Big) \\
&+ \frac{1}{(z-w)}\Big(-(8+3k) : G^+(w)\partial G^+(w) : +2(4+k)^2(10+3k) : L(w)\partial L(w) : \\
&+(8+3k) : G^-(w)\partial G^-(w) : -\frac{(2+k)(4+k)(4+3k)(10+3k)}{6}\partial^3 L(w)\Big),
\end{aligned}
$$

$$\begin{aligned} G^{\pm}(z)G^{\pm}(w) \sim & \pm \frac{(10+3k)(4+k)c_k}{2(z-w)^4} \\ & + \frac{1}{(z-w)^2}\left(\pm 2(4+k)(10+3k)L(w) - 4(3+k)G^{+}(w)\right) \\ & + \frac{1}{(z-w)}\left(\pm(4+k)(10+3k)\partial L(w) - 2(3+k)\partial G^{+}(w)\right), \\ G^{+}(z)G^{-}(w) \sim & \frac{4(3+k)}{(z-w)^2}G^{-}(w) + \frac{1}{(z-w)}\left(-2F(w) + 2(3+k)\partial G^{-}(w)\right), \end{aligned}$$

where

$$c_k = -\frac{4(k+2)(17+6k)}{4+k}.$$

To our knowledge, OPEs of the principal W-algebra $\mathcal{W}^k(G_2)$ were computed first in [**19**]. For completeness of the paper, we recall them in this section. The W-algebra $\mathcal{W}^k(G_2)$ is of type $(2,6)$, strongly generated by the fields $L(z)$ and $W(z)$, and satisfies the OPE:

$$\begin{aligned} W(z)W(w) \sim & \frac{-w_{0,k}c_k}{81(z-w)^{12}} + \frac{2w_{0,k}}{27(z-w)^{10}}L(w) + \frac{w_{0,k}}{27(z-w)^9}\partial L(w) \\ & + \frac{w_{1,k}}{108(z-w)^8}\left(-62(k+4) : L(w)^2 : +3\left(84k^2+579k+1000\right)\partial^2 L(w)\right) \\ & + \frac{w_{1,k}}{108(z-w)^7}\left(-62(k+4) : \partial L(w)L(w) : +\frac{2}{3}\left(84k^2+579k+1000\right)\partial^3 L(w)\right) \\ & + \frac{1}{(z-w)^6}(2w_{3,k}W(w) + w_{2,k}\Lambda^5(L)) + \frac{1}{(z-w)^5}(w_{3,k}\partial W(w) + w_{2,k}\Lambda^4(L)) \\ & + \frac{1}{(z-w)^4}(w_{5,k}\left(\frac{10}{9}\left(588k^2+3929k+6504\right)\partial^2 W(w) - \frac{1240}{3}(k+4) : L(w)W(w) :\right) \\ & \quad + w_{4,k}\Lambda^3(L)) \\ & + \frac{1}{(z-w)^3}(w_{5,k}\Omega^2(W,L) + w_{4,k}\Lambda^2(L)) + \frac{1}{(z-w)^2}(w_{7,k}\Omega^1(W,L) + w_{6,k}\Lambda^1(L)) \\ & + \frac{1}{(z-w)}(w_{7,k}\Omega^0(W,L) + w_{6,k}\Lambda^0(L)), \end{aligned}$$

where

$$c_k = -\frac{2(12k+41)(7k+24)}{4+k}.$$

and

$$\begin{aligned} w_{0,k} =& 2(k+4)^2(2k+5)(2k+7)(3k+10)(7k+22)(7k+23)(8k+27)(9k+34) \\ & (11k+40)(12k+37)(15k+52)(15k+53)(18k+65)(336k^2+2301k+3940) \\ & (588k^2+3991k+6752)) \\ w_{1,k} =& (k+4)^2(2k+5)(2k+7)(3k+10)(7k+22)(8k+27)(9k+34)(11k+40) \\ & (12k+37)(15k+52)(18k+65)\left(336k^2+2301k+3940\right)\left(588k^2+3991k+6752\right) \\ w_{2,k} =& (k+4)^2(2k+5)(2k+7)(3k+10)(7k+22)(8k+27)(9k+34)(11k+40) \\ & (12k+37)(15k+52)(18k+65) \\ w_{3,k} =& \frac{140}{9}(k+4)(2k+7)(3k+10)(7k+20)(12k+35)(13k+48)(24k+89) \\ & \left(3k^2+24k+47\right)\left(336k^2+2301k+3940\right)\left(588k^2+3991k+6752\right) \\ w_{4,k} =& (k+4)^2(2k+5)(2k+7)(3k+10)(9k+34)(11k+40)(12k+37) \end{aligned}$$

$$
\begin{aligned}
w_{5,k} =&(k+4)(2k+7)(3k+10)(7k+20)(24k+89)\left(3k^2+24k+47\right)\\
&\left(336k^2+2301k+3940\right)\left(588k^2+3991k+6752\right)\\
w_{6,k} =&(k+4)^2(2k+5)(2k+7)(3k+10)(9k+34)\\
w_{8,k} =&(k+4)\left(3k^2+24k+47\right)\left(336k^2+2301k+3940\right)\left(588k^2+3991k+6752\right),
\end{aligned}
$$

The fields $\Lambda^n(L)$ only depend on L while every summand in the fields $\Omega^n(W,L)$ depends on W. We refer to [**19**, Appendix] for their explicit definition.

4. Isomorphisms of W-algebras

In this section, we use the explicit description of W-algebras associated with Lie algebras of rank two to obtain remarkable isomorphisms of vertex algebras (Table 1).

4.1. Trivial simple W-algebras. Consider the W-algebra $\mathcal{W}^k(\mathfrak{g},f)$ associated with a nilpotent element $f\in\mathfrak{g}$. Denote $\mathcal{I}_k$ its maximal ideal – which depends on the level k. We have $\mathcal{W}_k(\mathfrak{g},f)=\mathcal{W}^k(\mathfrak{g},f)/\mathcal{I}_k$. At non-critical level, it follows from the conformal structure of $\mathcal{W}^k(\mathfrak{g},f)$ that the vertex algebra $\mathcal{W}_k(\mathfrak{g},f)$ is trivial, i.e. isomorphic to $\mathbb{C}|0\rangle$, if and only if the conformal vector generates $\mathcal{I}_k$. This happens only if the central charge c_k is zero, but this necessary condition is not sufficient. For instance, the conformal vector L of $\mathcal{W}_k(sp(4),f_{\min})$ does not belong to the maximal ideal when $k=-1$ whereas $c_{-1}=0$. Indeed, according to the OPEs described in Sect. 3.2, if L were in $\mathcal{I}_{-1}$ then J would be in $\mathcal{I}_{-1}$ as well as the vacuum $|0\rangle$.

Adding a supplementary condition, we obtain a complete criterion. We enjoy this opportunity to mention that the proofs of Propositions 2.2.2 and 2.2.7 in [**12**] are incomplete. We correct them in this paper. It does not impact the results presented in the thesis.

Proposition 4.1. *Let $J^{(1)},\ \ldots,\ J^{(n)}$ be a set of strong generators of $\mathcal{W}^k(\mathfrak{g},f)$ with conformal weights $w_1\ldots w_n\in\mathbb{R}_{>0}$ respectively. The simple quotient $\mathcal{W}_k(\mathfrak{g},f)$ is trivial if and only if for any $1\leq i,j\leq n$ the pole of degree w_i+w_j of the OPE $J^{(j)}(z)J^{(i)}(w)$ vanishes, that is $(J^{(j)})_{w_i}J^{(i)}=0$.*

Proof. Since the W-algebra $\mathcal{W}^k(\mathfrak{g},f)$ is positively graded by the conformal vector, its simple quotient is trivial if and only if all the strong generators are in the maximal ideal $\mathcal{I}_k$. For all $1\leq i,j\leq n$, the vector $(J^{(j)})_{w_i}J^{(i)}$ has conformal weight zero. It vanishes since $\mathcal{W}^k(\mathfrak{g},f)$ is conical, i.e. $\mathcal{W}^k(\mathfrak{g},f)_0=\mathbb{C}|0\rangle$.

Conversely, let $1\leq i,j\leq n$. For $m>0$, $(J^{(j)})_mJ^{(i)}$ has conformal weight w_i-m and it is a linear combination of normally ordered products and derivatives of strong generators $J^{(\ell)}$ of conformal weight w_ℓ lower than $w_i-m<w_i$. In particular, $(J^{(j)})_{w_i}J^{(i)}$ is a multiple of $|0\rangle$: it exists $c_{i,j}\in\mathbb{C}$ such that

$$(J^{(j)})_{w_i}J^{(i)}=c_{i,j}|0\rangle.$$

It follows that $J^{(i)}$ is in $\mathcal{I}_k$ if all strong generators $J^{(\ell)}$ of conformal weight lower than w_i are in $\mathcal{I}_k$ and if $c_{i,j}=0$ for all $1\leq j\leq n$. Note that if $J^{(i)}$ has the lowest conformal weight, $(J^{(j)})_mJ^{(i)}=0$ for all $m\neq w_i$ and so $J^{(i)}$ is in $\mathcal{I}_k$ if and only if $(J^{(j)})_{w_i}J^{(i)}=0$.

Therefore, if $(J^{(j)})_{w_i}J^{(i)}=0$ for all $1\leq i,j\leq n$, then all the strong generators are in $\mathcal{I}_k$ and thus $\mathcal{W}_k(\mathfrak{g},f)\simeq\mathbb{C}|0\rangle$. □

We obtain the following classification.

Corollary 4.2.

(1) *Consider the Lie algebra $\mathfrak{g}=sl(3)$, then $\mathcal{W}_k(\mathfrak{g},f)\simeq\mathbb{C}$ if and only if*

$$(f,k)\in\{(f_{\min},-3/2),(f_{\text{reg}},-5/3),(f_{\text{reg}},-9/4)\}.$$

(2) *Consider the Lie algebra* $\mathfrak{g} = sp(4)$, *then* $\mathcal{W}_k(\mathfrak{g}, f) \simeq \mathbb{C}$ *if and only if*

$$(f, k) \in \{(f_{\min}, -1/2), (f_{\text{reg}}, -13/6), (f_{\text{reg}}, -12/5)\}.$$

(3) *Consider the Lie algebra* $\mathfrak{g} = G_2$, *then* $\mathcal{W}_k(\mathfrak{g}, f) \simeq \mathbb{C}$ *if and only if*

$$(f, k) \in \{(f_{\min}, -5/3), (f_{\text{subreg}}, -17/6), (f_{\text{subreg}}, -2), (f_{\text{reg}}, -41/12), (f_{\text{reg}}, -24/7)\}.$$

Most of the pairs listed in the Corollary 4.2 have already been established in [**3, 7**]. However, we get new isomorphisms such as the one involving $\mathcal{W}_{-2}(G_2, f_{\text{subreg}})$. It provides a new example of rational W-algebra at non admissible level.

COROLLARY 4.3. *The simple* W*-algebra* $\mathcal{W}_{-2}(G_2, f_{\text{subreg}})$ *is* C_2*-cofinite and rational.*

The principal W-algebras $\mathcal{W}_k(sl(3))$, $\mathcal{W}_k(sp(4))$ and $\mathcal{W}_k(G_2)$, and the subregular W-algebra $\mathcal{W}_k(G_2, f_{\text{subreg}})$ are one-dimensional if and only if $c_k = 0$. More generally, we conjecture the following:

CONJECTURE 4.4. *If* f *is distinguished, i.e.* $\mathfrak{g}_0^f = 0$, *then* $\mathcal{W}_k(\mathfrak{g}, f) \simeq \mathbb{C}|0\rangle$ *if and only if* $c_k = 0$.

We check this conjecture in many cases. In particular, we verify in [**7**] that, when f is distinguished and k is an admissible level which vanishes the central charge, $\mathcal{W}_k(\mathfrak{g}, f) \simeq \mathbb{C}|0\rangle$.

In another direction, the Corollary 4.3 gives more evidence for the following conjecture (see [**12**, Chap. 6.3] for additional details). We will explore in depth this topic in our future works.

CONJECTURE 4.5. *The associated variety of the simple vertex algebra* $L_{-2}(G_2)$ *is the closure of the subregular nilpotent orbit of* G_2*:*

$$X_{L_{-2}(G_2)} = \overline{\mathbb{O}}_{\text{subreg}}.$$

4.2. Simple quotients isomorphic to Virasoro vertex algebras. From now on, we consider non-trivial simple W-algebras. In particular, the conformal vector does not belong to the maximal ideal $\mathcal{I}_k$ of $\mathcal{W}^k(\mathfrak{g}, f)$. If the other strong generators are in $\mathcal{I}_k$, then $\mathcal{W}_k(\mathfrak{g}, f)$ is isomorphic to the Virasoro vertex algebra Vir_{c_k} with central charge c_k.

PROPOSITION 4.6. *Let* $J^{(1)}, \ldots, J^{(n)}$ *be a set of strong generators of* $\mathcal{W}^k(\mathfrak{g}, f)$ *with conformal weights* $w_1 \ldots w_n \in \mathbb{R}$ *with respect to the conformal vector* $L := J^{(n)}$. *The simple quotient* $\mathcal{W}_k(\mathfrak{g}, f)$ *is isomorphic to the simple Virasoro vertex algebra* Vir_{c_k} *if and only if for any* $1 \leq i \leq n-1$ *and* $1 \leq j \leq n$, $(J^{(j)})_{w_i} J^{(i)} = 0$ *and* $(J^{(j)})_{w_i-2} J^{(i)}$ *does not depend on* L.

PROOF. The proof is similar to the one of Proposition 4.1. The simple quotient $\mathcal{W}_k(\mathfrak{g}, f)$ is isomorphic to Vir_{c_k} if and only if it is strongly generated only by the conformal vector L. Hence, $\mathcal{W}_k(\mathfrak{g}, f) \simeq \text{Vir}_{c_k}$ if and only if $J^{(1)}, \ldots, J^{(n-1)} \in \mathcal{I}_k$ and $L \notin \mathcal{I}_k$. Since $J^{(1)}, \ldots, J^{(n-1)} \in \mathcal{I}_k$, then $(J^{(j)})_{w_i} J^{(i)} = 0$ for all $1 \leq i \leq n-1$ and $1 \leq j \leq n$. Moreover, $(J^{(j)})_{w_i-2} J^{(i)}$ is a linear combination of normally ordered products and derivatives of strong generators $J^{(\ell)}$ of conformal weight lower than 2. There exist $c_{i,j} \in \mathbb{C}$ and a linear combination $R_{i,j}$ of normally ordered products and derivatives of the strong generators $J^{(\ell)}$, $1 \leq \ell \leq n-1$, such that

$$(J^{(j)})_{w_i-2} J^{(i)} = c_{i,j} L + R_{i,j}. \tag{4.1}$$

Since $(J^{(j)})_{w_i-2} J^{(i)}$ and $R_{i,j}$ belongs to $\mathcal{I}_k$, we deduce that $c_{i,j} = 0$. Thus, $(J^{(j)})_{w_i-2} J^{(i)}$ does not depend on L.

Conversely, let $1 \leq i \leq n-1$ and $1 \leq j \leq n$. For $m > 0$, $(J^{(j)})_m J^{(i)}$ is a sum of normally ordered products and derivatives of strong generators $J^{(\ell)}$ of conformal weight lower than $w_i - m < w_i$. In particular, $(J^{(j)})_{w_i} J^{(i)}$ is a multiple of $|0\rangle$. In addition,

L can appear only as a summand in the decomposition of $(J^{(j)})_{w_i-2}J^{(i)}$. We have a decomposition (4.1). Since $L \notin \mathcal{I}_k$, $J^{(i)}$ is in the maximal ideal $\mathcal{I}_k$ if $(J^{(j)})_{w_i}J^{(i)} = 0$ and $(J^{(j)})_{w_i-2}J^{(i)}$ does not depend on L for all $1 \le j \le n$. Therefore, if this condition is satisfied for all $1 \le i \le n-1$, then all the strong generators except L are in $\mathcal{I}_k$ and $\mathcal{W}_k(\mathfrak{g}, f) \simeq \mathrm{Vir}_{c_k}$. □

We deduce the following classification from the OPEs presented in Sect. 3.

COROLLARY 4.7. *Consider the simple Lie algebra $\mathfrak{g}$ of rank 2.*

(1) *The simple W-algebra $\mathcal{W}_k(sl(3))$ is isomorphic to Vir_{c_k} if and only if $k = -4/3$ or $k = -12/5$.*
(2) *If $\mathfrak{g} = sp(4)$, then $\mathcal{W}_k(\mathfrak{g}, f) \simeq \mathrm{Vir}_{c_k}$ if and only if*
  - *$f = f_{\mathrm{subreg}}$ and $k = -2$, or*
  - *$f = f_{\mathrm{reg}}$ and $k \in \{-3/2, -8/3, -11/5, -11/6, -18/7, -19/8\}$.*
(3) *If $\mathfrak{g} = G_2$, then $\mathcal{W}_k(\mathfrak{g}, f) \simeq \mathrm{Vir}_{c_k}$ if and only if*
  - *$f = f_{\tilde{A}_1}$ and $k = -3/2$, or*
  - *$f = f_{\mathrm{subreg}}$ and $k = -10/3$, or*
  - [**19**, Proposition 5.3] *$f = f_{\mathrm{reg}}$ and*

$$k \in \{-\tfrac{5}{2}, -\tfrac{7}{2}, -\tfrac{10}{3}, -\tfrac{22}{7}, -\tfrac{23}{7}, -\tfrac{27}{8}, -\tfrac{34}{9}, -\tfrac{40}{11}, -\tfrac{37}{12}, -\tfrac{52}{15}, -\tfrac{53}{15}, -\tfrac{65}{18}\}.$$

The OPEs of $\mathcal{W}^k(G_2, f_{\mathrm{subreg}})$ also admit simplifications at level $k = -16/5$ (see Sect. 3.3). Indeed, the strong generator F belongs to the maximal ideal $\mathcal{I}_{-16/5}$. This induces relations between the three other strong generators and their derivatives in the simple W-algebra. Using the OPEs, the authors of [**19**] proved that the simple W-algebra corresponds with the tensor product of three copies of the Virasoro vertex algebra of central charge $-22/5$:

$$\mathcal{W}_{-16/5}(G_2, f_{\mathrm{subreg}}) \simeq (\mathrm{Vir}_{-22/5})^{\otimes 3}.$$

Since $\mathrm{Vir}_{-22/5}$ is C_2-cofinite and rational, see for instance [**5**, **22**], the tensor product $\mathcal{W}_{-16/5}(G_2, f_{\mathrm{subreg}})$ is also C_2-cofinite and rational [**1**, **11**].

4.3. Other isomorphisms. For several additional particular levels, the OPEs give a complete or partial description of the maximal ideal $\mathcal{I}_k$ and so the simple W-algebras. For instance, we recover some isomorphisms from [**2**] and [**7**, Theorem 10.10]:

$$\mathcal{W}_{-1}(sl(3), f_{\min}) \simeq M(1), \qquad \mathcal{W}_{-2}(sp(4), f_{\min}) \simeq L_{-3/2}(sl(2)),$$
$$\mathcal{W}_{-4/3}(G_2, f_{\min}) \simeq L_1(sl(2)), \qquad \mathcal{W}_{-17/6}(G_2, f_{\tilde{A}_1}) \simeq L_{-4/3}(sl(2)),$$

where $M(1)$ denotes the Heisenberg vertex algebra of rank 1 and $L_k(\mathfrak{g})$ is the simple affine vertex algebra associated with $\mathfrak{g}$ at level k. We are also able to determine other isomorphisms of vertex algebras.

PROPOSITION 4.8. *We have the following isomorphisms:*

$$\mathcal{W}_{-10/3}(G_2, f_{\tilde{A}_1}) \simeq L_{-11/6}(sl(2)), \qquad \mathcal{W}_{-2}(G_2, f_{\tilde{A}_1}) \simeq L_{-\frac{1}{2}}(sl(2)),$$
$$\mathcal{W}_{-1}(sp(4), f_{\mathrm{subreg}}) \simeq M(1).$$

PROOF. We detail the proof of the isomorphism $\mathcal{W}_{-1}(sp(4), f_{\mathrm{subreg}}) \simeq M(1)$. Other cases are similar. Using the OPEs, we check that when $k = -1$, the vector G^+ and G^- generate a proper ideal. Hence, they belongs to the maximal ideal of $\mathcal{W}^{-1}(sp(4), f_{\mathrm{subreg}})$. It follows that $G^+_{(n)}G^- = 0$ for all $n \ge 0$. In particular, $G^+_{(1)}G^- = 0$ gives a relation between the conformal field $L(z)$ and the field $:J(z)^2:$

$$L(z) = \frac{1}{2} : J(z)^2 : .$$

Thus, L coincides with the conformal vector of the Heisenberg vertex algebra generated by J. Since J does not belong to the maximal ideal of $\mathcal{W}^{-1}(sp(4), f_{\mathrm{subreg}})$, we get the isomorphism. □

TABLE 1. Isomorphisms for W-algebras of rank two.

$\mathcal{W} = \mathcal{W}_k(\mathfrak{g}, f)$	affine vertex algebra $L = L_{k'}(\mathfrak{g}')$	k such that $\mathcal{W} \simeq L$	$\mathcal{V}$	k such that $\mathcal{W} \simeq \mathcal{V}$
$\mathcal{W}_k(sl(3), f_{\min})$	$M(1)$	-1	$\mathbb{C}$	$-\frac{3}{2}$
$\mathcal{W}_k(sl(3))$	$\mathbb{C}$	$-\frac{5}{3}, -\frac{9}{4}$	Vir_{c_k}	$-\frac{4}{3}, -\frac{12}{5}$
$\mathcal{W}_k(sp(4), f_{\min})$	$L_{-3/2}(sl(2))$	-2	$\mathbb{C}$	$-\frac{1}{2}$
$\mathcal{W}_k(sp(4), f_{\mathrm{subreg}})$	$M(1)$	-1	Vir_{c_k}	-2
$\mathcal{W}_k(sp(4))$	$\mathbb{C}$	$-\frac{13}{6}, -\frac{12}{5}$	Vir_{c_k}	$-\frac{3}{2}, -\frac{8}{3}, -\frac{11}{5}, -\frac{11}{6}, -\frac{18}{7}, -\frac{19}{8}$
$\mathcal{W}_k(G_2, f_{\min})$	$L_1(sl(2))$	$-\frac{4}{3}$	$\mathbb{C}$	$-\frac{5}{3}$
$\mathcal{W}_k(G_2, f_{\tilde{A}_1})$	$L_{p(k)}(sl(2))$ with $p(k) = -\frac{92+81k+18k^2}{52+42k+9k^2}$	$-\frac{17}{6}, -\frac{10}{3}, -2$	Vir_{c_k}	$-\frac{3}{2}$
$\mathcal{W}_k(G_2, f_{\mathrm{subreg}})$	$\mathbb{C}$	$-\frac{17}{6}, -2$	Vir_{c_k}	$-\frac{10}{3}$
$\mathcal{W}_k(G_2)$	$\mathbb{C}$	$-\frac{41}{12}, -\frac{24}{7}$	Vir_{c_k}	$-\frac{5}{2}, -\frac{7}{2}, -\frac{10}{3}, -\frac{22}{7}, -\frac{23}{7}, -\frac{27}{8}, -\frac{34}{9}, -\frac{40}{11}, -\frac{37}{12}, -\frac{52}{15}, -\frac{53}{15}, -\frac{65}{18}$

We summarize in Table 1 the isomorphisms of W-algebras associated with simple Lie algebras of rank two we mentioned in Sect. 4.

REMARK 4.9. In the Table 1, the levels appearing in the third column are called *collapsing* [**2**, **3**, **7**]. The simple affine vertex algebra $L_{k'}(\mathfrak{g}')$ in the second column corresponds to $L_{k^\natural}(\mathfrak{g}^\natural)$ where $\mathfrak{g}^\natural$ is the centralizer of the $sl(2)$-triple (e, h, f) in $\mathfrak{g}$ and $k^\natural$ is defined as in [**18**].

Acknowledgments

This paper is based on the author's Ph.D. thesis [**12**]. The author would like to thank her advisors A. Moreau and T. Arakawa for suggesting the problem. She is very grateful to D. Adamović for useful discussions on collapsing levels. She also thanks A. Milas for the interesting feedback and exchanges.

References

[1] Toshiyuki Abe, Geoffrey Buhl, and Chongying Dong, *Rationality, regularity, and C_2-cofiniteness*, Trans. Amer. Math. Soc. **356** (2004), no. 8, 3391–3402, DOI 10.1090/S0002-9947-03-03413-5. MR2052955

[2] Dražen Adamović, Victor G. Kac, Pierluigi Möseneder Frajria, Paolo Papi, and Ozren Perše, *Conformal embeddings of affine vertex algebras in minimal W-algebras II: decompositions*, Jpn. J. Math. **12** (2017), no. 2, 261–315, DOI 10.1007/s11537-017-1621-x. MR3694933

[3] Dražen Adamović, Victor G. Kac, Pierluigi Möseneder Frajria, Paolo Papi, and Ozren Perše, *Conformal embeddings of affine vertex algebras in minimal W-algebras I: structural results*, J. Algebra **500** (2018), 117–152, DOI 10.1016/j.jalgebra.2016.12.005. MR3765450

[4] Tomoyuki Arakawa, *Representation theory of superconformal algebras and the Kac-Roan-Wakimoto conjecture*, Duke Math. J. **130** (2005), no. 3, 435–478, DOI 10.1215/S0012-7094-05-13032-0. MR2184567

[5] Tomoyuki Arakawa, *A remark on the C_2-cofiniteness condition on vertex algebras*, Math. Z. **270** (2012), no. 1-2, 559–575, DOI 10.1007/s00209-010-0812-4. MR2875849

[6] Tomoyuki Arakawa, *Rationality of Bershadsky-Polyakov vertex algebras*, Comm. Math. Phys. **323** (2013), no. 2, 627–633, DOI 10.1007/s00220-013-1780-4. MR3096533

[7] Tomoyuki Arakawa, Jethro van Ekeren, and Anne Moreau, *Singularities of nilpotent Slodowy slices and collapsing levels of W-algebras*, 2022.
[8] Michael Bershadsky, *Conformal field theories via Hamiltonian reduction*, Comm. Math. Phys. **139** (1991), no. 1, 71–82. MR1116410
[9] David H. Collingwood and William M. McGovern, *Nilpotent orbits in semisimple Lie algebras: An introduction*, Van Nostrand Reinhold Mathematics Series, Van Nostrand Reinhold Co., New York, 1993. MR1251060
[10] Thomas Creutzig and Andrew R. Linshaw, *Cosets of the* $\mathcal{W}^k(\mathfrak{sl}_4, f_{\mathrm{subreg}})$*-algebra*, Vertex algebras and geometry, Contemp. Math., vol. 711, Amer. Math. Soc., [Providence], RI, 2018, pp. 105–117, DOI 10.1090/conm/711/14301. MR3831569
[11] Chongying Dong, Geoffrey Mason, and Yongchang Zhu, *Discrete series of the Virasoro algebra and the moonshine module*, Algebraic groups and their generalizations: quantum and infinite-dimensional methods (University Park, PA, 1991), Proc. Sympos. Pure Math., vol. 56, Amer. Math. Soc., Providence, RI, 1994, pp. 295–316, DOI 10.1090/pspum/056.2/1278737. MR1278737
[12] Justine Fasquel, *Geometry and new rational W-algebras*, PhD thesis, 2022.
[13] Justine Fasquel, *Rationality of the exceptional* $\mathcal{W}$*-algebras* $\mathcal{W}_k(\mathfrak{sp}_4, f_{subreg})$ *associated with subregular nilpotent elements of* $\mathfrak{sp}_4$, Comm. Math. Phys. **390** (2022), no. 1, 33–65, DOI 10.1007/s00220-021-04294-6. MR4381184
[14] Boris Feigin and Edward Frenkel, *Quantization of the Drinfeld-Sokolov reduction*, Phys. Lett. B **246** (1990), no. 1-2, 75–81, DOI 10.1016/0370-2693(90)91310-8. MR1071340
[15] Edward Frenkel and David Ben-Zvi, *Vertex algebras and algebraic curves*, 2nd ed., Mathematical Surveys and Monographs, vol. 88, American Mathematical Society, Providence, RI, 2004, DOI 10.1090/surv/088. MR2082709
[16] Victor Kac, *Introduction to vertex algebras, Poisson vertex algebras, and integrable Hamiltonian PDE*, Perspectives in Lie theory, Springer INdAM Ser., vol. 19, Springer, Cham, 2017, pp. 3–72. MR3751122
[17] Victor Kac, Shi-Shyr Roan, and Minoru Wakimoto, *Quantum reduction for affine superalgebras*, Comm. Math. Phys. **241** (2003), no. 2-3, 307–342, DOI 10.1007/s00220-003-0926-1. MR2013802
[18] Victor G. Kac and Minoru Wakimoto, *Quantum reduction and representation theory of superconformal algebras*, Adv. Math. **185** (2004), no. 2, 400–458, DOI 10.1016/j.aim.2003.12.005. MR2060475
[19] Antun Milas, Michael Penn, and Christopher Sadowski, S_3*-permutation orbifolds of Virasoro vertex algebras*, J. Pure Appl. Algebra **227** (2023), no. 10, Paper No. 107378, 25, DOI 10.1016/j.jpaa.2023.107378. MR4572460
[20] A. M. Polyakov, *Gauge transformations and diffeomorphisms*, Internat. J. Modern Phys. A **5** (1990), no. 5, 833–842, DOI 10.1142/S0217751X90000386. MR1035397
[21] Krivono Thielemans, *A Mathematica package for computing operator product expansions*, Internat. J. Modern Phys. C **2** (1991), no. 3, 787–798, DOI 10.1142/S0129183191001001. MR1130315
[22] Weiqiang Wang, *Rationality of Virasoro vertex operator algebras*, Internat. Math. Res. Notices **7** (1993), 197–211, DOI 10.1155/S1073792893000212. MR1230296
[23] Alexandre Zamolodchikov, *Infinite extra symmetries in two-dimensional conformal quantum field theory* (Russian, with English summary), Teoret. Mat. Fiz. **65** (1985), no. 3, 347–359. MR829902

School of Mathematics and Statistics, University of Melbourne, Parkville, Victoria 3010, Australia
Email address: justine.fasquel@unimelb.edu.au

Contemporary Mathematics
Volume **813**, 2025
https://doi.org/10.1090/conm/813/16283

Fusion rings acting on vertex operator algebras: First steps

Andrew Riesen

ABSTRACT. By using tensor categorical techniques, we take the first steps towards defining what it means for a hypergroup, or a "non-invertible symmetry", to act on a vertex operator algebra. In particular, we show for finite extensions of rational vertex operator algebras $\mathcal{W} \subset \mathcal{V}$ there is a hypergroup K that acts on $\mathcal{V}$ such that the fixed points are $\mathcal{W}$. We calculate this associated hypergroup for certain VOA extensions associated to affine $\mathfrak{sl}(2)$, and discuss what the general theory of hypergroup actions should look like.

1. Introduction

Quantum field theory has been a source of inspiration for many mathematical ideas. For example, vertex operator algebras first defined by Borcherds in [**7**] describe 1 space + 1 time dimensional quantum field theories called conformal field theories. More precisely, it describes the chiral half of a conformal field theory. The representation theory of vertex operator algebras, or VOAs for short, is particularly interesting as in sufficiently nice cases the category of modules of a VOA $\mathcal{V}$ will form what is called a modular fusion category, or MFC for short. Modular fusion categories are a very special type of category, they essentially have two main properties:

(1) There are direct sums, they have only finitely many simples up to isomorphism, there is a tensor product, and a notion of dual objects. Every object in the category will decompose into finitely many simple objects and Schur's Lemma will also hold.
(2) There is a projective representation of the modular group related to the aforementioned categorical data.

The specific contents of the second property for a particular MFC is called the *modular data*, and while it is certainly interesting, requiring the second property restricts us from considering many categories which naturally satisfy property one, such as representations of a finite group G over $\mathbb{C}$. If instead we require that our category $\mathcal{D}$ only satisfies the first property, then $\mathcal{D}$ is called a *fusion category* and the combinatorial data formed by looking at the isomorphism classes of simple objects and how they behave under the tensor product is referred to as the *fusion ring* of $\mathcal{D}$.

2020 *Mathematics Subject Classification.* Primary 17B69, 18M20.

Fusion rings are generalizations of groups, and there has been a breakthrough in the physics community that suggests we should think of the "symmetries" of a quantum field theory as not only just including invertible elements, which would form a group, but we should also include "non-invertible symmetries", which would form a fusion ring [**19**][**2**][**3**]. This should also be extremely exciting for the mathematical community as it seems that physics is again suggesting a new idea, namely there should be a notion of what it means for a fusion ring to act on a VOA and restricting ourselves to thinking of "symmetries" as only coming from groups is a holdover from 20th and 19th century mathematics.

In this paper, we take the first step towards describing what it means for a fusion ring to act on a vertex operator algebra. More specifically, we describe what it means in specific cases for hypergroups, think of these as "normalized fusion rings", to act on VOAs. In particular we show that for sufficiently nice VOA extensions $\mathcal{W} \subset \mathcal{V}$, called finite extensions, there is a hypergroup K which "acts" on $\mathcal{V}$ as $\mathcal{W}$-module homomorphisms and the fixed points give $\mathcal{W}$, that is $\mathcal{V}^K = \mathcal{W}$. We say "act" as an intrinsic definition for what it means for a fusion ring to act on a VOA is currently being developed and will be presented in future work [**21**]. The reason we need K to be a hypergroup and not just a fusion ring is in general it is not possible to take quotients of fusion rings, but you can take quotients of hypergroups. K will be a quotient of hypergroups associated to two fusion rings, and so one should think of K as describing the effective part of a fusion ring action on $\mathcal{V}$. Stating this result as a theorem, the main theoretical result of this paper is:

THEOREM 1. *Suppose that $\mathcal{W} \subset \mathcal{V}$ is a finite extension of rational VOAs. Then there exists a hypergroup K which acts on $\mathcal{V}$ as $\mathcal{W}$-module maps, and for which:*

$$\mathcal{V}^K = \mathcal{W} \tag{1.1}$$

In the case that G is a finite group acting on $\mathcal{V}$, and $\mathcal{W} := \mathcal{V}^G$, we recover that $K = G$.

Conformal nets are an alternative approach to conformal field theory, and significant work has been done by Marcel Bischoff in [**5**] where he defines what it means for a hypergroup to act on a conformal net, which we now briefly recall. Firstly, a conformal net $\mathcal{A}$ is a collection of von Neumann algebras, type III_1 factors to be precise, indexed by open subsets of the circle such that they form a net with respect to the topology of $\mathbb{S}^1$ and they are required to satisfy several other axioms. A hypergroup action of K on a type III_1 factor M is defined as a unital map $\phi : K \to \mathrm{Ex}(\mathrm{UCP}(M))$ such that $\phi(x * y) = \phi(x) \circ \phi(y)$ for $x, y \in K$, where $\mathrm{Ex}(\mathrm{UCP}(M))$ are the extremal points of the unital completely positive endomorphisms of M. An action of a hypergroup on a conformal net is then defined as a compatible collection of actions of K on all the associated Type III_1 factors.

One drawback the conformal net approach has is that to come up with an example requires constructing infinitely many type III_1 factors that are also compatible. For this reason, constructing and working with conformal nets explicitly can be difficult, compared to VOAs for which there are several known constructions that can be worked with explicitly. The trade off conformal nets have for being less accessible is developing their theory is more straightforward when compared to their VOA counterpart. For example, it is easier to define what it means for a hypergroup to act on a conformal net. Continuing this trend, we show that the hypergroup associated to a VOA extension and the way the hypergroup acts can be explicitly calculated in many cases. We carry out this calculation for all finite VOA extensions of nice VOAs $(\widehat{\mathfrak{sl}(2)})_k$ associated to affine $\mathfrak{sl}(2)$.

The paper is structured as follows. Section 2 covers the relevant background on étale algebras, graphical calculus, and hypergroups/fusion rings. In Section 3 we prove Theorem 1 and prove how the elements of the hypergroup can be calculated using the modular data of Rep $\mathcal{W}$. In Section 4 we calculate the hypergroups associated to all finite VOA extensions of $(\widehat{\mathfrak{sl}(2)})_k$ using the A-D-E classification. Section 5 is devoted to describing many of the ideas that are in development for [**21**], and is more speculative in nature. It is our hope that this section will not only be helpful to convey how we think about hypergroup actions, but also serve as a primer for [**21**] to make the contents more digestible.

2. Background

In this paper, we will be working over $\mathbb{C}$. By a modular fusion category we mean a modular tensor category as defined in [**15**, Chapter 8]. For sake of brevity we will denote a MFC $(\mathcal{C}, \otimes, 1_{\mathcal{C}}, (\alpha, \ell, r), R, \theta)$ simply by the underlying category $\mathcal{C}$ when the context is clear. If $\mathcal{C}$ is a fusion category, we denote its set of simple objects by $\mathcal{O}(\mathcal{C})$ and its fusion ring by $\mathcal{K}^0(\mathcal{C})$.

For the basics about VOAs see [**30**]. To use the structure of MFCs we use the main result of [**24**] which means we will be working with simple VOAs $\mathcal{V} = (V, Y, \omega, |0\rangle)$ that satisfy the following conditions:

(1) $V_{(n)} = 0$ for $n < 0$ and $V_{(0)} = \mathbb{C}|0\rangle$
(2) The $\mathcal{V}$-module $(V', Y_{V'})$ is isomorphic to $\mathcal{V}$ as a $\mathcal{V}$-module, where V' is the restricted dual of V.
(3) Every grading-restricted generalized $\mathcal{V}$-module is a finite direct sum of irreducible $\mathcal{V}$-modules
(4) $\mathcal{V}$ is C_2-cofinite.

By a nice VOA, we mean a VOA satisfying these conditions.

2.1. Étale algebras and VOA extensions.

Without loss of generality we assume all fusion categories are strict.

DEFINITION 2.1. *Let $\mathcal{C}$ be a balanced braided fusion category. A* commutative $\mathcal{C}$ algebra *is a tuple (A, μ, ι) where A is an object in $\mathcal{C}$, $\mu : A \otimes A \to A$ and $\iota : 1_{\mathcal{C}} \to A$ are maps such that the following hold*

(Associativity) $\quad \mu \circ (\mathrm{Id}_A \otimes \mu) = \mu \circ (\mu \otimes \mathrm{Id}_A)$

(Unital) $\quad \mu \circ (\iota \otimes \mathrm{Id}_A) = \mu \circ (\mathrm{Id}_A \otimes \iota) = \mathrm{Id}_A$

(Commutative) $\quad \mu \circ R_{A,A} = \mu$

If A has the property that:

(Uniqueness of Unit) $\quad \dim_{\mathbb{C}} \mathrm{Hom}_{\mathcal{C}}(1_{\mathcal{C}}, A) = 1$

we say that it is a haploid algebra. *If (A, μ, ι) is a haploid algebra, choose $\epsilon : A \to 1_{\mathcal{C}}$ such that $\epsilon \circ \iota = \mathrm{Id}_{1_{\mathcal{C}}}$. We say that a haploid algebra is an* étale algebra[1] *if $\epsilon \circ \mu : A \otimes A \to 1_{\mathcal{C}}$ defines a non-degenerate pairing and* $\dim_{\mathcal{C}}(A) \neq 0$.

[1] In [**29**] this is refereed to as a *rigid algebra*.

When the context makes it clear, we will denote a $\mathcal{C}$-algebra (A, μ_A, ι_A) simply by A. Similar as to how an algebra in $\mathcal{C}$ categorfies the notion of an algebra, one can do the same thing for modules of an algebra.

DEFINITION 2.2. *Let (A, μ_A, ι_A) be a $\mathcal{C}$-algebra. An A-module is an object $M \in \mathcal{C}$, and a morphism $\mu_M : A \otimes M \to M$ such that the following holds:*

$$\text{(Associativity)} \qquad \mu_M \circ (\mu_A \otimes \mathrm{Id}_M) = \mu_M \circ (\mathrm{Id}_A \otimes \mu_M)$$

$$\text{(Unital)} \qquad \mu_M \circ (\iota_A \otimes \mathrm{Id}_M) = \mathrm{Id}_M$$

We denote a module by (M, μ_M) and when the module structure is clear from context just as M. A morphism of A-modules $f : (M, \mu_M) \to (N, \mu_N)$ is a morphism $f : M \to N$ such that

$$f \circ \mu_M = \mu_N \circ (\mathrm{Id}_A \otimes f) \tag{2.1}$$

With these notions, we can consider the category of A-modules denoted by $\mathrm{Rep}_{\mathcal{C}} A$. In the case that we have an étale algebra it was proven by Ostrik and Kirillov in [**29**] that $\mathrm{Rep}_{\mathcal{C}} A$ will be a fusion category. Note the tensor product on $\mathrm{Rep}_{\mathcal{C}} A$ will be different then the one on $\mathcal{C}$.

A concept that will be key for our calculations is *Frobenius reciprocity* for étale algebras. Namely, if A is an étale algebra in $\mathcal{C}$, then $\mathrm{Ind}(X) := (A \otimes X, \mu_A \otimes \mathrm{Id}_X)$ is a left A-module. If $f : X \to Y$ is a morphism in $\mathcal{C}$, then $\mathrm{Ind}(f) := \mathrm{Id}_A \otimes f : \mathrm{Ind}(X) \to \mathrm{Ind}(Y)$ is a morphism of left A-modules. It can be shown that $\mathrm{Ind} : \mathcal{C} \to \mathrm{Rep}_{\mathcal{C}} A$ is a tensor functor which will be left and right adjoint to the restriction functor $\mathrm{Res}(X, \mu_X) = X$. In symbols, we have as $\mathbb{C}$-linear spaces:

$$\mathrm{Hom}_{\mathrm{Rep}_{\mathcal{C}} A}(\mathrm{Ind}(X), (Y, \mu_Y)) \cong \mathrm{Hom}_{\mathrm{Rep}_{\mathcal{C}} A}((Y, \mu_Y), \mathrm{Ind}(X)) \cong \mathrm{Hom}_{\mathcal{C}}(X, Y) \tag{2.2}$$

A fusion sub-category of $\mathrm{Rep}_{\mathcal{C}} A$ that will be important is the category of local modules $\mathrm{Rep}_{\mathcal{C}}^{loc} A$.

DEFINITION 2.3. *Let $(M, \mu_M) \in \mathrm{Rep}_{\mathcal{C}} A$. We say that (M, μ_M) is a local module if*

$$\mu_M \circ R_{M,A} \circ R_{A,M} = \mu_M \tag{2.3}$$

The category of local-modules is denoted by $\mathrm{Rep}_{\mathcal{C}}^{loc} A$.

In the étale algebra case this will be a braided fusion category [**29**, Theorem 1.10]. If in addition we assume $\mathcal{C}$ is a modular fusion category and the étale algebra has a trivial twist, then $\mathrm{Rep}_{\mathcal{C}}^{loc} A$ is also a MFC [**29**, Theorem 4.5].

If $\mathcal{W}$ is a nice VOA, then the following theorems establishes a bijection between étale algebras and certain VOA extensions.

THEOREM 2 ([**25**, Theorem 3.6]). *Let $\mathcal{W}$ be a nice VOA, and $\mathcal{C} := \mathrm{Rep}\,\mathcal{W}$. Assume the weights of every simple $\mathcal{W}$-module M are non-negative, with $M_{(0)} \neq 0$ if and only if M is equivalent to $\mathcal{W}$. Then the following are equivalent:*

1. *Finite VOA extensions $\mathcal{W} \subset \mathcal{V}$ such that $\mathcal{V}$ is a nice VOA, the weights of a simple $\mathcal{V}$-module M are non-negative, with $M_{(0)} \neq 0$ if and only if M is equivalent to $\mathcal{V}$.*
2. *Étale algebras A in $\mathrm{Rep}\,\mathcal{W}$ such that $\theta_A = \mathrm{Id}_A$.*

It should be noted that the proof given in [**25**] has a small gap in that they assume $\dim_{\mathcal{C}}(V) \neq 0$. This can be filled using [**11**, Lemma 4.2].

THEOREM 3 ([**9**, Theorem 3.65]). *With all the assumptions of the previous theorem if $\mathcal{V}$ is a finite VOA extension of the prescribed type, then considering $A := \mathcal{V}$ as a $\mathcal{W}$-module there is an equivalence of MFCs:*

$$\mathrm{Rep}^{loc}_{\mathcal{C}} A \cong \mathrm{Rep}\,\mathcal{V} \tag{2.4}$$

Therefore, if we want to study VOA extensions of the type prescribed above, it suffices to look at étale algebras with trivial twist in $\mathrm{Rep}\,\mathcal{W}$. From now on whenever we refer to a finite extension $\mathcal{W} \subset \mathcal{V}$ we mean one satisfying the conditions given in Theorem 2

REMARK 1. A natural question to ask is if we have a finite extension $\mathcal{W} \subset \mathcal{V}$, we know what the local-modules of A will be in terms of $\mathcal{V}$-modules. Is there any way to describe the other objects of $\mathrm{Rep}_{\mathcal{C}} A$ as some generalized $\mathcal{V}$-module? In the case that G is a finite group acting on $\mathcal{V}$ and $\mathcal{W} = \mathcal{V}^G$ the answer to this is yes. If we assume that rationality descends, then it turns out that $\mathrm{Rep}_{\mathcal{C}} A$ will be all G-twisted modules. What we are proposing in this paper is that there is a similar description in general for all finite extensions. Namely if $\mathcal{W} \subset \mathcal{V}$ is a finite extension there is some fusion ring K, more precisely a hypergroup, acting on $\mathcal{V}$ such that $\mathrm{Rep}_{\mathcal{C}} A$ are the K-twisted $\mathcal{V}$-modules and $\mathcal{W} = \mathcal{V}^K$.

2.2. Graphical calculus. We represent the morphism $f : X \to Y$ by the following coupon oriented bottom to top:

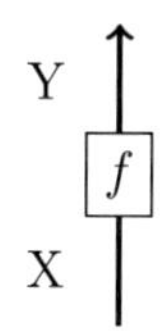

FIGURE 1. Morphism

Composition of morphisms $f : X \to Y$, $g : Y \to Z$ is given by stacking coupons:

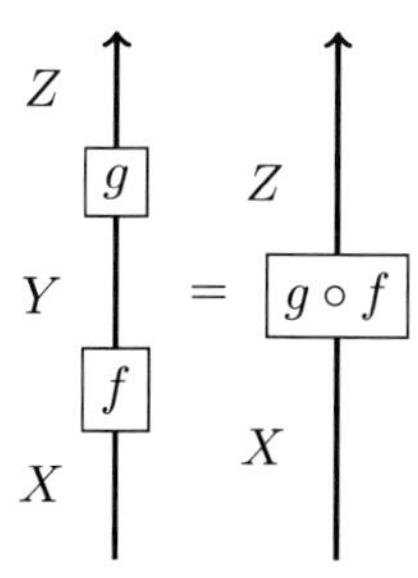

FIGURE 2. Composition

A straight line labelled by an object X should simply be interpreted as Id_X. If $f : X \to Y$, $g : W \to Z$ are morphisms, then the tensor product $f \otimes g : X \otimes W \to Y \otimes Z$ is depicted by:

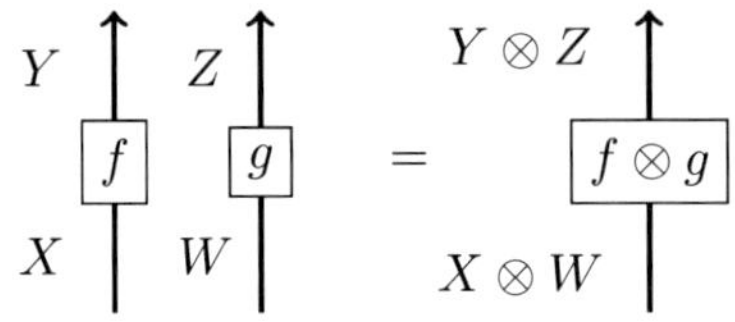

FIGURE 3. Tensor Product of Morphisms

The braiding of $\mathcal{C}$ and its inverse are depicted respectively by:

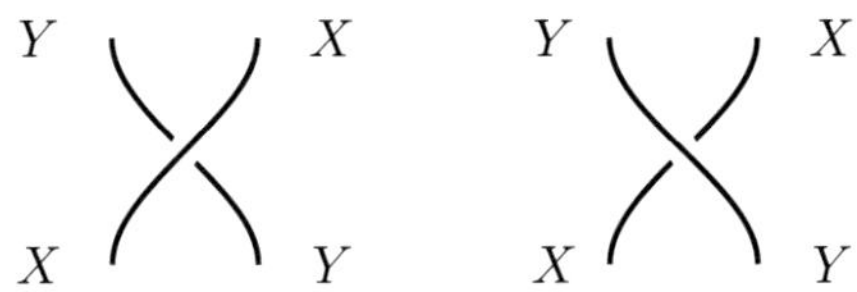

FIGURE 4. Braiding

Since we are assuming that $\mathcal{C}$ is a modular fusion category, this means it is rigid and pivotal. For every $X \in \mathrm{Obj}(\mathcal{C})$ we denote the left evaluation morphisms $\mathrm{ev}_X : X^* \otimes X \to 1$ and co-evaluation morphisms $\mathrm{coev}_X : 1 \to X \otimes X^*$ by:

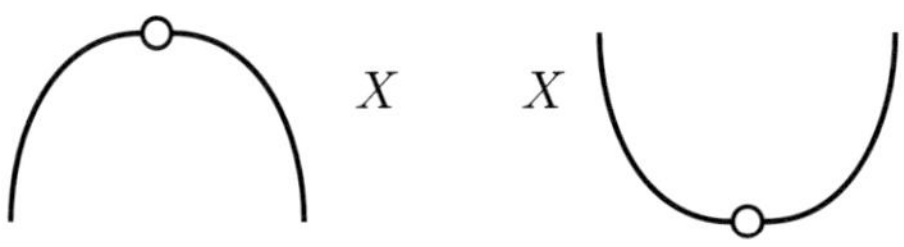

FIGURE 5. Evaluation and Co-evaluation

As an example, using graphical calculus we can write the morphism

$$(\mathrm{ev}_X \otimes \mathrm{Id}) \circ (\mathrm{Id}_{X^*} \otimes R^{-1}_{X,X^{**}}) \circ (\mathrm{coev}_{X^*} \otimes \mathrm{Id}) \tag{2.5}$$

as the following figure:

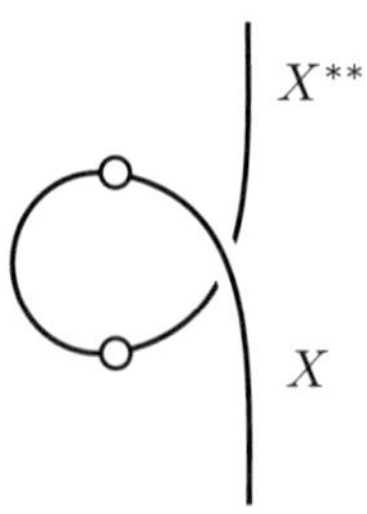

FIGURE 6.

In fact, using the ribbon structure θ of $\mathcal{C}$ we can write the pivotal structure and its inverse as:

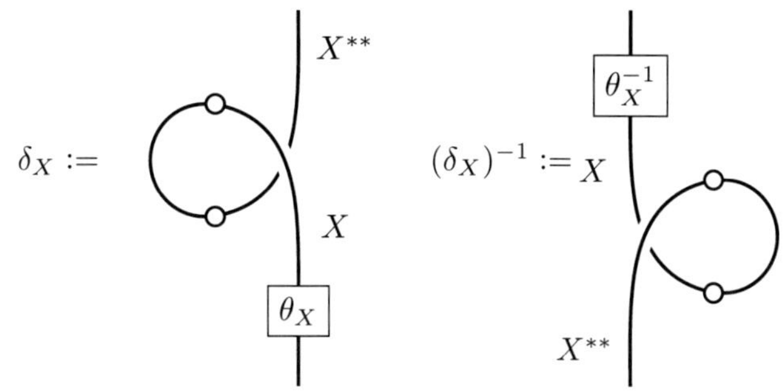

FIGURE 7. Pivotal Structures

see the appendix of [**23**] for details. We will need these diagrams in the proof of Lemma 4, but besides this we will simply suppress the pivotal structure in our graphical calculus throughout the rest of the paper. For example, under this convention the categorical dimension of $X \in \mathcal{C}$ will simply be written as an X-ring:

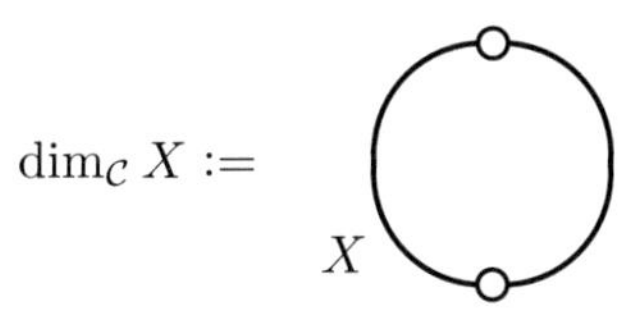

FIGURE 8. X-Ring

Now suppose that we have an algebra (A, μ_A, ι) in $\mathcal{C}$, throughout this paper we will be working with only one algebra at a time, and so morphisms from A to other objects in $\mathcal{C}$ are depicted by dashed lines and the domain is suppressed. The multiplication μ_A and unit ι_A are depicted respectively as:

FIGURE 9. Multiplication and Unit

With this notation the axioms of an algebra simply are:

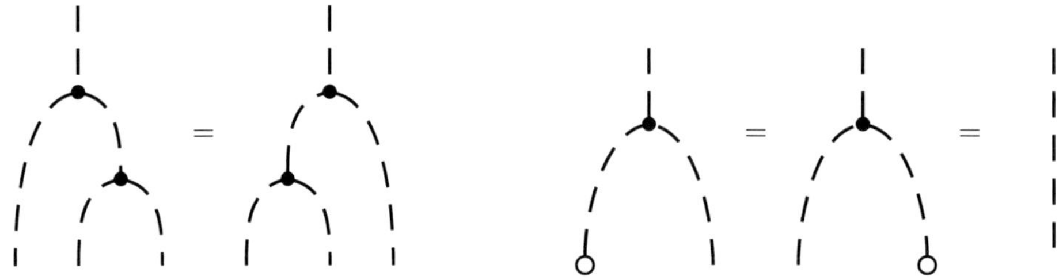

FIGURE 10. Associativity and Unit Axioms

If (M, μ_M) is a left A-module, and it is clear from the context that we are only considering the module structure on M given by μ_M, we depict μ_M as:

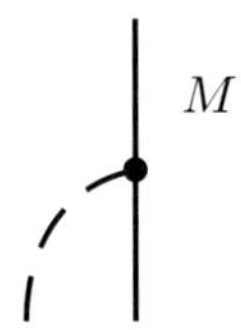

FIGURE 11. Module Structure

In [**29**] it is shown that if (A, μ_A, ι_A) is an étale algebra, then $\mathrm{Rep}_{\mathcal{C}}A$ is semi-simple, rigid and pivotal. We denote the categorical dimension of an object in $\mathrm{Rep}_{\mathcal{C}}\, A$ by $\dim_A(-)$. As a morphism in $\mathcal{C}$, we can depict the left evaluation morphisms and left co-evaluation morphisms of (M, μ_M) respectively as:

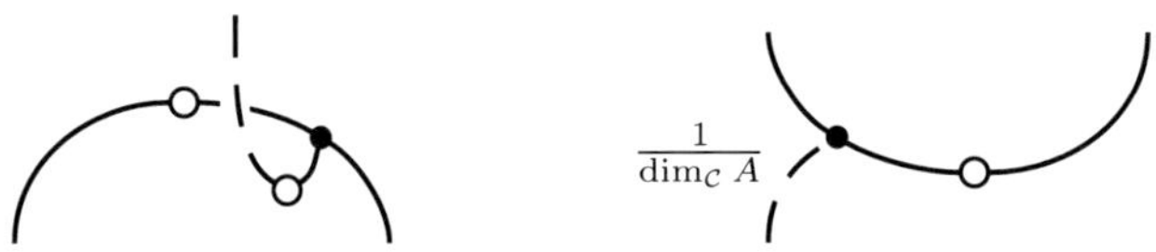

FIGURE 12. Evaluation and Co-Evaluation in $\mathrm{Rep}_{\mathcal{C}}\, A$

Using the fact that $\mathrm{Rep}_{\mathcal{C}}\, A$ is pivotal we can calculate the categorical dimension in $\mathrm{Rep}_{\mathcal{C}}A$:

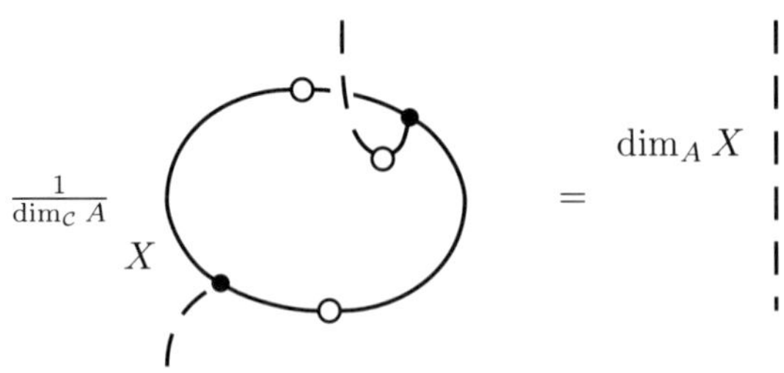

FIGURE 13. Dimension in $\mathrm{Rep}_{\mathcal{C}}\, A$

Additionally, using this graphical calculus one can show that $\dim_{\mathcal{C}}(\mathrm{Ind}(X)) = \dim_A X$ for all $X \in \mathcal{C}$, see [**29**, Theorem 1.18]. The following identities from [**29**] involving the left evaluation and co-evaluation morphisms in $\mathcal{C}$ will be essential to proving many of the results here:

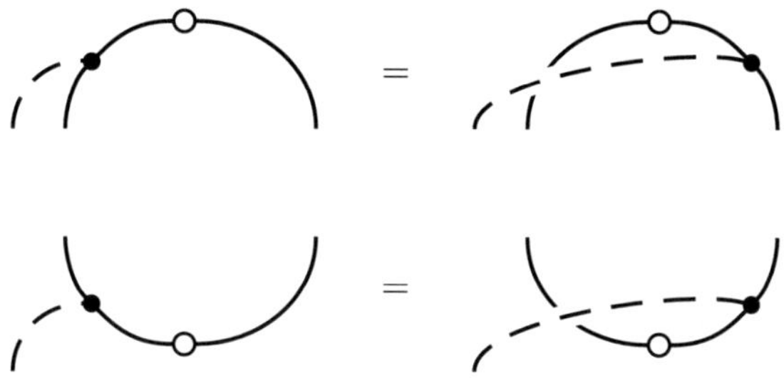

FIGURE 14. Module Evaluation and Co-evaluation Identity

If (A, μ_A, ι_A) is étale, then for every $(X, \mu_X), (Y, \mu_Y) \in \mathrm{Rep}_{\mathcal{C}} A$ we know that $X \otimes_A Y$, as an object in $\mathcal{C}$, is a direct summand of $X \otimes Y$. We have a projection of this given by:

$Q_{X,Y} = \frac{1}{\dim_{\mathcal{C}} A}$

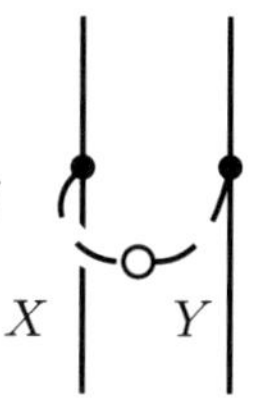

FIGURE 15. Projection of $X \otimes Y$ into $X \otimes_A Y$

Lastly, we will need the following identities [**29**, Lemma 1.12, Lemma 1.14]:

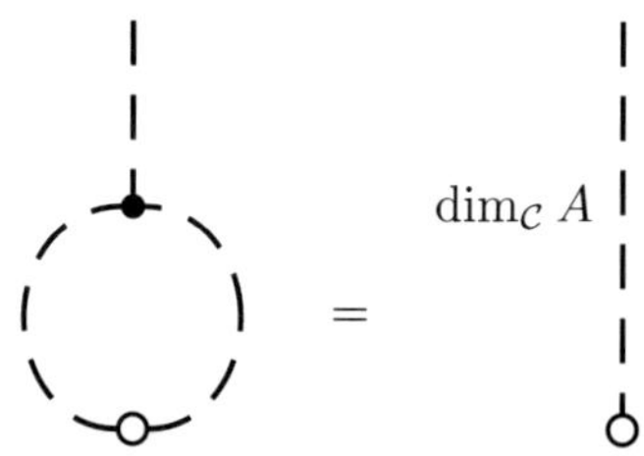

FIGURE 16. A-Ring Reduction

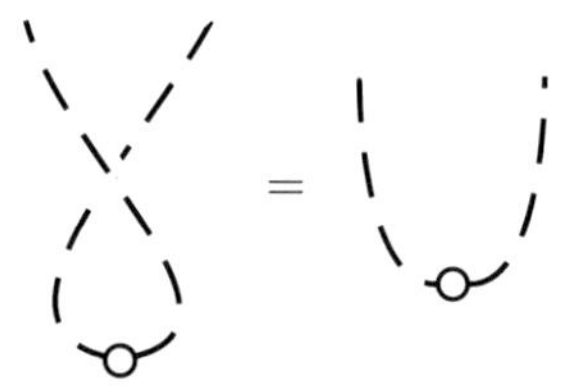

FIGURE 17.

2.3. Hypergroups and fusion rings. First, we recall the following from [**5**].

DEFINITION 2.4. *A hypergroup is a finite set* $K = \{r_i\}_{i\in I}$ *with a multiplication to the convex combinations of elements of* K, $\cdot : K \times K \to \mathrm{Conv}(K)$*:*

$$r_i \cdot r_j = \sum_{k\in I} C^k_{i,j} r_k \quad C^k_{i,j} \geq 0 \quad \sum_{k\in I} C^k_{i,j} = 1 \tag{2.6}$$

such that $\mathbb{C}[K]$ *is an associative algebra, and there is a distinguished element* $r_0 \in K$ *that acts as the identity making* $\mathbb{C}[K]$ *unital. Furthermore, there exists an involution* $i \mapsto i^*$ *of* K *that induces a unital anti-automorphism of* $\mathbb{C}[K]$ *such that:*

$$C^0_{i,j} = C^0_{j,i} \text{ and } C^0_{i,j} > 0 \text{ if and only if } j = i^* \tag{2.7}$$

When it is clear from the context we will usually refer to a hypergroup just by its set K. A sub-hypergroup is a subset $H \subset K$ that is closed under multiplication, involution and has the identity. Given a hypergroup K, the weight of a basis element r_i is defined as:

$$w_i := (C^0_{i,i^*})^{-1} \tag{2.8}$$

The weight of the hypergroup is defined as:

$$D(K) := \sum_{i\in I} w_i \tag{2.9}$$

Let $\mathbb{Z}_+ := \{n \in \mathbb{Z} : n \geq 0\}$. Strongly related to the concept of a hypergroup is the notion of a fusion ring and their modules [**15**, Chapter 3]:

DEFINITION 2.5 (Fusion Ring). *Let* R *be a free unital associative* $\mathbb{Z}$*-algebra. We say that it is a fusion ring if it has the following properties:*

(1) *There exists a* $\mathbb{Z}$*-basis* $\{x_i\}_{i\in I}$ *where* $|I| < \infty$ *such that:*

$$x_i \cdot x_j = \sum_{k\in I} N^k_{i,j} x_k \quad N^k_{i,j} \in \mathbb{Z}_+ \tag{2.10}$$

(2) *R has a unit* 1*, and there exists* $i_0 \in I$ *such that* $1 = x_{i_0}$.
(3) *There exists an involution* $i \mapsto i^*$ *of* I *such that the map* $\sum_{i\in I} c_i x_i \mapsto \sum_{i\in I} c_i x_{i^*}$ *gives a unital anti-automorphism of* R*, and* $N^0_{i,j} = N^0_{j,i} = \delta_{j,i^*}$.

Elements of $\{x_i\}_{i\in I}$ are referred to as *simple* objects of R.

DEFINITION 2.6 (Based Left $\mathbb{Z}_+$-Module over R). *Let* R *be a fusion ring with basis* $\{x_i\}_{i\in I}$*. A* based left $\mathbb{Z}_+$-module over R *is a unital* R*-module* M *with a* $\mathbb{Z}$*-basis* $\{m_\ell\}_{\ell\in L}$ *such that* $|L| < \infty$, $x_i \cdot m_\ell = \sum_{j\in L} b_j m_j$ *where* $b_j \geq 0$*, and* $(x_i)^*$ *acts as the transpose of* x_i *with respect to the basis* $\{m_\ell\}_{\ell\in L}$.

Similarly, a based right $\mathbb{Z}_+$- module over R and based $\mathbb{Z}_+$ bi-module over R can be defined. Just as fusion rings are the combinatorial data inspired by fusion categories, a based left $\mathbb{Z}_+$-module is the combinatorial data inspired by left module categories.

DEFINITION 2.7 (Sub-Fusion Ring). *Let R be as above, then a sub-ring $S \subset R$ is a sub-fusion ring if there exists a subset $J \subset I$ closed under the involution of I and containing* 1 *such that $S = \mathbb{Z}[x_i : i \in J]$.*

DEFINITION 2.8 (Sub Based Left $\mathbb{Z}_+$-Module over R). *Let M be a based left $\mathbb{Z}_+$-module over R with basis $\{m_\ell\}_{\ell \in L}$. A sub based left $\mathbb{Z}_+$-module N, or sub-module N for short, is an R sub-module such that there exists a subset $F \subset L$ for which $N = \mathbb{Z}[m_\ell : \ell \in F]$.*

DEFINITION 2.9 (Frobenius-Perron Dimension). *Let R be a fusion ring. The Frobenius-Perron dimension of x_i is defined to be the largest real eigenvalue of the matrix $(N_{i,j}^k)_{j,k \in I}$. We denote this value by d_{x_i} for every $i \in I$.*

Three properties of the Frobenius-Perron dimensions we will need are:

(1) $d_{x_i \cdot x_j} = d_{x_i} \cdot d_{x_j}$
(2) $d_{x_i} = d_{x_{i^*}}$
(3) $d_{x_0} = 1$

See [**15**] for details. As a short hand we write $d_i := d_{x_i}$ for $i \in I$.

A fusion ring R induces a hypergroup by setting $K := \{\frac{x_i}{d_i}\}_{i \in I}$, letting multiplication on $\mathbb{C}[K]$ be the one induced from R, and defining the involution of K by $\frac{x_i}{d_i} \mapsto \frac{x_{i^*}}{d_i}$. Notice that in this case the weight of $\frac{x_i}{d_i}$ will be d_i^2. As pointed out in [**5**] a hypergroup K will come from a fusion ring if and only if:

$$\forall i, j, k \in I \quad \sqrt{\frac{w_i w_j}{w_k}} C_{i,j}^k \in \mathbb{Z}_+ \tag{2.11}$$

This means the generic hypergroup will not come from a fusion ring, see the example below. One advantage hypergroups have over fusion rings is you can take quotients of them. To talk about quotients of hypergroups we need the notion of a Haar element

DEFINITION 2.10. *A Haar element of a hypergroup $K = \{r_i\}_{i \in I}$ is an element $\omega_K \in \mathrm{Conv}(K)$ such that $\omega_K^* = \omega_K = \omega_K^2$ and $r_i \cdot \omega_K = \omega_K \cdot r_i = \omega_K$ for all $i \in I$.*

PROPOSITION 1. *A Haar element of a hypergroup $K = \{r_i\}_{i \in I}$ can be constructed as:*

$$\omega_K = \frac{1}{D(K)} \sum_{i \in I} w_i r_i \tag{2.12}$$

Using this construction on a hypergroup coming from a fusion ring with basis $\{x_i\}_{i \in I}$ we get.

$$\omega_K := \frac{1}{D(K)} \sum_{i \in I} d_i \cdot x_i \tag{2.13}$$

PROOF. See [**5**] for details. □

Notice that in the fusion ring case $D(K) = \sum_{i \in I} d_i^2$.

Let K be a hypergroup. For $r, s \in K$ we say that $t \in r \cdot s$ if $C_{r,s}^t > 0$.

NOTATION. *If $A, B \subset K$, then $A \cdot B = \{t \in K : \exists a \in A, b \in B\ t \in a \cdot b\}$*

DEFINITION 2.11 (Right H-Cosets). *Let H be a sub-hypergroup of K. The quotient K/H will consist of cosets $\{x\} \cdot H$ for $x \in K$.*

Similarly we can define left H-cosets. For the sake of brevity we denote $\{x\} \cdot H$ by xH. The Haar element is a useful way to check if two H-cosets are equal:

LEMMA 1 ([**6**, Proposition 1.5.10]). *The right H-cosets of K define an equivalence relation on K, and $xH = yH$ if and only if $x \cdot \omega_H = y \cdot \omega_H$*

DEFINITION 2.12 (Double H-cosets). *Let H be a sub-hypergroup of K. The double quotient $K /\!\!/ H$ consists of double cosets $H \cdot \{x\} \cdot H$ for $x \in K$*

LEMMA 2 ([**6**, Proposition 1.5.14]). *The double H-cosets of K define an equivalence relation on K, and $HxH = HyH$ if and only if $\omega_H \cdot x \cdot \omega_H = \omega_H \cdot y \cdot \omega_H$.*

Define a quotient map:

$$p_H : K \to K /\!\!/ H \quad p_H(x) := HxH \tag{2.14}$$

This extends by linearity to $p_H : \mathbb{C}[K] \to \mathbb{C}[K /\!\!/ H]$.

THEOREM 4. *Let H and K be as before. $K /\!\!/ H$ can be given the structure of a hypergroup with multiplication defined by:*

$$(HxH) \cdot (HyH) := p_H(x \cdot \omega_H \cdot y)$$

The involution is defined as:

$$(HxH)^* := Hx^*H$$

PROOF. See [**6**] for details. □

For the sake of simplicity we will use the notation $[x] := HxH$. When K forms a group, and H is a normal subgroup then the double coset construction just gives the regular quotient group. However, if H is not normal we do not necessarily get a group structure on $K /\!\!/ H$ as the next example shows.

EXAMPLE 1. *Consider $H = \{1, (12)\}$ in S_3, this is not normal. By calculating $\omega_H \cdot \sigma \cdot \omega_H$ for all $\sigma \in S_3$, one can check that there are two double cosets with representatives given by*

$$[\text{Id}], [(13)]$$

and

$$[(13)] \cdot [(13)] = \frac{1}{2}[\text{Id}] + \frac{1}{2}[(13)]$$

Consider now a fusion ring R and a sub-fusion ring S. Clearly R will be a based $\mathbb{Z}_+$ bi-module over S. If y is a simple object of R denote by $\mathbb{Z}_+[SyS]$ the based $\mathbb{Z}_+$ bi-module over S generated by y. We can decompose R into based $\mathbb{Z}_+$ bi-modules over S:

$$R \cong \bigoplus_{k \in L} M_k \tag{2.15}$$

Let K be the associated hypergroup to R and H the associated hypergroup to S. H will be a sub-hypergroup of K.

PROPOSITION 2. *The indexing set L is in bijection with $K /\!/ H$ as follows. For every double coset, choose a representative $\frac{x_i}{d_i}$. Define $\sigma : K /\!/ H \to L$ by letting $\sigma(i) \in L$ be the element of L labelling the indecomposable based $\mathbb{Z}_+$ bi-module $\mathbb{Z}_+[Sx_iS]$. The inverse $\sigma^{-1} : L \to K /\!/ H$ is defined by first choosing a simple object $x_k \in M_k$ for every $k \in L$, and setting $\sigma^{-1}(k) = \left[\frac{x_k}{d_k}\right]$*

PROOF. First, let's show that if x_i is a simple object of R, then $\mathbb{Z}_+[Sx_iS]$ is an indecomposable based $\mathbb{Z}_+$ bi-module over S. Suppose not, then there exists non-zero based $\mathbb{Z}_+$ bi-modules N_0, N_1 such that $\mathbb{Z}_+[Sx_iS] = N_0 \oplus N_1$. By definition, N_0 and N_1 must be the span of simple objects of R. Therefore, x_i is either in N_0 or N_1 contradicting the definition of $\mathbb{Z}_+[Sx_iS]$. To see that σ is well-defined notice that $\omega_H \cdot \frac{x_i}{d_i} \cdot \omega_H = \omega_H \cdot \frac{x_j}{d_j} \cdot \omega_H$ if and only if there exists some $t, s \in S$ such that x_j is a sub-object of $t \cdot x_i \cdot s$. This implies that $\mathbb{Z}_+[Sx_jS] \subset \mathbb{Z}_+[Sx_iS]$. It is not difficult to show a based $\mathbb{Z}_+$ bi-module is indecomposable iff it is irreducible. Thus, we see that $\mathbb{Z}_+[Sx_iS] = \mathbb{Z}_+[Sx_jS]$ so σ is well-defined. Using the same observation, we see that σ is injective and hence bijective. A similar argument works for showing σ^{-1} is as described above. □

Let $R = \{x_i\}_{i \in I}$ be a fusion ring with associated hypergroup K, and $S = \{x_i\}_{i \in J}$ a sub-fusion ring with associated hypergroup H. When the context makes it clear we will refer to the double coset $K /\!/ H$ as $R /\!/ S$. The following is easy to show:

PROPOSITION 3. *Let $\rho : \mathbb{C}[K] \to \mathrm{End}_{\mathbb{C}}(V)$ be a unital algebra homomorphism. Suppose that for all $i \in J$ we have $\rho(x_i) = d_i \cdot \mathrm{Id}_V$. Then there is an induced action:*

$$\tilde{\rho} : K /\!/ H \to \mathrm{End}_{\mathbb{C}}(V) \qquad \tilde{\rho}([x]) := \rho(x) \tag{2.16}$$

Observe that the x are in the hypergroup associated to the fusion ring. This means that we have $x = \frac{x_i}{d_i}$ for some simple $x_i \in R$. If $\ker(\rho) := \mathbb{Z}[\{x_i : \rho(x_i) = d_i \cdot \mathrm{Id}_V\}]$ is a sub-fusion ring of R, then we say the *effective action of ρ* is given by the induced action $\tilde{\rho} : R /\!/ \ker(\rho) \to \mathrm{End}_{\mathbb{C}}(V)$. Lastly, if $\mathcal{C}, \mathcal{D}$ are fusion categories and $\mathcal{D}$ is a sub-fusion category of $\mathcal{C}$ then we have an inclusion of fusion rings $\mathcal{K}^0(\mathcal{D}) \subset \mathcal{K}^0(\mathcal{C})$ and so we we can form the double coset $\mathcal{K}^0(\mathcal{C}) /\!/ \mathcal{K}^0(\mathcal{D})$. When it is clear from context we will denote $\mathcal{K}^0(\mathcal{C}) /\!/ \mathcal{K}^0(\mathcal{D})$ simply as $\mathcal{C} /\!/ \mathcal{D}$.

3. Fusion ring action associated to a VOA extension

In this section let $\mathcal{W}$ be a nice VOA, $\mathcal{C} := \mathrm{Rep}\,\mathcal{W}$ and A an étale algebra with trivial twist in $\mathcal{C}$ corresponding to a VOA extension $\mathcal{V}$.

DEFINITION 3.1 (Hypergroup Action On an Étale Algebra). *We say that a hypergroup K acts on A if there is a unital algebra homomorphism $\rho_- : \mathbb{C}[K] \to \mathrm{End}_{\mathcal{C}}(A)$ such that $(\rho_r)^* = \rho_{r^*}$. Here $(\rho_X)^*$ denotes the dual morphism of ρ_X. Since $A^* = A$ this makes sense.*

We will show that the hypergroup $K := \mathrm{Rep}_{\mathcal{C}} A /\!/ \mathrm{Rep}^{loc}_{\mathcal{C}} A$ acts on A through A-open Hopf links depicted in Figure 18 for $(X, \mu_X) \in \mathrm{Rep}_{\mathcal{C}} A$. To the author's knowledge these diagrams were first used by Kirillov in the group orbifold case [**28**], and indeed in the case that $\mathcal{W} = \mathcal{V}^G$ one finds that $K = G$.

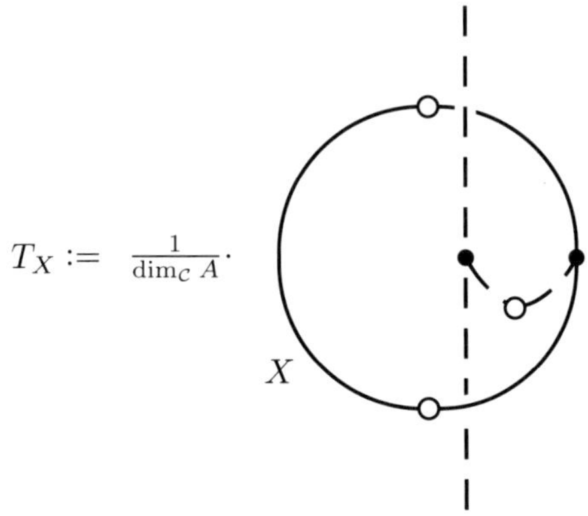

FIGURE 18. A-Open Hopf Link with X

3.1. The associated hypergroup. The following Lemma will be required to prove the main result of this subsection.

LEMMA 3. *The following holds for all* $(W, \mu_W), (Z, \mu_Z) \in \mathrm{Rep}_{\mathcal{C}} A$*:*

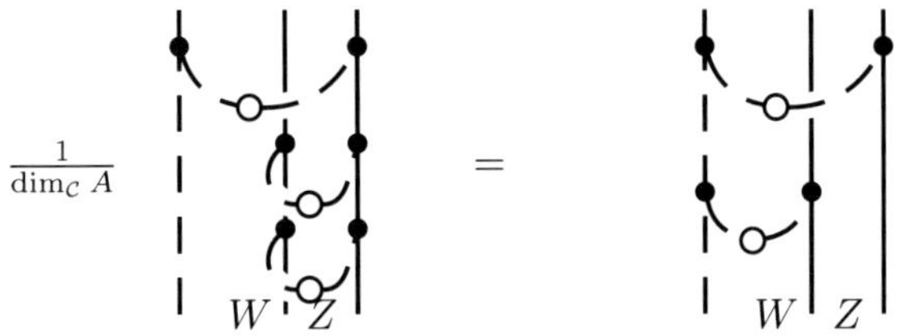

FIGURE 19.

PROOF. Using the fact that $Q_{W,Z}^2 = Q_{W,Z}$ we may simplify the left hand side of Figure 19 to the the first sub-figure on the left. The rest is completed through the following steps outline in the sub-figures

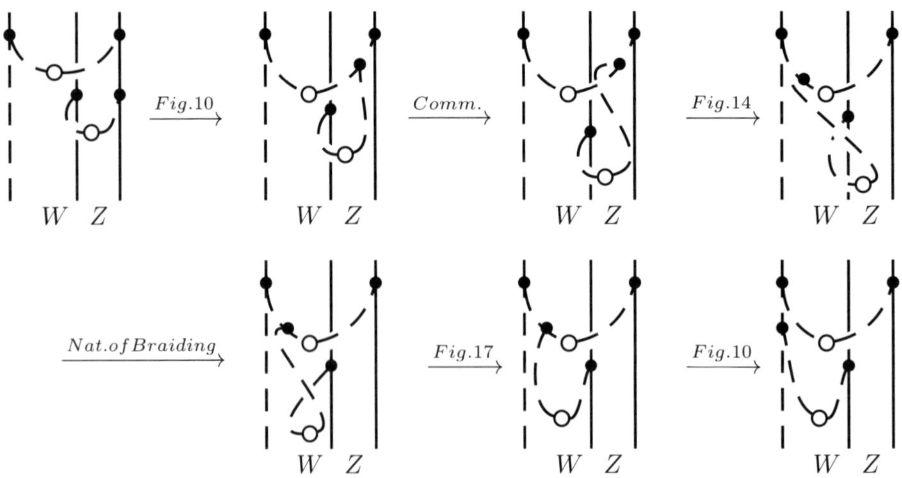

□

LEMMA 4. *For $(X, \mu_X), (Y, \mu_Y) \in \mathrm{Rep}_{\mathcal{C}} A$ the following holds:*

$$Q^*_{X,Y} = Q_{Y^*,X^*} \tag{3.1}$$

PROOF. Writing out $\dim_{\mathcal{C}}(X \otimes_A Y) \cdot Q^*_{X,Y}$ in graphical calculus we obtain the first sub-figure, and then by moving the A-string to the left side of the diagram using the moves given in Figures 12, 17 we obtain the second sub-figure.

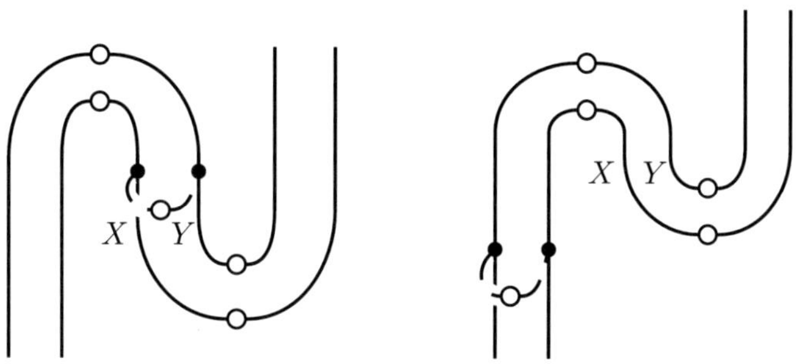

The rigidity axioms imply that we can straighten out the second subfigure to obtain $\dim_{\mathcal{C}}(Y^* \otimes_A X^*)Q_{Y^*,X^*}$. Using the spherical structure of $\mathcal{C}$ we see that:

$$\dim_{\mathcal{C}}(Y^* \otimes_A X^*) = \dim_{\mathcal{C}}((X \otimes_A Y)^*) = \dim_{\mathcal{C}}(X \otimes_A Y) \tag{3.2}$$

Therefore, we see that $\dim_{\mathcal{C}}(Y^* \otimes_A X^*)Q_{Y^*,X^*} = \dim_{\mathcal{C}}(X \otimes_A Y)Q^*_{X,Y}$ completing the proof. □

The following result was mentioned in [**28**] for the group orbifold case with only a sketch of a proof. Their proof idea works in the general setting which we now present in full detail.

THEOREM 5. *If $(X, \mu_X), (Y, \mu_Y) \in \mathrm{Rep}_{\mathcal{C}} A$, then $T_{X \otimes_A Y} = T_X \circ T_Y$.*

PROOF. First, observe that $X \otimes Y$ is a left A-module with A action given by $\mu_X \otimes \mathrm{Id}_Y$. With this module structure, the morphism $Q_{X,Y}$ is a morphism in $\mathrm{Rep}_{\mathcal{C}} A$. By Lemma 1.21 of [**29**] we see then that as a left A-module we have:

$$(X \otimes Y, \mu_X \otimes \mathrm{Id}_Y) \cong (\ker Q_{X,Y}) \oplus (X \otimes_A Y) \tag{3.3}$$

Now, denote the corresponding inclusion and projection A-module morphisms by:

$$\begin{gathered} \iota_{X \otimes_A Y} : (X \otimes_A Y, \mu_{X \otimes_A Y}) \to (X \otimes Y, \mu_X \otimes \mathrm{Id}) \\ \pi_{X \otimes_A Y} : (X \otimes Y, \mu_X \otimes \mathrm{Id}) \to (X \otimes_A Y, \mu_{X \otimes_A Y}) \\ \iota_{X \otimes_A Y} \circ \pi_{X \otimes_A Y} = Q_{X,Y} \qquad \pi_{X \otimes_A Y} \circ \iota_{X \otimes_A Y} = \mathrm{Id}_{X \otimes_A Y} \end{gathered} \tag{3.4}$$

By using the following identities as well as the pivotal structure we may depict $\dim_{\mathcal{C}} A \cdot T_{X \otimes_A Y}$ as Figure 20.

$$\begin{gathered} ev_{X \otimes_A Y} = ev_{X \otimes Y} \circ ((\pi_{X \otimes_A Y})^* \otimes \iota_{X \otimes_A Y}) \\ coev_{X \otimes_A Y} = (\pi_{X \otimes_A Y} \otimes (\iota_{X \otimes_A Y})^*) \circ coev_{X \otimes Y} \\ \mathrm{ev}_{(X \otimes_A Y)^*} = \mathrm{ev}_{(X \otimes Y)^*} \circ ((\iota_{X \otimes_A Y})^{**} \otimes (\pi_{X \otimes_A Y})^*) \end{gathered} \tag{3.5}$$

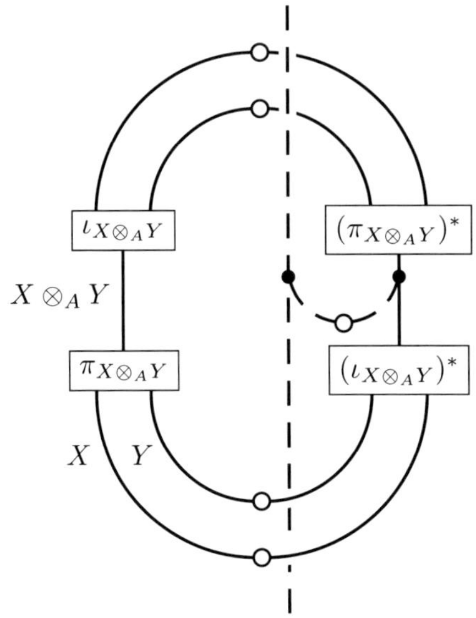

FIGURE 20.

Notice, in the last identity we use the fact that $(\text{coev}_X)^* = \text{ev}_{X^*}$. We can transform Figure 20 into Figure 21 by using the fact that $\pi^*_{X\otimes_A Y}$ is an A-module morphism from $(X \otimes_A Y, \mu_{X\otimes_A Y})^*$ to $(X \otimes Y, \mu_X \otimes \text{Id})^*$ and the pivotal structure.

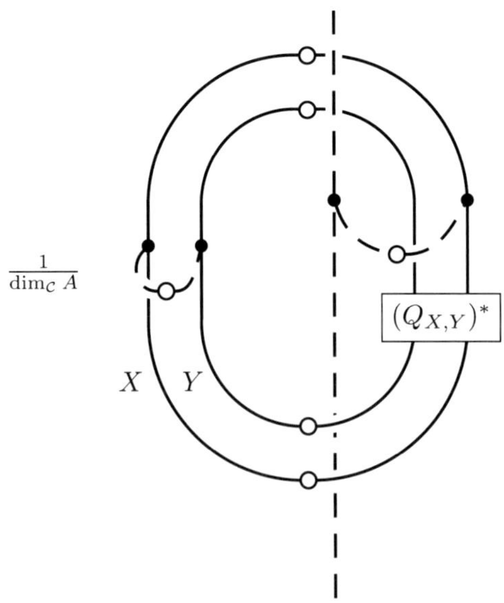

FIGURE 21.

Invoking Lemma 4 and moving the left A-string along the bottom of the $X\otimes Y$-ring we obtain Figure 22a.

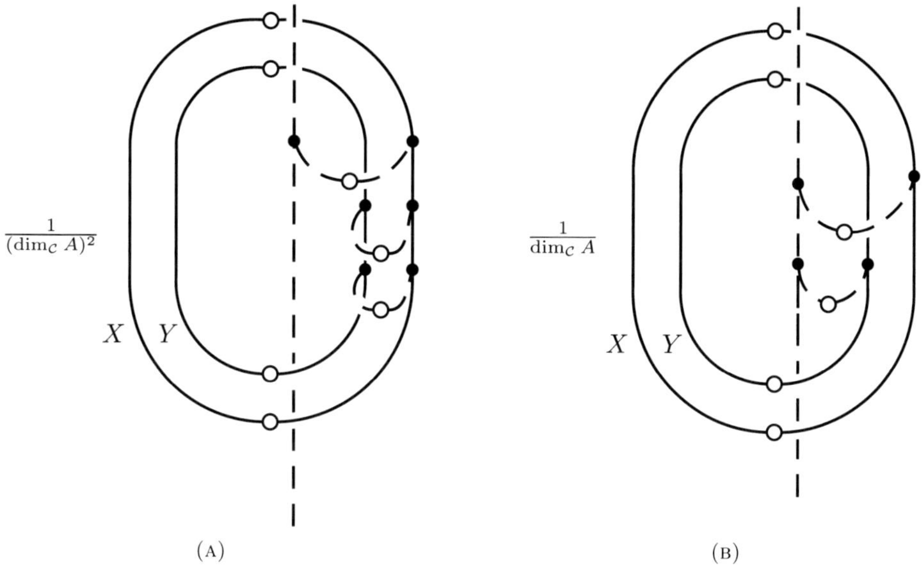

FIGURE 22.

Using Lemma 3 we see that this equals Figure 22b. We can now pull the A-open Hopf link with Y down through the A-open Hopf link with X to get $\dim_{\mathcal{C}} A \cdot (T_X \circ T_Y)$, completing the proof. □

Notice that the map $X \mapsto T_X$ induces a representation of the associative algebra $\mathcal{K}^0(\operatorname{Rep}_{\mathcal{C}} A)$ on $\operatorname{End}_{\mathcal{C}}(A)$ by acting through composition. The associated hypergroup will act on A through the set:

$$\{\frac{1}{\dim_A X} T_X : X \in \mathcal{O}(\operatorname{Rep}_{\mathcal{C}} A)\} \tag{3.6}$$

Since $\operatorname{Rep}_{\mathcal{C}} A$ is a semi-simple fusion category we have that $\dim_A X \neq 0$ for all $X \in \mathcal{O}(\operatorname{Rep}_{\mathcal{C}} A)$ [**15**, Proposition 4.8.4].

To calculate the effective action of this hypergroup, we calculate the kernel.

LEMMA 5. *Let* $(X, \mu_X) \in \operatorname{Rep}_{\mathcal{C}} A$. $(X, \mu_X) \in \operatorname{Rep}_{\mathcal{C}}^{loc} A$ *if and only if* $T_X = (\dim_A X) \cdot \operatorname{Id}_A$.

PROOF. First, notice that T_X can be re-written as Figure 23 by moving the bottom A-string along to the X-ring using the co-evaluation identity for A and then wrapped around to the left hand side of the X-ring by applying the co-evaluation identity for X.

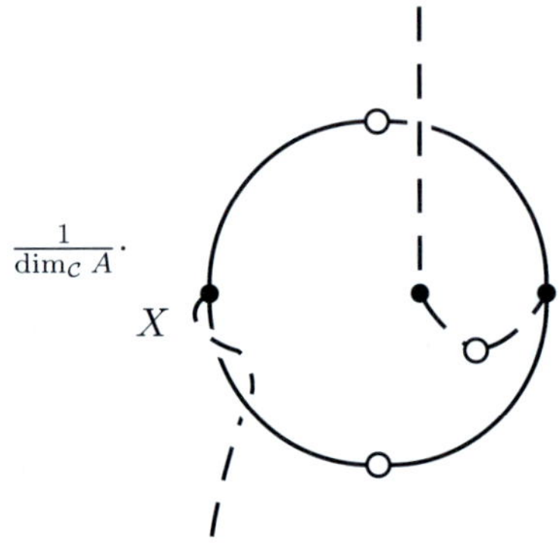

FIGURE 23.

By definition, a local module is a left A-module for which:

$$\mu_A \circ R_{X,A} \circ R_{A,X} = \mu_A$$

In the language of graphical calculus this means that we can straighten out the left A-string, and so we see that T_X will just be Figure 13. Therefore, if (X, μ_X) is local, we see that $T_X = (\dim_A X) \cdot \mathrm{Id}$.

Suppose now that $T_X = \dim_A(X) \cdot \mathrm{Id}_A$ for some simple $(X, \mu_X) \in \mathrm{Rep}_{\mathcal{C}} A$. For the sake of contradiction assume that $(X, \mu_X) \notin \mathrm{Rep}_{\mathcal{C}}^{loc} A$. If we take the trace of T_X we see that we get:

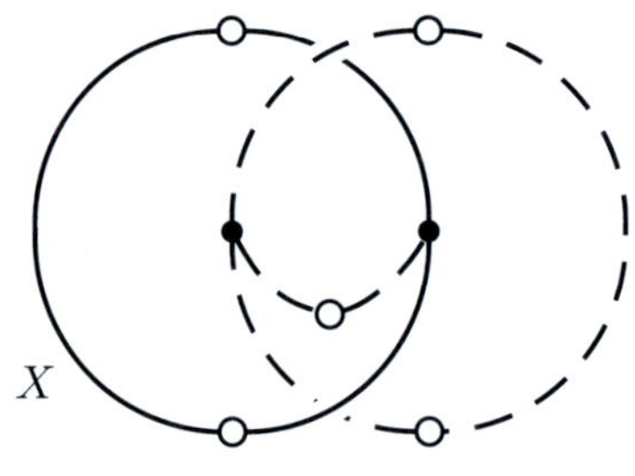

FIGURE 24.

Using the same reasoning as in [**29**, Lemma 4.4] we see that this will be zero. Since $\dim_{\mathcal{C}}(A) \neq 0$ this means that $\dim_A(X) = 0$, but this is impossible. □

We are now ready to prove the main result of the paper:

THEOREM 6. *A-open Hopf links T_- give an action of the fusion ring of* $\mathrm{Rep}_{\mathcal{C}} A$ *on A. The kernel of this action is* $\mathrm{Rep}_{\mathcal{C}}^{loc} A$, *and so the hypergroup* $K := \mathrm{Rep}_{\mathcal{C}} A \mathbin{/\!/} \mathrm{Rep}_{\mathcal{C}}^{loc} A$ *gives the effective action. If $\mathcal{W} = \mathcal{V}^G$ for some finite group G that acts on $\mathcal{V}$, then $K = G$.*

PROOF. By Lemma 5, Theorem 5, and Proposition 3 the last thing left to prove is that $(T_X)^* = T_{X^*}$. To that end notice that we can write $\dim_{\mathcal{C}}(A) \cdot (T_X)^*$ as the diagram below and apply the following manipulations:

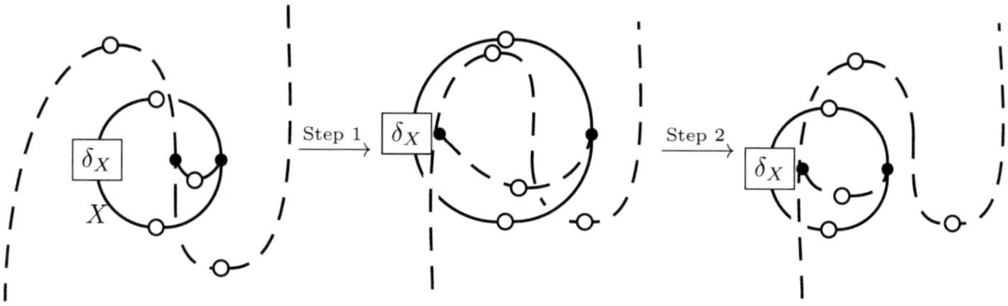

FIGURE 25.

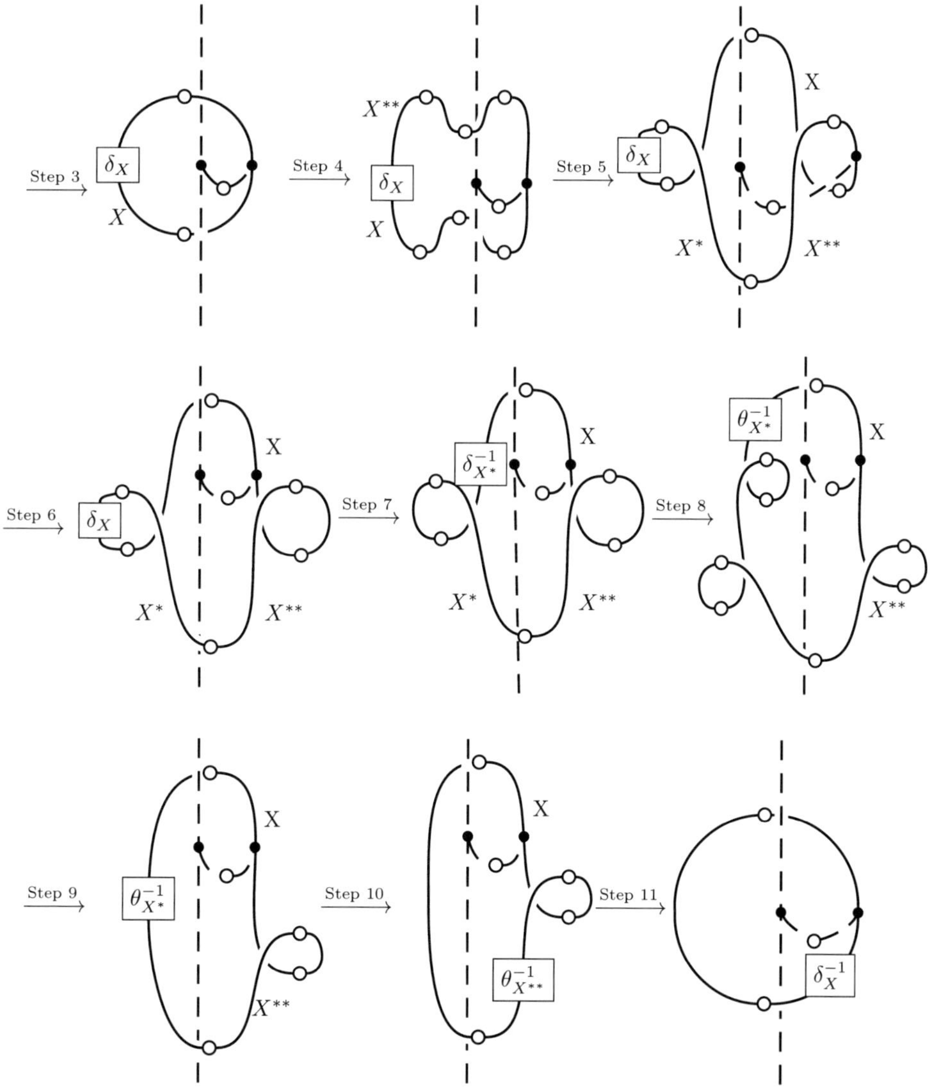

FIGURE 26.

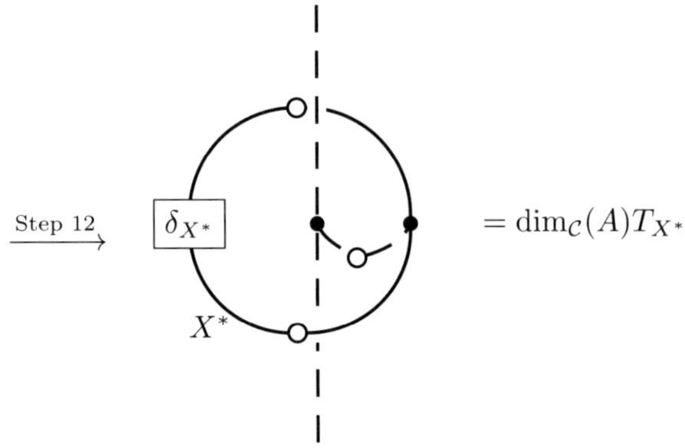

FIGURE 27.

Steps 1 and 2 follow from the naturality of the braiding and using the fact that A is commutative and the identities in Figure 14. Steps 3 and 4 follow from the rigidity axioms. Step 5 follows from pulling down the top of the ring past the bottom of the ring and using the naturality of the braiding. Step 6 follows from the identity in Figure 14. Step 7 follows from the fact that $\delta_{X^*}^{-1} = \delta_X^*$ and the identity $(\delta_X \otimes \mathrm{Id}) \circ \mathrm{coev}_X = (\mathrm{Id} \otimes (\delta_X)^*) \circ \mathrm{coev}_{X^{**}}$ (to see this just write out the diagram). Steps 8 and 9 follow from recalling the graphical description of the pivotal structure given in Figure 7. Step 10 follows from using the fact that $\theta_{X^{**}} = (\theta_{X^*})^*$ and the previous co-evaluation identity. Step 11 follows from Figure 7. Step 12 follows from the fact that δ_X is an A-module homomorphism [**29**, Theorem 1.17], and that $\mathrm{ev}_X \circ(\mathrm{Id} \otimes \delta_X^{-1}) = \mathrm{ev}_{X^*} \circ((\delta_X^{-1})^* \otimes \mathrm{Id})$ (to see this just write out the diagram). Therefore, we see that $(T_X)^* = T_{X^*}$. Recovering the group case follows from [**28**, Theorem 4.7]. □

We have a proof that K may be identified with Eq. (3.6) which will be presented in [**21**].

3.2. The fixed points of the hypergroup.

LEMMA 6. *Let ι be the inclusion of $1_{\mathcal{C}}$ into A. Then* $\frac{1}{\dim_A X}(T_X) \circ \iota = \iota$ *for all* $X \in \mathrm{Rep}_{\mathcal{C}} A$

PROOF. This follows from the sequence of figures given below. Note that the last equality holds because of the identity given in Figure 13.

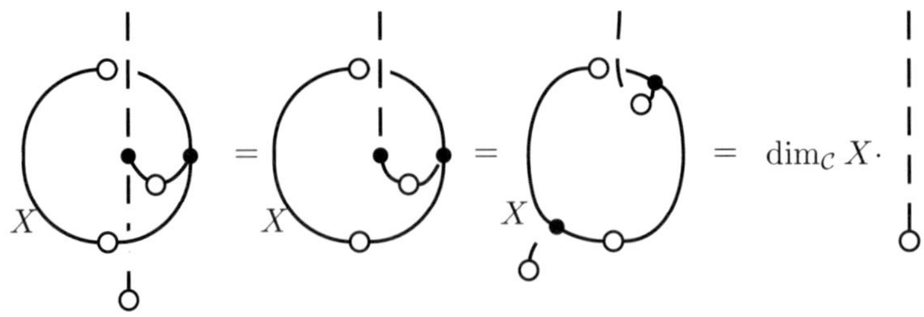

FIGURE 28.

Since $\dim_A X = \frac{\dim_{\mathcal{C}} X}{\dim_{\mathcal{C}} A}$ we get the desired result. □

Let K be as in Eq. (3.6). Define A^K to be the largest sub-object of A such that each $\frac{1}{\dim_A X} T_X$ restricts to the identity. By Lemma 6 we know that A^K will not be the zero object. Denote the open Hopf link of X with A by $\tilde{T}_X$.

LEMMA 7. *Let X be any object of $\mathcal{C}$. Then*

$$T_{\mathrm{Ind}(X)} = \tilde{T}_X \tag{3.7}$$

PROOF. Notice that $(A \otimes X)^* \cong X^* \otimes A$, and in particular

$$(A \otimes X, \mu_A \otimes \mathrm{Id}_X)^* \cong (X^* \otimes A, \mathrm{Id}_{X^*} \otimes \mu_A \circ R_{A,X^*} \otimes \mathrm{Id}_A)$$

Therefore, we can realize $\dim_{\mathcal{C}} A \cdot T_{\mathrm{Ind}(X)}$ as the following figure:

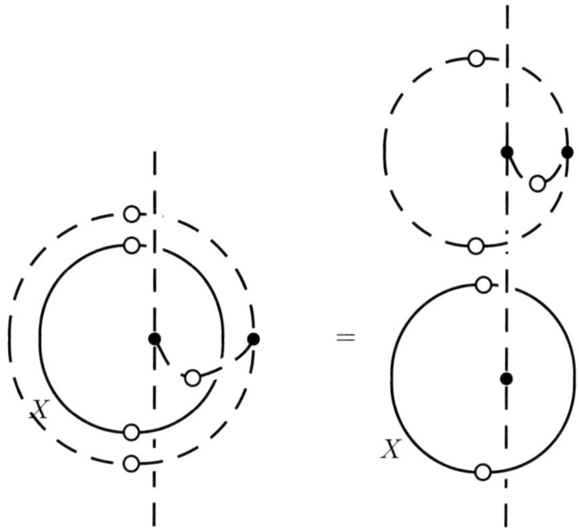

FIGURE 29.

By pulling the ring of X through we obtain the second picture, and evaluating the top morphism to $\dim A$ we see that $\dim_{\mathcal{C}} A \cdot T_{\mathrm{Ind}(X)} = \dim_{\mathcal{C}} A \cdot \tilde{T}_X$. □

Therefore, we see that induced modules act on A by their open Hopf links with A. As it will sometimes be the case that induction is simple this will turn out to be quite a useful tool for calculation as we will explain in the next sub-section.

THEOREM 7. *Let $\mathcal{W} \subset \mathcal{V}$ be an extension of nice VOAs such that $\mathcal{V}$ is an étale algebra in $\mathcal{C}$, and K as before. Then $\mathcal{V}^K = W$.*

PROOF. Let $r : A^K \to A, h : A \to A^K$ be the inclusion and projection respectively of A^K within A. Then by assumption $h \circ (\frac{1}{\dim_A Y} T_Y) \circ r = \mathrm{Id}_{A^K}$ for all $(Y, \mu_Y) \in \mathcal{O}(\mathrm{Rep}_{\mathcal{C}} A)$. On the other hand, we know that for every $Y, X \in \mathcal{O}(\mathrm{Rep}_{\mathcal{C}} A)$ there exist $N_{X,Y} \in \mathbb{Z}_{\geq 0}$ such that:

$$(A \otimes X, \mu_A \otimes \mathrm{Id}_X) \cong \bigoplus_{(Y,\mu_Y) \in \mathcal{O}(\mathrm{Rep}_{\mathcal{C}} A)} (Y, \mu_Y)^{\oplus N_{X,Y}} \tag{3.8}$$

This means that $T_{\mathrm{Ind}(X)} = \sum_{Y\in\mathcal{O}(\mathrm{Rep}_{\mathcal{C}}A)} N_{X,Y}T_Y$, and so

$$\begin{aligned} h\circ T_{\mathrm{Ind}(X)}\circ r &= \sum_{Y\in\mathcal{O}(\mathrm{Rep}_{\mathcal{C}}A)} N_{X,Y}(h\circ T_Y\circ r) \\ &= \sum_{Y\in\mathcal{O}(\mathrm{Rep}_{\mathcal{C}}A)} N_{X,Y}\dim_A Y\cdot \mathrm{Id}_{A^K} = \dim_A(\mathrm{Ind}(X))\cdot\mathrm{Id}_{A^K} \end{aligned} \tag{3.9}$$

By Lemma 7 we see then that $\frac{1}{\dim_{\mathcal{C}} X} h\circ\tilde{T}_X\circ r = \mathrm{Id}_{A^K}$ as $\dim_{\mathcal{C}}(X) = \dim_A(\mathrm{Ind}(X))$. But it is easy to see that $h\circ\tilde{T}_X\circ r$ is the open Hopf link of X with A^K. By closing this open Hopf link we see that

$$\frac{s_{X,A^K}}{\dim_{\mathcal{C}} X} = \dim_{\mathcal{C}} A^K \Rightarrow \frac{s_{X,A^K}}{\dim_{\mathcal{C}} A^K} = \dim_{\mathcal{C}} X \tag{3.10}$$

Write $A^K \cong \bigoplus_{Z\in\mathcal{O}(\mathcal{C})} Z^{\oplus N_{A^K,Z}}$. Then for all $X\in\mathcal{O}(\mathcal{C})$:

$$\frac{s_{X,A^K}}{\dim_{\mathcal{C}} A^K} = \sum_{Z\in\mathcal{O}(\mathcal{C})} \frac{N_{A^K,Z}}{\dim_{\mathcal{C}} A^K} s_{X,Z} = \dim_{\mathcal{C}} X = \frac{s_{X,1}}{s_{1,1}} \tag{3.11}$$

Since $\mathcal{C}$ is a modular fusion category we see that $\frac{N_{A^K,Z}}{\dim_{\mathcal{C}} A^K} s_{1,1} = \delta_{Z,1}$. This implies that $N_{Z,A^K} = 0$ if $Z\neq 1$. Since $N_{A^K,1_{\mathcal{C}}} = 1$, as A is haploid, we see that $A^K\cong 1$, or $A^K = 1$ as sub-objects of A. □

3.3. Calculating A-open Hopf links and the regular representation. In the last subsection we saw that $T_{\mathrm{Ind}(X)}$ is the open Hopf link of X with A. If we already understand $\mathcal{C}$, then this provides a useful tool to calculate the A-open Hopf links.

PROPOSITION 4. *Decompose* $A\cong\bigoplus_{Z\in\mathcal{O}(\mathcal{C})} Z^{\oplus N_{Z,A}}$, *then*

$$T_{\mathrm{Ind}(X)} = \bigoplus_{Z\in\mathcal{O}(\mathcal{C})} \frac{s_{X,Z}}{\dim_{\mathcal{C}} Z}\cdot p_Z \tag{3.12}$$

where $p_Z : A\to A$ *is the projection of* A *onto* $Z^{\oplus N_{Z,A}}$.

PROOF. This follows from Lemma 7, and the fact that if Z is simple, then the open Hopf link of X with Z is $\frac{s_{X,Z}}{\dim_{\mathcal{C}} Z}\cdot\mathrm{Id}_Z$. □

As we will see in the examples section, it is often possible to calculate K if one knows the fusion rules and modular data of $\mathcal{C}$.

If we have a group G acting faithfully on a nice VOA $\mathcal{V}$, then conjecturally $\mathcal{V}^G$ will be nice and under this assumption we have a Schur-Weyl duality [**12**] given by decomposing $\mathcal{V}$ as a $G\times\mathcal{V}^G$ module:

$$\mathcal{V}\cong\bigoplus_{\lambda\in\widehat{G}} M_\lambda\otimes V_\lambda \tag{3.13}$$

where M_λ is the irreducible G-module corresponding to the character λ, and V_λ is an irreducible $\mathcal{V}^G$-module. If we now have a finite extension $\mathcal{W}\subset\mathcal{V}$ we expect something like this to hold for the associated hypergroup K. We will say more about this in the Section 5, but in the case where $\mathcal{V}$ is multiplicity free as a $\mathcal{W}$-module we have the following:

PROPOSITION 5 (The Regular Representation in the Multiplicity Free Case). *Let $\mathcal{W} \subset \mathcal{V}$, K, A be as before, and assume that A is multiplicity free in $\mathcal{C}$. Identify K with Eq.* (3.6). *Then:*

$$\mathbb{C}[K] = \mathrm{End}_{\mathcal{C}}(A) \tag{3.14}$$

Irreducible representations of K correspond to irreducible $\mathcal{W}$ sub-modules of $\mathcal{V}$:

$$\mathcal{V} \cong \bigoplus_{\lambda \in \widehat{K}} V_\lambda \tag{3.15}$$

PROOF. Since $\mathcal{C}$ is a modular fusion category we know for every simple $Z_0 \in \mathcal{C}$ there exists some $X \in \mathbb{C}[\mathcal{C}]$ such that $s_{Z,X} = \delta_{Z,Z_0}$. Fixing such an X for every Z_0 in the decomposition of A, we see that $T_{\mathrm{Ind}(X)} = \frac{1}{\dim_{\mathcal{C}} Z_0} p_{Z_0}$, (notice the abuse of notation for the virtual object X). Since $T_{\mathrm{Ind}(X)} \in \mathbb{C}[K]$ we see that the projection $p_{Z_0} \in \mathbb{C}[K]$ for every Z_0 in A. Since A is multiplicity free these projections provide a basis of $\mathrm{End}_{\mathcal{C}}(A)$, this implies that $\mathbb{C}[K] = \mathrm{End}_{\mathcal{C}}(A)$. Now, notice that since A is multiplicity free $\mathrm{End}_{\mathcal{C}}(A)$ is a commutative ring. Therefore, K is commutative as well. This implies that all representations of K are one-dimensional and these correspond to the projection maps of simples in A, giving the second part of the theorem. □

4. Examples

In this section we work out the hypergroup associated to the finite extensions of nice VOAs $(\widehat{\mathfrak{sl}(2)})_k$ associated to affine $\mathfrak{sl}(2)$. We calculate them by only assuming the ADE classification of $\widehat{\mathfrak{sl}(2)}$ extensions and a result concerning modular invariants. See [**33**] for the categorical definition of modular invariants.

For the sake of brevity we denote the irreducible $(\widehat{\mathfrak{sl}(2)})_k$ module of highest weight r simply as r, and let $\mathcal{C} := \mathrm{Rep}(\widehat{\mathfrak{sl}(2)})_k$. The simple objects of $\mathcal{C}$ are then $\{0, \cdots, k\}$ and the fusion rules are given by [**17**]:

$$N_{a,b}^{c} = \begin{cases} 1 & c \equiv a+b \mod 2 \text{ and } |a-b| \leq c \leq \min\{a+b, 2k-a-b\} \\ 0 & \text{otherwise} \end{cases} \tag{4.1}$$

See also [**20**]. If $\mathcal{V}$ is a nice extension of $(\widehat{\mathfrak{sl}(2)})_k$, then this will correspond to an étale algebra A in $\mathcal{C}$ by Theorem 2. Therefore, Frobenius reciprocity given in Equation (2.2) holds. Our method of calculation for extensions of $(\widehat{\mathfrak{sl}(2)})_k$ requires the following description, well known to experts, of the associated modular invariant.

PROPOSITION 6. *Let $X, Y \in \mathcal{C}$ be simple, and let $\mathrm{Ind}(X)^{loc}$ denote the local-component of the left A-module $\mathrm{Ind}(X)$. The modular invariant matrix Z satisfies:*

$$Z_{X,Y} = \dim_{\mathbb{C}} \mathrm{Hom}_{\mathrm{Rep}_{\mathcal{C}} A}(\mathrm{Ind}(X)^{loc}, \mathrm{Ind}(Y)^{loc}) \tag{4.2}$$

PROOF. Note by semi-simplicity it suffices to show that:

$$Z_{X,Y} = \dim_{\mathbb{C}} \mathrm{Hom}_{\mathrm{Rep}_{\mathcal{C}} A}(\mathrm{Ind}(X)^{loc}, \mathrm{Ind}(Y))$$

By Frobenius reciprocity:

$$\dim_{\mathbb{C}} \mathrm{Hom}_{\mathrm{Rep}_{\mathcal{C}} A}(\mathrm{Ind}(X)^{loc}, \mathrm{Ind}(Y)) = \dim_{\mathbb{C}} \mathrm{Hom}_{\mathcal{C}}(\mathrm{Res}(\mathrm{Ind}(X)^{loc}), Y)$$

By [**18**, Proposition 3.6] this equals $Z_{X,Y}$ and we are done. □

Notice that for the fusion ring of $\operatorname{Rep}_{\mathcal{C}}^{loc} A$ the restriction functor

$$\operatorname{Res} : \mathcal{K}^0(\operatorname{Rep}_{\mathcal{C}}^{loc} A) \to \mathcal{K}^0(\mathcal{C}) \tag{4.3}$$

is a linear map of fusion rings. In particular, we can write down Res on the level of fusion rings as a $|\mathcal{K}^0(\mathcal{C})| \times |\mathcal{K}^0(\operatorname{Rep}_{\mathcal{C}}^{loc} A)|$ matrix. If the extension is clear from context we denote this matrix by B. Now, with everything in place our method of calculation for when the extension comes from a conformal embedding $(\widehat{\mathfrak{sl}(2)})_k \subset \mathcal{V}$ is as follows:

(1) Using Lie theory we have a complete description of $\operatorname{Rep} \mathcal{V}$, and we can determine how each simple $\mathcal{V}$-module decomposes as a $(\widehat{\mathfrak{sl}(2)})_k$ module. That is, we can calculate B.
(2) An easy corollary of Proposition 6 is that the modular invariant associated to the module category $\operatorname{Rep}_{\mathcal{C}} A$ is $Z = BB^t$. In particular by [**33**] we have:

$$\operatorname{Rnk}(\operatorname{Rep}_{\mathcal{C}} A) = \operatorname{Tr}(Z) \tag{4.4}$$

(3) From abstract nonsense we know that $\operatorname{Rep} \mathcal{V}$ is a fusion subcategory of $\operatorname{Rep}_{\mathcal{C}} A$, where $A := \mathcal{V}$ as a $(\widehat{\mathfrak{sl}(2)})_k$-module. Every simple object of $\operatorname{Rep}_{\mathcal{C}} A$ will be a sub-object of some $\operatorname{Ind}(L)$. By using Eq. (2.2) and Eq. (4.1) we determine the fusion rules of the simple objects of $\operatorname{Rep}_{\mathcal{C}} A$
(4) Once we know the fusion ring $\mathcal{K}^0(\operatorname{Rep}_{\mathcal{C}} A)$ we can calculate the double coset $\operatorname{Rep}_{\mathcal{C}} A \,/\!\!/\, \operatorname{Rep}_{\mathcal{C}}^{loc} A$

Using the A-D-E classification given in [**29**, Section 6] we only need to carry out this calculation for extensions of type D_{2m+2}, E_6, E_8.

4.1. D_{2m+2} **extensions.** In the type D we case we have an extension at level $k = 4m$ for all $m \in \mathbb{Z}_{>0}$, and in this case $A = 0 \oplus 4m$. In other words, we have a simple current extension. It is known that in this case there is a natural $\mathbb{Z}/2$ action on A such that $A^{\mathbb{Z}/2} = (\widehat{sl(2)})_{4m}$. To verify that $K = \mathbb{Z}/2$ we need to calculate $\operatorname{Rep}_{\mathcal{C}} A$ as a $\operatorname{Rep}_{\mathcal{C}}^0 A$ bi-module. We do this by brute force.

We know that every left A-module is a subobject of some $\operatorname{Ind}(i)$, $0 \leq i \leq 4m$. To see if $\operatorname{Ind}(i)$ is simple, by Frobenius-Reciprocity and rigidity we only need to check if:

$$\dim \operatorname{Hom}_{\mathcal{C}}(i \otimes i, 0 \oplus 4m) = 1 \tag{4.5}$$

It is easy to check that $\operatorname{Ind}(i) \cong \operatorname{Ind}(4m - i)$ for $0 \leq i \leq 4m$. This means we need to only check Equation (4.5) for $0 \leq i \leq 2m$.

Case 1. $[0 \leq i < 2m]$ In this case $2i < 8m - 2i$, so $N_{i,i}^c \neq 0$ if and only if c is even and $0 \leq c \leq 2i$. Therefore, $\operatorname{Ind}(i)$ will be simple.

Case 2. $[i = 2m]$ In this case we have $2i = 8m - 2i = 4m$, so both 0 and $4m$ are in $i \otimes i$. Therefore, $\dim_{\mathbb{C}} \operatorname{End}_{\operatorname{Rep}_{\mathcal{C}} A}(\operatorname{Ind}(2m)) = 2$ and so $\operatorname{Ind}(2m)$ is the direct sum of two simples.

From this case work we see that induction almost always gives a simple object. Denote the simples appearing in the decomposition of $\operatorname{Ind}(2m)$ as an object of $\operatorname{Rep}_{\mathcal{C}} A$ by a_{2m} and b_{2m}. An easy lemma states that $\operatorname{Ind}(i) \in \operatorname{Rep}_{\mathcal{C}}^{loc} A$ if and only if $R_{i,A} \circ R_{A,i} = \operatorname{Id}$. Using this one can calculate that:

$$\mathcal{O}(\operatorname{Rep}_{\mathcal{C}}{}^{loc}(A)) = \{a_{2m}, b_{2m}\} \cup \{\operatorname{Ind}(i) : 0 \leq i < 2m, i \equiv 0 \mod 2\} \tag{4.6}$$

$$\mathcal{O}(\operatorname{Rep}_{\mathcal{C}} A) = \mathcal{O}(\operatorname{Rep}_{\mathcal{C}}{}^{loc}(A)) \cup \{\operatorname{Ind}(i) : 0 \leq i < 2m, i \equiv 1 \mod 2\} \tag{4.7}$$

Using that induction is a tensor functor it is straightforward to check that $\{\mathrm{Ind}(i) : 0 \le i < 2m, i \equiv 1 \mod 2\}$ will give an irreducible $\mathrm{Rep}^{loc}_{\mathcal{C}} A$ bi-module. This implies that $|K| = 2$. Since $\mathrm{Ind}(1) \otimes_A \mathrm{Ind}(1) = \mathrm{Ind}(0) \oplus \mathrm{Ind}(2) \in \mathrm{Rep}^{loc}_{\mathcal{C}} A$ we see that indeed $K \cong \mathbb{Z}/2$.

4.2. E_6 extension. In this case we have the conformal embedding $(\widehat{\mathfrak{sl}(2)})_{10} \subset (\widehat{\mathfrak{sp}(4)})_1$. Let $A = (\widehat{\mathfrak{sp}(4)})_1$, from Lie theory we know that $\mathrm{Rep}(\widehat{\mathfrak{sp}(4)})_1$ will have three simple objects $(\widehat{\mathfrak{sp}(4)})_1, X, Y$ with fusion rules:

$$X \otimes X \cong (\widehat{\mathfrak{sp}(4)})_1 \oplus Y \tag{4.8}$$

$$X \otimes Y \cong X \tag{4.9}$$

$$Y \otimes Y \cong (\widehat{\mathfrak{sp}(4)})_1 \tag{4.10}$$

That is, the fusion ring of $(\widehat{\mathfrak{sp}(4)})_1$ gives a copy of Ising. We also know from Lie theory that:

$$\mathrm{Res}_{(\widehat{\mathfrak{sl}(2)})_{10}}((\widehat{\mathfrak{sp}(4)})_1) \cong 0 \oplus 6 \tag{4.11}$$

$$\mathrm{Res}_{(\widehat{\mathfrak{sl}(2)})_{10}}(X) \cong 3 \oplus 7 \tag{4.12}$$

$$\mathrm{Res}_{(\widehat{\mathfrak{sl}(2)})_{10}}(Y) \cong 4 \oplus 10 \tag{4.13}$$

Let B be the corresponding 11×3 restriction matrix. To calculate the trace of $Z = BB^t$ we need to only calculate:

$$\sum_{i=0}^{10} (\mathrm{Row}_i B) \cdot (\mathrm{Row}_i B) \tag{4.14}$$

This will only be non-zero when $i = 0, 3, 4, 6, 7, 10$ and in each case the inner product evaluates to one. Therefore, $\mathrm{Tr}(Z) = 6$. At this point one could mention some things about the indecomposable Ising module categories to determine the hypergroup, but let's take a more elementary approach. We know there are only three non-local simples. Based off the restrictions of the local modules a good guess for these three non-local simples are:

$$\mathrm{Ind}(1), \mathrm{Ind}(2), \mathrm{Ind}(9) \tag{4.15}$$

By Frobenius reciprocity and rigidity, to check they are simple we need to only calculate:

$$\dim \mathrm{Hom}_{\mathcal{C}}(1 \otimes 1, 0 \oplus 6), \dim \mathrm{Hom}_{\mathcal{C}}(2 \otimes 2, 0 \oplus 6), \dim \mathrm{Hom}_{\mathcal{C}}(9 \otimes 9, 0 \oplus 6) \tag{4.16}$$

It is easy to see that the first two dimensions will be 1, so $\mathrm{Ind}(1), \mathrm{Ind}(2)$ are simple. Since $9 \otimes 9 \cong 0 \oplus 2$ we see that $\mathrm{Ind}(9)$ is also simple. As induction is a tensor functor this implies that all the non-local simples are self-dual, and so the fusion ring $\mathcal{K}^0(\mathrm{Rep}_{\mathcal{C}} A)$ is commutative. By restriction one can check that none of these inductions coincide. Therefore, we have the three other simples. By using Frobenius reciprocity one can calculate how all other inductions decompose into simple left A-modules. In particular, we have that

$$\mathrm{Ind}(6) \cong A \oplus \mathrm{Ind}(2), \mathrm{Ind}(7) \cong X \oplus \mathrm{Ind}(1), \mathrm{Ind}(8) = \mathrm{Ind}(2), \mathrm{Ind}(10) \cong Y \tag{4.17}$$

We see then that:

$$\operatorname{Ind}(7)\otimes_A \operatorname{Ind}(1)\cong \operatorname{Ind}(6)\oplus \operatorname{Ind}(8)\cong A\oplus \operatorname{Ind}(2)\oplus \operatorname{Ind}(2) \tag{4.18}$$

$$\operatorname{Ind}(10)\otimes_A \operatorname{Ind}(1)\cong \operatorname{Ind}(9) \tag{4.19}$$

By similar calculations, one can calculate that $\operatorname{Ind}(2), \operatorname{Ind}(9)$ each generate a $\operatorname{Rep}_{\mathcal{C}}^{loc} A$ bi-module spanned by the three non-local simple objects. Therefore, they span an indecomposable $\operatorname{Rep}_{\mathcal{C}}^{loc} A$ bi-module. To make things look nicer let's denote $\operatorname{Ind}(1)$ by e_1, and its $\operatorname{Rep}_{\mathcal{C}} A$ dimension by d_1. By Proposition 3:

$$K=\left\{[A],\left[\frac{e_1}{d_1}\right]\right\} \tag{4.20}$$

The Haar element of $\operatorname{Rep}_{\mathcal{C}}^{loc} A$ is:

$$\omega_{\operatorname{Rep}_{\mathcal{C}}^{loc} A}=\frac{1}{4}A\oplus\frac{\sqrt{2}}{4}X\oplus\frac{1}{4}Y \tag{4.21}$$

Using the fact that induction is a tensor functor we can calculate:

$$\left[\frac{e_1}{d_1}\right]\cdot\left[\frac{e_1}{d_1}\right]=(2-\sqrt{3})[A]+(\sqrt{3}-1)\left[\frac{e_1}{d_1}\right] \tag{4.22}$$

Notice this implies that the weight of $[\frac{e_1}{d_1}]$ is:

$$w_{[\frac{e_1}{d_1}]}=\frac{1}{2-\sqrt{3}}=2+\sqrt{3}$$

A quick calculation shows that:

$$\sqrt{\frac{w^2_{[\frac{e_1}{d_1}]}}{w_{[\frac{e_1}{d_1}]}}C^{\frac{e_1}{d_1}}_{\frac{e_1}{d_1},\frac{e_1}{d_1}}}=\sqrt{2+\sqrt{3}(\sqrt{3}-1)}=\sqrt{2}\notin\mathbb{Z}$$

This implies that the hypergroup *cannot* come from a fusion ring. Lastly, the non-trivial part of the hypergroup will act on $\widehat{\mathfrak{sp}(4)}_1$ as:

$$\frac{1}{d_1}T_{e_1}=\operatorname{Id}_0\oplus(\sqrt{3}-2)\operatorname{Id}_6 \tag{4.23}$$

4.3. E_8 extension. In this case we have the conformal embedding $(\widehat{\mathfrak{sl}(2)})_{28}\subset(\widehat{G_2})_1$. Let $\mathcal{C}=\operatorname{Rep}(\widehat{\mathfrak{sl}(2)})_{28}$ and $A=(\widehat{G_2})_1$. From Lie theory we know that $\operatorname{Rep}(\widehat{G_2})_1\cong\operatorname{Rep}_{\mathcal{C}}^{loc} A$ will have two simples $(\widehat{G_2})_1, X$ with fusion rules:

$$X\otimes X\cong(\widehat{G_2})_1\oplus X \tag{4.24}$$

In other words $\operatorname{Rep}(\widehat{G_2})_1$ will be equivalent to Fib. Let ϕ denote the golden ratio, then this implies that:

$$\begin{aligned} &\dim_A((\widehat{G_2})_1)=1 && \dim_A(X)=\phi\\ &\dim(\operatorname{Rep}_{\mathcal{C}}^{loc} A)=2+\phi && \dim(\operatorname{Rep}_{\mathcal{C}} A)=(2+\phi)\cdot\dim_{\mathcal{C}}(A)\end{aligned} \tag{4.25}$$

As we know in this case $\dim(\operatorname{Rep}_{\mathcal{C}} A)=\dim_{\mathcal{C}}(A)\dim(\operatorname{Rep}_{\mathcal{C}}^{loc} A)$ (see [**29**, Theorem 4.5] or [**10**, Corollary 3.32]). We also know from Lie theory that:

$$\operatorname{Res}_{(\widehat{\mathfrak{sl}(2)})_{28}}((\widehat{G_2})_1)\cong 0\oplus 10\oplus 18\oplus 28 \tag{4.26}$$

$$\operatorname{Res}_{(\widehat{\mathfrak{sl}(2)})_{28}}(X)\cong 6\oplus 12\oplus 16\oplus 22 \tag{4.27}$$

Let B be the corresponding 29×2 restriction matrix. To calculate the trace of $Z := BB^t$ we need to only calculate

$$\sum_{i=0}^{28} (\mathrm{Row}_i B) \cdot (\mathrm{Row}_i B)$$

Evidently, each summand will only be non-zero when $i = 0, 6, 10, 12, 16, 18, 22, 28$ and in each case the summand will equal one. Therefore, we conclude that:

$$\mathrm{Rnk}(\mathrm{Rep}_{\mathcal{C}} A) = 8 \tag{4.28}$$

We already know two of these simples are $\widehat{(G_2)}_1, X$, so there are six other simples we have to determine. Consider $\mathrm{Rep}_{\mathcal{C}} A$ as a left $\mathrm{Rep}^{loc}_{\mathcal{C}} A$-module category. This will decompose into indecomposable $\mathrm{Rep}^{loc}_{\mathcal{C}} A$-module categories, but the only indecomposable Fib-module category is just the left regular action of Fib on itself. Since the rank of $\mathrm{Rep}_{\mathcal{C}} A = 8$ this implies there are 6 non-local simples I_j, X_j for $j = 1, 2, 3$ such that the pair I_j, X_j give a copy of Fib. That is $X \otimes_A I_j \cong X_j$, $X \otimes_A X_j \cong I_j \oplus X_j$. In particular, if $d_i := \dim_A(I_i)$ we have $\dim_A(X_i) = \phi \cdot d_i$. This means that:

$$\dim(\mathrm{Rep}_{\mathcal{C}} A) = (2 + \phi)(1 + d_1^2 + d_2^2 + d_3^2) \tag{4.29}$$

To determine how these I_j, X_j behave let's calculate some inductions using Frobenius reciprocity.

$$\dim \mathrm{Hom}_{\mathrm{Rep}_{\mathcal{C}} A}(\mathrm{Ind}(r), \mathrm{Ind}(r)) = \dim \mathrm{Hom}_{\mathcal{C}}(r, A \otimes r) = \dim \mathrm{Hom}_{\mathcal{C}}(r \otimes r, A) \tag{4.30}$$

But

$$1 \otimes 1 \cong 0 \oplus 2 \tag{4.31}$$

This implies that $\dim_{\mathbb{C}} \mathrm{Hom}_{\mathrm{Rep}_{\mathcal{C}} A}(\mathrm{Ind}(1), \mathrm{Ind}(1)) = 1$, and hence $\mathrm{Ind}(1)$ is simple. Since 1 doesn't appear in the restriction of any of the local modules we see it is non-local. Repeating this calculation for 2, 3, 4 we get.

$$2 \otimes 2 \cong 0 \oplus 2 \oplus 4 \tag{4.32}$$

$$3 \otimes 3 \cong 0 \oplus 2 \oplus 4 \oplus 6 \tag{4.33}$$

$$4 \otimes 4 \cong 0 \oplus 2 \oplus 4 \oplus 6 \oplus 8 \tag{4.34}$$

For the same reason as $\mathrm{Ind}(1)$ we see that $\mathrm{Ind}(2), \mathrm{Ind}(3), \mathrm{Ind}(4)$ are simple and non-local. Since induction is a tensor functor it will preserve dimension. Therefore, we see that:

$$\begin{aligned} \dim_A(\mathrm{Ind}(1)) &= \frac{\sin(\frac{2\pi}{30})}{\sin(\frac{\pi}{30})}, \quad \dim_A(\mathrm{Ind}(2)) = \frac{\sin(\frac{3\pi}{30})}{\sin(\frac{\pi}{30})}, \\ \dim_A(\mathrm{Ind}(3)) &= \frac{\sin(\frac{4\pi}{30})}{\sin(\frac{\pi}{30})}, \quad \dim_A(\mathrm{Ind}(4)) = \frac{\sin(\frac{5\pi}{30})}{\sin(\frac{\pi}{30})} \end{aligned} \tag{4.35}$$

Calculating the ratios of these dimensions we see only $\frac{\dim_A(\mathrm{Ind}(4))}{\dim_A(\mathrm{Ind}(2))} = \phi$, this implies that $\mathrm{Ind}(2), \mathrm{Ind}(4)$ will form an indecomposable $\mathrm{Rep}^{loc}_{\mathcal{C}} A$ module category, so let $I_2 = \mathrm{Ind}(2), X_2 = \mathrm{Ind}(4)$. We also see that $\mathrm{Ind}(1), \mathrm{Ind}(3)$ cannot be in the same module category. By fixing our labelling this means either $\mathrm{Ind}(1) = I_1$ or X_1 and $\mathrm{Ind}(3) = I_3$ or X_3. Combining Equations (4.25) and (4.29) we see that $1 + d_1^2 + d_2^2 + d_3^2 = \dim_{\mathcal{C}} A$. Since we know the dimensions of $\mathrm{Ind}(1), \mathrm{Ind}(3)$

and the relative dimensions in the module categories we can plug in the various possibilities for $\mathrm{Ind}(1), \mathrm{Ind}(3)$ into this equation and check that this will only hold when $\mathrm{Ind}(1) = I_1$ and $\mathrm{Ind}(3) = X_3$. This shows that the dimensions of all the simple objects in $\mathrm{Rep}_{\mathcal{C}} A$ are different, which implies that all simples are self-dual and so $\mathrm{Rep}_{\mathcal{C}} A$ has a commutative fusion ring.

By Proposition 2 we see that:

$$\mathrm{Rep}_{\mathcal{C}} A /\!/ \mathrm{Rep}_{\mathcal{C}}^{loc} A = [A] \cup \left[\frac{\mathrm{Ind}(1)}{d_1}\right] \cup \left[\frac{\mathrm{Ind}(2)}{d_2}\right] \cup \left[\frac{\mathrm{Ind}(3)}{\phi \cdot d_3}\right] \tag{4.36}$$

To calculate the hypergroup structure of the double coset, we use the Haar element of $\mathrm{Rep}(\widehat{G_2})_1$:

$$\omega_{\mathrm{Rep}_{\mathcal{C}}^{loc} A} = \frac{1}{2+\phi} A \oplus \frac{\phi}{2+\phi} X \tag{4.37}$$

To calculate $\left[\frac{\mathrm{Ind}(1)}{d_1}\right] \cdot \left[\frac{\mathrm{Ind}(1)}{d_1}\right]$ we simply need to evaluate

$$\mathrm{Ind}(1) \cdot \omega_{\mathrm{Rep}_{\mathcal{C}}^{loc} A} \cdot \mathrm{Ind}(1) = \frac{2}{5+\sqrt{5}} (\mathrm{Ind}(1) \otimes_A \mathrm{Ind}(1)) \oplus \frac{1}{\sqrt{5}} (\mathrm{Ind}(1) \otimes_A X \otimes_A \mathrm{Ind}(1)) \tag{4.38}$$

and then collect terms in the same $\mathrm{Rep}_{\mathcal{C}}^{loc} A$ bi-module.

Since $\mathrm{Ind}(1) \otimes_A \mathrm{Ind}(1) \cong \mathrm{Ind}(1 \otimes 1) \cong \mathrm{Ind}(0) \oplus \mathrm{Ind}(2) = A \oplus I_2$, and $X \otimes_A I_2 = X_2$ we can simplify to:

$$\mathrm{Ind}(1) \cdot \omega_{\mathrm{Rep}_{\mathcal{C}}^{loc} A} \cdot \mathrm{Ind}(1) = \frac{1}{2+\phi} A \oplus \frac{1}{2+\phi} I_2 + \frac{\phi}{2+\phi} (X \oplus X_2) \tag{4.39}$$

Normalizing so we get a well-defined hypergroup this becomes:

$$\begin{aligned} \left(\frac{I_1}{d_1}\right) \cdot \omega_{\mathrm{Rep}^{loc} A} \cdot \left(\frac{I_1}{d_1}\right) = \frac{1}{d_1^2} \bigg(& \frac{1}{2+\phi} A \oplus \frac{\phi^2}{2+\phi} \left(\frac{X}{\phi}\right) \\ & \oplus \frac{d_2}{2+\phi} \left(\frac{I_2}{d_2}\right) \oplus \frac{\phi^2 \cdot d_2}{2+\phi} \left(\frac{X_2}{\phi \cdot d_2}\right) \bigg) \end{aligned} \tag{4.40}$$

Gathering into equivalence classes we obtain:

$$\begin{aligned} \left[\frac{I_1}{d_1}\right] \cdot \left[\frac{I_1}{d_1}\right] &= \left(\frac{1}{d_1^2} \cdot \frac{1+\phi^2}{2+\phi}\right) [A] + \left(\frac{d_2}{d_1^2} \cdot \frac{1+\phi^2}{2+\phi}\right) \left[\frac{I_2}{d_2}\right] \\ &= \frac{1}{d_1^2} [A] + \frac{d_2}{d_1^2} \left[\frac{I_2}{d_2}\right] \\ &= \frac{1}{2} \left(4 - \sqrt{5} - \sqrt{3 \cdot (5 - 2\sqrt{5})}\right) [A] \\ &\quad + \frac{1}{2} \left(-2 + \sqrt{5} + \sqrt{3 \cdot (5 - 2\sqrt{5})}\right) \left[\frac{I_2}{d_2}\right] \end{aligned} \tag{4.41}$$

Notice that the fusion ring of $\mathrm{Rep}_{\mathcal{C}} A$ is commutative, and so the double coset hypergroup will be as well. To calculate the rest of the hypergroup we carry out a similar process, and this requires showing $\mathrm{Ind}(5) \cong X_1 \oplus I_3$, and $\mathrm{Ind}(6) \cong X \oplus I_2$ which can be done through similar induction/restriction arguments. For the sake of brevity we omit these calculations and just give the multiplication table:

TABLE 1. Multiplication Table for the E_8 Hypergroup

	$[A]$	$\left[\frac{I_1}{d_1}\right]$	$\left[\frac{I_2}{d_2}\right]$	$\left[\frac{I_3}{d_3}\right]$
$[A]$	$[A]$	$\left[\frac{I_1}{d_1}\right]$	$\left[\frac{I_2}{d_2}\right]$	$\left[\frac{I_3}{d_3}\right]$
$\left[\frac{I_1}{d_1}\right]$	$*$	$\frac{4-\alpha}{2}[A] + \frac{-2+\alpha}{2}\left[\frac{I_2}{d_2}\right]$	$\frac{-3+\tau}{4}\left[\frac{I_1}{d_1}\right] + \frac{7-\tau}{4}\left[\frac{I_3}{d_3}\right]$	$\left[\frac{I_2}{d_2}\right]$
$\left[\frac{I_2}{d_2}\right]$	$*$	$*$	$\frac{11+\beta}{4}[A] + \frac{-7-\beta}{4}\left[\frac{I_2}{d_2}\right]$	$\frac{4+\gamma}{2}\left[\frac{I_1}{d_1}\right] + \frac{-2-\gamma}{2}\left[\frac{I_3}{d_3}\right]$
$\left[\frac{I_3}{d_3}\right]$	$*$	$*$	$*$	$\frac{7-\zeta}{4}[A] + \frac{-3+\zeta}{4}\left[\frac{I_2}{d_2}\right]$

$$\alpha := \sqrt{5} + \sqrt{3\cdot(5-2\sqrt{5})} \qquad \beta := 3\sqrt{5} - 2\sqrt{\tfrac{75+33\sqrt{5}}{2}}$$
$$\gamma := \sqrt{5} - \sqrt{3\cdot(5+2\sqrt{5})} \qquad \zeta := \sqrt{5} + \sqrt{6\cdot(5-\sqrt{5})}$$
$$\tau := -\sqrt{5} + \sqrt{6\cdot(5+\sqrt{5})}$$

To calculate how the hypergroup $\mathrm{Rep}_{\mathcal{C}}\, A \,/\!\!/\, \mathrm{Rep}_{\mathcal{C}}^{loc}$ act on $(\widehat{G_2})_1$ we just need to look at:

$$T_{\mathrm{Ind}(0)}, T_{\mathrm{Ind}(1)}, T_{\mathrm{Ind}(2)}, T_{\mathrm{Ind}(3)} \tag{4.42}$$

By Proposition 4 we have:

$$T_{\mathrm{Ind}(i)} = \left(\frac{S_{0,i}}{S_{0,0}}\right)\cdot \mathrm{Id}_0 \oplus \left(\frac{S_{10,i}}{S_{10,0}}\right)\cdot \mathrm{Id}_{10} \oplus \left(\frac{S_{18,i}}{S_{18,0}}\right)\cdot \mathrm{Id}_{18} \oplus \left(\frac{S_{28,i}}{S_{28,0}}\right)\cdot \mathrm{Id}_{28} \tag{4.43}$$

Plugging this in we get:

$$T_{\mathrm{Ind}(0)} = \mathrm{Id}_A \tag{4.44}$$

(4.45)
$$T_{\mathrm{Ind}(1)} = \left(\frac{\sin(\frac{2\pi}{30})}{\sin(\frac{\pi}{30})}\right)\mathrm{Id}_0 \oplus \left(\frac{\sin(\frac{22\pi}{30})}{\sin(\frac{11\pi}{30})}\right)\mathrm{Id}_{10} \oplus \left(\frac{\sin(\frac{38\pi}{30})}{\sin(\frac{19\pi}{30})}\right)\mathrm{Id}_{18} \oplus \left(\frac{\sin(\frac{58\pi}{30})}{\sin(\frac{29\pi}{30})}\right)\mathrm{Id}_{28}$$

(4.46)
$$T_{\mathrm{Ind}(2)} = \left(\frac{\sin(\frac{3\pi}{30})}{\sin(\frac{\pi}{30})}\right)\mathrm{Id}_0 \oplus \left(\frac{\sin(\frac{33\pi}{30})}{\sin(\frac{11\pi}{30})}\right)\mathrm{Id}_{10} \oplus \left(\frac{\sin(\frac{57\pi}{30})}{\sin(\frac{19\pi}{30})}\right)\mathrm{Id}_{18} \oplus \left(\frac{\sin(\frac{87\pi}{30})}{\sin(\frac{29\pi}{30})}\right)\mathrm{Id}_{28}$$

(4.47)
$$T_{\mathrm{Ind}(3)} = \left(\frac{\sin(\frac{4\pi}{30})}{\sin(\frac{\pi}{30})}\right)\mathrm{Id}_0 \oplus \left(\frac{\sin(\frac{44\pi}{30})}{\sin(\frac{11\pi}{30})}\right)\mathrm{Id}_{10} \oplus \left(\frac{\sin(\frac{76\pi}{30})}{\sin(\frac{19\pi}{30})}\right)\mathrm{Id}_{18} \oplus \left(\frac{\sin(\frac{116\pi}{30})}{\sin(\frac{29\pi}{30})}\right)\mathrm{Id}_{28}$$

5. Future work

So far the results in this paper have given a glance at what the general theory and program should be for hypergroups acting on VOAs. In this section, we will outline what the general picture will be: results that we expect will be true and proven in the upcoming work [**21**].

5.1. The group orbifold triangle. To best illustrate these ideas, let's first review the group orbifold case and all the ingredients that go into understanding its category of representations. We will only be considering finite groups. First, recall that a group G acting on a VOA $\mathcal{V}$ is a representation $\rho : G \to \mathrm{End}_{\mathbb{C}}(\mathcal{V})$ such that the following conditions hold for each $g \in G$

$$\rho_g(\omega) = \omega \quad \rho_g(|0\rangle) = |0\rangle \tag{5.1}$$

$$\rho_g(v_n w) = (\rho_g(v))_n \rho_g(w) \tag{5.2}$$

The fixed points $\mathcal{V}^G$ will then form a VOA by restricting the state-field correspondence to $\mathcal{V}^G$ and taking the same vacuum vector and conformal vector. An area of active research over the past thirty years has been trying to understand how one can determine the categorical structure of $\mathrm{Rep}\,\mathcal{V}^G$ given only the categorical structure of $\mathrm{Rep}\,\mathcal{V}$ and the group action of G on $\mathrm{Rep}\,\mathcal{V}$ [**28**][**27**][**31**][**13**]. Recently, in joint work [**22**], and in [**35**] we have emphasized that to understand $\mathrm{Rep}\,\mathcal{V}^G$ an effective approach is to utilize the orbifold triangle.

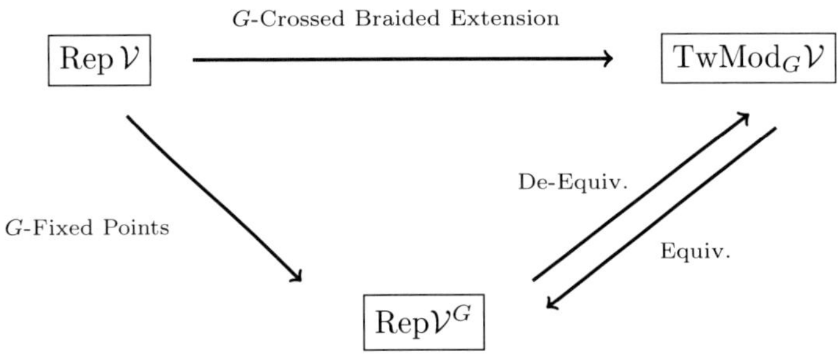

FIGURE 30. The Group Orbifold Triangle

To illustrate how this triangle is useful let's focus on each of the vertices and explain how to go to the adjacent ones.

Vertex: $\mathrm{Rep}\,\mathcal{V}$

There are two directions we would like to go: $\mathrm{Rep}\,\mathcal{V}^G$ and $\mathrm{TwMod}_G\mathcal{V}$. As mentioned above any group action on $\mathcal{V}$ will induce a VOA $\mathcal{V}^G$ by taking the fixed points, and from this we can look at its representations $\mathrm{Rep}\,\mathcal{V}^G$. While this is technically true, it is often difficult to understand the modules of $\mathcal{V}^G$ without using extra information. The extra information which allows us to understand $\mathrm{Rep}\,\mathcal{V}^G$ turns out to be the other vertex in the orbifold triangle the G-twisted $\mathcal{V}$-representations $\mathrm{TwMod}_G\mathcal{V}$. Since twisted modules are crucial to whole story we briefly review them and their categorical structure.

Firstly, if g is an automorphism then a g-twisted $\mathcal{V}$-module is a tuple (M, Y_M), which satisfies almost all of the usual conditions of a strongly-graded generalized $\mathcal{V}$-module except that we require that:

$$Y_M(g \cdot v, e^{2\pi i}z) = Y_M(v, z) \tag{5.3}$$

and the right hand side of the Jacobi identity is twisted in a suitable sense by g. Notice that we include $e^{2\pi i}$ in the LHS of equation (5.3) to indicate that we are taking a specific branch of log. We denote the collection of all g-twisted modules by $\mathrm{TwMod}_g\mathcal{V}$. In some cases, it is possible to construct and classify all g-twisted modules [**1**][**26**]. Denote the collection of all (direct sums) of g-twisted modules for $g \in G$ by $\text{TwMod}_G\mathcal{V}$. By definition, we have a G-grading:

$$\text{TwMod}_G\mathcal{V} = \bigoplus_{g\in G} \text{TwMod}_g\mathcal{V} \tag{5.4}$$

A group action on $\mathcal{V}$ will induce an action $T_g(M, Y_M) = (M, g(Y_M))$ for $g \in G$ on the twisted modules where:

$$g(Y_M)(v, z) := Y_M(\rho_{g^{-1}}(v), z) \tag{5.5}$$

It can be shown that this group action induces a categorical group action on $\text{TwMod}_G\mathcal{V}$. It will send $(M, Y_M) \in \text{TwMod}_h\mathcal{V}$ to $T_g(M, Y_M) \in \text{TwMod}_{ghg^{-1}}\mathcal{V}$. Furthermore, it can be shown that the G-twisted $\mathcal{V}$-modules will form a G-crossed braided fusion category when $\mathcal{V}$ is nice. See [**31**] for details. In summary, given a group G acting on $\mathcal{V}$, we can go to the $\text{TwMod}_G\mathcal{V}$ vertex by considering:

(1) What it means to have an object (M, Y_M) which $\mathcal{V}$ acts on in a g-twisted way.
(2) The categorical structure given by considering all g-twisted modules for $g \in G$

Vertex: $\text{TwMod}_G\,\mathcal{V}$

Just as before we would like to understand how to go to $\text{Rep}\,\mathcal{V}$ and $\text{Rep}\,\mathcal{V}^G$. The easiest direction to go to is $\text{Rep}\,\mathcal{V}$, simply take the Id_G-twisted modules or the identity graded component of $\text{TwMod}_G\mathcal{V}$. This means that $\text{TwMod}_G\mathcal{V}$ is a G-crossed graded extension, and as demonstrated in [**16**] such graded extensions are classified by two torsors, assuming certain obstructions vanish. In the case that $\text{Rep}\,\mathcal{V}$ is sufficiently simple [**16**] can provide a nice tool to determine the shape of $\text{TwMod}_G\mathcal{V}$'s categorical structure. See [**22**] for more details. To go from $\text{TwMod}_G\mathcal{V}$ to $\text{Rep}\,\mathcal{V}^G$ we need to take the categorical version of fixed points. This is the process of equivariantization, where objects $((M, Y_M), \{\tau_g\}_{g\in G}) \in (\text{TwMod}_G\mathcal{V})^G$ consist of $(M, Y_M) \in \text{TwMod}_G\mathcal{V}$ and isomorphisms $\tau_g : T_g(M, Y_M) \to (M, Y_M)$ for $g \in G$ such that they are compatible with the categorical G-action on $\text{TwMod}_G\mathcal{V}$. It is known that since we have a G-crossed braiding on the twisted modules, the equivariantization $(\text{TwMod}_G\mathcal{V})^G$ will be a braided fusion category equivalent to $\text{Rep}\,\mathcal{V}^G$ [**14**]. By defining G-twisted notions of ribbon twists it can be shown that this will actually be an equivalence of modular fusion categories. This equivalence can be made explicit by the functor [**27**]:

$$((M, Y_M), \{\tau_g\}_{g\in G}) \mapsto M^G \tag{5.6}$$

In summary, the group orbifold triangle says we can go from $\text{TwMod}_G\mathcal{V}$ to the other vertices by:

(1) Noting that the G-twisted modules of $\mathcal{V}$ are a G-crossed braided extension of $\text{Rep}\,\mathcal{V}$
(2) Equivariantization of the G-twisted modules is equivalent to $\text{Rep}\,\mathcal{V}^G$

Vertex: $\operatorname{Rep} \mathcal{V}^G$

Lastly, to go from $\operatorname{Rep} \mathcal{V}^G$ to the other vertices we need to use the concepts of VOA extensions and de-equivariantization. To this end consider $\mathcal{V}$ as a $\mathcal{V}^G$-module and denote it by A. It was shown in [**28**] that there is an embedding of $\operatorname{Rep}(G)$ inside of $\operatorname{Rep} \mathcal{V}^G$ as a braided fusion category, and under this embedding A will correspond to the regular representation $\mathbb{C}^G$. As previously mentioned, if $\mathcal{V}$ is a simple $\mathcal{V}$-module it is known that A will give an étale algebra in $\mathcal{C} := \operatorname{Rep} \mathcal{V}^G$. By looking at the modules of A in $\mathcal{C}$ we get a fusion category $\operatorname{Rep}_{\mathcal{C}} A$. The process of taking A-modules is called de-equivariantization, and it was shown [**28**] that this will carry the structure of a G-crossed braided fusion category. We can go to the vertex $\operatorname{Rep} \mathcal{V}$ by looking at the local-modules $\operatorname{Rep}_{\mathcal{C}}^{loc} A$. It was proved by McRae [**31**] that $\operatorname{Rep}_{\mathcal{C}} A$ will coincide with $\operatorname{TwMod}_G \mathcal{V}$ as a G-crossed braided fusion category, so by taking $\operatorname{Rep}_{\mathcal{C}} A$ we are able to go to the vertex $\operatorname{TwMod}_G \mathcal{V}$. See also [**9**]. In summary:

(1) $\mathcal{V}^G \subset \mathcal{V}$ is a finite extension, and this corresponds to an étale algebra in $\mathcal{C} = \operatorname{Rep} \mathcal{V}^G$ such that $\operatorname{Rep}_{\mathcal{C}}^{loc} A \cong \operatorname{Rep} \mathcal{V}$.
(2) The modules of A in $\mathcal{C}$ are naturally equivalent as a G-crossed braided fusion category to $\operatorname{TwMod}_G \mathcal{V}$.

5.2. Hypergroup actions and the hypergroup orbifold triangle. Now that we have reviewed the group orbifold triangle, we can anticipate what the story in the hypergroup case will be. More precisely, we will review some of the major results that are to appear in [**21**]. Firstly, the most important question that needs to be answered is:

> What does it mean for a hypergroup to act on a vertex operator algebra?

This will be answered in [**21**], but to the end of making the eventual material in [**21**] more digestible we will look at what is suggested from the examples worked out in this paper, and Bischoff's hypergroup story for conformal nets.

One natural candidate for a hypergroup action is to have the hypergroup act as VOA endomorphisms which are not necessarily invertible. Well, recall a endomorphism of a VOA $\mathcal{V}$ is a linear map $f : V \to V$ such that:

$$f(\omega) = \omega, \quad f(|0\rangle) = |0\rangle \tag{5.7}$$

$$\forall v, w \in V \quad (f \circ v_n)(w) = (f(v))_n(f(w)) \tag{5.8}$$

As we saw in Section 3 for any sufficiently nice VOA extension $\mathcal{W} \subset \mathcal{V}$ the hypergroup $\operatorname{Rep}_{\mathcal{C}} A \mathbin{/\!/} \operatorname{Rep}_{\mathcal{C}}^{loc} A$ will act on $\mathcal{V}$ as $\mathcal{W}$-module morphisms such that the fixed points give $\mathcal{W}$. One can show that in that in this case a $\mathcal{W}$-module endomorphism $f : V \to V$ satisfies equations (5.7) and (5.8) if and only if f is a $\mathcal{C}$-algebra endomorphism of A. If a hypergroup acts on $\mathcal{V}$ by VOA endomorphisms, then certainly all the hypergroups we have calculated should act VOA endomorphisms. One can check though that the linear maps from the E_6-hypergroup are not $\mathcal{C}$-algebra endomorphisms, and therefore are not VOA endomorphisms. This suggests that hypergroups should act on VOAs by maps which may not preserve their modes. In other words, we need something new.

As mentioned in the introduction Bischoff defined what it means for hypergroup to act on a conformal net in [**5**]. One of the reasons the theory of conformal nets is relatively more straightforward to develop is that conformal nets use unitarity, while VOAs in general do not. For example, the conformal net analogue of $\operatorname{Rep}\mathcal{V}$ will form a C^*-tensor category. Bischoff uses this categorical structure to define his hypergroup action. Despite this difference, *unitary* VOAs and conformal nets both attempt to describe the same natural phenomenon, and so there is an expected mathematical equivalence between them [**8**]. For this reason it might be useful to think about Bischoff's theory when trying to come up with the definition of a hypergroup acting on a VOA. Since we won't have access to unitarity for a general VOA, this will only give us a hint at what the definition is.

In [**5**] a hypergroup is defined to act as *stochastic maps*, these are extremal unital completely positive maps with a few more properties. In the language of VOAs, stochastic maps should translate to maps that preserve the n-point functions. Unital completely positive maps are particularly interesting as they have the following characterization:

THEOREM 8 ([**34**, Theorem 4.1], Stinespring's Dilation Theorem). *Let B be a unital C^*-algebra, $\mathcal{H}$ a Hilbert space and $\phi : B \to \mathcal{B}(\mathcal{H})$ a unital completely positive linear map. Then there exists a Hilbert space $\mathcal{K}$, a unital star homomorphism $\pi : B \to B(\mathcal{K})$, and an isometry $P_\phi : \mathcal{K} \to \mathcal{H}$ such that:*

$$\phi = \operatorname{Ad}(P_\phi)(\pi) \tag{5.9}$$

It can be shown that a unital completely positive map ϕ is multiplicative if and only if there exists a unitary element $V_\phi : \mathcal{H} \to \mathcal{H}$ such that $\phi = \operatorname{Ad}(V_\phi)$ [**32**, Lemma 2.1]. In the conformal net case, one should think of multiplicative maps as the VOA analogue of mode preserving maps. Multiplicative maps acting by conjugation is very similar to the automorphism formulation of Eq. (5.8):

$$fY(v,z)f^{-1} = Y(f(v),z) \tag{5.10}$$

This is no coincidence, as it was shown in [**8**, Theorem 6.9] that a multiplicative automorphism of a conformal net should correspond to an automorphism of a VOA. Now we don't know if hypergroups should act by invertible maps and we won't have access to adjoint maps when working with a non-unitary VOA, but the Stinespring dilation theorem seems to suggest that there should be some notion of *a mode transformation* $\pi : V\{z\} \to V\{z\}$ which will be related to how a hypergroup acts on a VOA. In the group case this transformation will be the identity map. In the finite extension case $\mathcal{W} \subset \mathcal{V}$ we can make this explicit by showing the hypergroup we introduced is the same as the "hypergroup" introduced in [**4**], and one can interpret their hypergroup operation as involving such a mode transformation. Although the mode transformation suggested by the work in [**4**] is dependent on the categorical structure of $\operatorname{Rep}\mathcal{W}$, and so we have not found it useful for defining a hypergroup action on a VOA.

In summary, from the hypergroups we have calculated it is clear that the type of linear maps a hypergroup can act on a VOA may not preserve its modes, and based off of the Stinespring dilation theorem Eq. (5.8) should be replaced by a more general condition of *mode transformation* instead of mode preservation.

Assuming now that there is a well-defined notion of a hypergroup/fusion ring acting on a VOA, what is the analogue of the group orbifold triangle? We propose the following:

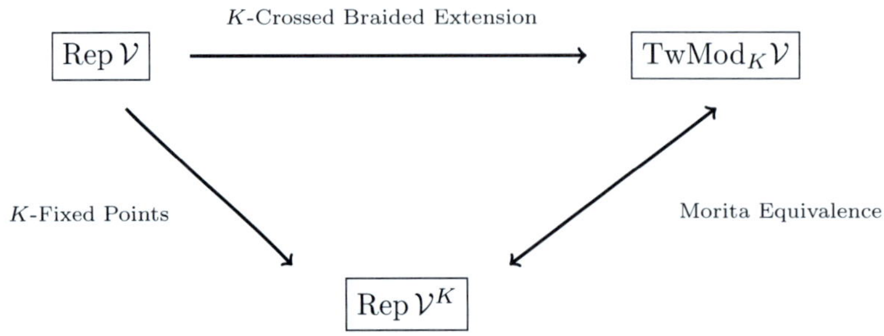

FIGURE 31. The Hypergroup Orbifold Triangle

To explain this diagram we do the same process as the group orbifold triangle, although much more briefly.

Vertex: Rep($\mathcal{V}$)

If we have a hypergroup K acting on $\mathcal{V}$, we can go to the vertex $\operatorname{Rep}\mathcal{V}^K$ by taking the fixed points. In general, there should be a notion of K-twisted modules that generalizes the notion of G-twisted modules. In the finite extension case $\mathcal{W} \subset \mathcal{V}$, we know that the indecomposable $\operatorname{Rep}^{loc}_{\mathcal{C}} A$ bi-module categories of $\operatorname{Rep}_{\mathcal{C}}(A)$ will be in bijection with the elements of the hypergroup $K := \operatorname{Rep}_{\mathcal{C}} A \,/\!\!/\, \operatorname{Rep}^{loc}_{\mathcal{C}} A$. If K is a group G, the associated indecomposable bi-module categories are the g-twisted modules $\operatorname{TwMod}_g\mathcal{V}$ for $g \in G$. This suggests that we refer to the left A-modules in the indecomposable $\operatorname{Rep}^{loc}_{\mathcal{C}} A$ bi-module category associated to $k \in K$ as k-twisted modules. We denote the collection of all k-twisted modules by $\operatorname{TwMod}_k\mathcal{V}$. Therefore, in the finite extension case we have a K-grading:

$$\operatorname{Rep}_{\mathcal{C}} A = \bigoplus_{k\in K} \operatorname{TwMod}_k(\mathcal{V}) \tag{5.11}$$

Since K is based off of the fusion ring of $\operatorname{Rep}_{\mathcal{C}} A$ this grading will be compatible with tensor products and duality in the sense that:

$$\text{If } X \in \operatorname{TwMod}_k(\mathcal{V}), Y \in \operatorname{TwMod}_r(\mathcal{V}), \text{ then } X \otimes_A Y \in \bigoplus_{s\in k\cdot r} \operatorname{TwMod}_s(\mathcal{V}) \tag{5.12}$$

$$\text{If } X \in \operatorname{TwMod}_k(\mathcal{V}), \text{ then } X^* \in \operatorname{TwMod}_{k^*}(\mathcal{V}) \tag{5.13}$$

Furthermore, in the multiplicity free case we can show that our definition of a hypergroup coincides with Bischoff's notion of a hypergroup for an étale algebra given in [**4**]. This means that the hypergroup K will correspond to irreducible sub A bi-modules of $A \otimes_{\mathcal{C}} A$. Label this decomposition as $A \otimes A = \bigoplus_{x\in K} M_x$. While each M_x will not be in $\operatorname{Rep}_{\mathcal{C}}(A)$ we have an induced functor given by:

$$T_x(Y, \mu_Y) := M_x \otimes_A (Y, \mu_Y), \quad T_x(f) := \operatorname{Id}_{M_x} \otimes_A f \tag{5.14}$$

Perhaps this will give a categorical action of K on $\operatorname{Rep}_{\mathcal{C}}(A)$. In any case there should be a categorical action of K on $\operatorname{Rep}_{\mathcal{C}} A$, and this should be compatible with the grading in the same way a G-crossed braided fusion category is. For any $x, y \in K$ there may be multiple elements of K in the convex combination of $x * y$ this implies that K will not always act through auto-equivalences as it will send simple objects to things which are not simple. In fact, when calculating the action described in Equation (5.14) for the E_6 hypergroup we found it is not even a unital functor.

While all these properties are interesting, it is worth noting that this requires we have a VOA extension, and therefore already have the hypergroup orbifold $\mathcal{V}^K$. Once we know what it means for a hypergroup to act on a VOA it should be possible to define what we mean by a K-twisted $\mathcal{V}$-module without reference to the orbifold VOA. From here, one should be able to define the categorical K-action on the twisted modules to be the hypergroup analogue of equation (5.5). Furthermore, the category of all K-twisted $\mathcal{V}$-modules $\mathrm{TwMod}_K\mathcal{V}$ should have most of the properties of G-crossed braided fusion categories naturally generalized to the hypergroup setting. We refer to such a hypothetical structure as a K-crossed braided fusion category. In summary, we can go from the vertex $\mathrm{Rep}\,\mathcal{V}$ to $\mathrm{TwMod}_K\mathcal{V}$ by considering:

(1) What it means to have an object (M, Y_M) on which $\mathcal{V}$ acts on in a k-twisted way.
(2) The K-crossed braided structure given by considering all k-twisted modules for $k \in K$.

Vertex: $\mathrm{TwMod}_K\,\mathcal{V}$

We can move to the vertex $\mathrm{Rep}\,\mathcal{V}$ by noting that $\mathrm{TwMod}_K\mathcal{V}$ will be a K-crossed braided extension of $\mathrm{Rep}\,\mathcal{V}$. It should be possible to do a similar classification for K-crossed braided extensions as there is for G-crossed braided extensions done in [**16**]. Although, K may need to be substituted for something which more naturally lifts to a bi-category. To go to $\mathrm{Rep}\,\mathcal{V}^K$ we need to shift our perspective on what equivariantization is. For if we define K-equivariantization in the same way as the group case we run into the issue that K a priori doesn't have a co-multiplication that is used crucially in the definition of the tensor product of G-equivariantization. Instead, notice that the categorical G-action on $\mathrm{TwMod}_G\mathcal{V}$ makes the G-twisted modules a module category of Vect_G. Similar to how we defined the left modules in $\mathcal{C}$ of an algebra A in $\mathcal{C}$ you can define the left modules in $\mathcal{M}$ of an algebra A in $\mathcal{D}$ when $\mathcal{M}$ is a left module category over $\mathcal{D}$. Denote this category by $\mathrm{Rep}_{\mathcal{M}}(A)$. Since $\mathrm{TwMod}_G\mathcal{V}$ is a left module category over Vect_G, we can consider the category:

$$\mathrm{Rep}_{\mathrm{TwMod}_G\mathcal{V}}(\mathbb{C}[G]) \tag{5.15}$$

As an Abelian category this will be equivalent to $(\mathrm{TwMod}_G\mathcal{V})^G$. Now in the finite extension case we can do something similar. Let $\mathcal{D}$ be the fusion category generated by A in $\mathcal{C} := \mathrm{Rep}\,\mathcal{W}$, in the group orbifold case this will be equivalent to $\mathrm{Rep}(G)$. $\mathcal{C}$ is naturally a left $\mathcal{D}$-module category. Let $\mathcal{M} = \mathrm{Rep}_{\mathcal{D}}(A)$. This is a right $\mathcal{D}$-module category. Taking the relative Deligne product (see [**16**] for details about this construction) we see as Abelian categories we have:

$$\mathrm{Rep}_{\mathcal{C}}(A) \cong \mathcal{M} \boxtimes_{\mathcal{D}} \mathcal{C} \tag{5.16}$$

Denote the category of A bi-modules in $\mathcal{D}$ by ${}_{\mathcal{D}}\,\mathrm{Rep}_{\mathcal{D}}(A)$. By abstract nonsense we know that $\mathcal{M}$ is also a left ${}_{\mathcal{D}}\,\mathrm{Rep}_{\mathcal{D}}(A)$-module category. Since we are in the finite extension case, we can consider $\mathrm{Rep}_{\mathcal{C}}(A)$ as the K-twisted modules for some hypergroup, denote $\mathrm{Rep}_{\mathcal{C}}(A)$ by $\mathrm{TwMod}_K\mathcal{V}$. We see from Eq. (5.16) that $\mathrm{TwMod}_K\mathcal{V}$ will be a left ${}_{\mathcal{D}}\,\mathrm{Rep}_{\mathcal{D}}(A)$-module category. It can be shown that $A \otimes_{\mathcal{D}} A$ is an algebra in ${}_{\mathcal{D}}\,\mathrm{Rep}_{\mathcal{D}}(A)$. In the group case $A \otimes_{\mathcal{D}} A$ will be isomorphic to $\mathbb{C}[G]$. Therefore, we believe the following may be an appropriate generalization of equivariantization for the hypergroup case:

$$\mathrm{Rep}_{\mathrm{TwMod}_K\mathcal{V}}(A \otimes_{\mathcal{D}} A) \tag{5.17}$$

We can show this is equivalent as an Abelian category to Rep $\mathcal{W}$. A nice aspect of this idea is it shows how equivariantization and de-equivariantization are really just a reflection of the Morita equivalence of Rep(G) and Vect_G through the module category Vect, and by recognizing this we can generalize to the hypergroup case. This idea was inspired by the physics done in [**3**]. One issue with this construction is it presupposes we already have the hypergroup orbifold. Therefore, to make this a true triangle we need to somehow make this intrinsic. Currently, it is not clear how to do this, although one way to do this may be to show that there is a tensor product structure on Rep(K) and that in the finite extension case $\mathcal{D}$ can be identified with this category.

There is another way to get Rep $\mathcal{V}^K$, which we state without explanation:

$$\mathrm{Rep}\,\mathcal{V}^K \cong (\mathrm{Rep}\,\mathcal{V})' \cong \mathrm{Rep}_{\mathcal{Z}(\mathrm{TwMod}_K\mathcal{V})\boxtimes \mathrm{Rep}\,\mathcal{V}}(1_{(\mathrm{Rep}\,\mathcal{V})'} \boxtimes T) \tag{5.18}$$

Here $(\mathrm{Rep}\,\mathcal{V})'$ is the Müger centralizer of Rep $\mathcal{V}$ in $\mathcal{Z}(\mathrm{TwMod}_K\mathcal{V})$ and T is the trivializing algebra of Rep $\mathcal{V}$ [**18**]. In summary we can go to the other vertices by:

(1) Taking the fixed points of the hypergroup action.
(2) Taking modules of an algebra in a dual category.

Vertex: Rep $\mathcal{V}^K$

In the finite extension case we know that $\mathcal{V}$ will be isomorphic as a $\mathcal{W}$-module to the regular representation of a hypergroup K, at least in the multiplicity free case. We expect that there should be a way to think about Rep(K) so that it has some tensor category structure, assuming that rationality descends to $\mathcal{V}^K$ it should also be the case that there is a copy of Rep(K) in Rep($\mathcal{V}^K$) and $\mathcal{V}$ will be equivalent to the regular representation of K just as in the group case. Using such results we will able to go to the vertex Rep $\mathcal{V}$. To go to the K-twisted modules we do the same thing as we did in the group case. Consider $A := \mathcal{V}$ as a $\mathcal{V}^K$-module. This should give an étale algebra in $\mathcal{C} := \mathrm{Rep}\,\mathcal{V}^K$ such that $\mathrm{Rep}_{\mathcal{C}} A$ has a structure of a K-crossed braided fusion category that is equivalent to $\mathrm{TwMod}_K\mathcal{V}$, and $\mathrm{Rep}^{loc}_{\mathcal{C}}(A)$ is equivalent to Rep $\mathcal{V}$ as a braided fusion category.

Acknowledgments

I would like to thank Terry Gannon for his support and encouragement to write this paper, as well as the referee for helpful remarks that significantly improved this paper.

References

[1] Bojko Bakalov and Victor G. Kac, *Twisted modules over lattice vertex algebras*, Lie theory and its applications in physics V, World Sci. Publ., River Edge, NJ, 2004, pp. 3–26, DOI 10.1142/9789812702562_0001. MR2172171

[2] Lakshya Bhardwaj, Sakura Schäfer-Nameki, and Jingxiang Wu, *Universal non-invertible symmetries*, Fortschr. Phys. **70** (2022), no. 11, Paper No. 2200143, 33, DOI 10.1002/prop.202200143. MR4528258

[3] Lakshya Bhardwaj and Yuji Tachikawa, *On finite symmetries and their gauging in two dimensions*, J. High Energy Phys. **3** (2018), 189, front matter+66, DOI 10.1007/jhep03(2018)189. MR3796103

[4] M. Bischoff and A. Davydov, *Hopf algebra actions in tensor categories*, Transform. Groups **26** (2021), no. 1, 69–80, DOI 10.1007/s00031-020-09560-w. MR4229659

[5] Marcel Bischoff, *Generalized orbifold construction for conformal nets*, Rev. Math. Phys. **29** (2017), no. 1, 1750002, 53, DOI 10.1142/S0129055X17500027. MR3595480

[6] Walter R. Bloom and Herbert Heyer, *Harmonic analysis of probability measures on hypergroups*, De Gruyter Studies in Mathematics, vol. 20, Walter de Gruyter & Co., Berlin, 1995, DOI 10.1515/9783110877595. MR1312826

[7] Richard E. Borcherds, *Vertex algebras, Kac-Moody algebras, and the Monster*, Proc. Nat. Acad. Sci. U.S.A. **83** (1986), no. 10, 3068–3071, DOI 10.1073/pnas.83.10.3068. MR843307

[8] Sebastiano Carpi, Yasuyuki Kawahigashi, Roberto Longo, and Mihály Weiner, *From vertex operator algebras to conformal nets and back*, Mem. Amer. Math. Soc. **254** (2018), no. 1213, vi+85, DOI 10.1090/memo/1213. MR3796433

[9] Thomas Creutzig, Shashank Kanade, and Robert McRae, *Tensor categories for vertex operator superalgebra extensions*, Mem. Amer. Math. Soc. **295** (2024), no. 1472, vi+181, DOI 10.1090/memo/1472. MR4720880

[10] Alexei Davydov, Michael Müger, Dmitri Nikshych, and Victor Ostrik, *The Witt group of non-degenerate braided fusion categories*, J. Reine Angew. Math. **677** (2013), 135–177, DOI 10.1515/crelle.2012.014. MR3039775

[11] Chongying Dong, Xiangyu Jiao, and Feng Xu, *Quantum dimensions and quantum Galois theory*, Trans. Amer. Math. Soc. **365** (2013), no. 12, 6441–6469, DOI 10.1090/S0002-9947-2013-05863-1. MR3105758

[12] Chongying Dong, Haisheng Li, and Geoffrey Mason, *Compact automorphism groups of vertex operator algebras*, Internat. Math. Res. Notices **18** (1996), 913–921, DOI 10.1155/S1073792896000566. MR1420556

[13] Chongying Dong, Li Ren, and Feng Xu, *On orbifold theory*, Adv. Math. **321** (2017), 1–30, DOI 10.1016/j.aim.2017.09.032. MR3715704

[14] Vladimir Drinfeld, Shlomo Gelaki, Dmitri Nikshych, and Victor Ostrik, *On braided fusion categories. I*, Selecta Math. (N.S.) **16** (2010), no. 1, 1–119, DOI 10.1007/s00029-010-0017-z. MR2609644

[15] Pavel Etingof, Shlomo Gelaki, Dmitri Nikshych, and Victor Ostrik, *Tensor categories*, Mathematical Surveys and Monographs, vol. 205, American Mathematical Society, Providence, RI, 2015, DOI 10.1090/surv/205. MR3242743

[16] Pavel Etingof, Dmitri Nikshych, and Victor Ostrik, *Fusion categories and homotopy theory*, Quantum Topol. **1** (2010), no. 3, 209–273, DOI 10.4171/QT/6. With an appendix by Ehud Meir. MR2677836

[17] Igor B. Frenkel and Yongchang Zhu, *Vertex operator algebras associated to representations of affine and Virasoro algebras*, Duke Math. J. **66** (1992), no. 1, 123–168, DOI 10.1215/S0012-7094-92-06604-X. MR1159433

[18] Jürg Fröhlich, Jürgen Fuchs, Ingo Runkel, and Christoph Schweigert, *Correspondences of ribbon categories*, Adv. Math. **199** (2006), no. 1, 192–329, DOI 10.1016/j.aim.2005.04.007. MR2187404

[19] Davide Gaiotto, Anton Kapustin, Nathan Seiberg, and Brian Willett, *Generalized global symmetries*, J. High Energy Phys. **2** (2015), 172, front matter+61, DOI 10.1007/JHEP02(2015)172. MR3321281

[20] Terry Gannon, *Modular data: the algebraic combinatorics of conformal field theory*, J. Algebraic Combin. **22** (2005), no. 2, 211–250, DOI 10.1007/s10801-005-2514-2. MR2164398

[21] Terry Gannon and Andrew Riesen, *Fusion rings acting on vertex operator algebras*, in propaération, 2023.

[22] Terry Gannon and Andrew Riesen, *Orbifolds of lattice vertex operator algebra I*, `https://arxiv.org/abs/2410.00809`, 2024.

[23] André Henriques, David Penneys, and James Tener, *Categorified trace for module tensor categories over braided tensor categories*, Doc. Math. **21** (2016), 1089–1149. MR3578212

[24] Yi-Zhi Huang, *Rigidity and modularity of vertex tensor categories*, Commun. Contemp. Math. **10** (2008), no. suppl. 1, 871–911, DOI 10.1142/S0219199708003083. MR2468370

[25] Yi-Zhi Huang, Alexander Kirillov Jr., and James Lepowsky, *Braided tensor categories and extensions of vertex operator algebras*, Comm. Math. Phys. **337** (2015), no. 3, 1143–1159, DOI 10.1007/s00220-015-2292-1. MR3339173

[26] Victor G. Kac and Ivan T. Todorov, *Affine orbifolds and rational conformal field theory extensions of* $W_{1+\infty}$, Comm. Math. Phys. **190** (1997), no. 1, 57–111, DOI 10.1007/s002200050234. MR1484548

[27] Alexander Kirillov Jr., *Modular categories and orbifold models*, Comm. Math. Phys. **229** (2002), no. 2, 309–335, DOI 10.1007/s002200200650. MR1923177

[28] Alexander Kirillov Jr., *Modular categories and orbifold models*, Comm. Math. Phys. **229** (2002), no. 2, 309–335, DOI 10.1007/s002200200650. MR1923177

[29] Alexander Kirillov Jr. and Viktor Ostrik, *On a q-analogue of the McKay correspondence and the ADE classification of* $\mathfrak{sl}_2$ *conformal field theories*, Adv. Math. **171** (2002), no. 2, 183–227, DOI 10.1006/aima.2002.2072. MR1936496

[30] James Lepowsky and Haisheng Li, *Introduction to vertex operator algebras and their representations*, Progress in Mathematics, vol. 227, Birkhäuser Boston, Inc., Boston, MA, 2004, DOI 10.1007/978-0-8176-8186-9. MR2023933

[31] Robert McRae, *Twisted modules and G-equivariantization in logarithmic conformal field theory*, Comm. Math. Phys. **383** (2021), no. 3, 1939–2019, DOI 10.1007/s00220-020-03882-2. MR4244264

[32] Constantin P. Niculescu, Anton Ströh, and László Zsidó, *Noncommutative extensions of classical and multiple recurrence theorems*, J. Operator Theory **50** (2003), no. 1, 3–52. MR2015017

[33] Victor Ostrik, *Module categories, weak Hopf algebras and modular invariants*, Transform. Groups **8** (2003), no. 2, 177–206, DOI 10.1007/s00031-003-0515-6. MR1976459

[34] Vern Paulsen, *Completely bounded maps and operator algebras*, Cambridge Studies in Advanced Mathematics, vol. 78, Cambridge University Press, Cambridge, 2002. MR1976867

[35] Andrew Riesen, *G-equivariant reconstruction and orbifold theory*, M.Sc. Thesis, `https://era.library.ualberta.ca/items/1dd7340d-9175-45de-a1be-309a379f9a84`, 2023.

Department of Mathematical and Statistical Sciences, University of Alberta, Edmonton, Alberta, Canada T6G 2J5

Email address: `riesen@ualberta.ca`

Contemporary Mathematics
Volume **813**, 2025
https://doi.org/10.1090/conm/813/16284

Ground-state degeneracy of twisted sectors of Conway moonshine SCFT

Alissa Furet and Theo Johnson-Freyd

Abstract. We calculate the ground state degeneracies of all twisted sectors in the "Conway Moonshine" holomorphic SCFT $V^{f\natural}$. We find that almost all sectors have ground states of only a single parity: specifically, 66 twisted sectors have nontrivial ground states of a single parity, 39 twisted sectors have spontaneous supersymmetry breaking, and only 6 twisted sectors have ground states of both parities. Although "nontrivial ground states, all of the same parity" is the expected behavior for a generic SQM model without symmetry protection, it is surprising in the presence of a large symmetry group, as is the case in $V^{f\natural}$. This surprise hints that there are as-yet-undiscovered features of Conway Moonshine.

1. Introduction

In a supersymmetric quantum mechanics (SQM) model — a super Hilbert space equipped with an odd self-adjoint operator $\hat{G}$ — the bosonic and fermionic modes almost exactly pair up. More precisely, in any eigenspace for the Hamiltonian $\hat{H} \propto \hat{G}^2$ (the precise proportionality factor depends on conventions) with eigenvalue $E \neq 0$, the supersymmetry $\hat{G}$ supplies an isomorphism between the bosonic and fermionic subspaces. But in the ground state — the eigenvectors with eigenvalue $E = 0$ — there are typically different numbers of bosons and fermions unless there is some extra symmetry identifying them. In the absence of any symmetry, one expects a generic SQM model to have all ground states of the same parity: a small deformation to $\hat{G}$ will typically annihilate ground states in boson-fermion pairs. For special models, there are no ground states at all: such models are said to have *spontaneous supersymmetry breaking*.

The story is more subtle in the presence of a symmetry group $\mathcal{S}$. In this case, the boson and fermion ground states are each (typically reducible) representations of $\mathcal{S}$, and a small $\mathcal{S}$-symmetric deformation to $\hat{G}$ can only annihilate pairs of states in isomorphic representations. As such, a generic SQM model enhanced with $\mathcal{S}$-symmetry will typically have both boson and fermion ground states, of different

2020 *Mathematics Subject Classification.* Primary 20D08, 17B69, 81Q60.

This article summarizes the main results of the first author's MSc thesis (Dalhousie University, 2022). We thank an anonymous referee for their suggestion to compare to the $N = 1$ extension of $E_{7,1}^2$ at the end of §2.2. The second author also thanks Ying Lin for related discussion. Both authors were supported by the NSERC Discovery Grant RGPIN-2021-02424 and by the Simons Collaboration on Global Categorical Symmetries.

numbers. We will refer to a symmetry-enhanced SQM model as *conspiratorial* when it lacks spontaneous supersymmetry breaking but nevertheless conspires to have only ground states of a single parity.

One interesting construction of SQM models with nontrivial symmetry groups starts with a highly-symmetric holomorphic 1+1D superconformal field theory (SQFT) V. Pick any supersymmetry-preserving automorphism g of V. Then the g-twisted Ramond sector of V, which we will denote $V_{\mathrm{R}g}$, comes with a supersymmetry generated by the zero mode of the supercurrent, and comes with a projective symmetry by the centralizer $C(g)$ of $g \in \mathrm{Aut}(V)$. Fairly often in these SQM models there are equal numbers of bosonic and fermionic ground states. For example, this happens whenever g commutes with a continuous family of supersymmetry-preserving automorphisms of V (see Lemma 2.2.1). The most important holomorphic SCFT without continuous automorphisms is the "Conway Moonshine" theory $V^{f\natural}$ of J. Duncan [**4**]. It is the unique $c \leq 12$ holomorphic SCFT without continuous automorphisms, and its (finite) group of automorphisms is a copy of Conway's largest sporadic simple group Co_1.

We decided to calculate the number of ground states of each parity for each of the g-twisted Ramond sectors of $V^{f\natural}$. We find the following:

MAIN THEOREM. *For exactly 66 of the 101 conjugacy classes* $[g]$ *in* Co_1, *the* g*-twisted Ramond sector* $V^{f\natural}_{\mathrm{R}g}$ *of* $V^{f\natural}$ *is conspiratorial: it has nontrivial ground states of a single parity. For exactly 39 conjugacy classes,* $V^{f\natural}_{\mathrm{R}g}$ *has spontaneous supersymmetry breaking. Exactly six conjugacy classes are nonconspiratorial, with ground states in* $V^{f\natural}_{\mathrm{R}g}$ *of both parities.*

The existence of six nonconspiratorial conjugacy classes means that there cannot be an easy universal explanation for why so many conjugacy classes in Co_1 are conspiratorial. Rather, we believe that explaining our result will shed new light on Conway moonshine. In lieu of an explanation, we offer simply an observation. The six nonconspiratorial classes are (in the standard ATLAS naming convention):

$$3C, \quad 4B, \quad 5C, \quad 6F, \quad 8D, \quad 9B.$$

These are some (but not all) of the small number of conjugacy classes in Co_1 whose Frame shapes are unbalanced (compare [**9**]). The latter three classes are some (but not all) of the square and cube roots of the first two classes: $6F^2 = 3C$, $8D^2 = 4B$, and $9B^3 = 3C$. Moreover, classes $3C$ and $5C$, together with class $2A$ (which is conspiratorial), are particularly special among conjugacy classes in Co_1: these are the only classes with centralizers of shape (extrapsecial).(almost simple); and these are the only classes g of prime order for which the (twisted) orbifold $V^{f\natural} /\!\!/ \langle g\rangle$ is the supersymmetric E_8 lattice VOA used in the original construction of $V^{f\natural}$ in [**5**].

The structure of this article is as follows. In Section 2, we briefly review some definitions and explain why g-twisted Ramond sectors of SCFTs are naturally SQMs. We moreover explain why typically these SQMs are far from conspiratorial: not only are there usually both bosonic and fermionic ground states, but the same number of them. In Section 3 we outline the methods we used to do the computations summarized in our Main Theorem. Finally, Section 4 tabulates the detailed outputs of those computations. Our computer code can be found in the Ancillary Files section of this article's arXiv posting.

2. Preliminaries

In this section, we briefly review some definitions and explain why g-twisted Ramond sectors of SCFTs are naturally SQMs.

2.1. SQM models and Witten index. A *quantum mechanics* (QM) model consists of a complex Hilbert space $\mathcal{H}$ equipped with a self-adjoint operator $\hat{H}$ called the *Hamiltonian.* We will abusively write simply $\mathcal{H}$ for the QM model $(\mathcal{H}, \hat{H})$. The Hamiltonian is typically unbounded, but we require that its spectrum is bounded below. In this case, the *Euclidean time-evolution operator*

$$U_s = e^{-s\hat{H}}, \quad s \in \mathbb{R}_{\geq 0}$$

is bounded for all $s \geq 0$. The QM model $\mathcal{H}$ is called *compact* if U_s is moreover trace-class for all $s > 0$. The *partition function* of a compact QM model is

$$Z(\mathcal{H}) : s \mapsto \mathrm{tr}_{\mathcal{H}}(U_s), \quad s > 0.$$

Note that it is common to work with an exponentiated coordinate $q := e^{-s}$, in which case compactness of $\mathcal{H}$ says that $Z(\mathcal{H})(q)$ converges for $|q| < 1$.

A *fermionic* QM model is just as above, but instead of an ordinary Hilbert space, $\mathcal{H} = \mathcal{H}_{\bar{0}} \oplus \mathcal{H}_{\bar{1}}$ is a super (aka $\mathbb{Z}_2$-graded) Hilbert space. The operator which is $+1$ on the even, aka bosonic, summand $\mathcal{H}_{\bar{0}}$ and -1 on the odd, aka fermionic, summand $\mathcal{H}_{\bar{1}}$ is called $(-1)^f$. The Hamiltonian in a fermionic QM model is required to be even, i.e. to commute with $(-1)^f$. A compact fermionic QM model has two partition functions:

$$Z^{\mathrm{NS}}(\mathcal{H}) : s \mapsto \mathrm{tr}_{\mathcal{H}}(U_s), \tag{2.1.1}$$

$$Z^{\mathrm{R}}(\mathcal{H}) : s \mapsto \mathrm{tr}_{\mathcal{H}}((-1)^f U_s). \tag{2.1.2}$$

The first is called the *Neveu–Schwarz partition function* and the second is called the *Ramond partition function.* Note that this trace is the trace on the underlying non-super Hilbert space of $\mathcal{H}$; the operation $X \mapsto \mathrm{tr}((-1)^f X)$ is often referred to as the *supertrace* $\mathrm{str}(X)$. One could more generally pick any automorphism $g \in \mathrm{Aut}(\mathcal{H})$, and study its character

$$Z^g(\mathcal{H}) : s \mapsto \mathrm{tr}_{\mathcal{H}}(gU_s),$$

(in which case, to match notation, we should simply write R for $(-1)^f$).

As mentioned already in the introduction, *supersymmetric* QM (aka SQM) models are equipped with a self-adjoint *odd* operator $\hat{G}$ whose square is proportional (by a fixed convention-dependent constant) to $\hat{H}$. More precisely, if there is one such chosen $\hat{G}$, then the model is said to have $N = 1$ *supersymmetry*; we will never use more than one supersymmetry, and so we will not write "$N = 1$" in the sequel. The following result is well-known:

LEMMA 2.1.1 ([**13**]). *In any compact SQM model $\mathcal{H}$, the Ramond partition function $Z^{\mathrm{R}}(\mathcal{H})$ is an integer (and independent of s).*

The integer Z^{R} is called the *Witten index* of the SQM model.

PROOF. Let $\mathcal{H}^\lambda \subset \mathcal{H}$ denote the eigenspace with eigenvalue $\hat{H} = \lambda$. Since $\mathcal{H}$ is a super Hilbert space and $\hat{H}$ preserves the super structure, $\mathcal{H}^\lambda$ is also a super space, and decomposes as a sum of bosonic and fermionic pieces, $\mathcal{H}^\lambda = \mathcal{H}^\lambda_{\bar 0} \oplus \mathcal{H}^\lambda_{\bar 1}$. The partition function is then

$$Z(\mathcal{H})(q) = \sum_\lambda q^\lambda(\dim(\mathcal{H}^\lambda_{\bar 0}) \pm \dim(\mathcal{H}^\lambda_{\bar 1})),$$

where addition is Neveu-Schwarz and subtraction is Ramond.

Since $\hat G$ commutes with $\hat H$, it preserves the eigenspaces; since it is odd, it maps $\hat G|_{\mathcal{H}^\lambda} : \mathcal{H}^\lambda_{\bar 0} \leftrightarrows \mathcal{H}^\lambda_{\bar 1}$. The composition of these two maps is $\hat H = \lambda$. In particular, if $\lambda \neq 0$ then $\hat G^2$ is an isomorphism so $\hat G$ is also an isomorphism. In particular $\dim(\mathcal{H}^\lambda_{\bar 0}) = \dim(\mathcal{H}^\lambda_{\bar 1})$ when $\lambda \neq 0$, and so

$$Z^{\mathrm{R}}(\mathcal{H}) = q^0\big(\dim(\mathcal{H}^{\lambda=0}_{\bar 0}) - \dim(\mathcal{H}^{\lambda=0}_{\bar 1})\big),$$

which is an s-independent integer. □

If $\hat G$ exists, then all eigenvalues of $\hat H$ are non-negative and so $Z^{\mathrm{NS}}(\mathcal{H})(q)$ is smooth at $q = 0$, equivalently $s = +\infty$. Moreover, its constant term $Z^{\mathrm{NS}}(\mathcal{H})(q = 0)$ counts (without sign) the number of states of eigenvalue 0. Thus we can retrieve the number of boson and fermion ground states via:

$$\#\text{bosons} = \frac{Z^{\mathrm{NS}}(\mathcal{H})(q = 0) + Z^{\mathrm{R}}(\mathcal{H})}{2}, \tag{2.1.3}$$

$$\#\text{fermions} = \frac{Z^{\mathrm{NS}}(\mathcal{H})(q = 0) - Z^{\mathrm{R}}(\mathcal{H})}{2}. \tag{2.1.4}$$

2.2. Holomorphic SCFTs. In order to produce interesting SQM models, we will work with a specific holomorphic superconformal field theory $V^{f\natural}$ first discovered by J. Duncan [**4, 5**]. We will not review the precise definition of "holomorphic conformal field theory." Suffice it to say that a *holomorphic CFT* is a (full) conformal field theory in which all vertex operators have only holomorphic dependence on the location of insertion. Holomorphic CFTs have a nice axiomatization in terms of vertex operator algebras (VOAs): they are precisely the (unitary and of CFT-type) VOAs V such that the category $\mathrm{Mod}(V)$ of vertex modules is semisimple with a unique simple object (namely V itself). We will without further comment allow our VOAs to have both bosonic and fermionic elements. Unitarity enforces the spin-statistics relation, which says that bosonic vertex operators have (non-negative) integral conformal weights whereas fermionic vertex operators have conformal weights in $\mathbb{N} + \frac{1}{2}$.

Let V be a VOA and $g \in \mathrm{Aut}(V)$ an automorphism of finite order. Then there is a well-studied category $\mathrm{Mod}_g(V)$ of *g-twisted modules* (we recommend the textbook [**12**]; the finite-order assumption can be dropped under some conditions on V [**7**], but we will not need that level of generality). Holomorphicity of V implies that all of the categories $\mathrm{Mod}_g(V)$ are semisimple with a unique simple module. This unique simple module is called the *g-twisted sector* of V, and we will denote it V_g. Note that the words "the" and "unique" are slightly abusive: V_g is well-defined up to non-canonical, possibly-odd isomorphism. (In particular, the overall parity of V_g is not well-defined.)

A special case is when $g = (-1)^f$ is the canonical parity operator on V. The $(-1)^f$-twisted sector is then conventionally called the *Ramond sector* V_{R} of V. (Thus, to match notation, we should write simply R for $(-1)^f$.) The untwisted ($g = 1$) sector is conventionally called the *Neveu–Schwarz sector* $V = V_{\mathrm{NS}}$. More generally, given $g \in \mathrm{Aut}(V)$, we will refer to $V_g = V_{\mathrm{NS}g}$ as the *g-twisted Neveu–Schwarz sector* and $V_{(-1)^f g} = V_{\mathrm{R}g}$ as the *g-twisted Ramond sector.*

The *zero-mode* of a bosonic vertex operator $X(-)$ is the operator $X_0 : V \to V$ defined by

$$X_0[Y] := \oint_{z\in U(1)} \frac{1}{2\pi i z} X(z)Y(0)\mathrm{d}z. \tag{2.2.1}$$

This formula makes sense on the g-twisted sector V_g provided that g preserves X. If X is fermionic, then $X(z)$ is anti-periodic if z goes around a circle with even spin structure, but periodic if z goes around a circle with odd spin structure; the upshot is that X_0 is defined on V_{R}, and on $V_{\mathrm{R}g}$ if g preserves X, but not on V_{NS}.

A special case is when $X = L$ is the conformal vector, which (by definition) is preserved by all symmetries $g \in \mathrm{Aut}(V)$. Its zero mode L_0 is then a reasonable choice of Hamiltonian for a quantum mechanics model with Hilbert space V_g. Actually, although it induces the correct time evolution on the projective space of states, L_0 is not quite the Hamiltonian usually used. Rather, the Hamiltonian on V_g is always taken to be $\hat{H} := L_0 - \frac{c}{24}$, where c is the central charge of the CFT V. The shift by the constant $\frac{c}{24}$ can be explained by carefully computing how the physics transforms when changing variables between rectilinear and polar coordinates (the buzzword is "Schwarzian derivative"); we refer the reader to [**3**] for further discussion.

Applying (2.1.1)–(2.1.2) to the Neveu–Schwarz and Ramond sectors supplies partition functions

$$Z_{\mathrm{NS}}^{\mathrm{NS}}(V) = \mathrm{tr}_{V_{\mathrm{NS}}}\left(q^{L_0-c/24}\right), \qquad Z_{\mathrm{NS}}^{\mathrm{R}}(V) = \mathrm{str}_{V_{\mathrm{NS}}}\left(q^{L_0-c/24}\right),$$
$$Z_{\mathrm{R}}^{\mathrm{NS}}(V) = \mathrm{tr}_{V_{\mathrm{R}}}\left(q^{L_0-c/24}\right), \qquad Z_{\mathrm{R}}^{\mathrm{R}}(V) = \mathrm{str}_{V_{\mathrm{R}}}\left(q^{L_0-c/24}\right).$$

More generally, we can use any twisting. We will be most interested in the g-twisted Ramond sectors:

$$Z_{\mathrm{R}g}^{\mathrm{NS\ or\ R}}(V) = Z^{\mathrm{NS\ or\ R}}(V_{\mathrm{R}g}) = (\mathrm{tr\ \ or\ \ str})_{V_{\mathrm{R}g}}\left(q^{L_0-c/24}\right).$$

We remind that $V_{\mathrm{R}g}$ is well-defined only up to a possibly-odd isomorphism. This does not affect the value of $Z_{\mathrm{R}g}^{\mathrm{NS}}$, but it means that $Z_{\mathrm{R}g}^{\mathrm{R}}$ is well-defined only up to a sign.

In Section 3.1, we will use more generally the *twisted and twined* Ramond–Ramond partition function, defined as follows. In general, for any $g, h \in \mathrm{Aut}(V)$, the action of $v \in V$ on V_g via $v \mapsto h(v)$ instead of the usual action writes V_g as a simple hgh^{-1}-twisted V-module. Thus h supplies an isomorphism $V_g \cong V_{hgh^{-1}}$, well defined up to a U(1)-phase ambiguity. Now suppose that g and h commute. Then $V_{hgh^{-1}}$ is literally equal to V_g, and so h supplies an automorphism of V_g, again well defined up to a U(1)-phase ambiguity. The character of this automorphism is the twisted and twined partition function:

$$Z_{\mathrm{R}g}^{\mathrm{R}h}(V) = \mathrm{str}_{V_{\mathrm{R}g}}\left(hq^{L_0-c/24}\right) = \mathrm{tr}_{V_{\mathrm{R}g}}\left((-1)^f hq^{L_0-c/24}\right). \tag{2.2.2}$$

A *supersymmetry* in a CFT of central charge c is a vertex operator $\hat{G}(-)$ of conformal weight $\frac{3}{2}$ solving the OPE

$$\hat{G}(z)\hat{G}(0) \sim \frac{\frac{2}{3}c}{z^3} + \frac{2L(0)}{z} \tag{2.2.3}$$

where $L(-)$ is the stress-energy vertex operator. Equation (2.2.3) implies that $\hat{G}_0^2 \propto L_0 - \frac{c}{24} = \hat{H}$. In other words, for each automorphism g of the SCFT $(V, \hat{G})$, the g-twisted Ramond sector $V_{\mathrm{R}g}$ is naturally an SQM model. Note that the supersymmetry is odd, and so *not* preserved by $(-1)^f$. In particular, $V_{\mathrm{R}g} \neq V_{\mathrm{NS}h}$ for any supersymmetry-preserving h; indeed, the latter is not naturally an SQM model.

Quite often the SQM model $V_{\mathrm{R}g}$ has vanishing Witten index:

LEMMA 2.2.1. *Suppose that V is a holomorphic SCFT and that $g \in \mathrm{Aut}(V)$ preserves a nonzero vertex operator $Y(-)$ of spin $1/2$. Then the g-twisted Ramond sector $V_{\mathrm{R}g}$ has an equal number of bosonic and fermionic ground states.*

We remark that such a $Y(-)$ supplies a continuous symmetry of V preserving the $N = 1$ structure. Indeed, it is well-known that if $X(-)$ is a vertex operator of spin 1, then X_0 generates a continuous symmetry of V which might not preserve the supersymmetry $\hat{G}$. If X_0 does indeed preserve $\hat{G}$, i.e. if $X_0[\hat{G}] = 0$, then the full OPE of X with $\hat{G}$ is

$$X(z)\hat{G}(0) \sim \frac{Y(0)}{z^2} + \frac{0}{z}$$

for some vertex operator Y of spin $1/2$. Moreover, (2.2.3) implies that

$$\hat{G}(z)Y(0) \sim \frac{2X(0)}{z}. \tag{2.2.4}$$

Conversely, any spin $1/2$ vertex operator Y selects a spin 1 operator X via (2.2.4), and the corresponding continuous symmetry fixes $\hat{G}$.

In the context of holomorphic VOAs, *Noether's Theorem* asserts that the map from spin-1 vertex operators to continuous symmetries is an isomorphism; it follows that the map from spin-1/2 vertex operators to susy-preserving continuous symmetries is also an isomorphism. This version of Noether's Theorem is widely believed, but does not seem to be established in the VOA literature.

PROOF. Since g preserve Y, the zero-mode Y_0 is defined on $V_{\mathrm{R}g}$. Because Y has spin $1/2$, its self-OPE with its complex conjugate $\bar{Y}$ is

$$Y(z)\bar{Y}(0) = \frac{\lambda}{z}$$

for some (positive, hence nonzero) constant $\lambda = \|Y\|^2$. But then $Y_0\bar{Y}_0 = \lambda \neq 0$, and since Y is odd, Y_0 provides an isomorphism between the bosonic and fermionic states, of any energy including 0, in $V_{\mathrm{R}g}$. □

Lemma 2.2.1 is not the only mechanism that can produce an equal number of bosonic and fermionic ground states. For example, symmetries with certain anomalies can produce the same behaviour [**2**]; for the g-twisted sector, the auxiliary anomalous symmetry must commute with g, but does not need to preserve the $N = 1$ supersymmetry.

In the introduction, we gave a heuristic argument that a typical SQM model with a large symmetry group $\mathcal{S}$ will typically have ground states of both parities provided supersymmetry is not spontaneously broken. It is worth testing if this heuristic applies to twisted sectors of holomorphic SCFTs other than $V^{f\natural}$. Unfortunately, there are not too many known holomorphic SCFTs with large symmetry groups, and most of those do have many continuous symmetries, triggering Lemma 2.2.1.

Even more severely, in an $N = 1$ SCFT V, the untwisted Ramond sector V_{R} might have spontaneous supersymmetry breaking. This happens for example in the holomorphic SCFT with bosonic part $E_{7,1}^2$ that is studied in [**8**, §6.1–§6.2]. Indeed, the central charge is $c = 14$, whereas the minimum spin of a Ramond-sector field is $3/4$, and $3/4 - 14/24 > 0$. (In particular, the spontaneous supersymmetry breaking holds for any choice of $N = 1$ structure; there are three possibly-inequivalent $N = 1$ structures found in [**8**].) A generalization of Lemma 3.1.1 below then implies that $Z^{\mathrm{R}}_{\mathrm{R}g}(V) = \mathrm{tr}_{\mathbb{R}^0}(\tilde{g}) = 0$ for any $g \in \mathrm{Aut}_{N=1}(V)$. This rules out any conspiratorial symmetries, and suggests (but does not imply) that almost all the twisted sectors will also have spontaneous supersymmetry breaking.

3. Calculations

We turn now to calculating the partition functions $Z^{\mathrm{NS}}_{\mathrm{R}g}(V^{f\natural}) = Z^{\mathrm{NS}}(V^{f\natural}_{\mathrm{R}g})$ and $Z^{\mathrm{R}}_{\mathrm{R}g}(V^{f\natural}) = Z^{\mathrm{R}}(V^{f\natural}_{\mathrm{R}g})$ of the g-twisted Ramond sector of $V^{f\natural}$, as these will supply a count of the ground states via (2.1.3)–(2.1.4).

If one forgets the supersymmetry, then the automorphism group of $V^{f\natural}$ is the group SO^+_{24}, which is by definition the image of Spin_{24} in the positive half-spin representation. This is a group with the same Lie algebra as SO_{24}, but a different global form: it is the unique, up to the outer automorphism, double cover of PSO_{24} other than SO_{24}; and it is the unique, up to the outer automorphism, $\mathbb{Z}_2$-quotient of Spin_{24} other than SO_{24}. In particular, twisted (and twined) partition functions of $V^{f\natural}$ make sense for all finite-order elements of SO^+_{24}. We will explain a little bit more about how SO^+_{24} arises in Section 3.2. Although the automorphism group of $V^{f\natural}$ is just SO^+_{24}, it is convenient to describe elements in terms of (either of their) lift(s) to Spin_{24}.

The main theorem of [**4**] states that $V^{f\natural}$ admits a unique-up-to-automorphism supersymmetry, and that the subgroup of SO^+_{24} that preserves this supersymmetry is a copy of Conway's largest sporadic group Co_1. To help orient the reader, we note that the double cover $\mathrm{Spin}_{24} \to \mathrm{SO}^+_{24}$ restricts over Co_1 to the double cover $\mathrm{Co}_0 \to \mathrm{Co}_1$, and that $\mathrm{Co}_0 \subset \mathrm{Spin}_{24}$ maps injectively along $\mathrm{Spin}_{24} \to \mathrm{SO}_{24} \subset \mathrm{O}_{24}$. As a subgroup of O_{24}, the group Co_0 arises as the group of automorphisms of the Leech lattice $\Lambda_{24} \subset \mathbb{R}^{24}$.

3.1. Frame shapes and conjugacy classes. Frame shapes (introduced in [**6**]) generalize cycle structures of permutations, where one writes a formal product like $1^2 2^1 3^1$ to indicate the conjugacy class of a permutation $(1)(2)(34)(567) \in S_7$. Frame shapes are defined as follows. Suppose that $g \in \mathrm{O}_n$ has finite order $o(g) < \infty$ and that there exists a basis in which the matrix for g has all entries in $\mathbb{Q}$ — in examples, this happens because g preserves a lattice $\mathbb{Z}^n \hookrightarrow \mathbb{R}^n$. Then the characteristic polynomial $p(x) = \det(x - g)$ has coefficients in $\mathbb{Q}$. On the other hand, because g has finite order, its eigenvalues are all roots of unity. It follows that the irreducible-over-$\mathbb{Q}$ factors of $p(x)$ are cyclotomic polynomials. Equivalent $p(x)$ factors (uniquely!) into a product of shape

$$p(x) = \prod_{d | o(g)} (1 - x^d)^{k_d},$$

where some of the exponents k_d may be negative. We remark for future reference that the x^1 term in this product is

$$\mathrm{tr}_{\mathbb{R}^n}(g) = k_1. \tag{3.1.1}$$

The *Frame shape* of g is then defined to be the formal symbol

$$\pi_g := \prod_{d|o(g)} d^{k_d}.$$

Because Frame shapes exactly encode characteristic polynomials, they are a convenient and compact way of recording O_n-conjugacy classes. In [**11**], the Frame shapes for elements of $\mathrm{Co}_0 \subset \mathrm{O}_{24}$ were calculated, resulting in 160 Frame shapes out of the 167 conjugacy classes.

We will need to modify this encoding in two ways. First, our formulas do not directly reference the Frame shape of g, but rather ($1/2\pi i$ times) the logarithms of the eigenvalues of g. These log-eigenvalues can be quickly extracted from the Frame shape as follows. A general element of SO_{24} can be "diagonalized" to a block-diagonal matrix of shape

$$\begin{bmatrix} \begin{bmatrix} \cos(2\pi\lambda_1) & \sin(2\pi\lambda_1) \\ -\sin(2\pi\lambda_1) & \cos(2\pi\lambda_1) \end{bmatrix} & & \\ & \ddots & \\ & & \begin{bmatrix} \cos(2\pi\lambda_{12}) & \sin(2\pi\lambda_{12}) \\ -\sin(2\pi\lambda_{12}) & \cos(2\pi\lambda_{12}) \end{bmatrix} \end{bmatrix},$$

where each block is a rotation matrix. The O_{24}-conjugacy class is determined by the total log-eigenvalue $\vec{\lambda} \in (\mathbb{R}/\mathbb{Z})^{12}$, up to permuations of the entries and sign-flips $\lambda_i \mapsto -\lambda_i \pmod 1$. The full list of log-eigenvalues $(\lambda_1, -\lambda_1, \lambda_2, \dots, -\lambda_{12})$ consists simply of the logarithms of the roots of the characteristic polynomial $\prod_{d|o(g)}(1-x^d)^{k_d}$. Each factor d^{k_d} in the Frame shape contributes k_d-many copies of each of the log-eigenvalues $0, \frac{1}{d}, \frac{2}{d}, \dots, \frac{d-1}{d}$ mod 1 to the full list. Naturally, negative exponents entries in the Frame shape contribute negatively many copies of the log-eigenvalues. After taking producing the full list of 24 numbers, we then record just 12 numbers by taking, for each i, whichever of λ_i is in the interval $[0, \frac{1}{2}] \subset \mathbb{R}/\mathbb{Z}$.

Second, given an element $g \in \mathrm{Co}_0$, our formulas depend not just on the O_{24}-conjugacy class of g, but on its conjugacy class in Spin_{24} — because $\mathrm{H}^1(\mathrm{Co}_0; \mathbb{Z}_2) = \mathrm{H}^2(\mathrm{Co}_0; \mathbb{Z}_2) = 0$, the inclusion $\mathrm{Co}_0 \subset \mathrm{O}_{24}$ lifts uniquely to an inclusion $\mathrm{Co}_0 \subset \mathrm{Spin}_{24}$. The fusion of conjugacy classes along $\mathrm{Spin}_{24} \to \mathrm{O}_{24}$ is typically four-to-one:

(1) Whereas the O_{24}-conjugacy class of $\vec{\lambda} = (\lambda_1, \dots, \lambda_{12})$ is unchanged by any sign flip $\lambda_i \mapsto -\lambda_i$, the SO_{24}-conjugacy class is unchanged only by an even number of sign flips. In other words, an O_{24}-conjugacy class parameterized by a vector $(\lambda_1, \dots, \lambda_{11}, \lambda_{12})$ could have come from the SO_{24}-conjugacy class parameterized by $(\lambda_1, \dots, \lambda_{11}, \lambda_{12})$ or from the conjugacy class parameterized by $(\lambda_1, \dots, \lambda_{11}, -\lambda_{12})$.

(2) Whereas the SO_{24}-conjugacy class depends on the vector $\vec{\lambda} \in (\mathbb{R}/\mathbb{Z})^{12} = \mathbb{R}^{12}/\mathbb{Z}^{12}$, the Spin_{24}-conjugacy class depends on a vector in $\mathbb{R}^{12}/D_{12}$, where $D_{12} \subset \mathbb{Z}^{12}$ is the coweight lattice of Spin_{12}, defined by

$$D_{12} = \{\vec{z} \in \mathbb{Z}^{12} : \sum z_i \in 2\mathbb{Z}\}.$$

In other words, an SO_{24}-conjugacy class $(\lambda_1,\dots,\lambda_{11},\lambda_{12})$ could have come from the Spin_{24}-conjugacy class $(\lambda_1,\dots,\lambda_{11},\lambda_{12})$ or from $(\lambda_1,\dots,\lambda_{11}, 1+\lambda_{12})$.

One reason Frame shapes, equivalently O_n-conjugacy classes, are convenient to work with is that they are comparatively quick to compute from character tables: to work out the characteristic polynomial of a representation requires knowing only the power operations, some elementary symmetric polynomial theory, and a desktop computer to actually do the computations. (That said, we did not do this computation directly; Kondo already did that work in [**11**], and we simply extracted the log-eigenvalues from the Frame shapes.) For elements of Co_0, it is in principle possible to in fact work out the Spin_{24}-conjugacy classes from a careful analysis of the characters of the spin representations of Co_0, but we did not pursue this route. Rather, we used a guess-and-check method guided by the fact that, for $V^{f\natural}$, we know that $Z^{\mathrm{R}}_{\mathrm{R}g}$ is an integer, and moreover we know which integer it is:

LEMMA 3.1.1. *Suppose that $\pi_g = \prod_d d^{k_d}$ is the Frame shape of any lift $\tilde g \in \mathrm{Co}_0$ of an element $g \in \mathrm{Co}_1$. Then $Z^{\mathrm{R}}_{\mathrm{R}g}(V^{f\natural}) = \pm k_1$.*

We remind that $k_1 = \mathrm{tr}_{\mathbb{R}^{24}}(\tilde g)$ by (3.1.1); that the two lifts $\tilde g$ of g, and hence their traces, differ only by a sign; and that $Z^{\mathrm{R}}_{\mathrm{R}g}(V^{f\natural})$ is only defined up to a sign.

PROOF. Zhu's Theorem states that if V is any rational VOA and M is any V-module, then $Z(M)$ is a scalar-valued modular function for some subgroup of the modular group $\Gamma = \mathrm{SL}_2(\mathbb{Z})$ [**14**], and moreover that the various $Z(M)$ are together a vector-valued modular form for the full modular group Γ. The main theorem by Carnahan and Miyamoto in [**1**] implies that the twisted and twined characters (for finite-order automorphisms) of any rational VOA also participate in vector-valued modular forms.

A special case says that, for any commuting elements g, h and any holomorphic VOA V, the S-transformation $S = \left(\begin{smallmatrix} 0 & 1 \\ -1 & 0 \end{smallmatrix}\right)$, which acts by $\tau \mapsto -1/\tau$ for $q = e^{-i\tau}$, takes

$$Z^{\mathrm{R}h}_{\mathrm{R}g}(V) \overset{S}{\longmapsto} Z^{\mathrm{R}g}_{\mathrm{R}h}(V)$$

up to an overall $\mathrm{U}(1)$-phase ambiguity; see (2.2.2) and the preceding discussion.

Specializing $V = V^{f\natural}$, and hence $c = 12$, and setting $h = 1$, we find that S takes

$$Z^{\mathrm{R}}_{\mathrm{R}g}(V^{f\natural}) \overset{S}{\longmapsto} Z^{\mathrm{R}g}_{\mathrm{R}}(V^{f\natural}) = \mathrm{tr}_{V^{f\natural}_{\mathrm{R}}}\big((-1)^f g q^{L_0 - c/24}\big).$$

But $Z^{\mathrm{R}}_{\mathrm{R}g}(V^{f\natural})$ is an integer by Lemma 2.1.1, and so is unchanged under S-transformations. On the other hand,

$$V^{f\natural}_{\mathrm{R}} = \mathbb{R}^{24} \oplus (\cdots)$$

where L_0 acts on $\mathbb{R}^{24}$ with eigenvalue $\frac{1}{2}$ and on $(\cdots)$ with higher eigenvalues, and where $\mathbb{R}^{24}$ is all of one parity. Thus

$$\mathrm{tr}_{V^{f\natural}_{\mathrm{R}}}\big((-1)^f g q^{L_0 - c/24}\big) = \mathrm{tr}_{\mathbb{R}^{24}}(g) + O(q).$$

The result follows. ☐

All together, the way our algorithm works is that it takes the Frame shapes as calculated by [**11**], and first extracts the vector $\vec\lambda$ that parameterizes the O_{24}-conjugacy class. It then tries all four possible variations $\lambda_{12} \mapsto \lambda_{12}, -\lambda_{12}, 1+\lambda_{12}, 1-\lambda_{12} \pmod 2$ in the formula for $Z^{\mathrm{R}}_{\mathrm{R}g}$. Remarkably, we find in each case that only one option produces the correct integer value. Our algorithm then uses this correct option in the formula for $Z^{\mathrm{NS}}_{\mathrm{R}g}$. We turn now to describing those formulas.

3.2. Lattices and θ-series. There are many constructions of $V^{f\natural}$. The most direct, and the most convenient for our purposes, is to recognize $V^{f\natural}$ as a lattice VOA for the D_{12}^+ lattice. By definition, D_{12}^+ is the weight (equivalently coweight, as it is self-dual) lattice of SO_{24}^+. It is the extension of $D_{12} = \{\vec{z} \in \mathbb{Z}^{12} : \sum_i z_i \in 2\mathbb{Z}\}$ by the coset through the vector $\vec{\varsigma} := (\frac{1}{2}, \frac{1}{2}, \dots, \frac{1}{2})$. The construction that inputs a lattice $\mathcal{L}$ and produces a VOA $V_{\mathcal{L}}$ is well-known, and surveyed for example in [**10**]. In general, $\mathrm{Aut}(V_{\mathcal{L}})$ is the Lie group with weight lattice $\mathcal{L}$. This is one way to see the equality $\mathrm{Aut}(V^{f\natural}) = \mathrm{SO}_{24}^+$.

For our purposes, we don't care about the precise vertex algebra structure on $V_{\mathcal{L}}$, but we do care about the grading by L_0, so we briefly review the basics. Let $\mathcal{L} \subset \mathbb{R}^r$ be a full-rank integral lattice. There is a vertex algebra $\mathrm{Bos}(r) = \mathrm{Bos}(1)^{\otimes r}$ that depends only on $\mathbb{R}^r$. As a vector space graded by L_0, there is a PBW-type isomorphism

$$\mathrm{Bos}(1) \cong \mathbb{C}[\phi, \partial\phi, \partial^2\phi, \dots] \cong \bigotimes_{n=1}^{\infty} \mathbb{C}[\partial^n\phi]$$

where "$\partial^n\phi$" is supposed to be understood as a single symbol, denoting the nth generator of this infinitely-generated polynomial ring. The L_0-grading is compatible with the polynomial ring structure, and $\partial^n\phi$ has grading $n+1$. The VOA $\mathrm{Bos}(r)$ has central charge $c = r$ and is naturally a (non-rational) sub-VOA of $V_{\mathcal{L}}$. Finally, there is also a PBW-type vector space isomorphism

$$V_{\mathcal{L}} \cong \mathrm{Bos}(r) \otimes \mathbb{C}[\mathcal{L}] = \bigoplus_{\vec{\ell} \in \mathcal{L}} \mathrm{Bos}(r)$$

where the component graded by $\vec{\ell}$ picks up an extra L_0-grading by $\frac{\|\vec{\ell}\|^2}{2}$.

The twisted sectors are just as easy to write down. The maximal torus of $\mathrm{Aut}(V_{\mathcal{L}})$ is $\mathbb{R}^r/\mathcal{L}$. Assuming $g \in \mathrm{Aut}(V_{\mathcal{L}})$ is in the identity component, it can be conjugated into the maximal torus, and so is parameterized by its log-eigenvalues $\vec{\lambda} \in \mathbb{R}^r/\mathcal{L}$. The corresponding twisted sector $(V_{\mathcal{L}})_g$ is then PBW-isomorphic to

$$(V_{\mathcal{L}})_g \cong \mathrm{Bos}(r) \otimes \mathbb{C}[\mathcal{L} + \vec{\lambda}] = \bigoplus_{\vec{\ell} \in \mathcal{L}+\vec{\lambda}} \mathrm{Bos}(r)$$

where of course $\mathcal{L} + \lambda$ just denotes the coset of $\mathcal{L}$ through $\vec{\lambda}$, and where the ℓ'th-component receives a contribution to its L_0-grading by $\frac{\|\vec{\ell}\|^2}{2}$.

These tensor products are compatible with the L_0-grading, and so the partition functions factor:

$$\begin{aligned} Z_g^{\mathrm{NS\ or\ R}}(V_{\mathcal{L}}) &= (\mathrm{tr\ \ or\ \ str})_{(V_{\mathcal{L}})_g}\left(q^{L_0 - c}\right) \\ &= (\mathrm{tr\ \ or\ \ str})_{\mathrm{Bos}(1)}\left(q^{L_0-c}\right)^r \times (\mathrm{tr\ \ or\ \ str})_{\mathbb{C}[\mathcal{L}+\vec{\lambda}]}\left(q^{L_0}\right) \\ &= Z(\mathrm{Bos}(1))^r \times \sum_{\vec{\ell} \in \mathcal{L}+\vec{\lambda}} \pm q^{\|\vec{\ell}\|^2/2}. \end{aligned} \tag{3.2.1}$$

Note that $\mathrm{Bos}(1)$ is purely bosonic and so the choice of trace or supertrace is irrelevant. The sign in the sum is all $+1$ in the Neveu–Schwarz case, and $(-1)^{\|\vec{\ell}\|^2}$ in Ramond case. Actually, different vectors $\vec{\lambda}$ representing the same class in $\mathbb{R}^r/\mathcal{L}$

can lead to different signs, but the difference is uniform in ℓ: it is just an overall sign-on Z^{R}, and in any case, Z^{R} is typically only well-defined up to an overall sign.

It remains to compute $Z(\mathrm{Bos}(1))$ and the sum. The former is straightforward. There is an overall factor of $q^{-1/24}$ coming from the shift by $c/24$. The rest is the graded dimension of a polynomial algebra, or more precisely an infinite product $\bigotimes \mathbb{C}[\partial^n \phi]$. But the nth factor in this product contributes

$$\operatorname{tr}_{\mathbb{C}[\partial^n\phi]}(q^{L_0}) = \sum_{m=0}^{\infty} \operatorname{tr}_{\mathbb{C}(\partial^n\phi)^m} q^{L_0} = \sum_{m=0}^{\infty} q^{m(n+1)} = \frac{1}{1-q^{n+1}},$$

and so

$$Z(\mathrm{Bos}(1)) = q^{-1/24} \prod_{n=0}^{\infty} \frac{1}{1-q^{n+1}} = \frac{1}{\eta} \tag{3.2.2}$$

where $\eta = q^{1/24}\prod_{n=0}^{\infty}(1-q^{n+1}) = \sqrt[24]{\Delta}$ is Dedekind's eta function.

The final factor depends on the lattice $\mathcal{L}$ and the vector $\vec{\lambda}$, and so doesn't have a closed form in general. It is called the $\vec{\lambda}$-*twisted Theta function* of $\mathcal{L}$:

$$\Theta^{\pm}_{\mathcal{L}+\vec{\lambda}}(q) := \sum_{\vec{\ell}\in\mathcal{L}} (\pm 1)^{\|\vec{\ell}\|^2} q^{\|\vec{\ell}+\vec{\lambda}\|^2/2}.$$

Note that the usual Theta function $\Theta_{\mathcal{L}}$ is the special case $\vec{\lambda} = 0$, and that we have included a sign $\pm$ in the notation since we are working with odd lattices: Θ^{+} appears in Z^{NS} and Θ^{-} appears in Z^{R}.

In the special case $\mathcal{L} = D_{12}^{+}$, we can write a closed formula. This is because D_{12}^{+} is almost the $\mathbb{Z}^{12}$-lattice, and so its Theta series almost factor as 12th powers of Theta series for $\mathbb{Z}$. The twisted Theta series of $\mathbb{Z}$ are the lower-case twisted theta series:

$$\theta^{\pm}_{\lambda}(q) := \sum_{\ell\in\mathbb{Z}} (\pm 1)^{\ell} q^{(\ell+\lambda)^2/2}.$$

The original Jacobi theta series arise when $\lambda = 0, \frac{1}{2}$:

$$\theta_1 = \theta^{-}_{1/2} = 0 \qquad \theta_2 = \theta^{+}_{1/2}$$
$$\theta_4 = \theta^{-}_{0} \qquad \theta_3 = \theta^{+}_{0}$$

Actually, the usual formula for θ_1 differs from our formula for $\theta^{-}_{1/2}$ by an overall phase, which in any case doesn't matter because the sum vanishes identically.

Suppose that we cared about $\mathcal{L} = \mathbb{Z}^{12}$. Then the twisted Theta function would factor:

$$\Theta^{\pm}_{\mathbb{Z}^{12}+\vec{\lambda}}(q) = \sum_{\vec{\ell}\in\mathbb{Z}^{12}} (\pm 1)^{\|\vec{\ell}\|^2} q^{\|\vec{\ell}+\vec{\lambda}\|^2/2} = \prod_{i=1}^{12} \theta^{\pm}_{\lambda_i}(q).$$

Observe that the signs of the $\vec{\ell}$th summand in Θ^{+} and Θ^{-} agree exactly when $\vec{\ell} \in D_{12}$ and exactly cancel when $\vec{\ell} \in \mathbb{Z}^{12} \smallsetminus D_{12}$. This means that the twisted Theta series for D_{12} is:

$$\Theta_{D_{12}+\vec{\lambda}}(q) = \frac{1}{2}\left(\Theta^{+}_{\mathbb{Z}^{12}+\vec{\lambda}}(q) + \Theta^{-}_{\mathbb{Z}^{12}+\vec{\lambda}}(q)\right) = \frac{1}{2}\left(\prod_{i=1}^{12}\theta^{+}_{\lambda_i}(q) + \prod_{i=1}^{12}\theta^{-}_{\lambda_i}(q)\right). \tag{3.2.3}$$

Note that D_{12} is even (every $\vec{\ell} \in D_{12}$ has $\|\vec{\ell}\|^2 \in 2\mathbb{Z}$), so there are no signs.

On the other hand, our desired lattice D_{12}^+ is nothing but D_{12} together with a shift thereof through $\vec{\varsigma} := (\frac{1}{2}, \dots, \frac{1}{2})$. Thus

$$\Theta^{\pm}_{D_{12}^+ + \vec{\lambda}} = \Theta_{D_{12}+\vec{\lambda}} \pm \Theta_{D_{12}+\vec{\varsigma}+\vec{\lambda}}. \tag{3.2.4}$$

The very last step is to understand which log-eigenvalues correspond to the canonical parity operator $\mathrm{R} = (-1)^f \in \mathrm{Aut}(V^{f\natural}) = \mathrm{SO}_{24}^+$. The answer is that R can be represented by any vector $\vec{\rho} \in \mathbb{R}^{12}$ which has integral dot product with D_{12} and half-integral dot product with $D_{12}^+ \smallsetminus D_{12} = D_{12} + \vec{\varsigma}$. An example of such a vector is $(0, \dots, 0, 1)$. Thus if g has log-eigenvalue $\vec{\lambda}$, then $\mathrm{R}g$ has log-eigenvalue $\vec{\lambda} + (0, \dots, 0, 1)$. Finally, note that such a shift acts very simply on twisted theta functions:

$$\theta^{\pm}_{\lambda+1}(q) = \pm\theta^{\pm}_{\lambda}(q) \tag{3.2.5}$$

Combining equations (3.2.1)–(3.2.5) then yields our final formula:

LEMMA 3.2.1. *Suppose (a lift to* Co_0 *of)* $g \in \mathrm{Co}_1$ *has log-eigenvalues* $\vec{\lambda} \in \mathbb{R}^{12}/D_{12}$. *Then*

$$Z^{\mathrm{NS}}_{\mathrm{R}g}(V^{f\natural}) = \frac{1}{2}\left(\prod_{i=1}^{12}\theta^+_{\lambda_i}(q) - \prod_{i=1}^{12}\theta^-_{\lambda_i}(q) + \prod_{i=1}^{12}\theta^+_{\lambda_i+\frac{1}{2}}(q) - \prod_{i=1}^{12}\theta^-_{\lambda_i+\frac{1}{2}}(q)\right)$$

$$Z^{\mathrm{R}}_{\mathrm{R}g}(V^{f\natural}) = \frac{1}{2}\left(\prod_{i=1}^{12}\theta^+_{\lambda_i}(q) - \prod_{i=1}^{12}\theta^-_{\lambda_i}(q) - \prod_{i=1}^{12}\theta^+_{\lambda_i+\frac{1}{2}}(q) + \prod_{i=1}^{12}\theta^-_{\lambda_i+\frac{1}{2}}(q)\right)$$

□

4. Results

The following table reports the results of our computations, which were done using Maple. We ran through all conjugacy classes $g \in \mathrm{Co}_0$. (The answer only depends on the corresponding class in Co_1, but we ran through all classes in Co_0 in order to double-check our results.) For each class, we used the Frame shape computed by **[11]** to produce log-eigenvalues to parameterize the class of g in O_{24}, and hence the log eigenvalues of the four possible lifts of g to Spin_{24}. For each of these lifts, we computed $Z^{\mathrm{R}}_{\mathrm{R}g}(V^{f\natural})$ from Lemma 3.2.1, and found that only one produced $\mathrm{tr}_{\mathbb{R}^{24}}(g)$. For that lift, we computed $Z^{\mathrm{NS}}_{\mathrm{R}g}(V^{f\natural})$ from Lemma 3.2.1. In the table, we have reported the Frame shapes and the computed values of $Z^{\mathrm{R}}_{\mathrm{R}g}(V^{f\natural})$ and $Z^{\mathrm{NS}}_{\mathrm{R}g}(V^{f\natural})$. We also record the names of $g \in \mathrm{Co}_0$ and of its image in Co_1 following the ATLAS naming convention: a conjugacy class is named nX where $n \in \mathbb{N}$ records the order of g, and $X \in \{\mathrm{A}, \mathrm{B}, \mathrm{C}, \dots\}$ sorts the conjugacy classes of a given order in terms of the size of their centralizers, starting from the largest.

The penultimate column of the table records whether g is conspiratorial (✓), nonconspiratorial (**X**), or spontaneously breaks supersymmetry (0) — as explained in Section 2, these are read off by comparing $Z^{\mathrm{R}}_{\mathrm{R}g}$ to the constant term of $Z^{\mathrm{NS}}_{\mathrm{R}g}$. As reported in the introduction, we find that the only conspiratorial Co_1-classes are

$$3C, \quad 4B, \quad 5C, \quad 6F, \quad 8D, \quad 9B.$$

In the final column of the table, included purely for the reader's convenience, we record from **[9]** whether the class is (✓) or is not ( ) anomalous. Although it was not to study the relationship between anomalies and spontaneous symmetry breaking, we find that a class in Co_1 is anomalous if and only if it spontaneously breaks supersymmetry. Actually, this is immediate from our observation in the introduction that the conspiratorial classes are all unbalanced. Indeed, if g is an anomalous symmetry of any holomorphic SCFT V, then $Z^{\mathrm{R}}_{\mathrm{R}g}(V) \propto Z^{\mathrm{R}g}_{\mathrm{R}}(V)$ must be a constant but must transform with nontrivial eigenvalue under T-transformations, and so must vanish; specializing to $V^{f\natural}$, we see that a necessary condition for an anomaly is that g must be traceless. On the other hand, we found by calculation that the only conspiratorial classes are unbalanced, and [**9**] found by calculation that all unbalanced classes are non-anomalous. So if g is balanced and traceless, then $Z^{\mathrm{NS}}_{\mathrm{R}g} = |Z^{\mathrm{R}}_{\mathrm{R}g}| + O(q) = 0 + O(q)$, and so $V_{\mathrm{R}g}$ has spontaneous supersymmetry breaking.

Co_0	Co_1	Frame shape	$Z^{\mathrm{R}}_{\mathrm{R}g}$	$Z^{\mathrm{NS}}_{\mathrm{R}g}$	Consp?	Anom?
1A	1A	1^{24}	24	$24 + 4096q + 98304q^2 + 1228800q^3 + 10747904q^4 + 74244096q^5 + 432144096q^6 + \dots$	✓	
2A	1A	$2^{24}1^{-24}$	-24	$24 + 4096q + 98304q^2 + 1228800q^3 + 10747904q^4 + 74244096q^5 + 432144096q^6 + \dots$	✓	
2B	2A	$1^8 2^8$	8	$8 + 256q^{1/2} + 2048q + 11264q^{3/2} + 49152q^2 + 183808q^{5/2} + 614400q^3 + 1882112q^{7/2} + \dots$	✓	
2C	2A	$2^{16}1^{-8}$	-8	$8 + 256q^{1/2} + 2048q + 11264q^{3/2} + 49152q^2 + 183808q^{5/2} + 614400q^3 + 1882112q^{7/2} + \dots$	✓	
4A	2B	$4^{12}2^{-12}$	0	$0 + 24q^{1/8} + 464q^{5/8} + 3192q^{9/8} + 16656q^{13/8} + 69040q^{17/8} + 251136q^{21/8} + \dots$	0	✓
2D	2C	2^{12}	0	$0 + 64q^{1/4} + 768q^{3/4} + 4992q^{5/4} + 24064q^{7/4} + 96576q^{9/4} + 340224q^{11/4} + \dots$	0	✓
3A	3A	$3^{12}1^{-12}$	-12	$12 + 24q^{1/3} + 440q^{2/3} + 1608q + 3192q^{4/3} + 13904q^{5/3} + 35688q^2 + 68160q^{7/3} + \dots$	✓	
6A	3A	$1^{12}6^{12}2^{-12}3^{-12}$	12	$12 + 24q^{1/3} + 440q^{2/3} + 1608q + 3192q^{4/3} + 13904q^{5/3} + 35688q^2 + 68160q^{7/3} + \dots$	✓	
3B	3B	$1^6 3^6$	6	$6 + 64q^{1/3} + 384q^{2/3} + 1344q + 4352q^{4/3} + 12672q^{5/3} + 32640q^2 + 79872q^{7/3} + \dots$	✓	
6B	3B	$2^6 6^6 1^{-6} 3^{-6}$	-6	$6 + 64q^{1/3} + 384q^{2/3} + 1344q + 4352q^{4/3} + 12672q^{5/3} + 32640q^2 + 79872q^{7/3} + \dots$	✓	
3C	3C	$3^9 1^{-3}$	-3	$5 + 72q^{1/3} + 360q^{2/3} + 1368q + 4392q^{4/3} + 12528q^{5/3} + 32760q^2 + 80064q^{7/3} + \dots$	X	
6C	3C	$1^3 6^9 2^{-3} 3^{-9}$	3	$5 + 72q^{1/3} + 360q^{2/3} + 1368q + 4392q^{4/3} + 12528q^{5/3} + 32760q^2 + 80064q^{7/3} + \dots$	X	
3D	3D	3^8	0	$0 + 16q^{1/9} + 128q^{4/9} + 576q^{7/9} + 2048q^{10/9} + 6304q^{13/9} + 17408q^{16/9} + 44416q^{19/9} + \dots$	0	✓
6D	3D	$6^8 3^{-8}$	0	$0 + 16q^{1/9} + 128q^{4/9} + 576q^{7/9} + 2048q^{10/9} + 6304q^{13/9} + 17408q^{16/9} + 44416q^{19/9} + \dots$	0	✓
4B	4A	$1^8 4^8 2^{-8}$	8	$8 + 16q^{1/4} + 128q^{1/2} + 448q^{3/4} + 1024q + 2272q^{5/4} + 5632q^{3/2} + 12672q^{7/4} + 24576q^2 + \dots$	✓	
4C	4A	$4^8 1^{-8}$	-8	$8 + 16q^{1/4} + 128q^{1/2} + 448q^{3/4} + 1024q + 2272q^{5/4} + 5632q^{3/2} + 12672q^{7/4} + 24576q^2 + \dots$	✓	
4D	4B	$4^8 2^{-4}$	0	$4 + 32q^{1/4} + 128q^{1/2} + 384q^{3/4} + 1024q + 2496q^{5/4} + 5632q^{3/2} + 12032q^{7/4} + 24576q^2 + \dots$	X	
4E	4C	$1^4 2^2 4^4$	4	$4 + 32q^{1/4} + 128q^{1/2} + 384q^{3/4} + 1024q + 2496q^{5/4} + 5632q^{3/2} + 12032q^{7/4} + 24576q^2 + \dots$	✓	
4F	4C	$2^6 4^4 1^{-4}$	-4	$4 + 32q^{1/4} + 128q^{1/2} + 384q^{3/4} + 1024q + 2496q^{5/4} + 5632q^{3/2} + 12032q^{7/4} + 24576q^2 + \dots$	✓	
4G	4D	$2^4 4^4$	0	$0 + 16q^{1/8} + 64q^{3/8} + 224q^{5/8} + 640q^{7/8} + 1616q^{9/8} + 3776q^{11/8} + 8288q^{13/8} + \dots$	0	✓
8A	4E	$8^6 4^{-6}$	0	$0 + 12q^{3/32} + 52q^{11/32} + 204q^{19/32} + 552q^{27/32} + 1456q^{35/32} + 3396q^{43/32} + 7560q^{51/32} + \dots$	0	✓
4H	4F	4^6	0	$0 + 8q^{1/16} + 48q^{5/16} + 168q^{9/16} + 496q^{13/16} + 1296q^{17/16} + 3072q^{21/16} + 6840q^{25/16} + \dots$	0	✓
5A	5A	$5^6 1^{-6}$	-6	$6 + 12q^{1/5} + 52q^{2/5} + 192q^{3/5} + 372q^{4/5} + 844q + 1584q^{6/5} + 3216q^{7/5} + 6388q^{8/5} + \dots$	✓	
10A	5A	$1^6 10^6 2^{-6} 5^{-6}$	6	$6 + 12q^{1/5} + 52q^{2/5} + 192q^{3/5} + 372q^{4/5} + 844q + 1584q^{6/5} + 3216q^{7/5} + 6388q^{8/5} + \dots$	✓	
5B	5B	$1^4 5^4$	4	$4 + 16q^{1/5} + 64q^{2/5} + 160q^{3/5} + 384q^{4/5} + 816q + 1664q^{6/5} + 3296q^{7/5} + 6144q^{8/5} + 1 \dots$	✓	
10B	5B	$2^4 10^4 1^{-4} 5^{-4}$	-4	$4 + 16q^{1/5} + 64q^{2/5} + 160q^{3/5} + 384q^{4/5} + 816q + 1664q^{6/5} + 3296q^{7/5} + 6144q^{8/5} + \dots$	✓	
5C	5C	$5^5 1^{-1}$	-1	$3 + 20q^{1/5} + 60q^{2/5} + 160q^{3/5} + 380q^{4/5} + 820q + 1680q^{6/5} + 3280q^{7/5} + 6140q^{8/5} + \dots$	X	
10C	5C	$1^1 10^5 2^{-1} 5^{-5}$	1	$3 + 20q^{1/5} + 60q^{2/5} + 160q^{3/5} + 380q^{4/5} + 820q + 1680q^{6/5} + 3280q^{7/5} + 6140q^{8/5} + \dots$	X	
6E	6A	$3^4 6^4 1^{-4} 2^{-4}$	-4	$4 + 16q^{1/6} + 8q^{1/3} + 112q^{1/2} + 232q^{2/3} + 224q^{5/6} + 792q + 1408q^{7/6} + 1576q^{4/3} + \dots$	✓	
6F	6A	$1^4 6^8 2^{-8} 3^{-4}$	4	$4 + 16q^{1/6} + 8q^{1/3} + 112q^{1/2} + 232q^{2/3} + 224q^{5/6} + 792q + 1408q^{7/6} + 1576q^{4/3} + \dots$	✓	
12A	6B	$2^6 12^6 4^{-6} 6^{-6}$	0	$0 + 12q^{1/8} + 40q^{7/24} + 24q^{11/24} + 192q^{5/8} + 360q^{19/24} + 384q^{23/24} + 1236q^{9/8} + \dots$	0	✓
6G	6C	$2^5 3^4 6^1 1^{-4}$	-4	$4 + 8q^{1/6} + 40q^{1/3} + 88q^{1/2} + 168q^{2/3} + 368q^{5/6} + 696q + 1216q^{7/6} + 2216q^{4/3} + \dots$	✓	
6H	6C	$1^4 2^1 6^5 3^{-4}$	4	$4 + 8q^{1/6} + 40q^{1/3} + 88q^{1/2} + 168q^{2/3} + 368q^{5/6} + 696q + 1216q^{7/6} + 2216q^{4/3} + \dots$	✓	
6I	6D	$1^5 3^1 6^4 2^{-4}$	5	$5 + 8q^{1/6} + 32q^{1/3} + 88q^{1/2} + 192q^{2/3} + 368q^{5/6} + 672q + 1216q^{7/6} + 2176q^{4/3} + \dots$	✓	
6J	6D	$2^1 6^5 1^{-5} 3^{-1}$	-5	$5 + 8q^{1/6} + 32q^{1/3} + 88q^{1/2} + 192q^{2/3} + 368q^{5/6} + 672q + 1216q^{7/6} + 2176q^{4/3} + \dots$	✓	
6K	6E	$1^2 2^2 3^2 6^2$	2	$2 + 16q^{1/6} + 32q^{1/3} + 80q^{1/2} + 192q^{2/3} + 352q^{5/6} + 672q + 1280q^{7/6} + 2176q^{4/3} + \dots$	✓	
6L	6E	$2^4 6^4 1^{-2} 3^{-2}$	-2	$2 + 16q^{1/6} + 32q^{1/3} + 80q^{1/2} + 192q^{2/3} + 352q^{5/6} + 672q + 1280q^{7/6} + 2176q^{4/3} + \dots$	✓	
6M	6F	$3^3 6^3 1^{-1} 2^{-1}$	-1	$3 + 12q^{1/6} + 36q^{1/3} + 84q^{1/2} + 180q^{2/3} + 360q^{5/6} + 684q + 1248q^{7/6} + 2196q^{4/3} + \dots$	X	
6N	6F	$1^1 6^6 2^{-2} 3^{-3}$	1	$3 + 12q^{1/6} + 36q^{1/3} + 84q^{1/2} + 180q^{2/3} + 360q^{5/6} + 684q + 1248q^{7/6} + 2196q^{4/3} + \dots$	X	
6O	6G	$2^3 6^3$	0	$0 + 8q^{1/12} + 24q^{1/4} + 48q^{5/12} + 128q^{7/12} + 264q^{3/4} + 480q^{11/12} + 944q^{13/12} + 1680q^{5/4} + \dots$	0	✓
12B	6H	$12^4 6^{-4}$	0	$0 + 8q^{5/72} + 16q^{17/72} + 56q^{29/72} + 112q^{41/72} + 248q^{53/72} + 464q^{65/72} + 896q^{77/72} + \dots$	0	✓

Co_0	Co_1	Frame shape	$Z^{\mathrm{R}}_{\mathrm{R}g}$	$Z^{\mathrm{NS}}_{\mathrm{R}g}$	Consp?	Anom?
6P	6I	6^4	0	$0+4q^{1/36}+16q^{7/36}+40q^{13/36}+96q^{19/36}+204q^{25/36}+400q^{31/36}+760q^{37/36}+\dots$	0	✓
7A	7A	$7^4 1^{-4}$	-4	$4+8q^{1/7}+16q^{2/7}+56q^{3/7}+104q^{4/7}+200q^{5/7}+328q^{6/7}+592q+976q^{8/7}+1560q^{9/7}+\dots$	✓	
14A	7A	$1^4 14^4 2^{-4} 7^{-4}$	4	$4+8q^{1/7}+16q^{2/7}+56q^{3/7}+104q^{4/7}+200q^{5/7}+328q^{6/7}+592q+976q^{8/7}+1560q^{9/7}+\dots$	✓	
7B	7B	$1^3 7^3$	3	$3+8q^{1/7}+24q^{2/7}+48q^{3/7}+104q^{4/7}+192q^{5/7}+336q^{6/7}+584q+984q^{8/7}+1608q^{9/7}+\dots$	✓	
14B	7B	$2^3 14^3 1^{-3} 7^{-3}$	-3	$3+8q^{1/7}+24q^{2/7}+48q^{3/7}+104q^{4/7}+192q^{5/7}+336q^{6/7}+584q+984q^{8/7}+1608q^{9/7}+\dots$	✓	
8B	8A	$8^4 2^{-4}$	0	$0+8q^{1/16}+8q^{3/16}+16q^{5/16}+48q^{7/16}+104q^{9/16}+152q^{11/16}+208q^{13/16}+400q^{15/16}+\dots$	0	✓
8C	8B	$2^4 8^4 4^{-4}$	0	$0+4q^{1/16}+16q^{3/16}+24q^{5/16}+32q^{7/16}+84q^{9/16}+176q^{11/16}+248q^{13/16}+352q^{15/16}+\dots$	0	✓
8D	8C	$1^4 8^4 2^{-2} 4^{-2}$	4	$4+4q^{1/8}+16q^{1/4}+32q^{3/8}+64q^{1/2}+120q^{5/8}+192q^{3/4}+320q^{7/8}+512q+788q^{9/8}+\dots$	✓	
8E	8C	$2^2 8^4 1^{-4} 4^{-2}$	-4	$4+4q^{1/8}+16q^{1/4}+32q^{3/8}+64q^{1/2}+120q^{5/8}+192q^{3/4}+320q^{7/8}+512q+788q^{9/8}+\dots$	✓	
8F	8D	$8^4 4^{-2}$	0	$2+8q^{1/8}+16q^{1/4}+32q^{3/8}+64q^{1/2}+112q^{5/8}+192q^{3/4}+320q^{7/8}+512q+808q^{9/8}+\dots$	**X**	
8G	8E	$1^2 2^1 4^1 8^2$	2	$2+8q^{1/8}+16q^{1/4}+32q^{3/8}+64q^{1/2}+112q^{5/8}+192q^{3/4}+320q^{7/8}+512q+808q^{9/8}+\dots$	✓	
8H	8E	$2^3 4^1 8^2 1^{-2}$	-2	$2+8q^{1/8}+16q^{1/4}+32q^{3/8}+64q^{1/2}+112q^{5/8}+192q^{3/4}+320q^{7/8}+512q+808q^{9/8}+\dots$	✓	
8I	8F	$4^2 8^2$	0	$0+4q^{1/32}+8q^{5/32}+20q^{9/32}+40q^{13/32}+72q^{17/32}+128q^{21/32}+220q^{25/32}+360q^{29/32}+\dots$	0	✓
9A	9A	$9^3 1^{-3}$	-3	$3+6q^{1/9}+8q^{2/9}+24q^{1/3}+42q^{4/9}+80q^{5/9}+120q^{2/3}+192q^{7/9}+296q^{8/9}+456q+\dots$	✓	
18A	9A	$1^3 18^3 2^{-3} 9^{-3}$	3	$3+6q^{1/9}+8q^{2/9}+24q^{1/3}+42q^{4/9}+80q^{5/9}+120q^{2/3}+192q^{7/9}+296q^{8/9}+456q+\dots$	✓	
9B	9B	$9^3 3^{-1}$	0	$2+6q^{1/9}+12q^{2/9}+24q^{1/3}+42q^{4/9}+72q^{5/9}+120q^{2/3}+192q^{7/9}+300q^{8/9}+456q+\dots$	**X**	
18B	9B	$3^1 18^3 6^{-1} 9^{-3}$	0	$2+6q^{1/9}+12q^{2/9}+24q^{1/3}+42q^{4/9}+72q^{5/9}+120q^{2/3}+192q^{7/9}+300q^{8/9}+456q+\dots$	**X**	
9C	9C	$1^3 9^3 3^{-2}$	3	$3+4q^{1/9}+12q^{2/9}+24q^{1/3}+44q^{4/9}+72q^{5/9}+120q^{2/3}+192q^{7/9}+300q^{8/9}+456q+\dots$	✓	
18C	9C	$2^3 3^2 18^3 1^{-3} 6^{-2} 9^{-3}$	-3	$3+4q^{1/9}+12q^{2/9}+24q^{1/3}+44q^{4/9}+72q^{5/9}+120q^{2/3}+192q^{7/9}+300q^{8/9}+456q+\dots$	✓	
10D	10A	$5^2 10^2 1^{-2} 2^{-2}$	-2	$2+8q^{1/10}+4q^{1/5}+16q^{3/10}+28q^{2/5}+56q^{1/2}+96q^{3/5}+112q^{7/10}+188q^{4/5}+\dots$	✓	
10E	10A	$1^2 10^4 2^{-4} 5^{-2}$	2	$2+8q^{1/10}+4q^{1/5}+16q^{3/10}+28q^{2/5}+56q^{1/2}+96q^{3/5}+112q^{7/10}+188q^{4/5}+\dots$	✓	
20A	10B	$2^3 20^3 4^{-3} 10^{-3}$	0	$0+6q^{3/40}+8q^{7/40}+18q^{11/40}+24q^{3/8}+32q^{19/40}+84q^{23/40}+114q^{27/40}+180q^{31/40}+\dots$	0	✓
20B	10C	$4^2 20^2 2^{-2} 10^{-2}$	0	$0+4q^{1/40}+4q^{1/8}+12q^{9/40}+16q^{13/40}+40q^{17/40}+56q^{21/40}+92q^{5/8}+136q^{29/40}+\dots$	0	✓
10F	10D	$2^3 5^2 10^1 1^{-2}$	-2	$2+4q^{1/10}+12q^{1/5}+16q^{3/10}+28q^{2/5}+52q^{1/2}+80q^{3/5}+128q^{7/10}+188q^{4/5}+\dots$	✓	
10G	10D	$1^2 2^1 10^3 5^{-2}$	2	$2+4q^{1/10}+12q^{1/5}+16q^{3/10}+28q^{2/5}+52q^{1/2}+80q^{3/5}+128q^{7/10}+188q^{4/5}+\dots$	✓	
10H	10E	$1^3 5^1 10^2 2^{-2}$	3	$3+4q^{1/10}+8q^{1/5}+16q^{3/10}+32q^{2/5}+52q^{1/2}+80q^{3/5}+128q^{7/10}+192q^{4/5}+\dots$	✓	
10I	10E	$2^1 10^3 1^{-3} 5^{-1}$	-3	$3+4q^{1/10}+8q^{1/5}+16q^{3/10}+32q^{2/5}+52q^{1/2}+80q^{3/5}+128q^{7/10}+192q^{4/5}+\dots$	✓	
10J	10F	$2^2 10^2$	0	$0+4q^{1/20}+8q^{3/20}+12q^{1/4}+24q^{7/20}+36q^{9/20}+64q^{11/20}+104q^{13/20}+152q^{3/4}+\dots$	0	✓
11A	11A	$1^2 11^2$	2	$2+4q^{1/11}+8q^{2/11}+12q^{3/11}+24q^{4/11}+36q^{5/11}+56q^{6/11}+88q^{7/11}+128q^{8/11}+\dots$	✓	
22A	11A	$2^2 22^2 1^{-2} 11^{-2}$	0	$0+2q^{1/44}+2q^{3/44}+2q^{5/44}+4q^{7/44}+4q^{9/44}+6q^{1/4}+8q^{13/44}+10q^{15/44}+12q^{17/44}+\dots$	0	✓
12C	12A	$2^4 3^4 12^4 1^{-4} 4^{-4} 6^{-4}$	-4	$4+8q^{1/6}+8q^{1/4}+8q^{1/3}+48q^{5/12}+56q^{1/2}+16q^{7/12}+104q^{2/3}+176q^{3/4}+112q^{5/6}+\dots$	✓	
12D	12A	$1^4 12^4 3^{-4} 4^{-4}$	4	$4+8q^{1/6}+8q^{1/4}+8q^{1/3}+48q^{5/12}+56q^{1/2}+16q^{7/12}+104q^{2/3}+176q^{3/4}+112q^{5/6}+\dots$	✓	
12E	12B	$2^2 12^4 4^{-4} 6^{-2}$	0	$0+8q^{1/12}+4q^{1/6}+8q^{1/4}+28q^{1/3}+16q^{5/12}+28q^{1/2}+96q^{7/12}+76q^{2/3}+\dots$	0	✓
12F	12C	$6^2 12^2 2^{-2} 4^2$	0	$0+4q^{1/24}+4q^{1/8}+8q^{5/24}+16q^{7/24}+20q^{3/8}+32q^{11/24}+56q^{13/24}+72q^{5/8}+\dots$	0	✓
12G	12D	$2^1 3^3 12^3 1^{-1} 4^{-1} 6^{-3}$	-1	$1+6q^{1/12}+6q^{1/6}+6q^{1/4}+18q^{1/3}+36q^{5/12}+42q^{1/2}+48q^{7/12}+90q^{2/3}+150q^{3/4}+\dots$	✓	
12H	12D	$1^1 12^3 3^{-3} 4^{-1}$	1	$1+6q^{1/12}+6q^{1/6}+6q^{1/4}+18q^{1/3}+36q^{5/12}+42q^{1/2}+48q^{7/12}+90q^{2/3}+150q^{3/4}+\dots$	✓	
12I	12E	$1^2 3^2 4^2 12^2 2^{-2} 6^{-2}$	2	$2+4q^{1/12}+8q^{1/6}+4q^{1/4}+16q^{1/3}+40q^{5/12}+40q^{1/2}+48q^{7/12}+96q^{2/3}+148q^{3/4}+\dots$	✓	
12J	12E	$4^2 12^2 1^{-2} 3^{-2}$	-2	$2+4q^{1/12}+8q^{1/6}+4q^{1/4}+16q^{1/3}+40q^{5/12}+40q^{1/2}+48q^{7/12}+96q^{2/3}+148q^{3/4}+\dots$	✓	
24A	12F	$4^3 24^3 8^{-3} 12^{-3}$	0	$0+6q^{5/96}+2q^{13/96}+6q^{7/32}+24q^{29/96}+18q^{37/96}+24q^{15/32}+84q^{53/96}+56q^{61/96}+\dots$	0	✓
12K	12G	$4^2 12^2 2^{-1} 6^{-1}$	0	$0+2q^{1/144}+4q^{13/144}+6q^{25/144}+12q^{37/144}+18q^{49/144}+28q^{61/144}+44q^{73/144}+\dots$	0	✓
12I	12H	$1^1 2^2 3^1 12^2 4^{-2}$	1	$1+4q^{1/12}+8q^{1/6}+12q^{1/4}+16q^{1/3}+24q^{5/12}+40q^{1/2}+64q^{7/12}+96q^{2/3}+132q^{3/4}+\dots$	✓	

Co_0	Co_1	Frame shape	$Z^{\mathrm{R}}_{\mathrm{R}g}$	$Z^{\mathrm{NS}}_{\mathrm{R}g}$	Consp?	Anom?
12M	12H	$2^3 6^1 12^2 1^{-1} 3^{-1} 4^{-2}$	-1	$1 + 4q^{1/12} + 8q^{1/6} + 12q^{1/4} + 16q^{1/3} + 24q^{5/12} + 40q^{1/2} + 64q^{7/12} + 96q^{2/3} + 132q^{3/4} + \dots$	✓	
12N	12I	$2^2 3^2 4^1 12^1 1^{-2}$	-2	$2 + 4q^{1/12} + 4q^{1/6} + 12q^{1/4} + 20q^{1/3} + 24q^{5/12} + 44q^{1/2} + 64q^{7/12} + 84q^{2/3} + 132q^{3/4} + \dots$	✓	
12O	12I	$1^2 4^1 6^2 12^1 3^{-2}$	2	$2 + 4q^{1/12} + 4q^{1/6} + 12q^{1/4} + 20q^{1/3} + 24q^{5/12} + 44q^{1/2} + 64q^{7/12} + 84q^{2/3} + 132q^{3/4} + \dots$	✓	
12P	12J	$2^1 4^1 6^1 12^1$	0	$0 + 4q^{1/24} + 4q^{1/8} + 8q^{5/24} + 16q^{7/24} + 20q^{3/8} + 32q^{11/24} + 56q^{13/24} + 72q^{5/8} + \dots$	0	✓
12Q	12K	$1^3 12^3 2^{-1} 3^{-1} 4^{-1} 6^{-1}$	3	$3 + 2q^{1/12} + 6q^{1/6} + 10q^{1/4} + 18q^{1/3} + 28q^{5/12} + 42q^{1/2} + 64q^{7/12} + 90q^{2/3} + 130q^{3/4} + \dots$	✓	
12R	12K	$2^2 3^1 12^3 1^{-3} 4^{-1} 6^{-2}$	-3	$3 + 2q^{1/12} + 6q^{1/6} + 10q^{1/4} + 18q^{1/3} + 28q^{5/12} + 42q^{1/2} + 64q^{7/12} + 90q^{2/3} + 130q^{3/4} + \dots$	✓	
24B	12L	$24^2 12^{-2}$	0	$0 + 4q^{11/288} + 4q^{35/288} + 8q^{59/288} + 12q^{83/288} + 24q^{107/288} + 32q^{131/288} + 52q^{155/288} + \dots$	0	✓
12S	12M	12^2	0	$0 + 2q^{1/144} + 4q^{13/144} + 6q^{25/144} + 12q^{37/144} + 18q^{49/144} + 28q^{61/144} + 44q^{73/144} + \dots$	0	✓
13A	13A	$13^2 1^{-2}$	-2	$2 + 4q^{1/13} + 4q^{2/13} + 8q^{3/13} + 12q^{4/13} + 24q^{5/13} + 32q^{6/13} + 48q^{7/13} + 68q^{8/13} + \dots$	✓	
26A	13A	$1^2 26^2 2^{-2} 13^{-2}$	2	$2 + 4q^{1/13} + 4q^{2/13} + 8q^{3/13} + 12q^{4/13} + 24q^{5/13} + 32q^{6/13} + 48q^{7/13} + 68q^{8/13} + \dots$	✓	
28A	14A	$2^2 28^2 4^{-2} 14^{-2}$	0	$0 + 4q^{3/56} + 4q^{1/8} + 8q^{11/56} + 8q^{15/56} + 20q^{19/56} + 16q^{23/56} + 32q^{27/56} + 48q^{31/56} + \dots$	0	✓
14C	14B	$1^1 2^1 7^1 14^1$	1	$1 + 4q^{1/14} + 4q^{1/7} + 8q^{3/14} + 12q^{2/7} + 16q^{5/14} + 24q^{3/7} + 36q^{1/2} + 52q^{4/7} + 68q^{9/14} + \dots$	✓	
14D	14B	$2^2 14^2 1^{-1} 7^{-1}$	-1	$1 + 4q^{1/14} + 4q^{1/7} + 8q^{3/14} + 12q^{2/7} + 16q^{5/14} + 24q^{3/7} + 36q^{1/2} + 52q^{4/7} + 68q^{9/14} + \dots$	✓	
15A	15A	$1^3 15^3 3^{-3} 5^{-3}$	3	$3 + 2q^{1/15} + 6q^{1/5} + 18q^{4/15} + 6q^{1/3} + 24q^{2/5} + 36q^{7/15} + 12q^{8/15} + 78q^{3/5} + 92q^{2/3} + \dots$	✓	
30A	15A	$2^3 3^3 5^3 30^3 1^{-3} 6^{-3} 10^{-3} 15^{-3}$	-3	$3 + 2q^{1/15} + 6q^{1/5} + 18q^{4/15} + 6q^{1/3} + 24q^{2/5} + 36q^{7/15} + 12q^{8/15} + 78q^{3/5} + 92q^{2/3} + \dots$	✓	
15B	15B	$3^2 15^2 1^{-2} 5^{-2}$	-2	$2 + 4q^{1/15} + 8q^{1/5} + 12q^{4/15} + 4q^{1/3} + 28q^{2/5} + 40q^{7/15} + 16q^{8/15} + 68q^{3/5} + 88q^{2/3} + \dots$	✓	
30B	15B	$1^2 5^2 6^2 30^2 2^{-2} 3^{-2} 10^{-2} 15^{-2}$	2	$2 + 4q^{1/15} + 8q^{1/5} + 12q^{4/15} + 4q^{1/3} + 28q^{2/5} + 40q^{7/15} + 16q^{8/15} + 68q^{3/5} + 88q^{2/3} + \dots$	✓	
15C	15C	$15^2 3^{-2}$	0	$0 + 4q^{2/45} + 4q^{1/9} + 4q^{8/45} + 8q^{11/45} + 8q^{14/45} + 24q^{17/45} + 28q^{4/9} + 32q^{23/45} + \dots$	0	✓
30C	15C	$3^2 30^2 6^{-2} 15^{-2}$	0	$0 + 4q^{2/45} + 4q^{1/9} + 4q^{8/45} + 8q^{11/45} + 8q^{14/45} + 24q^{17/45} + 28q^{4/9} + 32q^{23/45} + \dots$	0	✓
15D	15D	$1^1 3^1 5^1 15^1$	1	$1 + 4q^{1/15} + 4q^{2/15} + 4q^{1/5} + 12q^{4/15} + 12q^{1/3} + 20q^{2/5} + 32q^{7/15} + 36q^{8/15} + 52q^{3/5} + \dots$	✓	
30D	15D	$2^1 6^1 10^1 30^1 1^{-1} 3^{-1} 5^{-1} 15^{-1}$	-1	$1 + 4q^{1/15} + 4q^{2/15} + 4q^{1/5} + 12q^{4/15} + 12q^{1/3} + 20q^{2/5} + 32q^{7/15} + 36q^{8/15} + 52q^{3/5} + \dots$	✓	
15E	15E	$1^2 15^2 3^{-1} 5^{-1}$	2	$2 + 2q^{1/15} + 4q^{2/15} + 6q^{1/5} + 10q^{4/15} + 14q^{1/3} + 20q^{2/5} + 28q^{7/15} + 40q^{8/15} + 54q^{3/5} + \dots$	✓	
30E	15E	$2^2 3^1 5^1 30^2 1^{-2} 6^{-1} 10^{-1} 15^{-2}$	-2	$2 + 2q^{1/15} + 4q^{2/15} + 6q^{1/5} + 10q^{4/15} + 14q^{1/3} + 20q^{2/5} + 28q^{7/15} + 40q^{8/15} + 54q^{3/5} + \dots$	✓	
16A	16A	$2^2 16^2 4^{-1} 8^{-1}$	0	$0 + 2q^{1/32} + 4q^{3/32} + 4q^{5/32} + 8q^{7/32} + 10q^{9/32} + 12q^{11/32} + 20q^{13/32} + 24q^{15/32} + \dots$	0	✓
16B	16B	$1^2 16^2 2^{-1} 8^{-1}$	2	$2 + 2q^{1/16} + 4q^{1/8} + 4q^{3/16} + 8q^{1/4} + 12q^{5/16} + 16q^{3/8} + 24q^{7/16} + 32q^{1/2} + 42q^{9/16} + \dots$	✓	
16C	16B	$2^1 16^2 1^{-2} 8^{-1}$	-2	$2 + 2q^{1/16} + 4q^{1/8} + 4q^{3/16} + 8q^{1/4} + 12q^{5/16} + 16q^{3/8} + 24q^{7/16} + 32q^{1/2} + 42q^{9/16} + \dots$	✓	
18D	18A	$9^1 18^1 1^{-1} 2^{-1}$	-1	$1 + 4q^{1/18} + 2q^{1/9} + 4q^{1/6} + 4q^{2/9} + 8q^{5/18} + 12q^{1/3} + 16q^{7/18} + 22q^{4/9} + 28q^{1/2} + \dots$	✓	
18E	18A	$1^1 18^2 2^{-2} 9^{-1}$	1	$1 + 4q^{1/18} + 2q^{1/9} + 4q^{1/6} + 4q^{2/9} + 8q^{5/18} + 12q^{1/3} + 16q^{7/18} + 22q^{4/9} + 28q^{1/2} + \dots$	✓	
18F	18B	$1^2 9^1 18^1 2^{-1} 3^{-1}$	2	$2 + 2q^{1/18} + 2q^{1/9} + 4q^{1/6} + 6q^{2/9} + 8q^{5/18} + 12q^{1/3} + 16q^{7/18} + 22q^{4/9} + 28q^{1/2} + \dots$	✓	
18G	18B	$2^1 3^1 18^2 1^{-2} 6^{-1} 9^{-1}$	-2	$2 + 2q^{1/18} + 2q^{1/9} + 4q^{1/6} + 6q^{2/9} + 8q^{5/18} + 12q^{1/3} + 16q^{7/18} + 22q^{4/9} + 28q^{1/2} + \dots$	✓	
18H	18C	$2^2 9^1 18^1 1^{-1} 6^{-1}$	-1	$1 + 2q^{1/18} + 4q^{1/9} + 4q^{1/6} + 6q^{2/9} + 8q^{5/18} + 12q^{1/3} + 16q^{7/18} + 20q^{4/9} + 28q^{1/2} + \dots$	✓	
18I	18C	$1^1 2^1 18^2 6^{-1} 9^{-1}$	1	$1 + 2q^{1/18} + 4q^{1/9} + 4q^{1/6} + 6q^{2/9} + 8q^{5/18} + 12q^{1/3} + 16q^{7/18} + 20q^{4/9} + 28q^{1/2} + \dots$	✓	
20C	20A	$2^2 5^2 20^2 1^{-2} 4^{-2} 10^{-2}$	-2	$2 + 4q^{1/10} + 4q^{3/20} + 4q^{1/5} + 4q^{1/4} + 8q^{3/10} + 20q^{7/20} + 12q^{2/5} + 8q^{9/20} + 28q^{1/2} + \dots$	✓	
20D	20A	$1^2 20^2 4^{-2} 5^{-2}$	2	$2 + 4q^{1/10} + 4q^{3/20} + 4q^{1/5} + 4q^{1/4} + 8q^{3/10} + 20q^{7/20} + 12q^{2/5} + 8q^{9/20} + 28q^{1/2} + \dots$	✓	
20E	20B	$4^1 20^1$	0	$0 + 2q^{1/80} + 2q^{1/16} + 2q^{9/80} + 4q^{13/80} + 4q^{17/80} + 8q^{21/80} + 10q^{5/16} + 12q^{29/80} + \dots$	0	✓
20F	20C	$2^2 5^1 20^1 1^{-1} 4^{-1}$	-1	$1 + 2q^{1/20} + 2q^{1/10} + 4q^{3/20} + 6q^{1/5} + 6q^{1/4} + 8q^{3/10} + 12q^{7/20} + 14q^{2/5} + 18q^{9/20} + \dots$	✓	
20G	20C	$1^1 2^1 10^1 20^1 4^{-1} 5^{-1}$	1	$1 + 2q^{1/20} + 2q^{1/10} + 4q^{3/20} + 6q^{1/5} + 6q^{1/4} + 8q^{3/10} + 12q^{7/20} + 14q^{2/5} + 18q^{9/20} + \dots$	✓	
21A	21A	$1^2 21^2 3^{-2} 7^{-2}$	2	$2 + 4q^{2/21} + 4q^{1/7} + 8q^{5/21} + 8q^{2/7} + 4q^{1/3} + 20q^{8/21} + 24q^{3/7} + 8q^{10/21} + 32q^{11/21} + \dots$	✓	
42A	21A	$2^2 3^2 7^2 42^2 1^{-2} 6^{-2} 14^{-2} 21^{-2}$	-2	$2 + 4q^{2/21} + 4q^{1/7} + 8q^{5/21} + 8q^{2/7} + 4q^{1/3} + 20q^{8/21} + 24q^{3/7} + 8q^{10/21} + 32q^{11/21} + \dots$	✓	
21B	21B	$7^1 21^1 1^{-1} 3^{-1}$	-1	$1 + 2q^{1/21} + 4q^{2/21} + 2q^{1/7} + 2q^{4/21} + 8q^{5/21} + 4q^{2/7} + 10q^{1/3} + 16q^{8/21} + 18q^{3/7} + \dots$	✓	
42B	21B	$1^1 3^1 14^1 42^1 2^{-1} 6^{-1} 7^{-1} 21^{-1}$	1	$1 + 2q^{1/21} + 4q^{2/21} + 2q^{1/7} + 2q^{4/21} + 8q^{5/21} + 4q^{2/7} + 10q^{1/3} + 16q^{8/21} + 18q^{3/7} + \dots$	✓	

Co_0	Co_1	Frame shape	Z^{R}_{Rg}	Z^{NS}_{Rg}	Consp?	Anom?
21C	21C	$3^1 21^1$	0	$0 + 2q^{1/63} + 2q^{4/63} + 2q^{1/9} + 4q^{10/63} + 4q^{13/63} + 6q^{16/63} + 8q^{19/63} + 12q^{22/63} + + \dots$	0	✓
42C	21C	$6^1 42^1 3^{-1} 21^{-1}$	0	$0 + 2q^{1/63} + 2q^{4/63} + 2q^{1/9} + 4q^{10/63} + 4q^{13/63} + 6q^{16/63} + 8q^{19/63} + 12q^{22/63} + + \dots$	0	✓
22BC	22A	$2^1 22^1$	0	$0 + 2q^{1/44} + 2q^{3/44} + 2q^{5/44} + 4q^{7/44} + 4q^{9/44} + 6q^{1/4} + 8q^{13/44} + 10q^{15/44} + 12q^{17/44} + \dots$	0	✓
23AB	23AB	$1^1 23^1$	1	$1 + 2q^{1/23} + 2q^{2/23} + 2q^{3/23} + 4q^{4/23} + 4q^{5/23} + 6q^{6/23} + 8q^{7/23} + 10q^{8/23} + 12q^{9/23} + \dots$	✓	
46AB	23AB	$2^1 46^1 1^{-1} 23^{-1}$	-1	$1 + 2q^{1/23} + 2q^{2/23} + 2q^{3/23} + 4q^{4/23} + 4q^{5/23} + 6q^{6/23} + 8q^{7/23} + 10q^{8/23} + 12q^{9/23} + \dots$	✓	
24C	24A	$2^2 24^2 6^{-2} 8^{-2}$	0	$0 + 4q^{1/16} + 4q^{5/48} + 4q^{3/16} + 8q^{11/48} + 8q^{5/16} + 16q^{17/48} + 8q^{19/48} + 20q^{7/16} + \dots$	0	✓
24D	24B	$2^1 3^2 4^1 24^2 1^{-2} 6^{-1} 8^{-2} 12^{-1}$	-2	$2 + 2q^{1/24} + 2q^{1/8} + 4q^{1/6} + 8q^{1/4} + 12q^{7/24} + 4q^{1/3} + 14q^{3/8} + 16q^{5/12} + 4q^{11/24} + \dots$	✓	
24E	24B	$1^2 4^1 6^1 24^2 2^{-1} 3^{-2} 8^{-2} 12^{-1}$	2	$2 + 2q^{1/24} + 2q^{1/8} + 4q^{1/6} + 8q^{1/4} + 12q^{7/24} + 4q^{1/3} + 14q^{3/8} + 16q^{5/12} + 4q^{11/24} + \dots$	✓	
24F	24C	$8^1 24^1 2^{-1} 6^{-1}$	0	$0 + 2q^{1/48} + 2q^{1/16} + 4q^{5/48} + 2q^{3/16} + 8q^{11/48} + 4q^{13/48} + 4q^{5/16} + 12q^{17/48} + \dots$	0	✓
14G	24D	$12^1 24^1 4^{-1} 8^{-1}$	0	$0 + 2q^{1/32} + 4q^{7/96} + 4q^{5/32} + 4q^{19/96} + 10q^{9/32} + 12q^{31/96} + 4q^{35/96} + 16q^{13/32} + \dots$	0	✓
24H	24E	$2^1 6^1 8^1 24^1 4^{-1} 12^{-1}$	0	$0 + 2q^{1/48} + 2q^{1/16} + 4q^{7/48} + 6q^{3/16} + 4q^{11/48} + 4q^{13/48} + 8q^{5/16} + 12q^{17/48} + \dots$	0	✓
24I	24F	$2^1 3^1 4^1 24^1 1^{-1} 8^{-1}$	-1	$1 + 2q^{1/24} + 2q^{1/12} + 2q^{1/8} + 2q^{1/6} + 4q^{5/24} + 6q^{1/4} + 8q^{7/24} + 10q^{1/3} + 10q^{3/8} + \dots$	✓	
24J	24F	$1^1 4^1 6^1 24^1 3^{-1} 8^{-1}$	1	$1 + 2q^{1/24} + 2q^{1/12} + 2q^{1/8} + 2q^{1/6} + 4q^{5/24} + 6q^{1/4} + 8q^{7/24} + 10q^{1/3} + 10q^{3/8} + \dots$	✓	
52A	26A	$2^1 52^1 4^{-1} 26^{-1}$	0	$0 + 2q^{3/104} + 2q^{7/104} + 2q^{11/104} + 2q^{15/104} + 4q^{19/104} + 4q^{23/104} + 4q^{27/104} + 8q^{31/104} + \dots$	0	✓
28B	28A	$1^1 4^1 7^1 28^1 2^{-1} 14^{-1}$	1	$1 + 2q^{1/28} + 2q^{1/14} + 2q^{1/7} + 4q^{5/28} + 4q^{3/14} + 2q^{1/4} + 6q^{2/7} + 10q^{9/28} + 8q^{5/14} + \dots$	✓	
28C	28A	$4^1 28^1 1^{-1} 7^{-1}$	-1	$1 + 2q^{1/28} + 2q^{1/14} + 2q^{1/7} + 4q^{5/28} + 4q^{3/14} + 2q^{1/4} + 6q^{2/7} + 10q^{9/28} + 8q^{5/14} + \dots$	✓	
56AB	28B	$4^1 56^1 8^{-1} 28^{-1}$	0	$0 + 2q^{1/224} + 2q^{17/224} + 2q^{25/224} + 2q^{33/224} + 2q^{41/224} + 4q^{7/32} + 6q^{57/224} + 6q^{65/224} + \dots$	0	✓
30F	30A	$1^1 2^1 15^1 30^1 3^{-1} 5^{-1} 6^{-1} 10^{-1}$	1	$1 + 2q^{1/15} + 4q^{1/10} + 4q^{1/6} + 2q^{1/5} + 10q^{4/15} + 8q^{3/10} + 2q^{1/3} + 8q^{11/30} + 12q^{2/5} + \dots$	✓	
30G	30A	$2^2 3^1 5^1 30^2 1^{-1} 6^{-2} 10^{-2} 15^{-1}$	-1	$1 + 2q^{1/15} + 4q^{1/10} + 4q^{1/6} + 2q^{1/5} + 10q^{4/15} + 8q^{3/10} + 2q^{1/3} + 8q^{11/30} + 12q^{2/5} + \dots$	✓	
60A	30B	$2^1 10^1 12^1 60^1 4^{-1} 6^{-1} 20^{-1} 30^{-1}$	0	$0 + 2q^{1/40} + 4q^{11/120} + 2q^{1/8} + 4q^{23/120} + 6q^{9/40} + 8q^{7/24} + 8q^{13/40} + 4q^{43/120} + \dots$	0	✓
60B	30C	$6^1 60^1 12^{-1} 30^{-1}$	0	$0 + 2q^{1/360} + 2q^{5/72} + 2q^{49/360} + 4q^{61/360} + 4q^{73/360} + 4q^{17/72} + 4q^{97/360} + 4q^{109/360} + \dots$	0	✓
30H	30D	$1^1 6^1 10^1 15^1 3^{-1} 5^{-1}$	1	$1 + 2q^{1/30} + 2q^{1/10} + 2q^{2/15} + 2q^{1/6} + 4q^{1/5} + 4q^{7/30} + 4q^{4/15} + 6q^{3/10} + 8q^{1/3} + \dots$	✓	
30I	30D	$2^1 3^1 5^1 30^1 1^{-1} 15^{-1}$	-1	$1 + 2q^{1/30} + 2q^{1/10} + 2q^{2/15} + 2q^{1/6} + 4q^{1/5} + 4q^{7/30} + 4q^{4/15} + 6q^{3/10} + 8q^{1/3} + \dots$	✓	
30J	30E	$2^1 3^1 5^1 30^1 6^{-1} 10^{-1}$	0	$0 + 2q^{1/30} + 2q^{1/15} + 2q^{1/10} + 2q^{2/15} + 2q^{1/6} + 2q^{1/5} + 4q^{7/30} + 6q^{4/15} + 6q^{3/10} + \dots$	0	✓
30K	30E	$2^1 30^1 3^{-1} 5^{-1}$	0	$0 + 2q^{1/30} + 2q^{1/15} + 2q^{1/10} + 2q^{2/15} + 2q^{1/6} + 2q^{1/5} + 4q^{7/30} + 6q^{4/15} + 6q^{3/10} + \dots$	0	✓
33A	33A	$3^1 33^1 1^{-1} 11^{-1}$	-1	$1 + 2q^{1/33} + 2q^{1/11} + 2q^{4/33} + 4q^{2/11} + 4q^{7/33} + 6q^{3/11} + 8q^{10/33} + 2q^{1/3} + 10q^{4/11} + \dots$	✓	
66A	33A	$1^1 6^1 11^1 66^1 2^{-1} 3^{-1} 22^{-1} 33^{-1}$	1	$1 + 2q^{1/33} + 2q^{1/11} + 2q^{4/33} + 4q^{2/11} + 4q^{7/33} + 6q^{3/11} + 8q^{10/33} + 2q^{1/3} + 10q^{4/11} \dots$	✓	
35A	35A	$1^1 35^1 5^{-1} 7^{-1}$	1	$1 + 2q^{1/35} + 2q^{4/35} + 2q^{1/7} + 4q^{6/35} + 2q^{1/5} + 4q^{8/35} + 2q^{9/35} + 4q^{2/7} + 8q^{11/35} + \dots$	✓	
70A	35A	$2^1 5^1 7^1 70^1 1^{-1} 10^{-1} 14^{-1} 35^{-1}$	-1	$1 + 2q^{1/35} + 2q^{4/35} + 2q^{1/7} + 4q^{6/35} + 2q^{1/5} + 4q^{8/35} + 2q^{9/35} + 4q^{2/7} + 8q^{11/35} + \dots$	✓	
36A	36A	$2^1 9^1 36^1 1^{-1} 4^{-1} 18^{-1}$	-1	$1 + 2q^{1/18} + 2q^{1/12} + 2q^{1/9} + 2q^{1/6} + 4q^{7/36} + 2q^{2/9} + 2q^{1/4} + 4q^{5/18} + 8q^{11/36} + \dots$	✓	
36B	36A	$1^1 36^1 4^{-1} 9^{-1}$	1	$1 + 2q^{1/18} + 2q^{1/12} + 2q^{1/9} + 2q^{1/6} + 4q^{7/36} + 2q^{2/9} + 2q^{1/4} + 4q^{5/18} + 8q^{11/36} + \dots$	✓	
39AB	39AB	$1^1 39^1 3^{-1} 13^{-1}$	1	$1 + 2q^{2/39} + 2q^{1/13} + 2q^{5/39} + 2q^{2/13} + 4q^{8/39} + 4q^{3/13} + 6q^{11/39} + 6q^{4/13} + 2q^{1/3} + \dots$	✓	
78AB	39AB	$2^1 3^1 13^1 78^1 1^{-1} 6^{-1} 26^{-1} 39^{-1}$	-1	$1 + 2q^{2/39} + 2q^{1/13} + 2q^{5/39} + 2q^{2/13} + 4q^{8/39} + 4q^{3/13} + 6q^{11/39} + 6q^{4/13} + 2q^{1/3} + \dots$	✓	
40AB	40A	$2^1 40^1 8^{-1} 10^{-1}$	0	$0 + 2q^{3/80} + 2q^{1/16} + 2q^{7/80} + 4q^{13/80} + 2q^{3/16} + 4q^{19/80} + 4q^{21/80} + 6q^{23/80} + 4q^{5/16} + \dots$	0	✓
84A	42A	$4^1 6^1 14^1 84^1 2^{-1} 12^{-1} 28^{-1} 42^{-1}$	0	$0 + 2q^{1/168} + 2q^{3/56} + 2q^{1/8} + 2q^{25/168} + 4q^{11/56} + 4q^{37/168} + 4q^{15/56} + 6q^{7/24} + \dots$	0	✓
60C	60A	$1^1 4^1 6^1 10^1 15^1 60^1 2^{-1}$ $3^{-1} 5^{-1} 12^{-1} 20^{-1} 30^{-1}$	1	$1 + 2q^{1/60} + 2q^{1/10} + 2q^{3/20} + 2q^{1/6} + 2q^{1/5} + 4q^{13/60} + 2q^{1/4} + 4q^{4/15} + 4q^{3/10} + \dots$	✓	
60D	60A	$3^1 4^1 5^1 60^1 1^{-1} 12^{-1} 15^{-1} 20^{-1}$	-1	$1 + 2q^{1/60} + 2q^{1/10} + 2q^{3/20} + 2q^{1/6} + 2q^{1/5} + 4q^{13/60} + 2q^{1/4} + 4q^{4/15} + 4q^{3/10} + \dots$	✓	

References

[1] Scott Carnahan and Masahiko Miyamoto. *Regularity of fixed-point vertex operator subalgebras*, 2016.

[2] Diego Delmastro, Davide Gaiotto, and Jaume Gomis, *Global anomalies on the Hilbert space*, J. High Energy Phys. **11** (2021), Paper No. 142, 67, DOI 10.1007/jhep11(2021)142. MR4368302

[3] Philippe Di Francesco, Pierre Mathieu, and David Sénéchal, *Conformal field theory*, Graduate Texts in Contemporary Physics, Springer-Verlag, New York, 1997, DOI 10.1007/978-1-4612-2256-9. MR1424041

[4] John F. Duncan, *Super-moonshine for Conway's largest sporadic group*, Duke Math. J. **139** (2007), no. 2, 255–315, DOI 10.1215/S0012-7094-07-13922-X. MR2352133

[5] John Francis Robert Duncan, *Vertex operators, and three sporadic groups*, ProQuest LLC, Ann Arbor, MI, 2006. Thesis (Ph.D.)–Yale University. MR2708637

[6] J. S. Frame, *The characters of the Weyl group E_8*, Computational Problems in Abstract Algebra (Proc. Conf., Oxford, 1967), Pergamon, Oxford-New York-Toronto, Ont., 1970, pp. 111–130. MR269751

[7] Yi-Zhi Huang, *Generalized twisted modules associated to general automorphisms of a vertex operator algebra*, Comm. Math. Phys. **298** (2010), no. 1, 265–292, DOI 10.1007/s00220-010-0999-6. MR2657819

[8] Theo Johnson-Freyd, *Supersymmetry and the Suzuki chain*, Tunis. J. Math. **3** (2021), no. 2, 309–359, DOI 10.2140/tunis.2021.3.309. MR4190470

[9] Theo Johnson-Freyd and David Treumann, $\mathrm{H}^4(\mathrm{Co}_0;\mathbf{Z}) = \mathbf{Z}/24$, Int. Math. Res. Not. IMRN **21** (2020), 7873–7907.

[10] Victor Kac, *Vertex algebras for beginners*, 2nd ed., University Lecture Series, vol. 10, American Mathematical Society, Providence, RI, 1998, DOI 10.1090/ulect/010. MR1651389

[11] Takeshi Kondo, *The automorphism group of Leech lattice and elliptic modular functions*, J. Math. Soc. Japan **37** (1985), no. 2, 337–362, DOI 10.2969/jmsj/03720337. MR780666

[12] James Lepowsky and Haisheng Li, *Introduction to vertex operator algebras and their representations*, Progress in Mathematics, vol. 227, Birkhäuser Boston, Inc., Boston, MA, 2004, DOI 10.1007/978-0-8176-8186-9. MR2023933

[13] Edward Witten, *Supersymmetry and Morse theory*, J. Differential Geometry **17** (1982), no. 4, 661–692 (1983). MR683171

[14] Yongchang Zhu, *Modular invariance of characters of vertex operator algebras*, J. Amer. Math. Soc. **9** (1996), no. 1, 237–302, DOI 10.1090/S0894-0347-96-00182-8. MR1317233

Department of Mathematics and Statistics, Dalhousie University, Halifax, Nova Scotia, Canada

Email address: `alissa.furet@mail.mcgill.ca`

Perimeter Institute for Theoretical Physics, Waterloo, Ontario, Canada

Email address: `theojf@pitp.ca`

Contemporary Mathematics
Volume **813**, 2025
https://doi.org/10.1090/conm/813/16285

Combinatorial principles that enforce a group structure

Simon Lentner

ABSTRACT. We solve a system of equations on an abelian group G, using techniques from additive combinatorics. In essence, certain positivity assumptions lead somewhat unexpected to an additive group law. We obtain some related results that exhibit a similar phenomenon. The initial motivation for this work is to classify R-matrix for quantum groups.

1. Introduction

As we will discuss at the end of this introduction, the following elementary problem arises in an effort to classify R-matrices for quantum groups: Given a finite abelian group G (written additively) we would like to determine all families $(g(x,y))_{x,y\in G}$ of complex numbers not vanishing simultaneously such that for all $x,y\in G$ we have

$$g(x,y)=\sum_{y_1+y_2=y} g(x,y_1)g(x,y_2)=\sum_{x_1+x_2=x} g(x_1,y)g(x_2,y). \tag{1.1}$$

Let us call such a family $(g(x,y))_{x,y\in G}$ a *solution* for G in the sequel. As it turns out, this problem has an interesting structure in additive combinatorics. As final result we obtain

THEOREM 1.1. *Given a finite abelian group G a family $(g(x,y))_{x,y\in G}$ of complex numbers is a solution for G if and only if there are subgroups $H_1,H_2\leqslant G$ of the same order $d=|H_1|=|H_2|$ and a pairing $\omega\colon H_1\times H_2\longrightarrow\mathbb{C}^\times$ such that*

$$g(x,y)=\frac{1}{d}H_1(x)H_2(y)\omega(x,y) \tag{1.2}$$

holds for all $x,y\in G$.

Here ω being a pairing means the same as saying that ω be a homomorphism from the direct product $H_1\times H_2$ into the multiplicative group $\mathbb{C}^\times$, and in formula (1.2) we identified the sets H_1, H_2 with their characteristic functions. In particular this implies as expected

---

2020 *Mathematics Subject Classification.* Primary 05E10.

The author was partly supported by the DFG Priority Program 1388 "representation theory" and is currently supported by the Feyodor-Lyenen fellowship of the Humboldt foundation.

COROLLARY 1.2. *Given a finite abelian group G a family $(g(x,y))_{x,y\in G}$ of complex numbers is a solution of the equations*

$$\begin{aligned} g(x_1,y)\delta_{x_1,x_2} &= \sum_{y_1+y_2=y} g(x_1,y_1)g(x_2,y_2) \\ g(x,y_1)\delta_{y_1,y_2} &= \sum_{x_1+x_2=x} g(x_1,y_1)g(x_2,y_2) \\ \delta_{x,0} &= \sum_y g(x,y) \\ \delta_{y,0} &= \sum_x g(x,y) \end{aligned}$$

if and only if there are subgroups $H_1, H_2 \leqslant G$ and a nodegenerate pairing $\omega\colon H_1 \times H_2 \longrightarrow \mathbb{C}^\times$, which in particular implies $H_1 \cong H_2$, such that

$$g(x,y) = \frac{1}{d} H_1(x) H_2(y) \omega(x,y) \tag{1.3}$$

holds for all $x,y \in G$ with $d = |H_1| = |H_2|$.

The proof of Theorem 1.1 relies on methods from additive combinatorics, in particular the Fourier method [**8**]. The additive closure arises, very roughly, from the attaining of bounds in the triangle inequality. It seems interesting that an additive structure can arise in such a combinatorial way, solely from positivity. In our inital proof we have encountered some related results of a similar nature, which we record here, for completeness and because they may be interesting on their own:

THEOREM 1.3. *Let N be finite set, $\mathcal{P}(N)$ denote the power set of N and $\mathcal{E} \subset \mathcal{P}(N)$. Let $\mu\colon \mathcal{P}(N) \to [0,\infty]$ be a measure on $\mathcal{P}(N)$. Then the following is equivalent:*

(i) $a_C := (-1)^{|N|} \sum_{D\in\mathcal{E}} (-1)^{|C\cup D|} e^{\mu(C-D)} \geqslant 0$ *for all* $C \subset N$.
(ii) $\mathcal{E}$ *is* $\{D \subset N \mid D \supset A\}$ *for some* $A \subset N$.

Such a set $\mathcal{E}$ is called filter or ideal in N. This result and a generalized version with multiplicities gives an alternative way to prove our main result in the cyclic case $G = \mathbb{Z}_n$, where N is the set of prime factors of n

THEOREM 1.4. *A non-empty finite set U of complex roots of unity is a group if and only if for every nonnegative integer k the power sum $a_k = \sum_{\zeta\in U} \zeta^k$ is a nonnegative real number.*

We were made aware by a referee that this has been proven in [**7**]. Our proof relies on the polynomial $\sum_k a_k X^k$ and is quite different.

THEOREM 1.5. *A nonzero element ε/N of the group algebra $\mathbb{C}[\mathbb{Z}_N \times \mathbb{Z}_N]$, where ε has in the canonical basis $g^i \otimes g^j$ the representing matrix $\varepsilon_{ij} \in \{0,1\}$, is an idempotent if and only if there exists a subgroup $G \leqslant \mathbb{Z}_N \times \mathbb{Z}_N$ of order N such that ε is the characteristic function of G in the sense that*

$$\varepsilon_{ij} = \begin{cases} 1 & \text{if } (i,j) \in G \\ 0 & \text{else.} \end{cases} \tag{1.4}$$

This result can easily be rendered more explicit by classifying the possible groups G (see Proposition 5.1).

We now discuss the motivation behind our main result: Hopf algebras are algebras with additional structure, such that the category of representations admits a tensor product. A particular class of examples of Hopf algebras are the quantum groups $U_q(\mathfrak{g})$, which are deformations of the universal enveloping algebra $U(\mathfrak{g})$ of a finite-dimensional complex semisimple Lie algebra $\mathfrak{g}$, and their finite-dimensional quotients $u_q(\mathfrak{g})$ for q a root of unity. The representation theory of $U_q(\mathfrak{g})$ is related to the representation theory of $\mathfrak{g}$, while the representation theory of $u_q(\mathfrak{g})$ is related to the representation theory of $\mathfrak{g}$ over a field of positive characteristic. We refer to Lusztig's research monograph [**6**] for an introduction to this subject. Quantum groups and their representations are also important in theoretical physic, where they model for example spaces of solutions of certain differential equations.

An important feature of quantum groups is the commutativity of the tensor product $V \otimes W \to W \otimes V$ by $v \otimes w \mapsto R.(w \otimes v)$ for a special element $R \in U_q(\mathfrak{g}) \otimes U_q(\mathfrak{g})$ called R-matrix. The definition of $U_q(\mathfrak{g})$ comes with choices of a finite abelian group G, which corresponds to choices of center resp. fundamental group of an associated Lie group. For $G = 1$ and order of q not divisible by $2, 3$ it is shown in [**1**] that R is essentially unique. We are interested to construct and classify R-matrices for $G \neq 1$. This is in particular important for the case of arbitrary q, when considering the quotients $u_q(\mathfrak{g})$, where existence and nondegeneracy of R-matrices is more subtle, see [**2**] and references therein.

In [**6**, Sec. 32.1] Lusztig proposes an Ansatz for obtaining a family of R-matrices of the form $R_0\bar{\Theta}$, with $\bar{\Theta}$ is fully determined by the root system of $\mathfrak{g}$ and R_0 is contained in its toral part and is quasi-periodic in the root lattice. Hence, as we discussed in [**4**, Sec. 2], the choices R for a given G are determined uniquely up to the choice $g \in \mathbb{C}[G] \otimes \mathbb{C}[G]$ fulfilling the conditions in Theorem 1.1. Using the result of the present article, we are able to determine all possible R_0 for $u_q(\mathfrak{g})$ in [**4, 5**]. The improved results for arbitrary G allow for a uniform treatment of the case $\mathfrak{g} = D_{2n}$, the only one with non-cyclic fundamental group $\mathbb{Z}_2 \times \mathbb{Z}_2$, in which case we obtain unusual R-matrices with support $H_1 \neq H_2$, and for more general cases such as products of simple Lie algebras.

Organisation. The Theorems 1.1, 1.3, 1.4, and 1.5 are proved in Sections 2, 3, 4, and 5, respectively.

2. Discrete Fourier Analysis

This section is devoted to the proof of Theorem 1.1. The easy direction is to check that families defined by (1.2) are indeed solutions. For later use we establish this in the following slightly stronger form.

LEMMA 2.1. *Suppose that H_1, H_2 are two subgroups of a finite Abelian group G, that $\omega\colon H_1 \times H_2 \longrightarrow \mathbb{C}^\times$ is a pairing and that $d \in \mathbb{C}^\times$. The family $(g(x,y))_{x,y\in G}$ of complex numbers defined by*

$$g(x,y) = \frac{1}{d} H_1(x) H_2(y) \omega(x,y)$$

for all $x, y \in G$ is a solution for G if and only if $d = |H_1| = |H_2|$.

PROOF. Due to $g(1,1) = 1/d \neq 0$ the family $(g(x,y))_{x,y\in G}$ is non-degenerate. Now for arbitrary $x, y \in G$ a straightforward calculation yields

$$\begin{aligned} d^2 \sum_{y_1+y_2=y} g(x,y_1)g(x,y_2) &= \sum_{y_1+y_2=y} \omega(x,y_1)\omega(x,y_2)H_1(x)H_2(y_1)H_2(y_2) \\ &= \omega(x,y)H_1(x) \sum_{y_1+y_2=y} H_2(y_1)H_2(y_2) \\ &= |H_2|\omega(x,y)H_1(x)H_2(y) \\ &= d\,|H_2| \cdot g(x,y) \end{aligned}$$

and, therefore, the first equality in (1.1) is equivalent to $|H_2| = d$. Similarly, the equality of the first and third term in (1.1) is equivalent to $|H_1| = d$. □

The converse direction is less obvious. Our proof uses discrete Fourier analysis and we commence by fixing our notation. Given a finite abelian group G we write $\widehat{G}$ for its Pontryagin dual. So explicitly we set $\widehat{G} = \operatorname{Hom}(G, S^1)$, where $S^1 = \{z \in \mathbb{C} \colon |z| = 1\}$ denotes the unit circle in the complex plane, regarded as a multiplicative group. For an element $x \in G$ and a character $r \in \widehat{G}$ we usually write (r, x) instead of $r(x)$. It is well known that G and $\widehat{G}$ are (non-canonically) isomorphic.

The *Fourier transform* of a function $f \colon G \longrightarrow \mathbb{C}$ is defined to be the function $\widehat{f} \colon \widehat{G} \longrightarrow \mathbb{C}$ given by $\widehat{f}(r) = \sum_{x\in G} f(x)(r,x)$ for every $r \in \widehat{G}$. If $\widehat{f}$ is known one can compute f by means of the *inversion formula*

$$f(x) = \frac{1}{|G|} \sum_{r\in\widehat{G}} \widehat{f}(r)\overline{(r,x)}.$$

The *convolution* of two functions $f, g \colon G \longrightarrow \mathbb{C}$ is the function $(f \star g) \colon G \longrightarrow \mathbb{C}$ defined by $(f \star g)(x) = \sum_{x_1+x_2=x} f(x_1)g(x_2)$ for all $x \in G$. It is well known (and easy to confirm) that $\widehat{f \star g} = \widehat{f} \cdot \widehat{g}$, where the dot on the right side indicates pointwise multiplication. For further information on this topic we refer to the textbook by Tao and Vu [**8**].

PROOF OF THEOREM 1.1. The backwards implication is a trivial consequence of Lemma 2.1. Now suppose conversely that the family $(g(x,y))_{x,y\in G}$ is a solution for G. For each $x \in G$ let $g_x \colon G \longrightarrow \mathbb{C}$ be the function $y \longmapsto g(x,y)$. The first equality in (1.1) tells us that $g_x \star g_x = g_x$. Taking the Fourier transform, we infer $\widehat{g_x} = (\widehat{g_x})^2$ and thus there exists a set of characters $\widehat{B_x} \subseteq \widehat{G}$ whose characteristic function is $\widehat{g_x}$. Now the inversion formula yields

$$g(x,y) = \frac{1}{|G|} \sum_{r\in\widehat{G}} \widehat{B_x}(r)\overline{(r,y)}. \tag{2.1}$$

Substituting this back into the last sum occurring in (1.1) we learn

$$\begin{aligned} g(x,y) &= \sum_{x_1+x_2=x} g(x_1,y)g(x_2,y) \\ &= \frac{1}{|G|^2} \sum_{x_1+x_2=x} \sum_{r_1,r_2} \widehat{B_{x_1}}(r_1)\widehat{B_{x_2}}(r_2)\overline{(r_1+r_2,y)} \\ &= \frac{1}{|G|^2} \sum_{r} \left(\sum_{x_1+x_2=x} \sum_{r_1+r_2=x} \widehat{B_{x_1}}(r_1)\widehat{B_{x_2}}(r_2) \right) \overline{(r,y)}. \end{aligned}$$

For fixed x and variable y the last expression is the inverse Fourier transform of

$$r \longmapsto \frac{1}{|G|} \sum_{x_1+x_2=x} \sum_{r_1+r_2=x} \widehat{B_{x_1}}(r_1)\widehat{B_{x_2}}(r_2),$$

while $g(x, y)$ is the inverse transform of $\widehat{B_x}$. These considerations prove

$$\widehat{B_x}(r) = \frac{1}{|G|} \sum_{x_1+x_2=x} \sum_{r_1+r_2=r} \widehat{B_{x_1}}(r_1)\widehat{B_{x_2}}(r_2)$$

for all $x \in G$ and $r \in \widehat{G}$ and, in particular, we have

$$\widehat{B_{x_1}} + \widehat{B_{x_2}} \subseteq \widehat{B_{x_1+x_2}} \tag{2.2}$$

for all $x_1, x_2 \in G$. Consequently, the set

$$H_1 = \{x \in G \colon \widehat{B_x} \neq \varnothing\}$$

is closed under addition. Since G is finite, this implies that H_1 is either empty or a subgroup of G. But, in view of (2.1), $H_1 = \varnothing$ would imply $g(x, y) = 0$ for all $x, y \in G$, contrary to the demand that solutions be non-degenerate. Summarising this discussion, we have shown that H_1 is a subgroup of G.

In particular, we have $0 \in H_1$, i.e., the set $\widehat{B} = \widehat{B_0}$ is non-empty. By (2.2) $\widehat{B}$ is closed under addition as well and, hence, a subgroup of $\widehat{G}$. Moreover, (2.2) implies $\widehat{B} + \widehat{B_x} \subseteq \widehat{B_x}$ and $\widehat{B_x} + \widehat{B_{-x}} \subseteq \widehat{B}$ for every $x \in B$, wherefore each $\widehat{B_x}$ is a coset of $\widehat{B}$. So (2.2) reformulates as saying that $x \longmapsto \widehat{B_x}$ is a homomorphism from H_1 into the quotient group $\widehat{G}/\widehat{B}$.

Choose for every $x \in H_1$ an arbitrary representative $b_x \in \widehat{B_x}$. Now (2.1) rewrites as

$$g(x, y) = \frac{H_1(x)\overline{(b_x, y)}}{|G|} \sum_{r \in \widehat{B}} \overline{(r, y)}. \tag{2.3}$$

To simplify the sum appearing here we consider the subgroup

$$H_2 = \{y \in G \colon \text{ if } r \in \widehat{B} \text{ then } (r, y) = 1\}$$

of G and observe that

$$\sum_{r \in \widehat{B}} (r, y) = \begin{cases} |\widehat{B}| & \text{if } y \in H_2 \\ 0 & \text{else} \end{cases}$$

holds for every $y \in G$. Combined with (2.3) this reads

$$g(x, y) = \frac{|\widehat{B}|}{|\widehat{G}|} H_1(x)H_2(y)\overline{(b_x, y)}$$

for all $x, y \in G$. The map $\omega \colon H_1 \times H_2 \longrightarrow \mathbb{C}^\times$ given by $(x, y) \longmapsto \overline{(b_x, y)}$ is a group pairing: Indeed for fixed $y \in H_2$ the map $\omega(-, y)$ consists of the previous group homomorphism $H_1 \to \widehat{G}/\widehat{B}$ given by $x \mapsto \widehat{B}_x$, concatenated with the map $b_x \mapsto (b_x, y)$, which does by definition of H_2 not depend on the chosen $\widehat{B}$-coset representative b_x and thus gives a group homomorphism $\widehat{G}/\widehat{B} \to \mathbb{C}^\times$; on the other hand for fixed $x \in H_1$ the map $\omega(x, -)$ is clearly a group homomorphism $H_2 \to \mathbb{C}^\times$.

Finally, the forward implication of Lemma 2.1 applied to $d = |\widehat{G}|/|\widehat{B}|$ yields $d = |H_1| = |H_2|$ and our theorem is proved. □

3. A theorem on filters

In this section we prove Theorem 1.3. A straightforward calculation gives the values of the a_C for all $C \subset N$ if $\mathcal{E}$ is a given filter. It shows immediately the implication $(a) \to (b)$, but the precise value will also be crucial to the proof of $(b) \to (a)$ in what follows:

LEMMA 3.1. *Let $\mathcal{E}$ be a filter, i.e. $\mathcal{E} = \{D \subset N \mid D \supset A\}$ for some $A \subset N$. Then for any $C \subset N$ we have*

$$a_C := (-1)^{|N|} \sum_{D \in \mathcal{E}} (-1)^{|C \cup D|} e^{\mu(C-D)}$$
$$= \begin{cases} e^{\mu(N-A)} \prod_{p \in N-A} \left(1 + e^{-\mu(p)}\right), & C \cup A = N \\ 0, & else \end{cases}$$

PROOF. For a filter $\mathcal{E} = \{D \subset N \mid D \supset A\}$ generated by A the defining formula for a_C in Theorem 1.3 reads

$$a_C = (-1)^{|N|} \sum_{A \subset D \subset N} (-1)^{|C \cup D|} e^{\mu(C-D)}$$
$$= (-1)^{|N|} \sum_{D' \subset N-A} (-1)^{|C \cup A \cup D'|} e^{\mu(C-A-D')}$$
$$= (-1)^{|N-A|} \sum_{D' \subset N-A} (-1)^{|(C-A) \cup D'|} e^{\mu(C-A-D')}$$

This shows that the value of a_C for the filter generated by A in N is equal to the value of a_{C-A} for the filter generated by $\varnothing$ in $N - A$. Thus is suffices to show the claim for the filter $\mathcal{E} = \mathcal{P}(N)$ generated by $A = \varnothing$:

$$a_C = (-1)^{|N|} \sum_{D \subset N} (-1)^{|C \cup D|} e^{\mu(C-D)}$$
$$= (-1)^{|N|} \sum_{D_1 \subset C, D_2 \subset N-C} (-1)^{|C|+|D_2|} e^{\mu(C)-\mu(D_1)}$$
$$= (-1)^{|N|+|C|} e^{\mu(C)} \left(\sum_{D_2 \subset N-C} (-1)^{|D_2|} \right) \left(\sum_{D_1 \subset C} e^{-\mu(D_1)} \right)$$
$$= (-1)^{|N|+|C|} e^{\mu(C)} \left(\prod_{p \in N-C} (1-1) \right) \left(\prod_{p \in C} \left(1 + e^{-\mu(p)}\right) \right)$$
$$= \begin{cases} e^{\mu(N)} \prod_{p \in N} \left(1 + e^{-\mu(p)}\right), & C = N \\ 0, & else \end{cases}$$

The general formula for arbitrary A follows by again replacing N with $N - A$ and C by $C - A$. □

We use this result to show that if $\mathcal{E}$ is a small modification of a filter, the main assumption $a_C \geqslant 0$ for all $C \subset N$ usually fails to be true.

LEMMA 3.2.

(a) *Let $\mathcal{E} \neq \mathcal{P}(N)$ be a filter in $|N| > 1$, then $\mathcal{E} \cup \{\varnothing\}$ gives $a_C < 0$ for some $C \subset N$.*
(b) *Let $\mathcal{E} = \mathcal{P}(N)$ in $|N| > 1$, then $\mathcal{E} \setminus \{\varnothing\}$ gives $a_C < 0$ for some $C \subset N$.*

PROOF.

(a) By assumption $\mathcal{E}$ is a filter generated by some $A \neq \varnothing$ for $|N| > 1$. We wish to find a negative value of some $\tilde{a}_C$ for the set system $\tilde{\mathcal{E}} := \mathcal{E} \cup \{\varnothing\}$: Suppose first that also $A \neq N$ and choose some $p \in N - A$, then $a_{N-p} = 0$ by Lemma 3.1 and thus:

$$\begin{aligned}\tilde{a}_{N-p} &= (-1)^{|N|} \sum_{D \in \mathcal{E} \cup \{\varnothing\}} (-1)^{|C \cup D|} e^{\mu(C-D)} \\ &= a_{N-p} + (-1)^{|N|+|N-p|} e^{\mu(N-p)} \\ &= -e^{\mu(N-p)} < 0\end{aligned}$$

Suppose now that $A = N$ and choose some $q \in N$, then again by Lemma 3.1:

$$\begin{aligned}\tilde{a}_{N-q} &= (-1)^{|N|} \sum_{D \in \mathcal{E} \cup \{\varnothing\}} (-1)^{|C \cup D|} e^{\mu(C-D)} \\ &= a_{N-q} + (-1)^{|N|+|N-q|} e^{\mu(N-q)} \\ &= 1 - e^{\mu(N-q)} < 0\end{aligned}$$

(b) By assumption $\mathcal{E} = \mathcal{P}(N)$ for $|N| > 1$, so $\mathcal{E}$ is the filter generated by $A = \{\varnothing\}$. Then again by Lemma 3.1 $a_C = 0$ for $C \neq N$. Choose any $q \neq p \in N$, then we calculate $\tilde{a}_{N-\{p,q\}}$ for the filter $\tilde{\mathcal{E}} := \mathcal{E} - \{\varnothing\}$:

$$\begin{aligned}\tilde{a}_{N-\{p,q\}} &= (-1)^{|N|} \sum_{D \in \mathcal{E} - \{\varnothing\}} (-1)^{|C \cup D|} e^{\mu(C-D)} \\ &= a_{N-q} - (-1)^{|N|+|N-\{p,q\}|} e^{\mu(N-\{p,q\})} \\ &= 1 - e^{\mu(N-\{p,q\})} < 0\end{aligned}$$

□

We now proceed by introducing the induction step along $|N|$: Let $\mathcal{E}$ be any set system in N and $p \in N$, then we define a new set system for $N - p$ by

$$\mathcal{E}_p = \{D - p \mid p \in D, D \in \mathcal{E}\}$$

For $C \subset N - p$ we denote by a_C^p the corresponding sum over $\mathcal{E}_p$, i.e.

$$a_C^p = (-1)^{|N-p|} \sum_{D \in \mathcal{E}_p} (-1)^{|C \cup D|} e^{\mu(C-D)}$$

We will in the following only consider $\mathcal{E}_p$ for all p, such that there exists any $D \in \mathcal{E}$ with $p \in D$, so $\mathcal{E}_p$ is not empty.

We first wish to prove that our main assumption $a_C \geqslant 0$ implies $a_C^p \geqslant 0$ in $\mathcal{E}_p$:

LEMMA 3.3. *For any $p \in N$ we get for all $C \in \mathcal{E}_p$ (note that $p \notin C$):*

$$a_C^p = \frac{e^{\mu(p)}}{1 + e^{\mu(p)}} a_C + \frac{1}{1 + e^{\mu(p)}} a_{C \cup p}$$

In particular, $a_C \geqslant 0$ for all $C \in \mathcal{E}$ implies $a_C^p \geqslant 0$ for all $C \in \mathcal{E}_p$.

PROOF. We calculate the right hand side by splitting the sum over all $D \in \mathcal{E}$ into two summands for all $p \notin D$ resp. $p \in D$ and use $p \notin C$. The latter set of D

then correspond to $D' = D - p$ in $\mathcal{E}_p$:

$$\begin{aligned}
&\frac{e^{\mu(p)}}{1+e^{\mu(p)}}a_C + \frac{1}{1+e^{\mu(p)}}a_{C+p} \\
&= (-1)^{|N|}\frac{e^{\mu(p)}}{1+e^{\mu(p)}} \sum_{p\in D\in\mathcal{E}} (-1)^{|C\cup D|}e^{\mu(C-D)} \\
&\quad + (-1)^{|N|}\frac{1}{1+e^{\mu(p)}} \sum_{p\in D\in\mathcal{E}} (-1)^{|C\cup p\cup D|}e^{\mu((C\cup p)-D)} \\
&\quad + (-1)^{|N|}\frac{e^{\mu(p)}}{1+e^{\mu(p)}} \sum_{p\notin D\in\mathcal{E}} (-1)^{|C\cup D|}e^{\mu(C-D)} \\
&\quad + (-1)^{|N|}\frac{1}{1+e^{\mu(p)}} \sum_{p\notin D\in\mathcal{E}} (-1)^{|C\cup p\cup D|}e^{\mu((C\cup p)-D)} \\
&= (-1)^{|N|}\frac{e^{\mu(p)}}{1+e^{\mu(p)}} \sum_{p\in D\in\mathcal{E}} (-1)^{|C\cup D|}e^{\mu(C-D)} \\
&\quad + (-1)^{|N|}\frac{1}{1+e^{\mu(p)}} \sum_{p\in D\in\mathcal{E}} (-1)^{|C\cup D|}e^{\mu(C-D)} \\
&\quad + (-1)^{|N|}\frac{e^{\mu(p)}}{1+e^{\mu(p)}} \sum_{p\notin D\in\mathcal{E}} (-1)^{|C\cup D|}e^{\mu(C-D)} \\
&\quad + (-1)^{|N|}\frac{1}{1+e^{\mu(p)}} \sum_{p\notin D\in\mathcal{E}} (-1)^{|C\cup D|+1}e^{\mu(C-D)+\mu(p)} \\
&= \left(\frac{e^{\mu(p)}}{1+e^{\mu(p)}} + \frac{1}{1+e^{\mu(p)}}\right)\cdot(-1)^{|N|} \sum_{p\in D\in\mathcal{E}} (-1)^{|C\cup D|}e^{\mu(C-D)} \\
&\quad + \left(\frac{e^{\mu(p)}}{1+e^{\mu(p)}} - \frac{1}{1+e^{\mu(p)}}\cdot e^{\mu(p)}\right)\cdot(-1)^{|N|} \sum_{p\notin D\in\mathcal{E}} (-1)^{|C\cup D|}e^{\mu(C-D)} \\
&= (-1)^{|N|} \sum_{p\in D\in\mathcal{E}} (-1)^{|C\cup D|}e^{\mu(C-D)} \\
&= (-1)^{|N-p|} \sum_{D'\in\mathcal{E}_p} (-1)^{|C\cup D'|}e^{\mu(C-D')} = a_C^p
\end{aligned}$$

□

Thus if all $a_C \geqslant 0$ by induction hypothesis all $\mathcal{E}_p$ are filters. We now conclude the induction that $\mathcal{E}$ is a filter if all possible reductions $\mathcal{E}_p$ are filters. As induction step, we use the following lemma.

LEMMA 3.4. *Let $\mathcal{E}$ be a set system for N such that all $\mathcal{E}_p$ are filters generated by sets $A_p \subset N - p$.*

a) *One of the following cases applies:*
   (1) *There exists a $p \in N$ with $p \in A_q$ for all $p \neq q$.*
   (2) *For all $p \in N$ we have $A_p = \varnothing$.*
b) *If (1) applies, then $\mathcal{E}$ is the filter generated by $p \cup A_p$ or $\mathcal{E}$ is the set system consisting of this filter together with $D = \varnothing$.*
c) *If (2) applies, then $\mathcal{E} = \mathcal{P}(N)$ or $\mathcal{E} = \mathcal{P}(N) - \varnothing$.*

PROOF.

a) Assume there exists q' with $A_{q'} \neq \varnothing$ and let $p \in A_{q'}$, then we claim $p \in A_q$ for all $q \neq p$. We prove this by contradiction: if $p \notin A_q$ for some q then we consider $q' \cup A_q \in \mathcal{E}_q$ (since $\mathcal{E}_q$ is a filter) and hence $q \cup q' \cup A_q \in \mathcal{E}$ (by definition of $\mathcal{E}_q$). But then $q \cup A_q \in \mathcal{E}_{q'}$ and $A_{q'} \subset q \cup A_q$ (since $\mathcal{E}_{q'}$ is a filter). But this contradicts $p \notin A_q$.

b) Assume there exists $p \in A_q$ for all $q \neq p$. Let $D \supset p \cup A_p$ then $D \in \mathcal{E}_p$ (since $\mathcal{E}_p$ is a filter) and $D \in \mathcal{E}$ (by definition of $\mathcal{E}_p$). Let now conversely by $D \in \mathcal{E}$. If $p \in D$ then we have $D - p \in \mathcal{E}_p$ (by definition of $\mathcal{E}_p$) and hence also $D \supset A_p$ (since $\mathcal{E}_p$ is a filter), implying $D \supset p \cup A_p$. If $p \notin D$ then either $D = \varnothing$ or some $q \in D$. In the latter case $D - q \in \mathcal{E}_q$, hence $D - q \supset A_q \ni p$ which is a contradiction. So either $D \supset p \cup A_p$ or $D = \varnothing$ as asserted.

c) Assume $A_p = \varnothing$ for all $p \in N$, then for any set $D \neq \varnothing$ we may chose some $p \in D$ and yield $D - p \in \mathcal{E}_p$ and hence $D \in \mathcal{E}_q$. Hence any set with the possible exception of $D = \varnothing$ is in $\mathcal{E}$ as asserted. □

We can now conclude the inductive proof of the implication (i)→(ii) in Theorem 1.3: For $|N| = 0$ the only set system is $\mathcal{E} = \{\varnothing\}$ and is a filter. Let $|N| \geqslant 1$ and $\mathcal{E}$ such that all $a_C \geqslant 0$, then $a_C^p \geqslant 0$ for all $p \in N$ by Lemma 3.3. Thus by induction hypothesis all $\mathcal{E}_p$ are filters. Then by Lemma 3.4 we have that either $\mathcal{E}$ is a filter (in which case the induction step is finished) or some filter $\mathcal{E} \neq \mathcal{P}(N)$ together with $\varnothing$ or $\mathcal{E} = \mathcal{P}(N) - \{\varnothing\}$. By Lemma 3.2 these two cases can only fulfill $a_C \geqslant 0$ for $|N| = 1$ where both are filters. This concludes the proof of Theorem 1.3.

4. A theorem about roots of unity

In this section we prove Theorem 1.4. The forward direction follows from the well-known fact that if U denotes the set of all N-th root of unity, then

$$a_k = \begin{cases} N & \text{if } k \mid N \\ 0 & \text{else.} \end{cases} \tag{4.1}$$

Now assume conversely that U has the property that $a_0, a_1, \dots$ are nonnegative real numbers. Due to $U \neq \varnothing$ we know that $a_0 = |U|$ is positive. Now let N denote the least natural number such that every $\zeta \in U$ is an N-th root of unity and consider the polynomial

$$P(z) = a_0 + a_1 z + \cdots + a_{N-1} z^{N-1}.$$

For every N-th root of unity ξ we have

$$P(\xi) = \sum_{r=0}^{N-1} \sum_{\zeta \in U} (\zeta\xi)^r = \sum_{\zeta \in U} \sum_{r=0}^{N-1} (\zeta\xi)^r \overset{(4.1)}{=} \sum_{\zeta \in U} N\delta_{\zeta\xi,1} = NU(\xi^{-1}),$$

where U is identified with its characteristic function.

Since our hypothesis and $a_0 > 0$ yield $P(1) > 0$, it follows that $1 \in U$ and

$$P(1) = \sum_{r=0}^{N-1} a_r = N. \tag{4.2}$$

Now assume (reductio ad absurdum) that there exists some positive $r^\star < N$ such that $a_{r^\star} \neq 0$. For every $\zeta \in U$ we have $P(\zeta^{-1}) = N$ and, therefore, the triangle inequality entails

$$N = \Big| \sum_{r=0}^{N-1} a_r \zeta^{-r} \Big| \leqslant \sum_{r=0}^{N-1} |a_r \zeta^{-r}| = \sum_{r=0}^{N-1} a_r.$$

Actually, (4.2) tells us that this holds with equality, which due to $a_0, a_{r^\star} > 0$ is only possible if $\zeta^{-r^\star} = 1$. In other words, every member of U is an $r^\star$-th root of unity, contrary to our minimal choice of N.

We have thereby proved $a_r = 0$ for every positive $r < N$ and a final application of (4.2) discloses $|U| = a_0 = N$. Thus U is the set of all N-th root of unity, which concludes the proof of Theorem 1.4.

5. Idempotents and $\{0,1\}$-matrices

We begin this section by proving Theorem 1.5. For an arbitrary $\{0,1\}$-matrix $(\varepsilon_{ij})_{0\leqslant i,j<N}$ the element

$$\varepsilon = \sum_{i,j} \varepsilon_{ij} g^i \otimes g^j$$

of $\mathbb{C}[\mathbb{Z}_N \times \mathbb{Z}_N]$ has the property that ε/N is idempotent if and only if

$$\frac{1}{N^2} \sum_{i',i''} \sum_{j',j''} \varepsilon_{i'j'} \varepsilon_{i''j''} (g^{i'} \otimes g^{j'})(g^{i''} \otimes g^{j''}) = \frac{1}{N} \sum_{i,j} \varepsilon_{ij} g^i \otimes g^j.$$

By comparing coefficients one sees that this is in turn equivalent to

$$\sum_{i'+i''=i} \sum_{j'+j''=j} \varepsilon_{i'j'} \varepsilon_{i''j''} = N \varepsilon_{ij} \tag{5.1}$$

for all $i, j \in \mathbb{Z}_N$. This system of equations is clearly satisfied if (ε_{ij}) is the characteristic function of a subgroup of $\mathbb{Z}_N \times \mathbb{Z}_N$ of order N in the sense of (1.4).

Now suppose conversely that the equations (5.1) hold for some $\varepsilon \neq 0$. Clearly the set

$$G = \{(i, j) \in \mathbb{Z}_N \times \mathbb{Z}_N : \varepsilon_{ij} = 1\}$$

is non-empty. Moreover, if (i', j') and (i'', j'') are in G, then (5.1) reveals $\varepsilon_{i'+i'',j'+j''} > 0$ and thus the sum $(i', j') + (i'', j'')$ belongs to G as well. This proves that G is closed under addition and, hence, a subgroup of $\mathbb{Z}_N \times \mathbb{Z}_N$. Finally, by plugging $i = j = 0$ into (5.1) we find $|G| = N$. Thereby Theorem 1.5 is proved.

For the sake of completeness we mention the following classification of the groups G relevant to this result.

PROPOSITION 5.1. *For every subgroup $G \leqslant \mathbb{Z}_N \times \mathbb{Z}_N$ of order N there exists a unique pair (d, t) consisting of a divisor $d \mid N$ and a nonnegative integer $t < d$ such that*

$$G = \big\{(i, j) \in \mathbb{Z}_N \times \mathbb{Z}_N : \tfrac{N}{d} \mid i \text{ and } d \mid j - t\tfrac{i}{N/d}\big\} \tag{5.2}$$

is generated by $(0, d)$ and $(N/d, t)$. Conversely, for every such pair (d, t) this formula defines a subgroup of $\mathbb{Z}_N \times \mathbb{Z}_N$ of order N.

PROOF. Let $G \leqslant \mathbb{Z}_N \times \mathbb{Z}_N$ be a subgroup of order N. The projection

$$A = \big\{i \in \mathbb{Z}_N : \text{there is some } j \in \mathbb{Z}_N \text{ such that } (i, j) \in G\big\}$$

is a subgroup of $\mathbb{Z}_N$ and, hence, there exists a divisor $d \mid N$ such that A consists of the multiples of N/d. Now for each $i \in A$ we consider the fibre $B_i = \{j \in \mathbb{Z}_N : (i, j) \in G\}$. It is clear that any two of these sets have the same cardinality.

In fact, due to $|G| = N$ and $|A| = d$ they all must have size N/d. Since B_0 is a subgroup of $\mathbb{Z}_N$, this proves that B_0 consists of all multiples of d.

Now $B_{N/d}$ is a translate of B_0 and thus there exists an integer t such that $0 \leqslant t < d$ and $B_{N/d} = B_0 + t$. It follows easily that $B_{kN/d} = B_0 + kt$ holds for every integer k and this establishes (5.2). The remaining parts of Proposition 5.1 are left to the reader. $\square$

EXAMPLE 5.2. For $N = 12$, $d = 4$, and $t = 2$ we obtain the subgroup $G \leqslant \mathbb{Z}_{12} \times \mathbb{Z}_{12}$ generated by $(0,4)$ and $(3,2)$, and Theorem 1.5 leads to the matrix

$$\begin{array}{c} \\ 0 \\ \\ \\ N/d \\ \\ \\ \bar{d} \\ \\ \\ \\ \\ \\ \end{array}
\begin{array}{c} \begin{array}{cccccccccccc} 0 & & t & & d & & & & & & & \end{array} \\
\begin{pmatrix}
1 & 0 & 0 & 0 & 1 & 0 & 0 & 0 & 1 & 0 & 0 & 0 \\
0 & . & . & . & . & . & . & . & . & . & . & . \\
0 & . & . & . & . & . & . & . & . & . & . & . \\
0 & 0 & 1 & 0 & 0 & 0 & 1 & 0 & 0 & 0 & 1 & 0 \\
0 & . & . & . & . & . & . & . & . & . & . & . \\
0 & . & . & . & . & . & . & . & . & . & . & . \\
1 & 0 & 0 & 0 & 1 & 0 & 0 & 0 & 1 & 0 & 0 & 0 \\
0 & . & . & . & . & . & . & . & . & . & . & . \\
0 & . & . & . & . & . & . & . & . & . & . & . \\
0 & 0 & 1 & 0 & 0 & 0 & 1 & 0 & 0 & 0 & 1 & 0 \\
0 & . & . & . & . & . & . & . & . & . & . & . \\
0 & . & . & . & . & . & . & . & . & . & . & .
\end{pmatrix} \end{array}$$

Acknowledgments

This article builds on previous work with the PhD student D. Nett in the cyclic case [**3**]. I am indebted to Mihalis Kolountzakis and Christian Reiher for suggesting the methods that significantly improve our original approach, Christian Reiher for supplying the interesting alternative proof of 1.4, I thank an anonymous referee for pointing out [**7**] and an anonymous person for the beautiful typing.

References

[1] A. Davydov, P. Etingof, and D. Nikshych, *Autoequivalences of tensor categories attached to quantum groups at roots of 1*, Lie groups, geometry, and representation theory, Progr. Math., vol. 326, Birkhäuser/Springer, Cham, 2018, pp. 109–136, DOI 10.1007/978-3-030-02191-7_5. MR3890207 ↑119

[2] A. Gainutdinov, S. Lentner, and T. Ohrmann, *Modularization of small quantum groups*, Preprint, `arXiv:1809.02116`, 2018. ↑119

[3] S. Lentner and D. Nett, *A theorem on roots of unity and a combinatorial principle*, Preprint, `arXiv:1409.5822`, 2014. ↑127

[4] S. Lentner and D. Nett, *New R-matrices for small quantum groups*, Algebr. Represent. Theory **18** (2015), no. 6, 1649–1673, DOI 10.1007/s10468-015-9555-6. MR3435106 ↑119

[5] S. Lentner and T. Ohrmann, *Factorizable R-matrices for small quantum groups*, SIGMA Symmetry Integrability Geom. Methods Appl. **13** (2017), Paper No. 076, 25, DOI 10.3842/SIGMA.2017.076. MR3705499 ↑119

[6] G. Lusztig, *Introduction to quantum groups*, Progress in Mathematics, vol. 110, Birkhäuser Boston, Inc., Boston, MA, 1993. MR1227098 ↑119

[7] M. Newman, *Nonnegative sums of roots of unity*, J. Res. Nat. Bur. Standards Sect. B **80B** (1976), no. 1, 1–4. MR414515 ↑118, 127

[8] T. Tao and V. Vu, *Additive combinatorics*, Cambridge Studies in Advanced Mathematics, vol. 105, Cambridge University Press, Cambridge, 2006, DOI 10.1017/CBO9780511755149. MR2289012 ↑118, 120

Fachbereich Mathematik, Universität Hamburg, Hamburg, Germany
Email address: `Simon.Lentner@uni-hamburg.de`

Contemporary Mathematics
Volume **813**, 2025
https://doi.org/10.1090/conm/813/16286

On near-group centers and super-modular categories

Eric C. Rowell, Hannah Solomon, and Qing Zhang

Abstract. The construction and classification of super-modular categories is an ongoing project, of interest in algebra, topology and physics. In a recent paper, Cho, Kim, Seo and You produced two mysterious families of super-modular data, with no known realization. We show that these data are realized by modifying the Drinfeld centers of near-group fusion categories associated with the groups $\mathbb{Z}/6$ and $\mathbb{Z}/2 \times \mathbb{Z}/4$. The methods we develop have wider applications and we describe some of these, with a view towards understanding when near-group centers provide super-modular categories.

1. Introduction

Besides their interest in algebraic category theory and topology, modular (resp. super-modular) categories are important in condensed matter physics as they describe bosonic (resp. fermionic) topological phases of matter in two spacial dimensions (see eg. [**6, 30**]). This connection goes back to the mathematical study of conformal field theory in [**1, 26**]. It is of substantial interest to classify these categories both for these applications and their intrinsic beauty.

It is known [**9, 22**] that for any r there are finitely many modular (resp. super-modular) categories with precisely r isomorphism classes of simple objects, i.e. **rank** r. Although a complete classification of modular (resp. super-modular) categories is probably out of reach without some general structure theorems, for small r, classifications are known [**7, 8, 10, 28, 30**], at least up to *modular data.*

From any modular category $\mathcal{C}$ one obtains a (projective) representation of the modular group $\mathrm{SL}(2,\mathbb{Z})$ as the mapping class group of the torus. This representation is determined by a pair of matrices (S,T) called the modular data of $\mathcal{C}$. The matrices S and T satisfy a number of remarkable constraints, including the key result of [**29**] that says that the $\mathrm{SL}(2,\mathbb{Z})$ representation factors through $\mathrm{SL}(2,\mathbb{Z}/N)$ for some minimal N, called the **level** of the representation, and that N is the (finite) order of the T-matrix. The paper [**28**] introduced a computational method for classifying (low) rank r modular categories roughly as follows:

Step 1 Construct all irreducible prime-power-level representations of $\mathrm{SL}(2,\mathbb{Z})$ of dimension at most r, with the property that the image of $\mathfrak{s} := \begin{pmatrix} 0 & -1 \\ 1 & 0 \end{pmatrix}$

2020 *Mathematics Subject Classification.* Primary 18M20.

The first and second authors were partially supported by NSF grants DMS-2000331 and DMS-2205962.

is symmetric and the image of $\mathfrak{t} := \begin{pmatrix} 1 & 1 \\ 0 & 1 \end{pmatrix}$ is diagonal. That is, all such representations factoring over $\mathrm{SL}(2, \mathbb{Z}/p^k)$ for primes p and $k \geq 1$.

Step 2 Construct all finite-level representations ρ of dimension r by considering direct sums of tensor products of representations from Step 1.

Step 3 Construct possible modular data (S, T) of rank r by studying matrices of the form $S = U^{-1}\rho(\mathfrak{s})U$ where U is orthogonal [**28**, Theorem 3.4] and commutes with $\rho(\mathfrak{t})$ for each ρ from Step 2.

Step 4 Use the numerous constraints on modular data to eliminate as many pairs (S, T) as possible.

Step 5 For each remaining pair (S, T) find a modular category $\mathcal{C}$ with this modular data.

In [**28**] this approach was successfully applied to classify rank 6 modular categories up to modular data.[1] The number of cases to be considered proliferates rapidly, which requires significant assistance from computational software.

Super-modular categories are slight generalizations of modular categories. Modular categories $\mathcal{C}$ are non-degenerate in the sense that the symmetric center $\mathrm{Sym}(\mathcal{C})$ is the trivial category Vec, while super-modular categories $\mathcal{C}$ are **slightly degenerate** [**14**]: $\mathrm{Sym}(\mathcal{C})$ is equivalent to the braided fusion category sVec of super-vector spaces. A technical, but inessential, additional assumption is that a super-modular category should also be unitary. While super-modular categories have S and T matrices, they do not immediately yield a representation of a group, since S will be degenerate. However, after an appropriate reduction one obtains a representation of the index 3 subgroup $\Gamma_\theta \subset \mathrm{SL}(2, \mathbb{Z})$ generated by $\mathfrak{s}$ and $\mathfrak{t}^2$. The idea is that both S and T^2 have a well-defined tensor-decomposition into $S = \hat{S} \otimes \frac{1}{\sqrt{2}} \begin{pmatrix} 1 & 1 \\ 1 & 1 \end{pmatrix}$ and $T^2 = \hat{T}^2 \otimes \begin{pmatrix} 1 & 0 \\ 0 & 1 \end{pmatrix}$ where $\hat{S}$ and $\hat{T}^2$ give a projective representation of Γ_θ. The pair $(\hat{S}, \hat{T}^2)$ is called the **super-modular data** of $\mathcal{C}$ – note that $\hat{T}$ is only defined up to sign choices.

A natural problem is to extend the above-described approach for classifying low rank modular categories to super-modular categories. The crucial Step 1 is justified by the recently proved Minimal Modular Extension (MME) Theorem of [**21**] and [**4**] which together show that the representations of Γ_θ coming from super-modular categories factor over the finite group $\Gamma_\theta \cap \mathrm{SL}(2, \mathbb{Z}/N)$ for some N. Then one can prove appropriate modifications of the steps above for the super-modular setting. As a more modest goal one might hope to produce new super-modular data to aid in the classification.

Under certain assumptions, Steps 1-4 were used in [**12**] to carry out a partial classification of super-modular data of rank 8 and rank 10. The expectation is that the rank 8 classification is complete, as it coincides with the partial classification in [**10**]. The authors of [**12**] produced 2 families of super-modular data $(\hat{S}, \hat{T}^2)$ of rank 10 that were not known to have a realization, see (3.1) and (3.2).

The main motivation for this article is to realize the super-modular data (3.1) and (3.2) by finding super-modular categories with these super-modular data. To do this we must find modular categories containing a fermion f (i.e. **spin** modular

[1] In principle there could be inequivalent categories with the same modular data, but they would of course have the same fusion rules so the ambiguity is modest.

categories [**6**]) so that the centralizer $\mathcal{C}_{\mathcal{C}}(\langle f\rangle)$ of f (also denoted by $\langle f\rangle'$) is a super-modular category with the given data. Here by a **fermion** we mean an object f with $f\otimes f\cong \mathbf{1}$ that has self-braiding $c_{f,f} = -\operatorname{Id}_{f\otimes f}$. If such a spin modular category exists there is a 16-fold ambiguity, since there are precisely 16 minimal modular extensions of any given super-modular category [**6, 25**]. Our main insight is that the super-modular data in [**12**] bears some similarity with modular data found in [**19**], which is conjecturally the modular data associated with the Drinfeld centers of near-group fusion categories [**32**] (see Section 2.2). In particular, we find that near-group categories associated with the groups $\mathbb{Z}/6$ and $\mathbb{Z}/4\times\mathbb{Z}/2$ yield such categories, after taking their Drinfeld centers, condensing a boson in the second case, and discarding pointed modular factors. The work of [**18**] and [**20**] lay the groundwork for our approach, with the main difficulty being producing the modular data of these Drinfeld centers – this involves solving a large system of non-linear equations. We have the following:

THEOREM.

(a) *Let $\mathcal{C}$ be the Drinfeld center of a near-group category of type $\mathbb{Z}/6+6$. Then $\mathcal{C}\cong\mathcal{D}\boxtimes\mathcal{C}(\mathbb{Z}/3,q)$, where $\mathcal{D}$ is a spin modular category, and q is its associated quadratic form restricted to $\mathbb{Z}/3$. Moreover, the Müger centralizer of the fermion f in $\mathcal{D}$ is super-modular and either itself or one of its Galois conjugates has the same super-modular data as in* (3.1).

(b) *Let $\mathcal{C}$ be the Drinfeld center of a near-group category of type $\mathbb{Z}/2\times\mathbb{Z}/4+8$. Then $\left[\mathcal{C}_{\mathbb{Z}/2}\right]_0\cong\mathcal{D}\boxtimes\mathcal{C}(\mathbb{Z}/2,q)$, where $\mathcal{D}$ is a spin modular category and q is the associated quadratic form restricted to $\mathbb{Z}/2$. Moreover, the Müger centralizer of the fermion f in $\mathcal{D}$ is super-modular and either itself or one of its Galois conjugates has the same super-modular data as in* (3.2).

Along the way we noticed that our approach is quite general: we can often construct spin and hence super-modular categories from Drinfeld centers of near-group categories (for groups of even order). In regards to case (a), see Example 2.6, where it is shown that the Drinfeld center of a near-group category of type $G+2m$ with $|G|=2m$ for m odd has an analogous factorization, with a spin-modular factor.

Also, we remark that for case (b) we must first *condense a boson*, i.e. take the $\mathbb{Z}/2$-de-equivariantization with respect to the symmetric Tannakian category $\langle b\rangle$ where b is a **boson**: an object b so that $b\otimes b\cong\mathbf{1}$ and the self-braiding satisfies $c_{b,b}=\operatorname{Id}_{b\otimes b}$.

2. Preliminaries

We shall need a few standard, but technical notions from the theory of braided fusion categories. Throughout this paper, we use the notation $\zeta_n := e^{2\pi i/n}$, $\mathbb{T}=\{x\in\mathbb{C};|x|=1\}$, and $\chi_n^m = m+\sqrt{n}$. We occasionally employ the shorthand $XY := X\otimes Y$ for notational convenience, and also write $X\oplus Y$ as $X+Y$. We denote by $e(r)=exp(2\pi i r)$ for any rational number r.

A braiding on a fusion category $\mathcal{C}$ is a natural isomorphism $c_{X,Y}: X\otimes Y\to Y\otimes X$ satisfying the hexagon equations [**17**]. A braided fusion category is a fusion category equipped with a braiding. We call a braided fusion category $\mathcal{C}$ **symmetric** if $c_{Y,X}c_{X,Y}=\operatorname{Id}_{X\otimes Y}$ for all $X,Y\in\mathcal{C}$. Let $\mathcal{C}$ be a braided fusion category and $\mathcal{D}\subset\mathcal{C}$ a fusion subcategory. The Müger centralizer $\mathcal{C}_{\mathcal{C}}(\mathcal{D})$ of $\mathcal{D}$ in $\mathcal{C}$ is the symmetric fusion subcategory of $\mathcal{C}$ generalized by $X \in \mathcal{C}$ such that $c_{Y,X}c_{X,Y} = \mathrm{Id}_{X\otimes Y}$ for all $Y \in \mathcal{D}$. We will often use the shorthand notation $\mathcal{D}' = C_{\mathcal{C}}(\mathcal{D})$ when no confusion can arise. The Müger center of $\mathcal{C}$ is the fusion subcategory $C_{\mathcal{C}}(\mathcal{C}) = \mathcal{C}'$, which is also called the symmetric center and sometimes denoted $\mathrm{Sym}(\mathcal{C})$. A premodular category is a spherical braided fusion category (in other notation, a ribbon fusion category). A premodular category $\mathcal{C}$ is called **modular** if $\mathrm{Sym}(\mathcal{C}) \cong \mathrm{Vec}$, that is, $\mathrm{Sym}(\mathcal{C})$ is equivalent to the category of finite-dimensional vector spaces over $\mathbb{C}$. For a premodular category $\mathcal{C}$, the $\tilde{S}$ matrix is defined by $\tilde{S}_{X,Y} = \mathrm{Tr}(c_{Y,X^*}c_{X^*,Y})$ and the T matrix by $T_{X,Y} = \delta_{X,Y}\theta_X$, where θ_X is the (scalar form of the) ribbon twist. The S-matrix is a normalized version of $\tilde{S}$, namely $S := \frac{\tilde{S}}{\sqrt{\dim(\mathcal{C})}}$. An alternative definition of a modular category is a premodular category whose S matrix is non-degenerate. A **super-modular** category $\mathcal{B}$ is a premodular category with $\mathrm{Sym}(\mathcal{B})$ equivalent to the category sVec of super-vector spaces. In this case, the S matrix is degenerate. For any super-modular category it is more convenient to consider $(\hat{S}, \hat{T}^2)$, where $S = \hat{S} \otimes \frac{1}{\sqrt{2}}\begin{pmatrix} 1 & 1 \\ 1 & 1 \end{pmatrix}$ and $T^2 = \hat{T}^2 \otimes \begin{pmatrix} 1 & 0 \\ 0 & 1 \end{pmatrix}$. Then, $\hat{S}$ is non-degenerate. A super-modular category $\mathcal{C}$ is called **split** if there is a modular category $\mathcal{D}$ such that $\mathcal{C} \cong \mathcal{D} \boxtimes \mathrm{sVec}$ as braided fusion categories. Clearly non-split super-modular categories are of greatest interest. The following recent result is crucial:

THEOREM 2.1 ([**21**]). *Let $\mathcal{C}$ be any super-modular category. Then there exists a (psuedo-unitary) modular category $\mathcal{D}$ such that $\mathcal{C} \subset \mathcal{D}$ and* $\dim(\mathcal{D}) = 2\dim(\mathcal{C})$.

As the minimal dimension of a (pseudo-unitary) modular category containing a super-modular category $\mathcal{C}$ is $2\dim(\mathcal{C})$, such a category is called a **minimal modular extension** of $\mathcal{C}$. A modular category $\mathcal{D}$ with a fermion f is called a spin modular category. In this case $\mathcal{D} = \mathcal{D}_0 \oplus \mathcal{D}_1$ is $\mathbb{Z}/2$-graded, where the trivial component $\mathcal{D}_0 = C_{\mathcal{D}}(\langle f \rangle)$. By results of [**6, 25**] there are exactly 16 minimal modular extensions of any super-modular category. Clearly if $\mathcal{C} = \mathcal{D} \boxtimes \mathrm{sVec}$ is a split super-modular category then $\mathcal{D} \boxtimes \mathcal{K}$ is a minimal modular extension where $\mathcal{K}$ is one of the 16 minimal modular extensions of sVec [**24**]. Thus a super-modular category $\mathcal{C}$ is split if and only if any (hence, every) minimal modular extension factors in this way.

We recall some results and notation from [**17**, Section 8.4]. A braided fusion category is **pointed** if every simple object is invertible. In a pointed braided fusion category, the isomorphism classes of simple objects form a finite abelian group. For a finite abelian group G, a **quadratic form** on G is a map $q : G \to \mathbb{F}^\times$ such that $q(g) = q(-g)$ and the map $\langle\ ,\ \rangle : G \times G \to \mathbb{F}^\times$ with $\langle g, h \rangle = \frac{q(g+h)}{q(g)q(h)}$ is a symmetric bicharacter. Let G be a finite abelian group and $q : G \to \mathbb{F}^\times$ a quadratic form on G. The pair (G, q) is called a **pre-metric group**. Consider the pointed braided fusion category with (isomorphism classes of) simple objects labeled by $g \in G$ and braiding $c_{g,h}$. We can define a quadratic form $q : G \to \mathbb{F}^\times$ on G by sending g to $c_{g,g} \in \mathbb{F}^\times$. Then, (G, q) is a pre-metric group. In fact, up to braided equivalence, for each pre-metric group (G, q), there is a unique pointed braided fusion category, denoted by $\mathcal{C}(G, q)$. By equipping $\mathcal{C}(G, q)$ with the spherical structure $\theta_g = q(g)$, we get a premodular category with a symmetric S matrix defined by $S_{g,h} = \frac{1}{\sqrt{|G|}}\overline{\langle g, h \rangle}$ and a diagonal T matrix defined by $T_{g,g} = q(g)$ (we use the conventions of [**19**]). If

$\langle\ ,\ \rangle$ is a non-degenerate bicharacter, then q is a **non-degenerate** quadratic form. In this case, the pair (G, q) is called a **metric group**. The corresponding pointed category for a metric group, $\mathcal{C}(G, q)$, is then a modular category with S and T as its modular data.

2.1. ***G*-de-equivariantization and Boson condensation.** Suppose that $\mathcal{C}$ is a braided fusion category with a subcategory equivalent to $\text{Rep}(G)$. In [**5, 15, 27**], three related concepts are described: modularization, modules over an algebra object and de-equivariantization. The G-de-equivariantization $\mathcal{C}_G$ of $\mathcal{C}$ is G-graded, and the trivial component $[\mathcal{C}_G]_0$ is again braided. Boson condensation is the term used in the physical literature for the process $\mathcal{C} \rightsquigarrow [\mathcal{C}_G]_0$. This has a non-unique reverse process known as gauging [**13**]. There is an alternative description of boson condensation in terms of algebra objects that is sometimes more useful: the object $A = \text{Fun}(G) \in \text{Rep}(G) \subset \mathcal{C}$ has the structure of an algebra object in $\mathcal{C}$. The category of A-modules in $\mathcal{C}$, denoted $\mathcal{C}_A$ is equivalent to $\mathcal{C}_G$, and the category of so-called *local* A-modules $\mathcal{C}_A^0$ coincides with $[\mathcal{C}_G]_0$. If $\mathcal{C}$ is modular then so is $[\mathcal{C}_G]_0$. Moreover, in this case $\text{Sym}(\text{Rep}(G)') = \text{Rep}(G)$ and $[\mathcal{C}_G]_0$ is equivalent to $(\text{Rep}(G)')_G$: the G-de-equivariantization of the centralizer of $\text{Rep}(G)$ inside $\mathcal{C}$. This is called the **modularization** of $\text{Rep}(G)'$ [**5**]. We remark that many algebra objects not of the form $\text{Fun}(G)$ exist and leads to a more general construction.

Here we describe a practical computational approach to boson condensation in the special case of $G = \mathbb{Z}/2$ which is all we will need. The non-trivial simple object in $\text{Rep}(\mathbb{Z}/2)$ is a boson, which we denote b. The corresponding algebra $A = \mathbf{1}+b$ is an object. In this case for any simple object X we have either $bX \cong X$ or $bX \cong Y$ and $bY \cong X$ for some simple object $Y \ncong X$. Let F be the $\mathbb{Z}/2$-de-equivariantization functor restricted to the subcategory $\langle b\rangle'$ (i.e. the centralizer of the boson b) so that we have $F : \langle b\rangle' \to \mathcal{C}_A^0$. For an $X \in \langle b\rangle'$, if $bX \cong Y \ncong X$ then $F(X) \cong F(Y)$ is a simple object in $\mathcal{C}_A^0$ with dimension $\dim(X)$. We can denote the simple for $F(X) \cong F(Y)$ by α, and call it type I. If $bX \cong X$ then $F(X) \cong \alpha_1 \oplus \alpha_2$ where α_i are simple objects in $\mathcal{C}_A^0$ with dimension $\dim(X)/2$. This is called type II. For $X \in \langle b\rangle'$ we have that $\theta_{F(X)} = \theta_X$.

We would like to compute the $\tilde{S}$-matrix entries of $\mathcal{C}_A^0$ in the special case that $A = \mathbf{1} + b$ for a boson b. More general results are known, see also the recent thesis [**2**] and [**23**]. For our purposes the following somewhat *ad hoc* approach will suffice.

Recall that the balancing equation, for simple objects X, Y is

$$\theta_X\theta_Y\tilde{S}_{X,Y} = \sum_{Z\in\text{Irr}(\mathcal{C})} N_{X^*,Y}^Z\theta_Z d_Z \tag{2.1}$$

where the sum is over simple objects Z, $N_{X^*,Y}^Z = \dim\text{Hom}(X^* \otimes Y, Z)$ and d_Z is the categorical dimension of Z. From this we can infer further information about the $\tilde{S}$-matrix of $\mathcal{C}_A^0$. The following result is well-known to experts but we include a proof for completeness:

THEOREM 2.2. *Suppose $b \in \mathcal{C}$ is a boson and F is the condensation functor $F : \langle b\rangle' \to \mathcal{C}_A^0$. Let $X, Y \in \langle b\rangle'$.*

(a) *Suppose that $F(X) = \alpha$ and $F(Y) = \beta$ are simple, i.e., type I simple objects. Then $\tilde{S}_{\alpha,\beta} = \tilde{S}_{X,Y}$.*
(b) *Suppose that $F(X) = \alpha$ is simple and $F(Y) \cong \beta_1 + \beta_2$ with β_1, β_2 simple, i.e., X is type I and Y is type II. Then $\tilde{S}_{\alpha,\beta_1} + \tilde{S}_{\alpha,\beta_2} = \tilde{S}_{X,Y}$.*

PROOF. Recall [**5**, **27**] that

$$\operatorname{Hom}(F(X)\otimes F(Y),F(Z))=\operatorname{Hom}(X\otimes Y,Z\otimes(\mathbf{1}+b)).$$

Suppose that $bX\ncong X$ and $bY\ncong Y$ are simple objects in $\langle b\rangle'$, so that $F(X)=\alpha$ and $F(Y)=\beta$ are simple objects in $\mathcal{C}_A^0$. Consider the $\tilde{S}$-matrix entry $\tilde{S}_{\alpha,\beta}$. By (2.1) we have:

$$\tilde{S}_{\alpha,\beta}=\frac{1}{\theta_\alpha\theta_\beta}\sum_{\gamma\in\operatorname{Irr}(\mathcal{C}_A^0)}N_{(\alpha)^*,\beta}^{\gamma}d_\gamma\theta_\gamma.$$

There are two cases to consider:

(1) There is a simple $Z\in\mathcal{C}$ such that $F(Z)=\gamma$. In this case $bZ\ncong Z$, and $d_\gamma=d_Z$ and $\theta_\gamma=\theta_Z$.
(2) There is a simple $Z\in\mathcal{C}$ such that $F(Z)\cong\gamma_1+\gamma_2$ with γ_i simple. In this case $bZ\cong Z$ and we have $\theta_{\gamma_1}=\theta_{\gamma_2}=\theta_Z$ and $d_{\gamma_1}=d_{\gamma_2}=\frac{d_Z}{2}$.

In the first case we see that $N_{(\alpha)^*,\beta}^{\gamma}d_\gamma\theta_\gamma=N_{X^*,Y}^{Z}d_Z\theta_Z+N_{X^*,Y}^{bZ}d_{bZ}\theta_{bZ}$ with $Z\ncong bZ$. In the second case we find that $N_{(\alpha)^*,\beta}^{\gamma_1}=N_{(\alpha)^*,\beta}^{\gamma_2}=N_{X^*,Y}^{Z}$, so that $N_{X^*,Y}^{Z}d_Z\theta_Z=N_{(\alpha)^*,\beta}^{\gamma_1}d_{\gamma_1}\theta_{\gamma_1}+N_{(\alpha)^*,\beta}^{\gamma_2}d_{\gamma_2}\theta_{\gamma_2}$. Thus we find that

$$\tilde{S}_{\alpha,\beta}=\sum_{\gamma\in\operatorname{Irr}(\mathcal{C}_A^0)}N_{(\alpha)^*,\beta}^{\gamma}d_\gamma\frac{\theta_\gamma}{\theta_\alpha\theta_\beta}=\sum_{Z\in\operatorname{Irr}(\mathcal{C})}N_{X^*,Y}^{Z}d_Z\frac{\theta_Z}{\theta_X\theta_Y}=\tilde{S}_{X,Y}$$

proving (a).

Now consider the case of simple objects X,Y in $\langle b\rangle'$ such that $bX\ncong X$ and $bY\cong Y$. From this we have that $F(X)=\alpha$ and $F(Y)\cong\beta_1+\beta_2$ where β_i are non-isomorphic simple objects of dimension $\dim(\beta_i)=\frac{\dim(Y)}{2}$. We argue similarly as above to see the following:

(1) If $\gamma=F(Z)$ comes from an object $Z\in\operatorname{Irr}(\mathcal{C})$ such that $bZ\ncong Z$ we have:

$$N_{X^*,Y}^{Z}+N_{X^*,Y}^{bZ}=N_{\alpha^*,\beta_1}^{\gamma}+N_{\alpha^*,\beta_2}^{\gamma},$$

while
(2) if $\gamma_1+\gamma_2\cong F(Z)$ for $Z\in\operatorname{Irr}(\mathcal{C})$ we find that

$$N_{\alpha^*,\beta_1}^{\gamma_1}+N_{\alpha^*,\beta_2}^{\gamma_1}+N_{\alpha^*,\beta_1}^{\gamma_2}+N_{\alpha^*,\beta_2}^{\gamma_2}=2N_{X^*,Y}^{Z}.$$

Since $Z\in\operatorname{Irr}(\mathcal{C})$ is of type I if $F(Z)=\gamma$ is simple, i.e. if $bZ\ncong Z$, and type II if $F(Z)\cong\gamma_1+\gamma_2$, i.e. $bZ\cong Z$, we can partition $\operatorname{Irr}(\mathcal{C})=\operatorname{Irr}(\mathcal{C})_I\cup b\operatorname{Irr}(\mathcal{C})_I\cup\operatorname{Irr}(\mathcal{C})_{II}$ where we have chosen representatives of each $\mathbb{Z}/2$-orbit for objects of type I so that every type I object is in exactly one of $\operatorname{Irr}(\mathcal{C})_I$ or $b\operatorname{Irr}(\mathcal{C})_I$. Recall that $\theta_{\beta_i}=\theta_Y$,

$\theta_\alpha = \theta_X$, $\theta_\gamma = \theta_Z$, $d_\gamma = d_Z$, and $d_{\gamma_i} = \frac{d_Z}{2}$. We compute:

$$\begin{aligned}
\tilde{S}_{\alpha,\beta_1} + \tilde{S}_{\alpha,\beta_2} &= \sum_{\gamma \in \mathrm{Irr}(\mathcal{C}_A^0)} (N_{\alpha^*,\beta_1}^{\gamma} + N_{\alpha^*,\beta_2}^{\gamma}) \frac{d_Z \theta_Z}{\theta_X \theta_Y} \\
&= \sum_{Z \in \mathrm{Irr}(\mathcal{C})_I} (N_{X^*,Y}^{Z} + N_{X^*,Y}^{bZ}) \frac{d_Z \theta_Z}{\theta_X \theta_Y} \\
&\quad + \sum_{\substack{Z \in \mathrm{Irr}(\mathcal{C})_{II} \\ F(Z) = \gamma_1 + \gamma_2}} (N_{\alpha^*,\beta_1}^{\gamma_1} + N_{\alpha^*,\beta_2}^{\gamma_1} + N_{\alpha^*,\beta_1}^{\gamma_2} + N_{\alpha^*,\beta_2}^{\gamma_2}) \frac{d_Z \theta_Z}{2\theta_X \theta_Y} \\
&= \sum_{Z \in \mathrm{Irr}(\mathcal{C})_I} (N_{X^*,Y}^{Z} + N_{X^*,Y}^{bZ}) \frac{d_Z \theta_Z}{\theta_X \theta_Y} \\
&\quad + \sum_{Z \in \mathrm{Irr}(\mathcal{C})_{II}} (2N_{X^*,Y}^{Z}) \frac{d_Z \theta_Z}{2\theta_X \theta_Y} = \tilde{S}_{X,Y}
\end{aligned}$$

proving (b). □

2.2. Near-group categories. Let G be a finite group of order n and m a nonnegative integer. A **near-group category** of type $G+m$ is a fusion category with simple objects labeled by elements $g \in G$ and an extra simple object labeled by ρ such that the fusion rules are generalized by the group operation in G, $g\rho = \rho g = \rho$ for all $g \in G$, and $\rho^{\otimes 2} = m\rho + \sum_{g \in G} g$. The Tambara-Yamagami categories are the near-group categories with $m = 0$, which are fully classified in [**33**]. It is known [**18**, Theorem 2] that the only possible values of m are n, $n-1$ or $m = kn$ for some nonnegative integer k. This paper focuses on the cases when $m = n$ since these are related to the modular data we aim to realize.

THEOREM 2.3 ([**20**, Theorem 5.3], [**18**, Corollary 5]). *Let G be a finite abelian group of order n, $\langle\ ,\ \rangle$ a non-degenerate symmetric bicharacter on G and define $d = \dfrac{n + \sqrt{n^2+4n}}{2}$. Let $c \in \mathbb{T}$, $a : G \to \mathbb{T}$, $b : G \to \mathbb{C}$ be such that*

$$a(0) = 1, \quad a(x) = a(-x), \quad a(x+y)\langle x, y\rangle = a(x)a(y), \quad \sum_{a \in G} a(x) = \sqrt{n}c^{-3}, \tag{2.2}$$

$$b(0) = -\frac{1}{d}, \quad \sum_y \overline{\langle x, y\rangle b(y)} = \sqrt{n}c\overline{b(x)}, \quad a(x)b(-x) = \overline{b(x)}, \tag{2.3}$$

$$\sum_x b(x+y)\overline{b(x)} = \delta_{y,0} - \frac{1}{d}, \quad \sum_x b(x+y)b(x+z)\overline{b(x)} = \overline{\langle y, z\rangle b(y)b(z)} - \frac{c}{d\sqrt{n}}. \tag{2.4}$$

Then $\langle\ ,\ \rangle$, c, a, b determine a near-group fusion category of type $G+n$. Two such categories $\mathcal{C}_1$ and $\mathcal{C}_2$ determined by $\langle\ ,\ \rangle_1$, c_1, a_1, b_1 and $\langle\ ,\ \rangle_2$, c_2, a_2, b_2 are equivalent as fusion categories if and only if there is $\phi \in \mathrm{Aut}(G)$ such that $\langle x, y\rangle_1 = \langle \phi x, \phi y\rangle_2, a_1(x) = a_2(\phi x)$, $b_1(x) = b_2(\phi x)$ and $c_1 = c_2$.

2.3. Centers of near-group categories when $m = n$. The modular data for the center of a near-group of type $G+n$, for $|G| = n$, is given as follows [**20**].

First, we need to find all functions $\xi : G \to \mathbb{T}$ and values $\tau \in G, \omega \in \mathbb{T}$ which satisfy

$$\sum_g \xi(g) = \sqrt{n}\omega^2 a(\tau) c^3 - nd^{-1} \tag{2.5}$$

$$\bar{c}\sum_k b(g+k)\xi(k) = \omega^2 c^3 a(\tau)\overline{\xi(g+\tau)} - \sqrt{n}d^{-1} \tag{2.6}$$

$$\xi(\tau - g) = \omega c^4 a(g) a(\tau - g)\overline{\xi(g)} \tag{2.7}$$

$$\sum_k \xi(k) b(k-g) b(k-h) = c^{-2} b(g+h-\tau)\xi(g)\xi(h)\overline{a(g-h)} - c^2 d^{-1} \tag{2.8}$$

There are $n(n+3)/2$ triples $(\xi_i, \tau_i, \omega_i)$ that satisfy the above equations. The corresponding center has rank $n(n+3)$ with the following 4 subsets of simple objects:

(1) $\mathfrak{a}_g, g \in G$, $\dim(\mathfrak{a}_g) = 1$;
(2) $\mathfrak{b}_h, h \in G$, $\dim(\mathfrak{b}_h) = d+1$;
(3) $\mathfrak{c}_{l,k} = \mathfrak{c}_{k,l}, l, k \in G, l \neq k$, $\dim(\mathfrak{c}_{k,l}) = d+2$
(4) $\mathfrak{d}_j$, where j corresponds to a triple $(\xi_j, \tau_j, \omega_j)$, $\dim(\mathfrak{d}_j) = d$.

The T and S matrices are given by the following block form

$$T = \operatorname{diag}\left[\langle g, g\rangle, \langle h, h\rangle, \langle k, l\rangle, \omega_j\right] \tag{2.9}$$

(2.10)

$$S = \frac{1}{\lambda}\begin{bmatrix} \langle g, g'\rangle^{-2} & (d+1)\langle g, h'\rangle^{-2} & (d+2)\overline{\langle g, k'+l'\rangle} & d\langle g, \tau_{j'}\rangle \\ (d+1)\langle h, g'\rangle^{-2} & \langle h, h'\rangle^{-2} & (d+2)\overline{\langle h, k'+l'\rangle} & -d\langle h, \tau_{j'}\rangle \\ (d+2)\overline{\langle k+l, g'\rangle} & (d+2)\overline{\langle k+l, h'\rangle} & S_{(k,l),(k',l')} & \mathbf{0} \\ d\langle \tau_j, g'\rangle & -d\langle \tau_j, h'\rangle & \mathbf{0} & S_{j,j'} \end{bmatrix},$$

where[2]

$$\begin{aligned} S_{j,j'} &= \omega_j \omega_{j'} \sum_{g\in G} \langle \tau_j + \tau_{j'} + g, g\rangle \\ &+ d\omega_j \omega_{j'} c^6 a(\tau_j) a(\tau_{j'}) n^{-1} \sum_{g,h\in G} \overline{\xi_j(g)\xi_{j'}(h)\langle \tau_{j'} - \tau_j + h - g, h - g\rangle}, \end{aligned} \tag{2.11}$$

and

$$S_{(k,l),(k',l')} = (d+2)\left(\overline{\langle k, k'\rangle\langle l, l'\rangle} + \overline{\langle k, l'\rangle\langle l, k'\rangle}\right). \tag{2.12}$$

REMARK 2.4. As it may be useful to other researchers, here is our approach to finding solutions (ξ, τ, ω) to equations (2.5)–(2.8). We fix ω a root of unity and $\tau \in G$. Notice that equation (2.6) can be rephrased to simplify solving for ξ: equation (2.7) implies that

$$\overline{\xi(g+\tau)} = \xi(-g)\overline{\omega c^4 a(g+\tau) a(g)}.$$

Substituting this into equation (2.6), we have

$$\omega a(\tau)\overline{a(g+\tau)a(g)}\xi(-g) - \sum_k b(g+k)\xi(k) = \frac{c\sqrt{n}}{d}. \tag{2.13}$$

Let $C = C(\omega, \tau)$ be the matrix indexed by G with entries

$$C_{g,h}(\omega, \tau) = \omega a(\tau)\overline{a(g+\tau)a(g)}\delta_{g,-g}.$$

[2] There is a slight difference between our eqn. (2.11) and that of [**18**, eqn. (4.58)]: the τ_j and $\tau_{j'}$ are switched. This yielded consistent modular data in our setting.

Let B be the matrix with entries $B_{g,k} = b(g+k), g, k \in G$. Then equation (2.13) becomes the system:

$$(C(\omega,\tau) - B)\vec{\xi} = \vec{z} \tag{2.14}$$

where $\vec{z}$ has all entries equal to $\frac{c\sqrt{n}}{d}$.

We consider all pairs (ω, τ) where $\tau \in G$ and ω is a root of unity, of bounded degree [**9, 16**], and first solve the system (2.14). This gives an affine set consisting of vectors $\xi_h + \xi_p$ where ξ_p is a particular solution and $\xi_h \in \mathrm{Null}(C(\omega,\tau) - B)$. Depending on the dimension of $\mathrm{Null}(C(\omega,\tau) - B)$ we choose an appropriate number of equations (possibly 0) from (2.8) to determine the free parameters and obtain the vector form of ξ. We then test the corresponding triples (ξ, τ, ω) on the remaining equations of (2.5)–(2.8) to determine if the triple is a solution or not. When we have found exactly $n(n+3)/2$ triples (ξ, τ, ω) that satisfy all equations we stop and compute the S-matrix.

2.4. Leveraging the pointed part of near-group centers. Suppose $\mathcal{C}$ is a near-group category of type $G + n$ where $n = |G|$, corresponding to the non-degenerate symmetric bicharacter $\langle\ ,\ \rangle$. Using (2.10) and (2.9) we can glean a wealth of information about the center $\mathcal{Z}(\mathcal{C})$, by examining the pointed part.

The objects $\mathfrak{a}_g$ with $\mathfrak{a}_g^{\otimes 2} = \mathbf{1}$, i.e., $2g = 0$ as a group element, and $\langle g, g\rangle = 1$ are bosons. Generally, if $\mathfrak{a}_g$ is an invertible object then we can determine its centralizer. It is generated by:

(1) the invertibles $\mathfrak{a}_h$ of dimension 1 with $\langle g, h\rangle^{-2} = 1$,
(2) the simple objects $\mathfrak{b}_h$ of dimension $d+1$ with $\langle g, h\rangle^{-2} = 1$,
(3) the objects $\mathfrak{c}_{k,\ell}$ of dimension $d+2$ with $\langle g, k+\ell\rangle = 1$, and
(4) the objects $\mathfrak{d}_j = (\xi_j, \tau_j, \omega_j)$ of dimension d such that $\langle g, \tau_j\rangle = 1$.

Suppose that $|G|$ is even so that there is an element g with $2g = 0$. The corresponding object $\mathfrak{a}_g$ has $\mathfrak{a}_g^{\otimes 2} = \mathbf{1}$ and is either a boson or a fermion, depending on the value of $\langle g, g\rangle \in \{1, -1\}$ (recall that $\langle g, h\rangle$ is a bicharacter, so that the value of $\langle g, h\rangle = \pm 1$). Now if $b = \mathfrak{a}_g$ is a boson we can determine a significant portion of the condensation by $\langle b\rangle \cong \mathrm{Rep}(\mathbb{Z}/2)$. First note that $\langle g, h\rangle^{-2} = \langle 2g, h\rangle^{-1} = 1$ since $2g = 0$. Thus all of the objects of $\mathfrak{a}_h, \mathfrak{b}_h$ are centralized by b. Moreover, the objects $\mathfrak{b}_h$ must have $b \otimes \mathfrak{b}_h \not\cong \mathfrak{b}_h$. If not, then under the condensation functor the image of $\mathfrak{b}_h$ would be a sum of two simple objects of dimension $\frac{d+1}{2}$, which is not an algebraic integer: indeed $d+1$ is a unit in $\mathbf{Q}[d]$ so $\frac{d+1}{2}$ has norm $1/4$. On the other hand $\frac{d}{2}$ and $\frac{d+2}{2}$ are algebraic integers if and only if $4 \mid n$. So, the objects $\mathfrak{d}_j$ and $\mathfrak{c}_{k,l}$ could be fixed under tensoring with the boson if $4 \mid n$.

EXAMPLE 2.5. Now suppose that $G = \mathbb{Z}/m2^s$, where m is odd and $s \geq 2$. Suppose that there is a near-group fusion category $\mathcal{C}$ realizing the fusion rules of type $G + m2^s$. If we write G additively, the non-degenerate quadratic form a on G is given by

$$a(x,y) = exp(\frac{2\varepsilon_1 \pi i x^2}{m})exp(\frac{\varepsilon_2 \pi i y^2}{2^s})$$

where $\varepsilon_1 = \pm 1$ and $\varepsilon_2 \in \{\pm 1, \pm 5\}$. Consider the Drinfeld center $\mathcal{Z}(\mathcal{C})$. First notice that the $\mathcal{C}(\mathbb{Z}/m, q)$ part of the pointed subcategory in $\mathcal{Z}(\mathcal{C})$ is modular since the S-matrix entries are $\langle (x_1, 0), (x_2, 0)\rangle^{-2} = \left(\frac{a(x_1+x_2,0)}{a(x_1,0)a(x_2,0)}\right)^{-2}$ which is non-degenerate. The element $(0, 2^{s-1})$ has order 2 and $\langle (0, 2^{s-1}), (0, 2^{s-1})\rangle = 1$ and so

$\mathfrak{a}_{2^{s-1}} = b$ is a boson. Since none of the invertible objects are fixed by tensoring with b we find that the pointed part of $[\mathcal{Z}(\mathcal{C})_{\mathbb{Z}/2}]_0$ is the product of the modular $\mathcal{C}(\mathbb{Z}/m, q)$ above and a category $\mathcal{C}(\mathbb{Z}/2^{s-1}, A)$, with *non-degenerate* quadratic form $A(x) = \langle x, x\rangle = exp(\pi i \varepsilon_2 x^2/2^{s-1})$. Thus, the pointed part of $[\mathcal{Z}(\mathcal{C})_{\mathbb{Z}/2}]_0$ is modular so that $[\mathcal{Z}(\mathcal{C})_{\mathbb{Z}/2}]_0 \cong \mathcal{C}(\mathbb{Z}/m, q) \boxtimes \mathcal{C}(\mathbb{Z}/2^{s-1}, A) \boxtimes \mathcal{D}$ where $\mathcal{D}$ is a modular category with trivial pointed part.

In particular we see that no non-split super-modular categories are obtained from near-group categories associated with $\mathbb{Z}/m2^s$ for $s \geq 2$ and any odd m.

EXAMPLE 2.6. Now suppose that G is an abelian group such that $|G| = 2m$ with m odd and there is a near-group category $\mathcal{C}$ with fusion rules as type $G + 2m$. In this case a similar computation reveals that $\mathcal{Z}(\mathcal{C}) \cong \mathcal{C}(H, q) \boxtimes \mathcal{D}$ where $\mathcal{D}$ is spin modular and H is a group of order m. Indeed, the pointed subcategory of $\mathcal{Z}(\mathcal{C})$ of order m is non-degenerate since the square of a non-degenerate quadratic form on an abelian group of odd order is also non-degenerate. Moreover, $\mathcal{D}$ does not contain a minimal modular extension of sVec, and hence $\mathcal{D}_0$ is non-split super-modular with simple objects of dimension 1, $d = m + \sqrt{m^2 + 2m}$, $d + 1$ and $d + 2$. We shall see this explicitly in the case of $m = 3$ below.

3. Realizing modular data

In [**12**], two families of unrealized super-modular data are constructed from the congruence representations of the Γ_θ group, as outlined in Section 1. Below are two representatives of them, where $\chi_n^m = m + \sqrt{n}$.

$$\hat{S} = \frac{1}{\sqrt{30\chi_{15}^4}} \begin{bmatrix} 1 & \chi_{15}^4 & \chi_{15}^5 & \chi_{15}^3 & \chi_{15}^3 \\ \chi_{15}^4 & 1 & \chi_{15}^5 & -\chi_{15}^3 & -\chi_{15}^3 \\ \chi_{15}^5 & \chi_{15}^5 & -\chi_{15}^5 & 0 & 0 \\ \chi_{15}^3 & -\chi_{15}^3 & 0 & \frac{1}{2}\chi_5^1\chi_{15}^3 & -\frac{2\sqrt{30\chi_{15}^4}}{\chi_5^5} \\ \chi_{15}^3 & -\chi_{15}^3 & 0 & -\frac{2\sqrt{30\chi_{15}^4}}{\chi_5^5} & \frac{1}{2}\chi_5^1\chi_{15}^3 \end{bmatrix}, \tag{3.1}$$
$$\hat{T}^2 = \mathrm{diag}\left[1, 1, e^{2i\pi/3}, e^{4i\pi/5}, e^{-4i\pi/5}\right].$$

$$\hat{S} = \frac{1}{2\sqrt{6\chi_{24}^5}} \begin{bmatrix} 1 & \chi_{24}^5 & \chi_6^3 & \chi_6^3 & \chi_{24}^4 \\ \chi_{24}^5 & 1 & \chi_6^3 & \chi_6^3 & -\chi_{24}^4 \\ \chi_6^3 & \chi_6^3 & -\chi_6^3 - i\sqrt{6\chi_{24}^5} & -\chi_6^3 + i\sqrt{6\chi_{24}^5} & 0 \\ \chi_6^3 & \chi_6^3 & -\chi_6^3 + i\sqrt{6\chi_{24}^5} & -\chi_6^3 - i\sqrt{6\chi_{24}^5} & 0 \\ \chi_{24}^4 & -\chi_{24}^4 & 0 & 0 & \chi_{24}^4 \end{bmatrix}, \tag{3.2}$$
$$\hat{T}^2 = \mathrm{diag}[1, 1, -1, -1, e^{2i\pi/3}].$$

In [**12**, Table 6] we find a family of 4 different data related to (3.1) and a pair of data related to (3.2) with positive dimensions. They are related to each other by Galois automorphisms, so we will be satisfied with realizing one set of data for each family.

Observing that $\frac{6+\sqrt{6^2+24}}{2} = 3+\sqrt{15} = \chi_{15}^3$ and $\frac{8+\sqrt{8^2+32}}{2} = 4+\sqrt{24} = \chi_{24}^4$ it is reasonable to expect that this super-modular data is related to that of the centers of near-group fusion categories coming from groups of order 6 and 8. We will show that this is indeed the case: for $G = \mathbb{Z}/6$ and $G = \mathbb{Z}/2 \times \mathbb{Z}/4$. We point out that by Example 2.5 $G = \mathbb{Z}/8$ will not yield a non-split super modular category, whereas the near-group fusion rule associated with $G = (\mathbb{Z}/2)^3$ is not realizable by results of [**31**].

3.1. $G = \mathbb{Z}/6$. There are 4 inequivalent near-group fusion categories [**11**,**18**] of type $\mathbb{Z}/6+6$. Their basic data is summarized in Table 1. The nondegenerate symmetric bicharacter for a cyclic group $\mathbb{Z}/n$ is in the form of $\langle x, y\rangle = \exp(2\pi imxy/n)$, where $m \in \mathbb{Z}$ and $\gcd(m,n) = 1$. When n is even, the function a is in the form of $a(x) = \varepsilon^x \exp\left(-\pi imx^2/n\right)$, where $\varepsilon = \pm 1$. By taking $m \in \mathbb{Z}/2n$ and coprime to n, we can eliminate the potential factor of -1. This m is recorded in the second column of Table 1. As $b(0) = -\frac{1}{d}$, $b(-x) = \overline{a(x)b(x)}$ and $b(x) = \frac{1}{\sqrt{n}}\exp(ij(x))$, where $j(x) \in [-\pi,\pi]$, it suffices to list the values of $j(1), j(2)$ and $j(3)$ to determine $b(x)$.

TABLE 1. The data for near-group categories of type $\mathbb{Z}/6+6$.

#	m	c	$j(1), j(2), j(3)$
J_6^1	5	ζ_{24}^5	2.91503, −1.59091, 2.35619
$\overline{J_6^1}$	−5	ζ_{24}^{-5}	−2.91503, 1.59091, −2.35619
J_6^2	1	ζ_{24}^1	2.95526, 0.0553542, −0.785398
$\overline{J_6^2}$	−1	ζ_{24}^{-1}	−2.95526, −0.0553542, 0.785398

We consider the near-group J_6^1 in Table 1 and solve equations (2.5)–(2.8). The solutions $(\omega_j, \tau_j, \xi_j)$ are listed in Table 2. The ω_j's are equal to ζ_{60}^k for some integer k, and we list the values for k in the corresponding column. As $\xi_j(g) = \exp(i\theta_{j,g})$, we record the values $\theta_{j,g}$ in the column for ξ with the order $g = 0, \ldots, 5 \in \mathbb{Z}/6$.

The center of the near-group category J_6^1 has the following simple objects:

six invertible objects X_g, $g \in \mathbb{Z}/6$;
six χ_{15}^4-dimensional objects Y_h, $h \in \mathbb{Z}/6$;
fifteen χ_{15}^5-dimensional objects $Z_{k,l}$, $k,l \in \mathbb{Z}/6$, $k < l$; and
twenty-seven χ_{15}^3-dimensional objects $W_{\omega_i,\tau_i,\xi_i}$, where $(\omega_i, \tau_i, \xi_i)$ are solutions in Table 2.

TABLE 2. $(\omega_j, \tau_j, \xi_j)$ for J_6^1 in Table 1

#	ω	τ	ξ
1	12	0	−3.03687, 1.02812, −0.75497, 1.67552, −1.12999, 0.228519
2	18	0	0.418879, −0.476051, 1.75517, 1.98968, −3.0118, 2.36101
3	42	0	1.67552, −1.12999, 0.228519, −3.03687, 1.02812, −0.75497
4	48	0	1.98968, −3.0118, 2.36101, 0.418879, −0.476051, 1.75517
5	23	1	−2.58859, 1.33196, 1.6366, −0.961707, −0.29493, −0.798841
6	23	1	−0.961707, −0.29493, −0.798841, −2.58859, 1.33196, 1.6366
7	47	1	3.06462, −1.80798, −1.19626, −0.601743, 1.85838, −1.73589
8	47	1	−0.601743, 1.85838, −1.73589, 3.06462, −1.80798, −1.19626
9	35	1	2.69346, −2.69346, 1.0472, 2.69346, −2.69346, 1.0472
10	2	2	0.228519, −3.03687, 1.02812, −0.75497, 1.67552, −1.12999
11	8	2	2.36101, 0.418879, −0.476051, 1.75517, 1.98968, −3.0118
12	32	2	−0.75497, 1.67552, −1.12999, 0.228519, −3.03687, 1.02812
13	38	2	1.75517, 1.98968, −3.0118, 2.36101, 0.418879, −0.476051
14	3	3	1.6366, −0.961707, −0.29493, −0.798841, −2.58859, 1.33196
15	3	3	−0.798841, −2.58859, 1.33196, 1.6366, −0.961707, −0.29493
16	27	3	−1.73589, 3.06462, −1.80798, −1.19626, −0.601743, 1.85838
17	27	3	−1.19626, −0.601743, 1.85838, −1.73589, 3.06462, −1.80798
18	15	3	1.0472, 2.69346, −2.69346, 1.0472, 2.69346, −2.69346
19	2	4	1.02812, −0.75497, 1.67552, −1.12999, 0.228519, −3.03687
20	8	4	−0.476051, 1.75517, 1.98968, −3.0118, 2.36101, 0.418879
21	32	4	−1.12999, 0.228519, −3.03687, 1.02812, −0.75497, 1.67552
22	38	4	−3.0118, 2.36101, 0.418879, −0.476051, 1.75517, 1.98968
23	23	5	1.33196, 1.6366, −0.961707, −0.29493, −0.798841, −2.58859
24	23	5	−0.29493, −0.798841, −2.58859, 1.33196, 1.6366, −0.961707
25	47	5	1.85838, −1.73589, 3.06462, −1.80798, −1.19626, −0.601743
26	47	5	−1.80798, −1.19626, −0.601743, 1.85838, −1.73589, 3.06462
27	35	5	−2.69346, 1.0472, 2.69346, −2.69346, 1.0472, 2.69346

Applying the formulas for the modular data in Section 2.3, we have the following T- and S-matrices

$$T_{X_g} = e\left(\frac{5g^2}{6}\right), \quad T_{Y_h} = e\left(\frac{5h^2}{6}\right), \quad T_{Z_{k,l}} = e\left(\frac{5kl}{6}\right), g, h, k, l \in \mathbb{Z}/6, k < l,$$

$$T_{W\omega,\tau,\xi} = \mathrm{diag}[e(\tfrac{1}{5}), e(\tfrac{3}{10}), e(-\tfrac{3}{10}), e(-\tfrac{1}{5}), e(\tfrac{23}{60}), e(\tfrac{23}{60}), e(-\tfrac{13}{60}), e(-\tfrac{13}{60}), e(-\tfrac{5}{12}), e(\tfrac{1}{30}), e(\tfrac{2}{15}), e(-\tfrac{7}{15}), e(-\tfrac{11}{30}), e(\tfrac{1}{20}), e(\tfrac{1}{20}), e(\tfrac{9}{20}), e(\tfrac{9}{20}), e(\tfrac{1}{4}), e(\tfrac{1}{30}), e(\tfrac{2}{15}), e(-\tfrac{7}{15}), e(-\tfrac{11}{30}), e(\tfrac{23}{60}), e(\tfrac{23}{60}), e(-\tfrac{13}{60}), e(-\tfrac{13}{60}), e(-\tfrac{5}{12})]; \text{ and}$$

$$S = \frac{1}{\lambda_1}\begin{bmatrix} s_{X,X} & \chi_{15}^4 s_{X,X} & \chi_{15}^5 s_{X,Z} & \chi_{15}^3 s_{X,W} \\ \chi_{15}^4 s_{X,X} & s_{X,X} & \chi_{15}^5 s_{X,Z} & -\chi_{15}^3 s_{X,W} \\ \chi_{15}^5 s_{X,Z}^T & \chi_{15}^5 s_{X,Z}^T, & \chi_{15}^5 s_{Z,Z} & \mathbf{0} \\ \chi_{15}^3 s_{X,W}^T & -\chi_{15}^3 s_{X,W}^T & \mathbf{0} & s_{W,W} \end{bmatrix},$$

$$\lambda_1 = 6\sqrt{10\chi_{15}^4}, \quad s_{X_g,X_h} = e\left(\frac{gh}{3}\right),$$

$$s_{X_g,Z_{k,l}} = e\left(\frac{g(k+l)}{6}\right), \quad s_{Z_{k,l},Z_{k',l'}} = e\left(\frac{kk'+ll'}{6}\right) + e\left(\frac{kl'+k'l}{6}\right),$$

where $g, h, k, k', l, l' \in \mathbb{Z}/6, k \neq l, k' \neq l'$, and the entries for $s_{X,W}$ and $s_{W,W}$ can be obtained from equations (2.10)and (2.11) using the solutions in Table 2. We provide detailed modular data in the Mathematica notebook `CenterofJ61.nb` in the `arxiv` source.

THEOREM 3.1. *Let $\mathcal{C}$ be the Drinfeld center of a near-group category of type $\mathbb{Z}/6+6$. Then $\mathcal{C} \cong \mathcal{D} \boxtimes \mathcal{C}(\mathbb{Z}/3, q)$, where $\mathcal{D}$ is a spin modular category, and q is its associated quadratic form restricted to $\mathbb{Z}/3$. Moreover, the Müger centralizer of the fermion f in $\mathcal{D}$ is super-modular and either itself or one of its Galois conjugates has the same super-modular data as in* (3.1).

PROOF. We first consider the case J_6^1. Take the pointed modular category $\mathcal{C}(\mathbb{Z}/3, q)$ with $q = e^{2\pi i a^2/3}$, $a \in \mathbb{Z}/3$. Notice $\mathcal{C}$ has $\mathcal{C}(\mathbb{Z}/3, q)$ as a pointed modular subcategory, generated by the invertible objects X_0, X_2 and X_4. Also, one observes that the invertible object X_3 is a fermion. Thus $\mathcal{C} \cong \mathcal{D} \boxtimes \mathcal{C}(\mathbb{Z}/3, q)$, where $\mathcal{D}$ is a spin modular category with the fermion $f = X_3$. Moreover, $\mathcal{D} \cong \mathcal{D}_0 \oplus \mathcal{D}_1$, where $\mathcal{D}_0 = C_{\mathcal{D}}(\langle f \rangle)$ is a super-modular category.

Specifically, the spin modular category $\mathcal{D}$ is generated by the simple objects: $\mathbf{1}$, X_3, Y_0, Y_3, $Z_{0,1}$, $Z_{0,3}$, $Z_{0,5}$, $Z_{1,5}$, $Z_{2,4}$, $W_1, \ldots, W_9$. The 10 simple objects in $\mathcal{D}_0$ are $\mathbf{1}$, X_3, Y_0, Y_3, $Z_{1,5}$, $Z_{2,4}$, $W_1, \ldots, W_4$. The resulting super-modular data is

$$\hat{S} = \frac{1}{\sqrt{30\chi_{15}^4}} \begin{bmatrix} 1 & \chi_{15}^4 & \chi_{15}^5 & \chi_{15}^3 & \chi_{15}^3 \\ \chi_{15}^4 & 1 & \chi_{15}^5 & -\chi_{15}^3 & -\chi_{15}^3 \\ \chi_{15}^5 & \chi_{15}^5 & -\chi_{15}^5 & 0 & 0 \\ \chi_{15}^3 & -\chi_{15}^3 & 0 & \frac{1}{2}\chi_5^1\chi_{15}^3 & -\frac{2\sqrt{30\chi_{15}^4}}{\chi_5^5} \\ \chi_{15}^3 & -\chi_{15}^3 & 0 & -\frac{2\sqrt{30\chi_{15}^4}}{\chi_5^5} & \frac{1}{2}\chi_5^1\chi_{15}^3 \end{bmatrix}, \tag{3.3}$$
$$\hat{T}^2 = \text{diag}\left[1, 1, e^{2i\pi/3}, e^{4i\pi/5}, e^{-4i\pi/5}\right].$$

which coincides with (3.1).

The other cases in Table 1 can be similarly verified. Each of them provides a realization of the super-modular data in [**12**, Table 6] with the global dimension 472.379. □

3.2. $G = \mathbb{Z}/2 \times \mathbb{Z}/4$. Up to equivalence, there are 4 inequivalent near group categories of type $\mathbb{Z}/2 \times \mathbb{Z}/4 + 8$ listed in Table 3 [**18**, Proposition 6].

TABLE 3. The data for near-group categories of type $\mathbb{Z}/2 \times \mathbb{Z}/4 + 8$.

#	c	$\langle , \rangle$	a	$j(0,1), j(0,2), j(1,0), j(1,1), j(1,2)$
$J_{(2,4)}^1$	ζ_{12}^5	1	(1, 1)	$-0.992441, 1.5708, 0.785398, -1.42977, -0.785398$
$J_{(2,4)}^2$	ζ_{12}^{-5}	-1	$(-1, 1)$	$0.992441, -1.5708, -0.785398, 1.42977, 0.785398$
$J_{(2,4)}^3$	ζ_{12}^{-5}	1	$(1, -1)$	$1.42977, -1.5708, 0.785398, -1.77784, -0.785398$
$J_{(2,4)}^4$	ζ_{12}^5	-1	$(-1, -1)$	$-1.42977, 1.5708, -0.785398, 1.77784, 0.785398$

In Table 3, the bicharacter for $\langle , \rangle$ is of the form

$$\langle (x_1, y_1), (x_2, y_2) \rangle = \exp(2\pi i x_1 x_2/2) \exp(2\pi i m y_1 y_2/4),$$

where m is given in the $\langle , \rangle$ column. The functions a are given by $a(x,y) = s_1^x s_2^y \exp(-\pi i x^2/2) \exp(-\pi i m y^2/4)$, where $s_1, s_2 \in \{\pm 1\}$ are given in the column for a. Note $b(x) = \frac{1}{\sqrt{n}} \exp(ij(x))$ if $x \neq 0$, where $j(x) \in [-\pi, \pi]$, we list the values for $j(x)$ in Table 3 for the 5 values of x needed to determine $b(x)$: the remaining values can be obtained from $b(0) = -\frac{1}{d}$ and $b(-x) = \overline{a(x)b(x)}$.

We consider the near-group $J^1_{(2,4)}$ and solve for the triples $(\omega_j, \tau_j, \xi_j)$ using the equations (2.5)–(2.8). The solutions are listed in Table 4. The ω_j equals to ζ_{48}^k for some $k \in \mathbb{Z}$, which are recorded in the column of ω. Since $\xi_j(g) = \exp(i\theta_{j,g})$, the values of $\theta_{j,g}$ are listed in the column of ξ with the order

$$(0,0), \dots, (0,3), (1,0), \dots, (1,3) \in \mathbb{Z}/2 \times \mathbb{Z}/4.$$

We also provide the solutions to $(\omega_j, \tau_j, \xi_j)$ and further computation of modular data in the arxiv source `CenterofJ241.nb`.

The center of the near-group category $J^1_{(2,4)}$ has rank 88 with the following simple objects:

eight invertible objects X_g, $g \in \mathbb{Z}/2 \times \mathbb{Z}/4$;
eight χ_{24}^5-dimensional objects Y_h, $h \in \mathbb{Z}/2 \times \mathbb{Z}/4$;
twenty-eight $2\chi_6^3$-dimensional objects $Z_{k,l}$, $k, l \in \mathbb{Z}/2 \times \mathbb{Z}/4$, $k \neq l$; and
forty-four χ_{24}^4-dimensional objects $W_{\omega_i,\tau_i,\xi_i}$, where $(\omega_i, \tau_i, \xi_i)$ are solutions in Table 4.

The modular data is given as the following:

$$T_{X_g} = e\left(\frac{2g_1^2 + g_2^2}{4}\right), \quad T_{Y_h} = e\left(\frac{2h_1^2 + h_2^2}{4}\right), \quad T_{Z_{k,l}} = e\left(\frac{2k_1 l_1 + k_2 l_2}{4}\right),$$

where $g, h, k, l \in \mathbb{Z}/2 \times \mathbb{Z}/4, k \neq l$.

$$\begin{aligned} T_{W_{\omega,\tau,\xi}} = \operatorname{diag}\Big[& e\left(\tfrac{1}{3}\right), e\left(\tfrac{1}{3}\right), e\left(-\tfrac{1}{6}\right), e\left(-\tfrac{1}{6}\right), e\left(\tfrac{7}{16}\right), e\left(-\tfrac{1}{16}\right), e\left(\tfrac{13}{48}\right), e\left(\tfrac{13}{48}\right), \\ & e\left(-\tfrac{11}{48}\right), e\left(-\tfrac{11}{48}\right), e\left(\tfrac{1}{12}\right), e\left(\tfrac{1}{12}\right), e\left(-\tfrac{5}{12}\right), e\left(-\tfrac{5}{12}\right), e\left(\tfrac{7}{16}\right), e\left(-\tfrac{1}{16}\right), e\left(\tfrac{13}{48}\right), \\ & e\left(\tfrac{13}{48}\right), e\left(-\tfrac{11}{48}\right), e\left(-\tfrac{11}{48}\right), e\left(\tfrac{11}{24}\right), e\left(\tfrac{11}{24}\right), e\left(\tfrac{11}{24}\right), e\left(\tfrac{11}{24}\right), e\left(\tfrac{1}{8}\right), e\left(\tfrac{1}{8}\right), e\left(\tfrac{5}{16}\right), \\ & e\left(-\tfrac{3}{16}\right), e\left(\tfrac{7}{48}\right), e\left(\tfrac{7}{48}\right), e\left(-\tfrac{17}{48}\right), e\left(-\tfrac{17}{48}\right), e\left(-\tfrac{7}{24}\right), e\left(-\tfrac{7}{24}\right), e\left(-\tfrac{7}{24}\right), e\left(-\tfrac{7}{24}\right), \\ & e\left(\tfrac{3}{8}\right), e\left(\tfrac{3}{8}\right), e\left(\tfrac{5}{16}\right), e\left(-\tfrac{3}{16}\right), e\left(\tfrac{7}{48}\right), e\left(\tfrac{7}{48}\right), e\left(-\tfrac{17}{48}\right), e\left(-\tfrac{17}{48}\right)\Big]; \end{aligned}$$

$$S = \frac{1}{\lambda_2}\begin{bmatrix} s_{X,X} & \chi_{24}^5 s_{X,X} & 2\chi_6^3 s_{X,Z} & \chi_{24}^4 s_{X,W} \\ \chi_{24}^5 s_{X,X} & s_{X,X} & 2\chi_6^3 s_{X,Z} & -\chi_{24}^4 s_{X,W} \\ 2\chi_6^3 s_{X,Z}^T & 2\chi_6^3 s_{X,Z}^T, & 2\chi_6^3 s_{Z,Z} & \mathbf{0} \\ \chi_{24}^4 s_{X,W}^T & -\chi_{24}^4 s_{X,W}^T & \mathbf{0} & s_{W,W} \end{bmatrix},$$

where

$$\lambda_2 = 16\sqrt{3\chi_{24}^5}, \quad s_{X_g,X_h} = e\left(-\tfrac{2g_1h_1+g_2h_2}{2}\right), \quad s_{X_g,Z_{k,l}} = e\left(-\tfrac{2g_1(k_1+l_1)+g_2(k_2+l_2)}{4}\right),$$

$$s_{Z_{k,l},Z_{k',l'}} = e\left(-\tfrac{2k_1k_1'+k_2k_2'+2l_1l_1'+l_2l_2'}{4}\right) + e\left(-\tfrac{2k_1'l_1+k_2'l_2+2k_1l_1'+k_2l_2'}{4}\right),$$

$$g, h, k, k', l, l' \in \mathbb{Z}/2 \times \mathbb{Z}/4, k \neq l, k' \neq l',$$

and $s_{X,W}$ and $s_{W,W}$ can be computed from equations (2.10) and (2.11) using the solutions in Table 4.

THEOREM 3.2. *Let $\mathcal{C}$ be the Drinfeld center of a near-group category of type $\mathbb{Z}/2 \times \mathbb{Z}/4 + 8$. Then $\left[\mathcal{C}_{\mathbb{Z}/2}\right]_0 \cong \mathcal{D} \boxtimes \mathcal{C}(\mathbb{Z}/2, q)$, where $\mathcal{D}$ is a spin modular category and q is the associated quadratic form restricted to $\mathbb{Z}/2$. Moreover, the Müger centralizer of the fermion f in $\mathcal{D}$ is super-modular and either itself or one of its Galois conjugates has the same super-modular data as in* (3.2).

TABLE 4. $(\omega_j, \tau_j, \xi_j)$ for $J^1_{(2,4)}$ in Table 3.

#	ω	τ	ξ
1	16	(0,0)	$0, 0.4373, 0, -2.008, -1.571, 3.076, 1.571, -1.505$
2	16	(0,0)	$0, -2.008, 0, 0.4373, 1.571, -1.505, -1.571, 3.076$
3	40	(0,0)	$1.571, -1.505, -1.571, 3.076, 0, -2.008, 0, 0.4373$
4	40	(0,0)	$-1.571, 3.076, 1.571, -1.505, 0, 0.4373, 0, -2.008$
5	21	(0,1)	$-1.134, 1.003, -0.0339, 3.045, -0.0339, 3.045, -1.134, 1.003$
6	45	(0,1)	$-0.0339, 3.045, -1.134, 1.003, -1.134, 1.003, -0.0339, 3.045$
7	13	(0,1)	$0.2202, -1.398, -1.675, -2.644, -2.842, -1.477, 2.793, 2.312$
8	13	(0,1)	$2.793, 2.312, -2.842, -1.477, -1.675, -2.644, 0.2202, -1.398$
9	37	(0,1)	$-1.675, -2.644, 0.2202, -1.398, 2.793, 2.312, -2.842, -1.477$
10	37	(0,1)	$-2.842, -1.477, 2.793, 2.312, 0.2202, -1.398, -1.675, -2.644$
11	4	(0,2)	$3.076, 1.571, -1.505, -1.571, 0.4373, 0, -2.008, 0$
12	4	(0,2)	$-1.505, -1.571, 3.076, 1.571, -2.008, 0, 0.4373, 0$
13	28	(0,2)	$0.4373, 0, -2.008, 0, 3.076, 1.571, -1.505, -1.571$
14	28	(0,2)	$-2.008, 0, 0.4373, 0, -1.505, -1.571, 3.076, 1.571$
15	21	(0,3)	$1.003, -0.0339, 3.045, -1.134, 3.045, -1.134, 1.003, -0.0339$
16	45	(0,3)	$3.045, -1.134, 1.003, -0.0339, 1.003, -0.0339, 3.045, -1.134$
17	13	(0,3)	$2.312, -2.842, -1.477, 2.793, -2.644, 0.2202, -1.398, -1.675$
18	13	(0,3)	$-1.398, -1.675, -2.644, 0.2202, -1.477, 2.793, 2.312, -2.842$
19	37	(0,3)	$-2.644, 0.2202, -1.398, -1.675, 2.312, -2.842, -1.477, 2.793$
20	37	(0,3)	$-1.477, 2.793, 2.312, -2.842, -1.398, -1.675, -2.644, 0.2202$
21	22	(1,0)	$-2.241, -1.335, -0.835, 0.8973, 1.455, 3.03, 0.04963, -1.022$
22	22	(1,0)	$-0.835, 0.8973, -2.241, -1.335, 0.04963, -1.022, 1.455, 3.030$
23	22	(1,0)	$0.04963, -1.022, 1.455, 3.03, -0.835, 0.8973, -2.241, -1.335$
24	22	(1,0)	$1.455, 3.03, 0.04963, -1.022, -2.241, -1.335, -0.835, 0.8973$
25	6	(1,0)	$2.235, 2.673, 2.235, 2.673, 1.168, -0.8402, 1.168, -0.8402$
26	6	(1,0)	$1.168, -0.8402, 1.168, -0.8402, 2.235, 2.673, 2.235, 2.673$
27	15	(1,1)	$0.6315, -3.119, 2.979, -2.324, 0.6315, -3.119, 2.979, -2.324$
28	39	(1,1)	$2.979, -2.324, 0.6315, -3.119, 2.979, -2.324, 0.6315, -3.119$
29	7	(1,1)	$2.65, 1.091, 2.736, -0.06957, 1.658, 0.09927, -0.3231, -3.129$
30	7	(1,1)	$1.658, 0.09927, -0.3231, -3.129, 2.65, 1.091, 2.736, -0.06957$
31	31	(1,1)	$2.736, -0.06957, 2.65, 1.091, -0.3231, -3.129, 1.658, 0.09927$
32	31	(1,1)	$-0.3231, -3.129, 1.658, 0.09927, 2.736, -0.06957, 2.65, 1.091$
33	34	(1,2)	$-1.022, 1.455, 3.03, 0.04963, 0.8973, -2.241, -1.335, -0.835$
34	34	(1,2)	$0.8973, -2.241, -1.335, -0.835, -1.022, 1.455, 3.03, 0.04963$
35	34	(1,2)	$3.03, 0.04963, -1.022, 1.455, -1.335, -0.835, 0.8973, -2.241$
36	34	(1,2)	$-1.335, -0.835, 0.8973, -2.241, 3.03, 0.04963, -1.022, 1.455$
37	18	(1,2)	$2.673, 2.235, 2.673, 2.235, -0.8402, 1.168, -0.8402, 1.168$
38	18	(1,2)	$-0.8402, 1.168, -0.8402, 1.168, 2.673, 2.235, 2.673, 2.235$
39	15	(1,3)	$-3.119, 2.979, -2.324, 0.6315, -3.119, 2.979, -2.324, 0.6315$
40	39	(1,3)	$-2.324, 0.6315, -3.119, 2.979, -2.324, 0.6315, -3.119, 2.979$
41	7	(1,3)	$1.091, 2.736, -0.06957, 2.65, 0.09927, -0.3231, -3.129, 1.658$
42	7	(1,3)	$0.09927, -0.3231, -3.129, 1.658, 1.091, 2.736, -0.06957, 2.65$
43	31	(1,3)	$-3.129, 1.658, 0.09927, -0.3231, -0.06957, 2.65, 1.091, 2.736$
44	31	(1,3)	$-0.06957, 2.65, 1.091, 2.736, -3.129, 1.658, 0.09927, -0.3231$

TABLE 5. Fusion rules of tensoring with b and centralization by b

dim	$b \otimes -$		in $\langle b\rangle'$?
1	$bX_{(i,j)} = bX_{(i,j+2)}$	$(i,j) \in \mathbb{Z}/2 \times \mathbb{Z}/4$	Yes
χ_{24}^5	$bY_{(i,j)} = bY_{(i,j+2)}$	$(i,j) \in \mathbb{Z}/2 \times \mathbb{Z}/4$	Yes
$2\chi_6^3$	$bZ_{(k_1,k_1),(l_1,l_2)} = Z_{(k_1,k_2),(l_1,l_2)}$	$(k_1,k_2),(l_1,l_2) =$ $(0,0),(0,2);(0,1),(0,3);$ $(1,0),(1,2);(1,1),(1,3)$	Yes
	$bZ_{(k_1,k_1),(l_1,l_2)} = Z_{(k_1,k_2+2),(l_1,l_2+2)}$	$(k_1,k_2),(l_1,l_2) =$ $(0,0),(1,0);(0,0),(1,2);$ $(0,1),(1,1);(0,1),(1,3)$	Yes
	$bZ_{(k_1,k_2),(l_1,l_2)} = Z_{(k_1,k_2+2),(l_1,l_2+2)}$	$(k_1,k_2),(l_1,l_2) =$ $(0,0),(0,1);(0,0),(1,1);$ $(0,0),(1,3);(0,1),(1,0);$ $(0,1),(1,2);(1,0),(1,1)$	No
	$bZ_{(k_1,k_2),(l_1,l_2)} = Z_{(k_1,k_2+1),(l_1,l_2-1)}$	$(k_1,k_2),(l_1,l_2) =$ $(0,0),(0,3);(1,0),(1,3)$	No
χ_{24}^4	$bW_i = W_i$	$i = 25, 26, 37, 38$	Yes
	$bW_i = W_{i+1}$	$i = 1, 3, 11, 13, 21, 23$	Yes
	$bW_i = W_{i+1}$	$i = 5, 15, 18, 27, 39, 42$	No
	$bW_i = W_{i+2}$	$i = 7, 8, 29, 30$	No
	$bW_i = W_{i+2}$	$i = 33, 34$	Yes
	$bW_i = W_{i+3}$	$i = 17, 41$	No

PROOF. First, we examine the case $J_{2,4}^1$ in Table 3. Notice that the invertible simple object $X_{(0,2)}$ has twist 1, thus it is a boson. We can apply the boson condensation described in Section 2.1 to obtain a modular category $\left[\mathcal{C}_{\mathbb{Z}/2}\right]_0$. Using the Verlinde formula with the S matrix associated to $J_{2,4}^1$, we can obtain the fusion rules of $\mathcal{C}$. We collect the fusion rules of simple objects tensoring with b and those centralized by b in Table 5. Therefore, $[\mathcal{C}_{\mathbb{Z}/2}]_0$ is a modular category with the 36 simple objects in Table 6.

As $\theta_{F(X_{(0,1)})} = i$, $\left[\mathcal{C}_{\mathbb{Z}/2}\right]_0$ has a modular subcategory generated by the semion $F(X_{(0,1)}) = F(X_{(0,3)})$, which is equivalent to $\mathcal{C}(\mathbb{Z}/2, q)$, where $q = i^{x^2}$, $x \in \mathbb{Z}/2$. Thus $[\mathcal{C}_{\mathbb{Z}/2}]_0 \cong \mathcal{D} \boxtimes \mathcal{C}(\mathbb{Z}/2, q)$, where $\mathcal{D}$ contains the simple object $F(X_{(1,0)}) = F(X_{(1,2)})$, which is a fermion since $\theta_{F(X_{1,0})} = -1$. We list the simple objects in the spin modular category $\mathcal{D}$ in Table 7.

The centralizer of the fermion $\mathcal{D}_0$ is generated by the following 10 simple objects: $F(\mathbf{1})$, $F(X_{(1,0)})$, $F(Y_{(0,0)})$, $F(Y_{(1,0)})$, $\left(Z_{(0,1),(0,3)}\right)_1$, $\left(Z_{(0,1),(0,3)}\right)_2$, $\left(Z_{(1,1),(1,3)}\right)_1$, $\left(Z_{(1,1),(1,3)}\right)_2$, $F(W_1)$, and $F(W_3)$.

Using Theorem 2.2 we find that the $\hat{S}$-matrix corresponding to $\mathcal{D}_0$ is:

$$\frac{1}{\hat{\lambda}_2}\begin{bmatrix} 1 & 5+2\sqrt{6} & 3+\sqrt{6} & 3+\sqrt{6} & 4+2\sqrt{6} \\ 5+2\sqrt{6} & 1 & t & r & -4-2\sqrt{6} \\ 3+\sqrt{6} & t & u_1 & u_2 & x \\ 3+\sqrt{6} & r & u_2 & u_3 & y \\ 4+2\sqrt{6} & -4-2\sqrt{6} & x & y & 4+2\sqrt{6} \end{bmatrix}$$

TABLE 6. Simple objects in $\left[\mathcal{C}_{\mathbb{Z}/2}\right]_0$

dim	Objects	Twists	Number Count
1	$F(\mathbf{1})$, $F\left(X_{(0,1)}\right)$, $F\left(X_{(1,0)}\right)$, $F\left(X_{(1,1)}\right)$	$1, i, -1, -i$	4
χ_{24}^5	$F\left(Y_{(0,0)}\right)$, $F\left(Y_{(0,1)}\right)$, $F\left(Y_{(1,0)}\right)$, $F\left(Y_{(1,3)}\right)$	$1, i, -1, -i$	4
$2\chi_6^3$	$F\left(Z_{(0,0),(1,0)}\right)$, $F\left(Z_{(0,0),(1,2)}\right)$, $F\left(Z_{(0,1),(1,1)}\right)$, $F\left(Z_{(0,1),(1,3)}\right)$	$1, 1, i, -i$	4
χ_6^3	$\left(Z_{(0,0),(0,2)}\right)_1$, $\left(Z_{(0,0),(0,2)}\right)_2$, $\left(Z_{(0,1),(0,3)}\right)_1$, $\left(Z_{(0,1),(0,3)}\right)_2$, $\left(Z_{(1,0),(1,2)}\right)_1$, $\left(Z_{(1,0),(1,2)}\right)_2$, $\left(Z_{(1,1),(1,3)}\right)_1$, $\left(Z_{(1,1),(1,3)}\right)_2$	$1, 1, -i, -i, -1, -1, i, i$	8
χ_{24}^4	$F(W_1)$, $F(W_3)$, $F(W_{11})$, $F(W_{13})$, $F(W_{21})$, $F(W_{23})$, $F(W_{33})$, $F(W_{34})$	$e^{\frac{2i\pi}{3}}, e^{-\frac{i\pi}{3}}, e^{\frac{i\pi}{6}}, e^{-\frac{5i\pi}{6}}, e^{\frac{11i\pi}{12}}, e^{\frac{11i\pi}{12}}, e^{-\frac{7i\pi}{12}}, e^{-\frac{7i\pi}{12}}$	8
χ_6^2	$(W_{25})_1$, $(W_{25})_2$, $(W_{26})_1$, $(W_{26})_2$, $(W_{37})_1$, $(W_{37})_2$, $(W_{38})_1$, $(W_{38})_2$	$e^{\frac{i\pi}{4}}, e^{\frac{i\pi}{4}}, e^{\frac{i\pi}{4}}, e^{\frac{i\pi}{4}}, e^{\frac{3i\pi}{4}}, e^{\frac{3i\pi}{4}}, e^{\frac{3i\pi}{4}}, e^{\frac{3i\pi}{4}}$	8

TABLE 7. Simple objects in $\mathcal{D}$

dim	Objects	Twists
1	$F(\mathbf{1})$, $F(X_{(1,0)})$	$1, -1$
χ_{24}^5	$F(Y_{(0,0)}), F(Y_{(1,0)})$	$1, -1$
$2\chi_6^3$	$F(Z_{(0,0),(1,0)}), F(Z_{(0,0),(1,2)})$	$1, 1$
χ_6^3	$\left(Z_{(0,1),(0,3)}\right)_1$, $\left(Z_{(0,1),(0,3)}\right)_2$, $\left(Z_{(1,1),(1,3)}\right)_1$, $\left(Z_{(1,1),(1,3)}\right)_2$	$-i, -i, i, i$
χ_{24}^4	$F(W_1), F(W_3), F(W_{21}), F(W_{23})$	$e^{\frac{2\pi i}{3}}, e^{-\frac{\pi i}{3}}, e^{\frac{11\pi i}{12}}, e^{\frac{11\pi i}{12}}$
χ_6^2	$(W_{25})_1$, $(W_{25})_2$, $(W_{26})_1$, $(W_{26})_2$	$e^{\frac{i\pi}{4}}, e^{\frac{i\pi}{4}}, e^{\frac{i\pi}{4}}, e^{\frac{i\pi}{4}}$

where $\hat{\lambda}_2 = 2\sqrt{6\chi_{24}^5}$, $t + r = 2(3 + \sqrt{6})$ and $x + y = 0$. For notational simplicity we define $\mathcal{S} = \hat{\lambda}_2 \hat{S}$. Orthogonality implies that $x = y = 0$ and $t = r = 3 + \sqrt{6}$, and $u_1 = u_3$, $u_2 = -2\sqrt{6} - u_3 - 6$.

Thus we have:

$$\mathcal{S} = \begin{bmatrix} 1 & 5+2\sqrt{6} & 3+\sqrt{6} & 3+\sqrt{6} & 4+2\sqrt{6} \\ 5+2\sqrt{6} & 1 & 3+\sqrt{6} & 3+\sqrt{6} & -4-2\sqrt{6} \\ 3+\sqrt{6} & 3+\sqrt{6} & u_3 & -2\sqrt{6}-u_3-6 & 0 \\ 3+\sqrt{6} & 3+\sqrt{6} & -2\sqrt{6}-u_3-6 & u_3 & 0 \\ 4+2\sqrt{6} & -4-2\sqrt{6} & 0 & 0 & 4+2\sqrt{6} \end{bmatrix}$$

Since we don't *a priori* know if the objects of dimension $3 + \sqrt{6}$ are a dual pair or self-dual, we must analyze both possibilities.

If they are self-dual orthogonality implies: $-2u_3^2 + \left(-4\sqrt{6} - 12\right) u_3 + 30 + 12\sqrt{6} = 0$ so that $u_3 = -\sqrt{6} - 3 \pm (3\sqrt{2} - 2\sqrt{3})$. Using the Verlinde formula for the naive fusion rules [**7**, Proposition 2.7] we compute all the $\hat{N}_{i,j}^k$ and find that these choices of u_3 yield negative fusion coefficients, a contradiction.

If they are a dual pair u_3 satisfies: $2u_3^2 + \left(4\sqrt{6}+12\right) u_3 + 90 + 36\sqrt{6} = 0$ which yields $u_3 = 3i\sqrt{2} + 2\,i\sqrt{3} - \sqrt{6} - 3$ or $u_3 = -3\,i\sqrt{2} - 2\,i\sqrt{3} - \sqrt{6} - 3$. These are complex conjugates of each other, so we get $\mathcal{S}$ is:

$$\begin{bmatrix} 1 & 5+2\sqrt{6} & 3+\sqrt{6} & 3+\sqrt{6} & 4+2\sqrt{6} \\ 5+2\sqrt{6} & 1 & 3+\sqrt{6} & 3+\sqrt{6} & -4-2\sqrt{6} \\ 3+\sqrt{6} & 3+\sqrt{6} & \pm i(3\sqrt{2}+2\sqrt{3})-\sqrt{6}-3 & \mp i(3\sqrt{2}+2\sqrt{3})-\sqrt{6}-3 & 0 \\ 3+\sqrt{6} & 3+\sqrt{6} & \mp i(3\sqrt{2}+2\sqrt{3})-\sqrt{6}-3 & \pm i(3\sqrt{2}+2\sqrt{3})-\sqrt{6}-3 & 0 \\ 4+2\sqrt{6} & -4-2\sqrt{6} & 0 & 0 & 4+2\sqrt{6} \end{bmatrix}$$

where the 4 signs are determined by any single sign, which we now calculate. We first note that without loss of generality (see [**7**]) the twists are: $[1, 1, i, i, e^{-\pi i/3}]$, with corresponding objects $\mathbf{1}, \beta, \gamma_1, \gamma_2$ and η of dimension $1, \chi_{24}^5, \chi_6^3, \chi_6^3$ and χ_{24}^4. We denote by f the fermion, so that the objects in $\mathcal{D}_0$ can be labeled by the above and $f, f\beta, f\gamma_1, f\gamma_2$ and $f\eta$. We aim to calculate $\hat{\lambda}_2\hat{S}_{\gamma_1,\gamma_1} = \pm i(3\sqrt{2}+2\sqrt{3})-\sqrt{6}-3$. Since $\gamma_1^* = \gamma_2$ the balancing equation gives us:

$$-\hat{S}_{\gamma_1,\gamma_1} = (\theta_{\gamma_1})^2\hat{S}_{\gamma_1,\gamma_1} = \frac{1}{\sqrt{2}\hat{\lambda}_2}\sum_{\psi} N_{\gamma_2,\gamma_1}^{\psi}\theta_{\psi}d_{\psi}$$

where the sum is over the 10 simple objects in $\mathcal{D}_0$. The Verlinde formula for $\hat{S}$ is invariant under complex conjugation so we obtain the (naive) fusion rules:

$N_{\gamma_2,\gamma_1}^{\mathbf{1}} = 1, N_{\gamma_2,\gamma_1}^{\beta} + N_{\gamma_2,\gamma_1}^{f\beta} = 2$ and $N_{\gamma_2,\gamma_1}^{\eta} + N_{\gamma_2,\gamma_1}^{f\eta} = 1$. One can now simply try each of the two possibilities for $\hat{S}_{\gamma_1,\gamma_1}$ with the fusion rules choices and check for consistency. The only possibility is that $N_{\gamma_2,\gamma_1}^{\beta} = N_{\gamma_2,\gamma_1}^{f\beta} = 1$ and $N_{\gamma_2,\gamma_1}^{\eta} = 1$, so that $\hat{\lambda}_2\hat{S}_{\gamma_1,\gamma_1} = i(3\sqrt{2}+2\sqrt{3})-\sqrt{6}-3$. Noting that $\hat{T}^2 = \mathrm{diag}[1, 1, -1, -1, e^{-2\pi i/3}]$ we see that we have the complex conjugate of the data in (3.2), which completes the proof. □

4. Conclusions and Discussion

We have now realized two previously unknown super-modular data using near-group categories. We hope to use these techniques to further find modular and super-modular categories, aiding in continuing the classification of these categories. Although there has been much work on understanding the S and T matrices of centers of near-group categories [**11, 18, 19**], some of these matrices are only conjectural. We hope to continue to expand the list of categories for which these conjectures have been verified, along with their modular/super-modular data.

4.1. Additional Examples. Below in Table 8 we have summarized some cases where a familiar category appears as a modular factor of a near-group center. As we do not provide proofs the reader may take these as speculations. Moreover, we do not specify a category of the given type in Table 8 so there is ambiguity in any case. The authors welcome comments and references for verifications/original sources for these statements.

A few comments on the notation of Table 8:

- We denote a pointed modular category by $\mathcal{C}(A, Q)$ where Q is an unspecified non-degenerate quadratic form.
- A G-gauging $\mathcal{D}$ of $\mathcal{B}$ [**13**] is denoted $\mathcal{B}_G^{\times,G}$, which is a convenient way of saying that the category $[\mathcal{D}_G]_0 \cong \mathcal{B}$.

TABLE 8. Familiar categories conjecturally related to centers of $G+n$ near-group categories.

Type of $\mathcal{F}$	Conj. form of $\mathcal{Z}(\mathcal{F})$	Notes
$\mathbb{Z}/1+1$	$\mathrm{Fib}\boxtimes\mathrm{Fib}^{\mathrm{rev}}$	$\mathcal{F}=\mathrm{Fib}$
$\mathbb{Z}/2+2$	$\mathrm{SU}(2)_{10}$	cf. [3]
$\mathbb{Z}/3+3$	$\mathrm{G}(2)_3\boxtimes\mathcal{C}(\mathbb{Z}/3,Q)$	$\mathrm{rank}(\mathrm{G}(2)_3)=6$ cf. [28]
$\mathbb{Z}/2\times\mathbb{Z}/2+4$	$\left(\left[[\mathrm{SU}(2)_6^{\boxtimes 2}]_{\mathbb{Z}/2}\right]_0\right)_{\mathbb{Z}/2}^{\times,\mathbb{Z}/2}$	cf. [7]
$\mathbb{Z}/4+4$	$[\mathrm{PSU}(3)_5\boxtimes\mathcal{C}(\mathbb{Z}/2,Q)]_{\mathbb{Z}/2}^{\times,\mathbb{Z}/2}$	$\mathrm{rank}(\mathrm{PSU}(3)_5)=7$
$\mathbb{Z}/5+5$	$\mathcal{B}\boxtimes\mathcal{C}(\mathbb{Z}/5,Q)$	$\mathrm{rank}(\mathcal{B})=8$
$\mathbb{Z}/6+6$	$\mathcal{D}\boxtimes\mathcal{C}(\mathbb{Z}/3,Q)$	$\mathcal{D}$ in Theorem 3.1
$\mathbb{Z}/7+7$	$\mathcal{B}\boxtimes\mathcal{C}(\mathbb{Z}/7,Q)$	$\mathrm{rank}(\mathcal{B})=10$
$\mathbb{Z}/8+8$	$[G(2)_4\boxtimes\mathcal{C}(\mathbb{Z}/4,Q)]_{\mathbb{Z}/2}^{\times,\mathbb{Z}/2}$	$\mathrm{rank}(G(2)_4)=9$
$\mathbb{Z}/2\times\mathbb{Z}/4+8$	$[\mathcal{D}\boxtimes\mathcal{C}(\mathbb{Z}/2,Q)]_{\mathbb{Z}/2}^{\times,\mathbb{Z}/2}$	$\mathcal{D}$ in Theorem 3.2

- The 4 rows corresponding to near-group categories associated with groups of order 4 and 8 can be understood as follows. The category $\mathcal{Z}(\mathcal{F})$ contains a boson, and after boson condensation one obtains the category $\mathcal{B}$ where the second column has $\mathcal{B}_{\mathbb{Z}/2}^{\times,\mathbb{Z}/2}$.

Acknowledgments

The authors thank T. Gannon, A. Schopieray, X. Cui, Y. Wang, A. Bagheri and P. Gustafson for enlightening conversations.

References

[1] Luis Alvarez-Gaumé, Gregory Moore, and Cumrun Vafa, *Theta functions, modular invariance, and strings*, Comm. Math. Phys. **106** (1986), no. 1, 1–40. MR853977

[2] Aaron Robert Bagheri, *On Condensation of Anyons and Applications*, ProQuest LLC, Ann Arbor, MI, 2023. Thesis (Ph.D.)–University of California, Santa Barbara. MR4691138

[3] Parsa Bonderson, Eric C. Rowell, and Zhenghan Wang, *On realizing modular data*, J. Math. Sci. Univ. Tokyo **27** (2020), no. 1, 65–79. MR4246625

[4] Parsa Bonderson, Eric C. Rowell, Zhenghan Wang, and Qing Zhang, *Congruence subgroups and super-modular categories*, Pacific J. Math. **296** (2018), no. 2, 257–270, DOI 10.2140/pjm.2018.296.257. MR3830835

[5] Alain Bruguières, *Catégories prémodulaires, modularisations et invariants des variétés de dimension 3* (French, with English summary), Math. Ann. **316** (2000), no. 2, 215–236, DOI 10.1007/s002080050011. MR1741269

[6] Paul Bruillard, César Galindo, Tobias Hagge, Siu-Hung Ng, Julia Yael Plavnik, Eric C. Rowell, and Zhenghan Wang, *Fermionic modular categories and the 16-fold way*, J. Math. Phys. **58** (2017), no. 4, 041704, 31, DOI 10.1063/1.4982048. MR3641612

[7] Paul Bruillard, César Galindo, Siu-Hung Ng, Julia Y. Plavnik, Eric C. Rowell, and Zhenghan Wang, *Classification of super-modular categories by rank*, Algebr. Represent. Theory **23** (2020), no. 3, 795–809, DOI 10.1007/s10468-019-09873-9. MR4109138

[8] Paul Bruillard, Siu-Hung Ng, Eric C. Rowell, and Zhenghan Wang, *On classification of modular categories by rank*, Int. Math. Res. Not. IMRN **24** (2016), 7546–7588, DOI 10.1093/imrn/rnw020. MR3632091

[9] Paul Bruillard, Siu-Hung Ng, Eric C. Rowell, and Zhenghan Wang, *Rank-finiteness for modular categories*, J. Amer. Math. Soc. **29** (2016), no. 3, 857–881, DOI 10.1090/jams/842. MR3486174

[10] Paul Bruillard, Julia Plavnik, Eric C. Rowell, and Qing Zhang, *On classification of super-modular categories of rank 8*, J. Algebra Appl. **20** (2021), no. 1, Paper No. 2140017, 36, DOI 10.1142/S021949882140017X. MR4209971

[11] Paul Garrett Budinski, *Exotic fusion categories and their modular data*, Thesis, 2021.

[12] Gil Young Cho, Hee-cheol Kim, Donghae Seo, and Minyoung You. *Classification of fermionic topological orders from congruence representations*, arXiv:2210.03681, 2022.

[13] Shawn X. Cui, César Galindo, Julia Yael Plavnik, and Zhenghan Wang, *On gauging symmetry of modular categories*, Comm. Math. Phys. **348** (2016), no. 3, 1043–1064, DOI 10.1007/s00220-016-2633-8. MR3555361

[14] Alexei Davydov, Dmitri Nikshych, and Victor Ostrik, *On the structure of the Witt group of braided fusion categories*, Selecta Math. (N.S.) **19** (2013), no. 1, 237–269, DOI 10.1007/s00029-012-0093-3. MR3022755

[15] Vladimir Drinfeld, Shlomo Gelaki, Dmitri Nikshych, and Victor Ostrik, *On braided fusion categories. I*, Selecta Math. (N.S.) **16** (2010), no. 1, 1–119, DOI 10.1007/s00029-010-0017-z. MR2609644

[16] Pavel Etingof, *On Vafa's theorem for tensor categories*, Math. Res. Lett. **9** (2002), no. 5-6, 651–657, DOI 10.4310/MRL.2002.v9.n5.a8. MR1906068

[17] Pavel Etingof, Shlomo Gelaki, Dmitri Nikshych, and Victor Ostrik, *Tensor categories*, Mathematical Surveys and Monographs, vol. 205, American Mathematical Society, Providence, RI, 2015, DOI 10.1090/surv/205. MR3242743

[18] David E. Evans and Terry Gannon, *Near-group fusion categories and their doubles*, Adv. Math. **255** (2014), 586–640, DOI 10.1016/j.aim.2013.12.014. MR3167494

[19] Pinhas Grossman and Masaki Izumi, *Infinite families of potential modular data related to quadratic categories*, Comm. Math. Phys. **380** (2020), no. 3, 1091–1150, DOI 10.1007/s00220-020-03710-7. MR4179724

[20] Masaki Izumi, *The structure of sectors associated with Longo-Rehren inclusions. II. Examples*, Rev. Math. Phys. **13** (2001), no. 5, 603–674, DOI 10.1142/S0129055X01000818. MR1832764

[21] Theo Johnson-Freyd and David Reutter, *Minimal nondegenerate extensions*, J. Amer. Math. Soc. **37** (2024), no. 1, 81–150, DOI 10.1090/jams/1023. MR4654609

[22] C. Jones, S. Morrison, D. Nikshych, and E. C. Rowell, *Rank-finiteness for G-crossed braided fusion categories*, Transform. Groups **26** (2021), no. 3, 915–927, DOI 10.1007/s00031-020-09576-2. MR4309554

[23] Alexander Kirillov Jr. and Viktor Ostrik, *On a q-analogue of the McKay correspondence and the ADE classification of $\mathfrak{sl}_2$ conformal field theories*, Adv. Math. **171** (2002), no. 2, 183–227, DOI 10.1006/aima.2002.2072. MR1936496

[24] Alexei Kitaev, *Anyons in an exactly solved model and beyond*, Ann. Physics **321** (2006), no. 1, 2–111, DOI 10.1016/j.aop.2005.10.005. MR2200691

[25] Tian Lan, Liang Kong, and Xiao-Gang Wen, *Modular extensions of unitary braided fusion categories and 2 + 1D topological/SPT orders with symmetries*, Comm. Math. Phys. **351** (2017), no. 2, 709–739, DOI 10.1007/s00220-016-2748-y. MR3613518

[26] Gregory Moore and Nathan Seiberg, *Classical and quantum conformal field theory*, Comm. Math. Phys. **123** (1989), no. 2, 177–254. MR1002038

[27] Michael Müger, *Galois theory for braided tensor categories and the modular closure*, Adv. Math. **150** (2000), no. 2, 151–201, DOI 10.1006/aima.1999.1860. MR1749250

[28] Siu-Hung Ng, Eric C. Rowell, Zhenghan Wang, and Xiao-Gang Wen, *Reconstruction of modular data from $\mathrm{SL}_2(\mathbb{Z})$ representations*, Comm. Math. Phys. **402** (2023), no. 3, 2465–2545, DOI 10.1007/s00220-023-04775-w. MR4630478

[29] Siu-Hung Ng and Peter Schauenburg, *Congruence subgroups and generalized Frobenius-Schur indicators*, Comm. Math. Phys. **300** (2010), no. 1, 1–46, DOI 10.1007/s00220-010-1096-6. MR2725181

[30] Eric Rowell, Richard Stong, and Zhenghan Wang, *On classification of modular tensor categories*, Comm. Math. Phys. **292** (2009), no. 2, 343–389, DOI 10.1007/s00220-009-0908-z. MR2544735

[31] Andrew Schopieray, *Categorification of integral group rings extended by one dimension*, J. Lond. Math. Soc. (2) **108** (2023), no. 4, 1617–1641, DOI 10.1112/jlms.12789. MR4655273

[32] Jacob Siehler, *Near-group categories*, Algebr. Geom. Topol. **3** (2003), 719–775, DOI 10.2140/agt.2003.3.719. MR1997336

[33] Daisuke Tambara and Shigeru Yamagami, *Tensor categories with fusion rules of self-duality for finite abelian groups*, J. Algebra **209** (1998), no. 2, 692–707, DOI 10.1006/jabr.1998.7558. MR1659954

Department of Mathematics, Texas A&M University, College Station, Texas 77843-3368

Email address: rowell@math.tamu.edu

Department of Mathematics, Texas A&M University, College Station, Texas 77843-3368

Email address: hmsolomon@tamu.edu

Department of Mathematics, Purdue University, West Lafayette, Indiana 47907

Email address: zhan4169@purdue.edu

Contemporary Mathematics
Volume **813**, 2025
https://doi.org/10.1090/conm/813/16287

Fixed-point-free fusion automorphisms

Andrew Schopieray

ABSTRACT. This is a study of fusion ring automorphisms leaving only the trivial element fixed. We prove that a variety of classical results on fixed-point-free automorphisms of finite groups are true in the generality of fusion rings. As a result, we show there are 8 Grothendieck equivalence classes of fusion categories of rank less than 9 with a fixed-point-free fusion automorphism of prime order, generalizing existing results about modular fusion categories of odd dimension.

1. Introduction

The study of fusion rings and categories in the 21st century is a continuation of the study of finite groups in the 20th century. Every finite group G can be realized as the fusion ring $\mathbb{Z}G$, its integral group ring, and as the fusion category (over $\mathbb{C}$) Vec_G^ω for any $\omega \in H^3(G, \mathbb{C}^\times)$. But not every fusion ring is isomorphic to $\mathbb{Z}G$ for some G, and not every fusion ring can be realized as the Grothendieck ring of a fusion category. So one of the most pervasive questions is where the boundaries between these mathematical objects lie.

For example, subgroups of a finite group G correspond to fusion subrings of $\mathbb{Z}G$, and thus fusion subcategories of Vec_G^ω. Frobenius-Perron dimension (FPdim) [**11**, Section 3.3] is the notion analogous to the order of a group for fusion rings and categories, and one might believe that results about orders of finite groups hold true in this more abstract setting. The classical result is that the order of a finite group is divisible by the order of any of its subgroups, and while it is true that the Frobenius-Perron dimension of a fusion category is divisible by the Frobenius-Perron dimension of any of its fusion subcategories [**11**, Theorem 7.17.6], the same is not true for fusion rings in general (Example 2.5). Even more, the obvious generalization of the Sylow theorems for finite groups in this analogy fails for both fusion rings and fusion categories, unless additional restrictions are imposed (e.g. [**4**]).

For a second example, solvability of finite groups has an analog for fusion categories [**11**, Definition 9.8.1] so that the fusion categories Vec_G^ω are solvable if and only if G is solvable [**11**, Proposition 9.8.5(ii)]. With this definition, many results about solvable groups also apply to fusion categories such as the fact that any fusion category of Frobenius-Perron dimension p^aq^b for primes $p, q \in \mathbb{Z}_{\geq 2}$ and nonnegative integers a, b is solvable [**11**, Theorem 9.15.9]. With the existence of additional

2020 *Mathematics Subject Classification.* Primary 18M20.

structures such as braidings, other classes of fusion categories of odd Frobenius-Perron dimension have been shown to be solvable [**10**, **27**]. It has been conjectured that all fusion categories of odd Frobenius-Perron dimension are solvable [**8**], and at least one notion of solvability has been proposed recently on the level of fusion rings and their generalizations [**37**].

Continuing in this thread, *fixed-point-free* automorphisms of finite groups have been studied for over a century. Of course, "fixed-point-free" is a misnomer since the multiplicative identity will be fixed by definition, but this term is firmly established in the group theory literature. One of the first observations about fixed-point-free automorphisms of finite groups was if G is a finite group and ϕ a fixed-point-free automorphism of G of order 2, then G is abelian of odd order and $\phi(g) = g^{-1}$ for all $g \in G$. In a result attributed to W. Burnside [**7**], it is noted that if G is a finite group with a fixed-point-free automorphism of order 3, then G is nilpotent of nilpotency class at most 2. More generally, in the doctoral thesis of J. G. Thompson [**36**], it is proven that any finite group with a fixed-point-free automorphism of prime order is nilpotent. If one utilizes the classification of finite simple groups, there are very condensed proofs of other related facts about fixed-point-free automorphisms of finite groups [**17**, Theorem 1.48][**31**][**38**].

In this manuscript we study how results about finite groups and their fixed-point-free automorphisms generalize to fusion rings and categories. There is currently very little written about fusion rings with assumptions about their automorphisms that does not fall under the purview of finite group theory, representations of Hopf algebras, or fusion rings related affine Lie algebras and their categorifications (e.g. [**16**, **32**]); we will attempt to recount the relevant pre-existing results here. Recall that for commutative fusion rings (R, B), the duality antiautomorphism $x \mapsto x^*$ for all $x \in B$ is furthermore a fusion automorphism of R. There are several pre-existing results on fusion categories for which this duality is fixed-point-free. In particular, it is known that if the Frobenius-Perron dimension of a fusion category $\mathcal{C}$ is odd, then $X \not\cong X^*$ for all simple objects of $\mathcal{C}$ [**28**, Corollary 8.2]. When $\mathcal{C}$ is commutative, this gives a fixed-point-free automorphism of the underlying fusion ring. This statement is false for commutative fusion rings in general with the smallest example having rank 4 (Example 2.5). Note that there is no reason to believe such an automorphism will correspond to a tensor autoequivalence of the fusion category in general (refer to [**9**], for example). Conversely, it was observed [**19**, Theorem 2.2] that modular fusion categories $\mathcal{C}$ such that $X \not\cong X^*$ for all simple X in $\mathcal{C}$ are integral, i.e. $\mathrm{FPdim}(X) \in \mathbb{Z}$ for all X in $\mathcal{C}$, and therefore $\mathrm{FPdim}(\mathcal{C})$ is an odd integer. We prove that this statement is true for arbitrary fusion rings with fixed-point-free automorphisms of prime order (Theorem 2.3(1)), which is to say if (R, B) is a fusion ring with a fixed-point free automorphism of prime order $p \in \mathbb{Z}_{\geq 2}$, then R is integral, and thus $\mathrm{FPdim}(R) \equiv 1 \pmod{p}$. This follows from the fact that fixed-point-free automorphisms of fusion rings act without fixed-points on their isomorphism classes of irreducible representations distinct from FPdim (Proposition 2.2).

Using these general results, we then analyze several special cases. First we prove that any fixed-point-free involution (order 2) of a commutative fusion ring must be the duality homomorphism (Corollary 3.2). This leaves the lingering and interesting question of whether the existence of a fixed-point-free automorphism of a fusion ring of order 2 implies commutativity as is the case for finite groups.

Next, we describe all fusion rings with fixed-point-free automorphisms ϕ of prime order such that R has 2 or 3 ϕ-orbits of basis elements. In the former case (Section 4.1), all examples are integral group rings $\mathbb{Z}G$ where $G \cong C_3$ is the cyclic group of order 3, or $G \cong C_2^n$ is an elementary abelian 2-group such that $2^n - 1$ is a Mersenne Prime. In particular, for any odd prime $p \in \mathbb{Z}_{\geq 3}$ such that $p+1$ is not a power of 2, there does not exist a fusion ring of rank $p+1$ with a fixed-point-free automorphism of any prime order. When there are exactly 3 ϕ-orbits (Section 4.2), R has a nontrivial fusion subring $\mathbb{Z}G$ where G is one of the groups from the previous case. This follows from the fact that an integral fusion ring whose basis elements have at most 3 distinct Frobenius-Perron dimensions must have a nontrivial pointed subring (Lemma 4.2).

Lastly, we demonstrate there are exactly 10 fusion rings (Figure 3) of rank less than 9 with fixed-point-free automorphisms of prime order $p \in \mathbb{Z}_{\geq 2}$ and integer formal codegrees, in the sense of [**29**, Section 2]. Integer formal codegrees are necessary for a fusion ring to be categorifiable, but it is not clear if integrality of formal codegrees is a consequence of the existence of a fixed-point-free automorphism. Eight of the 10 classified fusion rings possess categorifications and 3 do not; six of these 10 rings are not isomorphic to $\mathbb{Z}G$ for a finite group G, 3 of which are the character rings of the nonabelian groups of orders $3 \cdot 7$, $3 \cdot 13$, and $5 \cdot 11$. This is in contrast to more constrained results that state that all modular fusion categories of rank at most 11 [**6**, Theorem 4.5] and subsequently rank at most 15 [**8**, Theorem 6.3(a)] are pointed. Many questions are motivated from this initial study and deserve further analysis; we prioritize a few in the following list.

QUESTION 1.1. Are there finitely many fusion rings of a given rank admitting a fixed-point-free automorphism of prime order?

QUESTION 1.2. Does there exist a fusion ring (necessarily noncommutative) with a fixed-point-free automorphism of order 2 which is not duality?

QUESTION 1.3. Is there a nontrivial invertible basis element in every fusion ring with a fixed-point-free automorphism of prime order?

QUESTION 1.4. Does there exist a noncommutative fusion ring with a fixed-point-free automorphism of prime order whose rank is less than 64?

2. Basic definitions and results

A fusion ring (R, B) is an associative unital ring R which is free as a $\mathbb{Z}$-module with a distinguished basis $B := \{b_0 = 1_R, b_1, \ldots, b_n\}$ and anti-involution $b_i \mapsto b_i^* =: b_{i^*}$ of B (extended linearly to all of R) such that $b_i b_j = \sum_{j=0}^{n} c_{ij}^k b_k$ for some $c_{ij}^k \in \mathbb{Z}_{\geq 0}$ (the *fusion rules* of R) and $c_{ij}^0 = \delta_{ij^*}$. The order of B is denoted $\operatorname{rank}(R)$. Perhaps the most vital symmetry of a fusion ring is the cyclic invariance of the fusion rules [**11**, Proposition 3.1.6] which states that for all $x, y, z \in B$,

$$c_{x,y}^{z^*} = c_{y,z}^{x^*} = c_{z,x}^{y^*}. \tag{1}$$

By a fusion subring $(R', B') \subset (R, B)$, we mean (R', B') is a fusion ring in its own right and $B' \subset B$. We define the group $\operatorname{Aut}(R, B)$ as the set of permutations ϕ of B such that ϕ, extended linearly to R, is a unital ring automorphism. Of course, there exist many unital ring automorphisms of R which do not belong to $\operatorname{Aut}(R, B)$ in general. If $\phi \in \operatorname{Aut}(R, B)$, then $\phi(x)^* = \phi(x^*)$ for all $x \in B$.

Indeed, applying ϕ to xx^* yields $c_{\phi(x),\phi(x^*)}^{1_R} = 1$. One can replace most results about $\mathrm{Aut}(R, B)$ with analogous results about *anti*automorphisms, i.e. replacing the condition $\phi(xy) = \phi(x)\phi(y)$ with $\phi(xy) = \phi(y)\phi(x)$. We will only do so briefly in Section 3 in the context of fixed-point-free involutions,. i.e. $\phi \circ \phi = \mathrm{id}_R$.

EXAMPLE 2.1. There are myriad examples of fixed-point-free automorphisms of prime order of the integral group rings $\mathbb{Z}G$ for finite groups G. It is a well-known exercise to show that there exists a fixed-point-free automorphism of order 2 of a finite group G if and only if G is abelian of odd order and the automorphism has $g \mapsto g^{-1}$ for all $g \in G$.

When the order of the fixed-point-free automorphism is an odd prime $p \in \mathbb{Z}_{\geq 3}$, there exist non-abelian examples G which we will list by `GAP` identification numbers $[|G|, \#]$. The smallest non-abelian examples have order 64, and all such groups were classified in [**22**]. In particular, there are 7 isomorphism classes of finite groups of order 64 which admit a fixed-point-free automorphism of prime order p; three of these are non-abelian groups: group $[64, 82]$ is an example for $p = 7$, and both $[64, 242]$ and $[64, 245]$ are examples for $p = 3$. We note that the order of $\mathrm{Aut}(G)$ for each of these groups is over 9000.

Let (R, B) be a fusion ring. Let $\mathrm{Irr}(R)$ be the set of irreducible complex representations of R up to isomorphism, i.e. unital algebra homomorphisms $\varphi : R \to \mathrm{End}(V)$ for some finite-dimensional complex vector space V. The only irreducible representation existing for all fusion rings (R, B) is the *Frobenius-Perron* representation $\mathrm{FPdim} : R \to \mathbb{R}_{\geq 1}$ which is the unique unital ring homomorphism $\rho : R \to \mathbb{C}$ such that $\rho(x) > 0$ for all $x \in B$ [**11**, Proposition 3.3.6(3)]. Returning to the motivation of the definition of $\mathrm{Aut}(R, B)$, the requirement that $\phi \in \mathrm{Aut}(R, B)$, versus ϕ being an arbitrary unital ring automorphism of R, is mild but ensures that for all $x \in R$, $\mathrm{FPdim}(\phi(x)) = \mathrm{FPdim}(x)$ [**11**, Proposition 3.3.13]. If $\mathrm{FPdim}(x) = 1$ for some $x \in R$, then $x \in B$, $xx^* = 1_R$, and we call x *invertible*. The $\mathbb{Z}$-linear span of all invertible elements of R form a fusion subring R_{pt} of R, which is characteristic in R, in the sense that it is preserved under all $\phi \in \mathrm{Aut}(R, B)$. It is clear that the invertible objects of R form a finite group G, and $R = R_{\mathrm{pt}}$ if and only if $R \cong \mathbb{Z}G$. In general, the group G acts on R by left or right multiplication, and for each $x \in B$, the stabilizer subgroup $G_x \subset G$ is the subset of $g \in G$ such that $c_{x,x^*}^g = c_{x,x^*}^{g^{-1}} = c_{g,x}^x = 1$ by the cyclic invariance of the fusion rules (Equation (1)) and the fact that xx^* is self-dual.

The finite group $\mathrm{Aut}(R, B)$ acts on $\mathrm{Irr}(R)$; for all $\varphi \in \mathrm{Irr}(R)$ and $\phi \in \mathrm{Aut}(R, B)$, $\varphi \circ \phi \in \mathrm{Irr}(R)$ which may or may not be distinct from φ. The characters $\chi_\varphi := \mathrm{Tr}(\varphi) : R \to \mathbb{C}$ of $\varphi \in \mathrm{Irr}(R)$ satisfy the familiar orthogonality relation [**3**, Proposition 2.12] which is if $\varphi \not\cong \rho \in \mathrm{Irr}(R)$,

$$\sum_{x \in B} \chi_\varphi(x)\chi_\rho(x^*) = 0. \tag{2}$$

When $\rho = \mathrm{FPdim}$, or in general when χ_ρ is real-valued, one can ignore the duality in this formula, and we will do so in the future without note. It is known is much broader generality that the character values $\chi_\varphi(x)$ are algebraic integers for all $\varphi \in \mathrm{Irr}(R)$ and $x \in B$ [**3**, Proposition 2.17].

PROPOSITION 2.2. *Let (R, B) be a fusion ring and $\phi \in \mathrm{Aut}(R, B)$ of prime order $p \in \mathbb{Z}_{\geq 2}$. If 1_R is the unique fixed-point of the action of ϕ on B, then* FPdim *is the unique fixed-point of the action of ϕ on* $\mathrm{Irr}(R)$.

PROOF. Let $\phi \in \mathrm{Aut}(R, B)$ be fixed-point-free of prime order $p \in \mathbb{Z}_{\geq 2}$. Assume to the contrary that φ is a fixed-point of the action of ϕ on $\mathrm{Irr}(R)$, which is to say that $\varphi \circ \phi^j = \varphi$, and thus $\chi_{\varphi \circ \phi^j} = \chi_\varphi$, for all $1 \leq j \leq p$. In particular, $\chi_\varphi(x) = \chi_\varphi(\phi^j(x))$ for all $1 \leq j \leq p$ and $x \in B$. Let Γ be a set of representatives of nontrivial ϕ-orbits in B, which all have order p by the assumption that p is prime. Then applying the orthogonality relation of Equation (2) to χ_φ and FPdim,

$$0 = \sum_{x \in B} \chi_\varphi(x)\mathrm{FPdim}(x^*) = 1 + p \sum_{x \in \Gamma} \chi_\varphi(x)\mathrm{FPdim}(x), \tag{3}$$

which erroneously implies $1/p$ is an algebraic integer. So we must conclude FPdim is the unique fixed-point of ϕ acting on $\mathrm{Irr}(R)$. □

A coarser, but more easily computable, invariant of a fusion ring (R, B) is its multi-set of formal codegrees, indexed by irreducible representations of R. For each irreducible $\varphi : R \to \mathrm{End}(V)$, the *formal codegree* $f_\varphi \in \mathbb{C}$ of φ can be defined in terms of the central element $\alpha := \sum_{b \in B} bb^*$ which satisfies $\varphi(\alpha) = \dim(\varphi) f_\varphi \mathrm{id}_V$ [**29**, Lemma 2.6]. Hence the multi-set of the formal codegrees f_φ for $\varphi \in \mathrm{Irr}(R)$ are the eigenvalues of (left) multiplication by α occurring with multiplicity $\dim(\varphi)^2$ [**30**, Remark 2.11] and so are algebraic integers closed under Galois conjugacy. Furthermore the Frobenius-Perron dimension $\mathrm{FPdim}(R) := f_{\mathrm{FPdim}} = \sum_{x \in B} \mathrm{FPdim}(x)^2$ is one of the formal codegrees of R. These numerical invariants are heavily controlled by the constraint [**30**, Proposition 2.10]

$$\sum_{\varphi \in \mathrm{Irr}(R)} \frac{\dim(\varphi)}{f_\varphi} = 1, \tag{4}$$

and the fact that they are all real numbers greater than or equal to 1 [**30**, Remark 2.12]. It is clear that for any $\phi \in \mathrm{Aut}(R, B)$, $\phi(\alpha) = \alpha$, and therefore $f_\varphi = f_{\varphi \circ \phi}$. The following theorem summarizes the above results for fixed-point-free automorphisms of prime order.

THEOREM 2.3. *Let (R, B) be a fusion ring and $\phi \in \mathrm{Aut}(R, B)$ of prime order $p \in \mathbb{Z}_{\geq 2}$. If ϕ is fixed-point-free, then*

1. $\mathrm{FPdim}(x) \in \mathbb{Z}$ *for all* $x \in B$ *and* $\mathrm{FPdim}(R) \equiv 1 \pmod{p}$;
2. $\mathrm{rank}(R)$ *and the number of* $\varphi \in \mathrm{Irr}(R)$ *with* $\dim(\varphi) = 1$ *are congruent to 1 modulo* p;
3. *the number of* $\varphi \in \mathrm{Irr}(R)$ *with* $\dim(\varphi) = k$ *for any* $k \in \mathbb{Z}_{\geq 2}$ *is divisible by* p;
4. *the formal codegrees of the* ϕ*-orbit of* $\varphi \in \mathrm{Irr}(R)$ *are equal.*

PROOF. Recall that $\mathrm{FPdim}(x)$ are algebraic integers for all $x \in B$ as eigenvalues of nonnegative integer matrices. Proposition 2.2 implies that FPdim is the unique character χ of R such that $\chi(x) = \chi(\phi(x))$ for all $x \in B$. But for any $\sigma \in \mathrm{Gal}(\overline{\mathbb{Q}}/\mathbb{Q})$ where $\overline{\mathbb{Q}}$ is the algebraic closure of $\mathbb{Q}$, the character $\sigma(\mathrm{FPdim})$ has this property since FPdim is preserved by ϕ, thus $\sigma(\mathrm{FPdim}) = \mathrm{FPdim}$. Therefore $\mathrm{FPdim}(x)$ are rational integers for all $x \in B$. Moreover $\mathrm{FPdim}(R) \equiv 1 \pmod{p}$ by collecting $x \in B$ by ϕ-orbits. The second and third results follow as any nontrivial ϕ-orbit in $\mathrm{Irr}(R)$ has order p, and the final claim is true for any $\phi \in \mathrm{Aut}(R, B)$ □

COROLLARY 2.4. *Let (R, B) be a commutative fusion ring. If 2 divides* $\mathrm{FPdim}(R)$, *then there exists nontrivial $x \in B$ with $x = x^*$.*

PROOF. Assume for all nontrivial $x \in B$, $x \neq x^*$. Then duality is a fixed-point-free automorphism of order 2 since R is commutative. The contrapositive of Theorem 2.3(1) then implies $2 \nmid \mathrm{FPdim}(R)$. □

EXAMPLE 2.5. The converse of Corollary 2.4 is false. Consider the commutative rank 4 fusion ring (R, B) with fusion rules

$$\begin{bmatrix} 1 & 0 & 0 & 0 \\ 0 & 1 & 0 & 0 \\ 0 & 0 & 1 & 0 \\ 0 & 0 & 0 & 1 \end{bmatrix} \quad \begin{bmatrix} 0 & 1 & 0 & 0 \\ 1 & 0 & 0 & 0 \\ 0 & 0 & 1 & 0 \\ 0 & 0 & 0 & 1 \end{bmatrix} \tag{5}$$

$$\begin{bmatrix} 0 & 0 & 1 & 0 \\ 0 & 0 & 1 & 0 \\ 1 & 1 & 1 & 0 \\ 0 & 0 & 0 & 2 \end{bmatrix} \quad \begin{bmatrix} 0 & 0 & 0 & 1 \\ 0 & 0 & 0 & 1 \\ 0 & 0 & 0 & 2 \\ 1 & 1 & 2 & 1 \end{bmatrix} \tag{6}$$

The Frobenius-Perron dimensions of $x \in B$ are $1, 1, 2, 3$ and $\mathrm{FPdim}(R) = 1^2 + 1^2 + 2^2 + 3^2 = 3 \cdot 5$. But every basis element is self-dual.

So we see the relationship between the parity of Frobenius-Perron dimensions and duality in fusion rings is subtle. For example, there exist fusion rings whose Frobenius-Perron dimensions of nontrivial basis elements are all odd, but the Frobenius-Perron dimension of the ring is even, e.g. $\mathbb{Z}G$ for a group of even order. But there cannot exist a fusion ring whose Frobenius-Perron dimensions of nontrivial basis elements are all even, and the Frobenius-Perron dimension of the ring is odd. Indeed, taking the Frobenius-Perron dimension of xx^* for any nontrivial basis element x in such a ring will lead to a contradiction. This reasoning provides a "local" version of Corollary 2.4 which does not require commutativity.

PROPOSITION 2.6. *Let (R, B) be a fusion ring. If 2 divides* $\mathrm{FPdim}(x)^2$ *for some $x \in B$, then there exists nontrivial $y \in B$ with $y = y^*$.*

PROOF. Assume to the contrary, that $y \neq y^*$ for all nontrivial $y \in B$. Since xx^* is self-dual, then $c_{x,x^*}^y = c_{x,x^*}^{y^*}$ for all $y \in B$. Let $\Gamma \subset B$ be representatives of the nontrivial orbits of B under the duality antiautomorphism. We compute

$$\mathrm{FPdim}(x)^2 = \mathrm{FPdim}(xx^*) \tag{7}$$

$$= 1 + 2\sum_{y \in \Gamma} c_{x,x^*}^y \mathrm{FPdim}(y) \tag{8}$$

and therefore upon rearrangement,

$$-\frac{1}{2} = \frac{-\mathrm{FPdim}(x)^2}{2} + \sum_{y \in \Gamma} c_{x,x^*}^y \mathrm{FPdim}(y). \tag{9}$$

The right-hand side of Equation (9) is an algebraic integer since 2 divides $\mathrm{FPdim}(x)^2$, but the left-hand side is not. □

The relationship between fixed-point-free automorphisms of order 2, parity of Frobenius-Perron dimensions, and duality is discussed in greater detail in Section 3.

Another consequence of Theorem 2.3(3) is that the rank of a noncommutative fusion ring with a fixed-point-free automorphism of prime order $p \in \mathbb{Z}_{\geq 2}$ cannot be too small relative to p. To achieve a reasonable lower bound for our purposes requires the following elementary lemma.

LEMMA 2.7. *There is no fusion ring (R, B) such that* $\mathrm{FPdim}(R) \in \mathbb{Z}_{\geq 1}$ *has multiplicity* 1 *as a formal codegree, and R has at exactly* 2 *distinct formal codegrees.*

PROOF. Assume that R has exactly two distinct formal codegrees, $f := \mathrm{FPdim}(R) \in \mathbb{Z}_{\geq 1}$ and $g \in \mathbb{Z}_{\geq 1}$, where integrality follows from the formal codegrees being closed under Galois conjugacy. Then with h being the sum of dimensions of irreducible representations of R distinct from FPdim, we have $1/f + h/g = 1$ by Equation (4). Therefore $h/g \leq (g-1)/g$ since $f, g, h \in \mathbb{Z}_{\geq 1}$. Thus if $g < f$, $1/f + h/g < 1$. Moreover no such fusion ring exists. □

PROPOSITION 2.8. *Let (R, B) be a fusion ring and $\phi \in \mathrm{Aut}(R, B)$ be fixed-point-free of prime order. If R is noncommutative, then* $\mathrm{rank}(R) \geq 11$.

PROOF. This proof is a brief computation by Theorem 2.3. In particular if $r := \mathrm{rank}(R)$, we seek to create a list of dimensions of irreducible representations $d_1, \ldots, d_\ell$, not all equal to 1, such that $r = \sum_{j=1}^{\ell} d_j^2$. Specifically, let $n_k \in \mathbb{Z}_{\geq 0}$ be the number of $\varphi \in \mathrm{Irr}(R)$ with $\dim(\varphi) = k \in \mathbb{Z}_{\geq 1}$. There must exist a prime $p \in \mathbb{Z}_{\geq 2}$ such that $r \equiv 1 \pmod{p}$ and $n_k \equiv 0 \pmod{p}$. All d_j, p, k, and n_k are bounded by r, so this is a finite computation which can be done by hand. One feasible solution exists with $r < 11$ which is the case $p = 2$ and $r = 9$ with $d_1 = 1$, $d_2 = d_3 = 2$. But this violates Lemma 2.7 since the two-dimensional irreducible representations share the same formal codegree by Theorem 2.3(4). □

Our last result of the section notes that for fusion rings of small rank, which are predominately commutative, many of the formal codegrees of R are determined by the group of invertible elements of R, since the action of of the invertible elements on the noninvertible basis elements is often trivial.

LEMMA 2.9. *Let (R, B) be a commutative fusion ring with $R \neq R_{\mathrm{pt}}$ such that G the group of invertible objects of R acts trivially on the set of noninvertible basis elements. Then R has $|G|$ as a formal codegree of multiplicity exactly $|G| - 1$.*

PROOF. Let $\varphi \in \mathrm{Irr}(G)$ be nontrivial. We can extend φ to a ring homomorphism $\tilde{\varphi} : R \to \mathbb{C}$ by assigning $\tilde{\varphi}(x) = 0$ for all noninvertible $x \in B$. Note that for noninvertible $x, y \in B$, xy either contains all invertible elements as summands in the case $y = x^*$, or none, by the cyclic invariance of the fusion rules (Equation (1)). Therefore the only nontrivial verification that $\tilde{\varphi}$ is a homomorphism is

$$\tilde{\varphi}(xx^*) = \sum_{g \in G} \tilde{\varphi}(g) + \sum_{y \in B \setminus G} \tilde{\varphi}(y) = \sum_{g \in G} \varphi(g), \tag{10}$$

vanishes, which follows since $\varphi \in \mathrm{Irr}(G)$ is nontrivial. Conversely, if $\varphi : R \to \mathbb{C}$ is a ring homomorphism and $\varphi(x) = 0$ for all noninvertible basis elements x, then the left-hand side of Equation (10) vanishes, thus $\varphi|_G \in \mathrm{Irr}(G)$ is nontrivial. Now if $\varphi : R \to \mathbb{C}$ is any ring homomorphism, then $f_\varphi \geq |G|$ by [**29**, Example 2.4] since $\varphi(x)$ is a root of unity for all invertible $x \in B$, with equality if and only if $\varphi(x) = 0$ for all noninvertible $x \in B$. Moreover $|G|$ is a formal codegree of multiplicity exactly $|G| - 1$. □

3. When $p = 2$

In the case when p is the unique even prime, more can be said about fixed-point-free $\phi \in \mathrm{Aut}(R, B)$. For example, when $R \cong \mathbb{Z}G$ for a finite group G, then the existence of such ϕ implies G is an abelian group of odd order, and ϕ is the inversion map $g \mapsto g^{-1}$.

3.1. Relation with duality.

PROPOSITION 3.1. *Let (R, B) be a fusion ring and ϕ be an antiautomorphism of R of order 2. Then ϕ is fixed-point-free if and only if $\phi(x) = x^* \neq x$ for all $x \in B$.*

PROOF. The converse half of the statement is trivial, so assume ϕ is fixed-point-free. For all $x \in B$, $x\phi(x)$ is ϕ-invariant, and by definition $\phi(x) = x^*$ if and only if $c^{1_R}_{x,\phi(x)} = 1$. Let $\Gamma_0 \subset B$ be a set of representatives of ϕ-orbits of B which appear as summands of $x\phi(x)$. If $\phi(x) \neq x^*$, then since ϕ preserves Frobenius-Perron dimensions [**11**, Proposition 3.3.9],

$$\mathrm{FPdim}(x)^2 = \mathrm{FPdim}(x\phi(x)) = 2\sum_{y\in\Gamma_0} c^y_{x,\phi(x)}\mathrm{FPdim}(y), \tag{11}$$

hence $2 \mid \mathrm{FPdim}(x)^2$ for any $x \in B$ with $\phi(x) \neq x^*$. Now consider xx^*, which is a self-dual object of R, hence its summands are closed under duality. Let $\Gamma_1 \subset B$ be all of the summands y of xx^* for which $\phi(y) \neq y^*$, and let $\Gamma_2 \subset B$ be a set of representatives of the ϕ-orbits of all nontrivial summands y of xx^* for which $\phi(y) = y^*$. It is evident from their definition that Γ_1 and Γ_2 are disjoint and contain all nontrivial summands of xx^*. We compute

$$\mathrm{FPdim}(xx^*) = 1 + \sum_{y\in\Gamma_1} c^y_{x,x^*}\mathrm{FPdim}(y)^2 + 2\sum_{z\in\Gamma_2} c^z_{x,x^*}\mathrm{FPdim}(z)^2, \tag{12}$$

and therefore upon rearrangement,

$$\frac{1}{2} = \frac{1}{2}\left(\mathrm{FPdim}(x)^2 - \sum_{y\in\Gamma_1} c^y_{x,x^*}\mathrm{FPdim}(y)^2\right) - \sum_{z\in\Gamma_2} c^z_{x,x^*}\mathrm{FPdim}(z)^2 \tag{13}$$

is an algebraic integer, which is false. Therefore $\phi(x) = x^*$ for all $x \in B$. □

COROLLARY 3.2. *Let (R, B) be a commutative fusion ring and $\phi \in \mathrm{Aut}(R, B)$ of order 2. If ϕ is fixed-point-free, then $\phi(x) = x^*$ for all $x \in B$.*

PROOF. The set of antiautomorphisms and automorphisms are the same when R is commutative. □

PROPOSITION 3.3. *Let (R, B) be a fusion ring and $f \in \mathrm{Aut}(R, B)$ be fixed-point-free of order 2. For all $x \in B$, if $x\phi(x) = \phi(x)x$, then $\phi(x) = x^*$.*

PROOF. The proof is identical to that of Proposition 3.1 since the element $x\phi(x)$ is ϕ-invariant under the assumption $x\phi(x) = \phi(x)x$. □

EXAMPLE 3.4. Basis elements of fusion rings need not commute with their duals. If they did, then Proposition 3.3 would immediately imply the only fixed-point-free involution is duality. This still may be true but the reason must be slightly deeper. For example, one of the fusion rings associated to the extended

Haagerup subfactor [**2**, Figure 8] possesses a basis element (labeled "B" in loc. cit.) with the following fusion rule, which does not commute with its dual.

$$\begin{bmatrix} 0&0&0&0&0&0&1&0\\ 0&0&0&0&0&1&0&0\\ 0&0&0&1&1&0&0&0\\ 0&0&1&1&1&1&0&1\\ 0&1&0&1&1&1&1&0\\ 0&0&1&1&1&1&0&0\\ 0&0&0&1&0&0&0&0\\ 1&0&0&0&0&1&0&0 \end{bmatrix} \tag{14}$$

3.2. Representation theory and formal codegrees.

PROPOSITION 3.5. *Let (R, B) be a commutative fusion ring and $\phi \in \mathrm{Aut}(R, B)$ which is fixed-point-free of order 2. For all $\varphi \in \mathrm{Irr}(R)$, if $\varphi \neq \mathrm{FPdim}$, then $\varphi(x) \notin \mathbb{R}$ for some $x \in B$.*

PROOF. Commutativity of R implies $\phi(x) = x^* \neq x$ for all $x \in B$, so if $\varphi(x) = \overline{\varphi(x)} = \varphi(x^*)$ for all $x \in B$, then $\varphi \neq \mathrm{FPdim}$ is a fixed-point of the action of ϕ on $\mathrm{Irr}(R)$, violating Proposition 2.2. □

PROPOSITION 3.6. *Let (R, B) be a commutative fusion ring and $\phi \in \mathrm{Aut}(R, B)$ have order 2. If ϕ is fixed-point-free, then 2 does not divide f_φ for any $\varphi \in \mathrm{Irr}(R)$.*

PROOF. Let $\alpha := \sum_{x\in B} xx^*$. The eigenvalues of the matrix of (left) multiplication by α are f_φ for $\varphi \in \mathrm{Irr}(R)$ since R is commutative, with multiplicity 1 [**30**, Remark 2.11]. Let Γ be an index set of representatives of nontrivial ϕ-orbits in B. As R is commutative, $\phi(x) = x^*$ for all $x \in B$ by Corollary 3.2. Then

$$\alpha = 1 + 2\sum_{x\in\Gamma} xx^*, \tag{15}$$

Therefore the matrix N of multiplication by α is the $|B| \times |B|$ identity added to a matrix entirely composed of even integers. Moreover the elements of the diagonal are odd integers. We claim the determinant of N is an odd integer as well. The case $|B| = 1$ is trivial, and we continue by induction. If $|B| = n$ for some $n > 1$, then the terms in the cofactor expansion along the first row are all even, except the first term which is odd by the inductive hypothesis. □

We expect the analogous result to Proposition 3.6 to be true for odd primes $p \in \mathbb{Z}_{\geq 3}$. When $p = 2$, small formal codegrees are highly restrictive.

LEMMA 3.7. *Let (R, B) be a commutative fusion ring and $\phi \in \mathrm{Aut}(R, B)$ be fixed-point-free of order 2. If 3 is a formal codegree of R, then $R_{\mathrm{pt}} \cong \mathbb{Z}C_3$.*

PROOF. Let $\varphi \in \mathrm{Irr}(R)$ such that $f_\varphi = 3$. If $R = R_{\mathrm{pt}}$ we are done, so we may assume that $\varphi \neq \mathrm{FPdim}$. We must have

$$[\mathbb{Q}(\varphi(y) : y \in B) : \mathbb{Q}] \in \{1, 2\} \tag{16}$$

as all Galois conjugates of φ will share the same formal codegree which can be computed as

$$3 = \sum_{y\in B} |\varphi(y)|^2 = 1 + \sum_{y\in B\setminus\{1_R\}} |\varphi(y)|^2. \tag{17}$$

We may conclude $\mathbb{Q}(\varphi(y) : y \in B)$ is a complex quadratic field by Proposition 3.5. But $|\varphi(y)|^2 = \varphi(y)\overline{\varphi(y)} \in \mathbb{Z}_{\geq 0}$ is the algebraic norm of $\varphi(y) \in \mathbb{Q}(\varphi(y) : y \in B)$, thus there exist distinct nontrivial $x_1, x_2 \in B$ such that $|\varphi(x_1)|^2 = |\varphi(x_2)|^2 = 1$. Therefore $\varphi(y) = 0$ for all $y \notin \{1_R, x_1, x_2\}$, and moreover $x := x_1 = x_2^*$. The only complex quadratic units which are not integers are $\zeta \in \{\pm\zeta_4, \pm\zeta_3, \pm\zeta_3^2\}$, so without loss of generality, assume $\varphi(x) = \zeta$ and thus $\varphi(x^*) = \zeta^{-1}$. Orthogonality of φ with FPdim gives

$$0 = 1 + (\zeta + \zeta^{-1})\mathrm{FPdim}(x), \tag{18}$$

which implies that $\zeta = \zeta_3$, up to a relabelling of x and x^*, and that $\mathrm{FPdim}(x) = \mathrm{FPdim}(x^*) = 1$. There can be no other nontrivial invertible basis elements $y \in B$ since necessarily $\varphi(y)$ would be nonzero, violating Equation (17). □

LEMMA 3.8. *Let (R, B) be a commutative fusion ring with $R \neq R_{\mathrm{pt}}$ and $\phi \in \mathrm{Aut}(R, B)$ be fixed-point-free of order 2. If 5 is a formal codegree of R, then 5 is a formal codegree of multiplicity 4.*

PROOF. Let $\varphi \in \mathrm{Irr}(R)$ such that $f_\varphi = 5$. By Proposition 2.2 and Theorem 2.3(4), the multiplicity of 5 as a formal codegree is either 2 or 4 since $R \neq R_{\mathrm{pt}}$; any higher even multiplicity violates Equation (4). Assume the multiplicity of 5 as a formal codegree is 2. Then $\mathbb{Q}(\varphi(y) : y \in B)$ is a complex quadratic field by Proposition 3.5. The definition of f_φ is $5 = 1 + \sum_{y \in B \setminus \{1_R\}} |\varphi(y)|^2$ with $|\varphi(y)|^2 \in \mathbb{Z}_{\geq 0}$. One possibility is that there exists a unique (up to duality) $x \in B$ such that $|\varphi(x)|^2 = |\varphi(x^*)|^2 = 2$. In this case orthogonality with FPdim gives

$$0 = 1 + (\varphi(x) + \varphi(x^*))\mathrm{FPdim}(x). \tag{19}$$

Since $\varphi(x) + \varphi(x^*) = 2\mathrm{Re}(\varphi(x)) \neq 0$ for this to be true, then $-1/(2\mathrm{Re}(\varphi(x))) = \mathrm{FPdim}(x)$. But $\varphi(x)$ is a complex quadratic integer so its real part must be a half integer, which forces $\mathrm{Re}(\varphi(x)) = -1/2$ so that $\mathrm{FPdim}(x) \in \mathbb{Z}_{\geq 1}$. Let $b \in \mathbb{Z}_{\neq 0}$ such that $\varphi(x) = (1/2)(-1 + b\sqrt{n})$ for some $n \in \mathbb{Z}_{<0}$. Then

$$2 = |\varphi(x)|^2 = (1/2)(-1 + b\sqrt{n})(1/2)(-1 - b\sqrt{n}) = (1/4)(1 - b^2 n). \tag{20}$$

Thus $b^2 n = -7$ and furthermore $b = \pm 1$ and $n = -7$. Equation (19) then implies $\mathrm{FPdim}(x) = 1$, which is a contradiction since $\varphi(x)$ is nonzero, and not a root of unity.

The only other possibility is that there exist unique (up to duality) nontrivial $x, y \in B$ such that $x \neq y^*$ and $|\varphi(x)|^2 = |\varphi(y)|^2 = 1$. If $\varphi(x) = a_x + b_x\sqrt{n}$ for some half-integers a_x, b_x and $n \in \mathbb{Z}_{<0}$, then $1 = |\varphi(x)|^2 = a_x^2 - b_x^2 n$. Since $-b_x^2 n \geq 0$, then there are only three cases:

$$\begin{aligned}
&(21) && a_x = 0 && \text{and} && -b_x^2 n = 1 \\
&\Rightarrow && b_x = \pm 1 && \text{and} && n = -1, \\
&(22) && a_x = \pm 1/2 && \text{and} && -b_x^2 n = 3/4 \\
&\Rightarrow && b_x = \pm 1/2 && \text{and} && n = -3, \text{ and} \\
&(23) && a_x = \pm 1 && \text{and} && b_x = 0 .
\end{aligned}$$

The same is true for $\varphi(y) = a_y + b_y\sqrt{n}$ for some half integers a_y, b_y. Orthogonality of φ and FPdim gives

$$-1/2 = a_x\mathrm{FPdim}(x) + a_y\mathrm{FPdim}(y). \tag{24}$$

If $a_x = 0$, then either $a_y = 0$ or $a_y = \pm 1$; neither satisfies Equation (24). If $a_x = \pm 1/2$, then $a_y = \pm 1/2$ or $a_y = \pm 1$. In the former case we may assume $a_x = 1/2$ and $a_y = -1/2$ without loss of generality. Thus $\text{FPdim}(y) - 1 = \text{FPdim}(x)$ which cannot occur since this implies at least one of $\text{FPdim}(x)$ or $\text{FPdim}(y)$ is even, violating Proposition 2.6. In the latter case, orthogonality of φ with $\overline{\varphi}$ gives

$$0 = 1 + 2\text{Re}(\varphi(x)^2) + 2\text{Re}(\varphi(y)^2) = 1 - 2 + 2 = 1. \tag{25}$$

So we must conclude there is no viable character φ and thus the multiplicity of 5 as a formal codegree is not 2, so it must be 4. □

PROPOSITION 3.9. *Let (R, B) be a commutative fusion ring and $\phi \in \text{Aut}(R, B)$ be fixed-point-free of order 2. If 5 is a formal codegree of R, then $\text{R}_{\text{pt}} \cong \mathbb{Z}C_5$.*

PROOF. If R is pointed, we are done. Otherwise R has 5 as a formal codegree of multiplicity 4 from Lemma 3.8. Let $\varphi_1, \overline{\varphi_1}, \varphi_2, \overline{\varphi_2}$ be the irreducible representations of R with formal codegree 5. Let $x \in B$ be such that $\varphi_1(x) \notin \mathbb{R}$ which exists by Proposition 3.5. As in the proof of Lemma 3.8, if φ_1 and φ_2 are not Galois conjugate, then $\mathbb{Q}(\varphi(x))$ is a complex quadratic field and $|\varphi(x)|^2 = |\varphi(x^*)|^2 \in \{1, 2\}$. If $|\varphi(x)|^2 = |\varphi(x^*)|^2 = 2$, then $\varphi(y) = 0$ for all $y \in B \backslash \{1_R, x, x^*\}$. Moreover a contradiction is reached in the exact manner as in the discussion around Equation (20). Thus $\varphi_1(x)$ is a complex quadratic unit, i.e. $\varphi_1(x) \in \{\pm\zeta_4, \pm\zeta_3, \pm\zeta_3^2\}$. By the same reasoning, there must exist a unique up to duality $y \in B \setminus \{1_R, x, x^*\}$ such that $\varphi(y) \in \{\pm\zeta_4, \pm\zeta_3, \pm\zeta_3^2\}$ as well. Orthogonality of φ_1 and $\overline{\varphi_1}$ insists $-1/2 = \text{Re}(\varphi_1(x)^2) + \text{Re}(\varphi_1(y)^2)$ and $\mathbb{Q}(\varphi(x)) = \mathbb{Q}(\varphi(y))$. A finite check shows there are no possible solutions, so we must conclude φ_1 and φ_2 are Galois conjugate.

With φ_1 and φ_2 Galois conjugate, assume first that $|\varphi_1(x)|^2$ and $|\varphi_2(x)|^2$ are less than or equal to 1. Then $\varphi_1(x), \varphi_2(x)$ are roots of unity [**11**, Lemma 3.3.14], therefore $|\varphi_1(x)|^2 = |\varphi_2(x)|^2 = 1$. Since $1 + 2|\varphi_1(x)|^2 < 5$, there must exist $y \in B \backslash \{1_R, x, x^*\}$ such that $\varphi_1(y) \neq 0$, and if $\varphi_1(y)$ is not a root of unity, then either $|\varphi_1(y)|^2$ or $|\varphi_2(y)|^2$ is strictly greater than 1, violating the fact that $f_{\varphi_1} = f_{\varphi_2} = 5$. So we conclude $\varphi_1(x), \varphi_1(y), \varphi_2(x), \varphi_2(y)$ are roots of unity and at least one has exactly 4 distinct Galois conjugates, say $\varphi_1(x)$ without loss of generality. In this case we must have $\varphi_1(y), \varphi_2(y) \subset \mathbb{Q}(\varphi_1(x))$ and $\varphi_1(z) = 0$ for all $z \notin \{1_R, x, x^*, y, y^*\}$. The possible primitive roots of unity $\varphi_1(x)$ with exactly 4 Galois conjugates have orders in the set $\{5, 8, 10, 12\}$. There are no solutions to the orthogonality relation of φ_1 and $\overline{\varphi_1}$: $-1/2 = \text{Re}(\varphi_1(x)^2) + \text{Re}(\varphi_1(y)^2)$ when $\varphi_1(x)$ and $\varphi_1(y)$ are 8th roots of unity, and when they are 12th roots of unity, the only feasible solutions have $2\text{Re}(\varphi_1(x)) = \pm\sqrt{3}$ and $\text{Re}(\varphi_1(y)) = 0$ without loss of generality. Orthogonality of φ_1 and FPdim would then imply $0 = 1 \pm \sqrt{3}\text{FPdim}(x)$, which cannot occur since $\text{FPdim}(x) \in \mathbb{Z}$. Moreover $\varphi_1(x)$ and $\varphi_1(y)$ are 10th roots of unity. For all solutions to $-1/2 = \text{Re}(\varphi(x)^2) + \text{Re}(\varphi(y)^2)$, we have $2\text{Re}(\varphi_1(x)) = \epsilon_x \alpha$ and $2\text{Re}(\varphi_1(y)) = \epsilon_y \alpha^{-1}$ where $\epsilon_x, \epsilon_y \in \{\pm 1\}$ and α is $(1/2)(1+\sqrt{5})$. Thus orthogonality of φ_1 and FPdim gives

$$-1 = \epsilon_x \alpha \text{FPdim}(x) + \epsilon_y \alpha^{-1} \text{FPdim}(y). \tag{26}$$

Clearly $\text{FPdim}(x) = \text{FPdim}(y)$, or else the right-hand side is irrational, thus $-1/(\epsilon_x\alpha + \epsilon_y\alpha^{-1}) = \text{FPdim}(x) = \text{FPdim}(y)$. The only selection of signs ϵ_x, ϵ_y producing a positive integer Frobenius-Perron dimension is $\epsilon_x = \epsilon_y = -1$ with $\text{FPdim}(x) = \text{FPdim}(y) = 1$. And as $\varphi_1(z) = 0$ for all $z \notin \{1_R, x, x^*, y, y^*\}$, then $R_{\text{pt}} \cong \mathbb{Z}C_5$ in this case.

Finally, we need to demonstrate the case $|\varphi_1(x)|^2 < 1$ and $|\varphi_2(x)|^2 > 1$ is not possible, without loss of generality. In this case $|\varphi_1(x)|^2$, and thus $|\varphi_1(y)|^2$ are totally positive quadratic integers, say $|\varphi_1(x)|^2 = a_x + b_x\sqrt{n}$ and $|\varphi_1(y)|^2 = a_y + b_y\sqrt{n}$ for some half integers a_x, a_y, b_x, b_y and $n \in \mathbb{Z}_{\geq 1}$ subject to the formal codegree conditions for φ_1 and φ_2:

$$2 = a_x + a_y \pm (b_x + b_y)\sqrt{n}. \tag{27}$$

Therefore $a_x + a_y = 2$, and to ensure total positivity, the real constraints are $a_x, a_y \in \{1/2, 1, 3/2\}$, and the complex constraints are $|b_x| < a_x/\sqrt{n}$ and $|b_y|\sqrt{n} < a_y/\sqrt{n}$. If a_x (or a_y) is $1/2$, there are no positive integers n such that $|b_x|$ (or $|b_y|$) can be a half integer. If a_x (or a_y) is 1, then $|b_x| = 1/2$ (or $|b_y| = 1/2$) for any positive square-free integer n, which forces $n \equiv 1 \pmod 4$. But for the smallest possible value $n = 5$, $|b_x| < 1/\sqrt{5} < 1/2$. Lastly if a_x (or a_y) is $3/2$, then $n \equiv 1 \pmod 4$, so the only possibility is that $n = 5$ and $|b_x| = 1/2$ (or $|b_y| = 1/2$). In conclusion, the only possible values for a_x and a_y are $3/2$, but this violates Equation (27). □

EXAMPLE 3.10. Proposition 3.9 is false for fusion rings with self-dual basis elements. Choose a sign $\pm$ and $a \in \mathbb{Z}$ with $a \pm 1 \geq 0$. Then the rank 4 commutative fusion ring with nontrivial fusion rules

$$\begin{bmatrix} 0 & 1 & 0 & 0 \\ 1 & a & a\pm 1 & a\pm 1 \\ 0 & a\pm 1 & a\pm 1 & a \\ 0 & a\pm 1 & a & a \end{bmatrix} \tag{28}$$

$$\begin{bmatrix} 0 & 0 & 1 & 0 \\ 0 & a\pm 1 & a\pm 1 & a \\ 1 & a\pm 1 & a & a\pm 1 \\ 0 & a & a\pm 1 & a \end{bmatrix} \tag{29}$$

$$\begin{bmatrix} 0 & 0 & 0 & 1 \\ 0 & a\pm 1 & a & a \\ 0 & a & a\pm 1 & a \\ 1 & a & a & a \end{bmatrix} \tag{30}$$

has formal codegrees 5, 5, and the roots of

$$t^2 - (27a^2 \pm 24a + 12)t + 45a^2 \pm 40a + 20. \tag{31}$$

One may convince themself that this is the smallest rank for which a counterexample exists as an exercise. Note that the formal codegrees are all integers if and only if the chosen sign is positive and $a = 0$; in this case the above fusion ring is the character ring R_{D_5} of the dihedral group of order 10. When the chosen sign is negative and $a = 1$, the above fusion ring is the direct product $F \times F$ where F is the rank 2 fusion ring with nontrivial basis element x such that $x^2 = 1_F + x$.

4. Examples of fixed-point-free automorphisms of prime order

Let (R, B) be a fusion ring with a fixed-point-free automorphism ϕ of prime order $p \in \mathbb{Z}_{\geq 2}$ for the remainder of this section. We organize our discussion around the number of ϕ-orbits of B. The unit 1_R is an orbit itself, so the trivial fusion ring is the only possible fusion ring with exactly 1 ϕ-orbit. But the identity automorphism does not have prime order so this case is not relevant.

4.1. Two orbits. Assume there is exactly one nontrivial ϕ-orbit of basis elements, hence $\text{rank}(R) = p + 1$. Then Theorem 2.3(1) implies there exists $d \in \mathbb{Z}_{\geq 1}$ such that $\text{FPdim}(x) \in \{1, d\}$ for all $x \in B$ and d satisfies $d^2 = nd + 1$ for some $n \in \mathbb{Z}_{\geq 0}$ by taking the Frobenius-Perron dimension of xx^* for any nontrivial $x \in B$. The only positive integer root of $d^2 - nd - 1$ is $d = 1$, when $n = 0$. Therefore $R \cong \mathbb{Z}G$ for a finite group of order $p + 1$ admitting a fixed-point-free automorphism of prime order p. We include a proof of the following well-known lemma, which characterizes such rings, for completeness.

LEMMA 4.1. *Let G be a finite group. If $\text{Aut}(G)$ acts transitively on $G \setminus \{e\}$, then G is an elementary abelian p-group for a prime $p \in \mathbb{Z}_{\geq 2}$.*

PROOF. Let $p \in \mathbb{Z}_{\geq 2}$ be any prime dividing $|G|$. Cauchy's theorem implies there exists an element $g \in G \setminus \{e\}$ of order p. If $h \in G \setminus \{e\}$, there exists $\phi \in \text{Aut}(G)$ such that $\phi(g) = h$, hence $h^p = \phi(g)^p = \phi(g^p) = \phi(e) = e$ as well. Moreover every nontrivial element of G has order p and therefore G is a p-group. As all p-groups are nilpotent, $Z(G)$ is nontrival. But $Z(G) \subset G$ is a characteristic subgroup (preserved under all automorphisms of G) in general, thus $G = Z(G)$ and G is abelian, and moreover an elementary abelian p-group. □

If $p = 2$, then $R \cong \mathbb{Z}C_3$, and if $p + 1$ is even, then $2^n - 1 = p$ is prime for some $n \in \mathbb{Z}_{\geq 2}$, i.e. $|G| - 1$ is a *Mersenne prime.* This currently provides 51 more examples; the largest known example has order just shy of 25 million digits, and the corresponding p is the largest known prime number at this point in history.

4.2. Three orbits. Assume there are exactly two nontrivial ϕ-orbits of basis elements, hence $\text{rank}(R) = 2p+1$. Then Theorem 2.3(1) implies there exist $d_1, d_2 \in \mathbb{Z}_{\geq 1}$ such that $\text{FPdim}(x) \in \{1, d_1, d_2\}$ for all $x \in B$.

LEMMA 4.2. *Let (R, B) be a nontrivial integral fusion ring. If there exist $d_1, d_2 \in \mathbb{Z}_{\geq 1}$ such that $\text{FPdim}(x) \in \{1, d_1, d_2\}$ for all $x \in B$, then R_{pt} is nontrivial.*

PROOF. If $R = R_{\text{pt}}$, we are done as R was assumed to be nontrivial. So consider the case $d_1 = d_2 \neq 1$. For any nontrivial $x \in B$, taking the Frobenius-Perron dimension of xx^* gives $d_1^2 = 1 + nd_1$ for some $n \in \mathbb{Z}_{\geq 0}$ which has no positive integer solutions different from 1. Therefore $d_1 > d_2$ without loss of generality if 1_R is the unique invertible element. Taking the Frobenius-Perron dimension of xx^* for $x \in B$ with $\text{FPdim}(x) = d_1$,

$$d_1^2 = 1 + n_1 d_1 + n_2 d_2 \tag{32}$$

$$\Rightarrow \qquad 1 = (d_1 - n_1)d_1 + (-n_2)d_2 \tag{33}$$

with $n_1, n_2 \in \mathbb{Z}_{\geq 0}$. A solution to such a linear Diophantine equation exists if and only if d_1, d_2 are coprime. Now let $x, y \in B$ with $x \neq y$ such that $\text{FPdim}(x) = \text{FPdim}(y) = d_2$. Taking the Frobenius-Perron dimension of y^*x,

$$d_2^2 = m_1 d_1 + m_2 d_2 \tag{34}$$

for some $m_1, m_2 \in \mathbb{Z}_{\geq 0}$. Hence $m_1 d_1 \equiv 0 \pmod{d_2}$. If $m_1 > 0$, then $m_1 = \ell d_2$ for some $\ell \in \mathbb{Z}_{\geq 1}$ since $\gcd(d_1, d_2) = 1$ which would imply $d_2 \geq \ell d_1$. As $d_2 < d_1$, we must conclude $m_1 = 0$. Thus $0 = c_{y^*,x}^{z^*} = c_{x,z}^{y}$ for any $z \in B$ with $\text{FPdim}(z) = d_1$. Since $y \neq x$ was arbitrary, $xz = d_1 x$ by comparing Frobenius-Perron dimensions, and moreover $c^z_{x,x^*} = d_1$ contradicting the fact that $d_1 > d_2$. So we must conclude R_{pt} is nontrivial. □

NOTE 4.3. Lemma 4.2 and its proof are attributed to [**1**, Proposition 3.2] whose conclusion is that R is not simple. As previously observed in loc. cit., this is an optimal result, in the sense that there exist integral fusion rings (R, B) with R_{pt} trivial, and with basis elements of 4 distinct Frobenius-Perron dimensions. The smallest such fusion ring is the character ring of the alternating group A_5, with Frobenius-Perron dimensions of basis elements $1, 3, 3, 4, 5$.

Back to the three-orbit problem, we may now assume $d_1 = 1$ without loss of generality. Assume further that $d_2 \neq 1$. Then by Section 4.1, $R_{\text{pt}} \cong \mathbb{Z}_{C_3}$ or $R_{\text{pt}} \cong \mathbb{Z}C_2^n$ for some $n \in \mathbb{Z}_{\geq 2}$ such that $2^n - 1$ is a Mersenne prime. In the former case, $p = 2$ and $\text{rank}(R) = 5$, thus R is commutative by Proposition 2.8. Moreover the automorphism ϕ is duality by Corollary 3.2. Therefore the two noninvertible basis elements x, y satisfy $x = y^*$ and are fixed by multiplication by the invertible elements, hence $d_2^2 = nd_2 + 3$ for some $n \in \mathbb{Z}_{\geq 0}$. But $d_2 \in \mathbb{Z}_{\geq 2}$, hence $d_2 = 3$ and $n = 2$. This forces $1 = c^x_{x,y} = c^y_{x,y} = c^x_{x,x}$ by the fact that xy is self-dual and the cyclic symmetry of the fusion rules. Moreover $c^y_{x,x} = 2$ by considering Frobenius-Perron dimensions. The fusion of this ring is then determined by the fusion subring $\mathbb{Z}C_3$ acting trivially on x, y, and the following fusion rule for x, since the fusion rule for y is its transpose.

$$\left[\begin{array}{c|cc|cc} 0 & 0 & 0 & 0 & 1 \\ \hline 0 & 0 & 0 & 0 & 1 \\ 0 & 0 & 0 & 0 & 1 \\ \hline 1 & 1 & 1 & 1 & 1 \\ 0 & 0 & 0 & 2 & 1 \end{array}\right] \tag{35}$$

One may recognize this as the character ring of $C_7 \rtimes C_3$, the unique non-abelian group of order $3 \cdot 7$ up to isomorphism.

Alternatively, $R_{\text{pt}} \cong \mathbb{Z}C_2^n$ for some $n \in \mathbb{Z}_{\geq 2}$ such that $p = 2^n - 1$ is a Mersenne prime. As p is odd, the ring R is self-dual, and the action of R_{pt} on the p non-invertible objects must be trivial, i.e. x^2 contains all invertible elements as summands. Moreover $\text{FPdim}(x)$ is a root of $x^2 - kx - 2^n$ for some $k \in \mathbb{Z}_{\geq 0}$. This bounds the fusion coefficients of R by $\text{FPdim}(x) \leq 2^n$, producing a very restricted set of examples for each Mersenne prime p.

EXAMPLE 4.4. Assume $n = 2$, thus $p = 3$ and $k \in \{0, 3\}$ in the notation introduced above. Consider first the case $k = 0$. As $\text{FPdim}(x) = 2$ and x^2 contains all invertible elements, then $0 = c^{\phi^2(x)}_{x,x} = c^x_{\phi(x),\phi(x)} = c^{\phi(x)}_{x,\phi(x)}$. Similarly $c^{\phi^2(x)}_{x,\phi^2(x)} = 0$, which determines the fusion rule for x (Figure (1a)), which in turn determines the fusion rules for R. We denote this fusion ring by S_2. The only remaining option is $k = 3$, thus $\text{FPdim}(x) = 4$. If we set $a := c^x_{x,x}$, $b := c^{\phi(x)}_{x,x} = c^x_{\phi^2(x),\phi^2(x)} = c^{\phi^2(x)}_{x,\phi^2(x)}$, then

$$3 - (a + b) = c^{\phi^2(x)}_{x,x} = c^x_{\phi(x),\phi(x)} = c^{\phi(x)}_{x,\phi(x)}. \tag{36}$$

Along with the fact that $c^{\phi^2(x)}_{\phi(x),\phi(x)} = c^{\phi(x)}_{x,x}$, the choice of nonnegative integers a, b with $a + b \leq 3$ determines the fusion rule for x, and the commutation relation of x and $\phi(x)$ implies $a = 0$ and $b = 1$. We denote this ring by S_4. The fusion rules for x in the cases $k = 0$ and $k = 3$ are given in Figure 1.

$$\left[\begin{array}{c|ccc|ccc}
0 & 0 & 0 & 0 & 1 & 0 & 0 \\ \hline
0 & 0 & 0 & 0 & 1 & 0 & 0 \\
0 & 0 & 0 & 0 & 1 & 0 & 0 \\
0 & 0 & 0 & 0 & 1 & 0 & 0 \\ \hline
1 & 1 & 1 & 1 & 0 & 0 & 0 \\
0 & 0 & 0 & 0 & 0 & 0 & 2 \\
0 & 0 & 0 & 0 & 0 & 2 & 0
\end{array}\right]$$

(1a) $k = 0$

$$\left[\begin{array}{c|ccc|ccc}
0 & 0 & 0 & 0 & 1 & 0 & 0 \\ \hline
0 & 0 & 0 & 0 & 1 & 0 & 0 \\
0 & 0 & 0 & 0 & 1 & 0 & 0 \\
0 & 0 & 0 & 0 & 1 & 0 & 0 \\ \hline
1 & 1 & 1 & 1 & 0 & 1 & 2 \\
0 & 0 & 0 & 0 & 1 & 2 & 1 \\
0 & 0 & 0 & 0 & 2 & 1 & 1
\end{array}\right]$$

(1b) $k = 3$

FIGURE 1. Fusion rule for x in the rings S_2 and S_4

NOTE 4.5. Observe that the fusion rings S_2 and S_4 do admit fixed-point free automorphisms of order 3. In particular, there exist fixed-point-free $\phi \in \mathrm{Aut}(C_2^2)$ from elementary group theory permuting the three nontrivial elements. Fix any such ϕ and label the noninvertible basis elements of S_2 and S_4 by x_1, x_2, x_3 as ordered in the fusion matrices in Figure 1. We may extend ϕ to a fixed-point-free fusion automorphism of S_2 or S_4 by defining $\phi(x_i) = x_{i+1}$ where indices will always be considered modulo 3 in the remainder of this argument. As the invertible elements act trivially on x_i in either ring, $\phi(gx_j) = \phi(x_j) = \phi(g)\phi(x_j)$ for any invertible g. In addition, for S_2, $\phi(x_ix_{i+1}) = \phi(2x_{i+2}) = 2x_i$, while $\phi(x_i)\phi(x_{i+1}) = x_{i+1}x_{i+2} = 2x_i$. Similarly, for S_4, $\phi(x_ix_{i+1}) = \phi(x_i + 2x_{i+1} + x_{i+2}) = x_{i+1} + 2x_{i+2} + x_i$, while $\phi(x_i)\phi(x_{i+1}) = x_{i+1}x_{i+2} = x_{i+1} + 2x_{i+2} + x_i$. Now note that in S_2, $\phi(x_i^2) = x_{i+1}^2 = \phi(x_i)\phi(x_i)$ since x_i^2 is the sum of invertible elements with multiplicity 1. Lastly for S_4, the invertible summunds of $\phi(x_i^2)$ and $\phi(x_i)\phi(x_i)$ coincide for the same reason as in S_2, while the noninvertible summands of $\phi(x_i^2)$ are $x_{i+2} + 2x_i$, which coincide with the noninvertible summands of $\phi(x_i)\phi(x_i) = x_{i+1}^2$.

Lastly, assume $R \cong \mathbb{Z}G$ is pointed. These groups are a very small subset of a larger family (finite or infinite) known as *almost homogenous* groups, i.e. those groups G such that the action of $\mathrm{Aut}(G)$ on G has at most 3 orbits. All such groups (finite or infinite) were classified in [**25**]. If $|G|$ is divisible by two distinct primes, p, q, then $|G| = pq$ and G is nonabelian with $q \equiv 1 \pmod{p}$ without loss of generality, or else G is cyclic and has elements with 4 distinct orders, which lie in distinct ϕ-orbits. The Sylow theorems imply the unique q-Sylow subgroup C_q is characteristic, hence $q = 3$ and ϕ must be the inverse map (duality). But $S_3 \cong C_3 \rtimes C_2$ has elements of order 2, so no such example exists. The only other possibility is that $|G| = q^n$ for some $n \in \mathbb{Z}_{\geq 1}$, and that $q^n = 2p + 1$ for some prime q. In other words, p is odd. By McKay's proof of the Cauchy Theorem, the number of elements $g \in G$ of order q is congruent to -1 modulo q, thus G is a q-group of exponent q. If G is abelian, then G is an elementary abelian q-group. If G is nonabelian, then G is a (generalized) Heisenberg group [**25**, Section 4] of exponent q.

The simplest examples of such finite groups G are cyclic groups of order $q = 2p+1$ where p, q are primes. Primes with this relation have a long history, and with this labelling q is called a *safe* prime and p is called a *Sophie Germain* prime. We have $\mathrm{Aut}(G) \cong C_{2p}$ and the fixed-point subgroup of any nontrivial automorphism is trivial since q is prime. There are exactly $p - 1$ automorphisms ϕ of C_{2p} of order p, each having 3 ϕ-orbits.

It should be noted that there is no known infinite family of groups of this type due to the condition $p = (1/2)(q^n - 1)$. For $q = 3$, the smallest four primes p are 13, 1093, 797161, and 3754733257489862401973357979128773.

5. Integer formal codegrees

Fusion rings with integer formal codegrees are highly restrictive by Equation (4). In particular, there exist only finitely many lists of positive integers of a given length satisfying such an equation by the analogous classical argument bounding the number of finite groups with a given number of conjugacy classes [**23**]. Therefore the Frobenius-Perron dimension of such a fusion ring of fixed rank is bounded. The result then follows since there exist only finitely many fusion rings of bounded Frobenius-Perron dimension. One can refer to [**5**, Lemma 3.14] for a proof of this fact which is stated in terms of fusion categories, but uses no categorical data in the proof. The objective of this section is to use these bounds to classify fusion rings with fixed-point-free automorphisms of prime order of rank less than 9, which are listed in Figure 3. Note that all such fusion rings must be commutative by Proposition 2.8, so this is an inoccuous assumption moving forward.

5.1. Bounds on FPdim. Assume (R, B) is a commutative fusion ring with integer formal codegrees and a fixed-point free automorphism ϕ of prime order $p \in \mathbb{Z}_{\geq 2}$. There are $n := (\text{rank}(R) - 1)/p$ nontrivial ϕ-orbits of $\text{Irr}(R)$ and thus positive integers $f_1, \ldots, f_n$ such that

$$\frac{1}{p\text{FPdim}(R)} + \sum_{j=1}^{n} \frac{1}{f_j} = \frac{1}{p}. \tag{37}$$

It is known [**35**] that in this case, $p\text{FPdim}(R)$, and moreover all f_j, are bounded by A_{n+1}, where $A_1 = p$ and inductively $A_{j+1} = A_j(A_j + 1)$ for $2 \leq j \leq n$. We will explain how this bounded provides a classification of fusion rings in a few of the cases already described above, to prepare for a novel classification in Section 5.2. For example, when $n = 1$, $p\text{FPdim}(\mathcal{C}) \leq p(p+1)$, thus $\text{FPdim}(\mathcal{C}) \leq p+1 = \text{rank}(R)$, thus $R \cong \mathbb{Z}G$ for a finite group G, as we proved without the assumption of integer formal codegrees or commutativity in Section 4.1.

For $n = 2$, $p\text{FPdim}(\mathcal{C}) \leq p(p+1)(p^2 + p + 1)$. Hence

$$\text{FPdim}(\mathcal{C}) \leq (p+1)(p^2 + p + 1).$$

When $p = 2$ this gives a bound of 21. The only possible formal codegrees by an exhaustive search are

$$[\text{FPdim}(R), f_1, f_2] \in \{[5, 5, 5],\ [9, 9, 3], [21, 7, 3]\}. \tag{38}$$

But there must also exist positive integer Frobenius-Perron dimensions a, b such that $\text{FPdim}(R) = 1 + 2a^2 + 2b^2$. There is a unique solution $a = b = 1$ for $\text{FPdim}(R) = 5$ which is realized by the fusion ring $\mathbb{Z}C_5$, a unique solution $a = 1, b = 3$ for $\text{FPdim}(R) = 21$ which is realized by the fusion ring $R_{C_7 \rtimes C_3}$, and no solution for a, b when $\text{FPdim}(R) = 9$. When $p = 3$, we have $\text{FPdim}(R) \leq 52$. The only possible formal codegrees by an exhaustive search are

$$[\text{FPdim}(R), f_1, f_2] \in \{[7, 7, 7], [10, 10, 5], [16, 16, 4], \tag{39}$$

$$[28, 14, 4], [40, 8, 5], [52, 13, 4]\}. \tag{40}$$

But there must also exist positive integer Frobenius-Perron dimensions a, b such that $\mathrm{FPdim}(R) = 1 + 3a^2 + 3b^2$. There is a unique solution $a = b = 1$ for $\mathrm{FPdim}(R) = 7$ realized by the fusion ring $\mathbb{Z}C_7$, a unique solution $a = 1, b = 2$ for $\mathrm{FPdim}(R) = 16$ realized by the fusion ring S_2 and a unique solution $a = 1, b = 4$ for $\mathrm{FPdim}(R) = 52$ realized by the fusion ring S_4. There is a unique, but spurious solution for $\mathrm{FPdim}(R) = 40$: $a = 2$ and $b = 3$, which violates Lemma 4.2 and so there are no fusion rules with these dimensions and formal codegrees. Again, this was known from Section 4.2 without the assumption of integral formal codegrees or commutativity.

5.2. Four ϕ-orbits; the case $n = 3$. In this case

$$\mathrm{FPdim}(\mathcal{C}) \leq (p+1)(p^2+p+1)(p^4+2p^3+2p^2+p+1). \tag{41}$$

For $p = 2$ this bound is 903. There are 14 distinct lists of feasible formal codegrees $[\mathrm{FPdim}(R), f_1, f_2, f_3]$, and checking for the existence of positive integers a, b, c such that $\mathrm{FPdim}(R) = 1 + 2(a^2 + b^2 + c^2)$, we find 11 distinct pairings of feasible formal codegrees and triples a, b, c. Of those for which a, b, c are not distinct, we can discard any solution for which one of a, b, c is not 1 by Lemma 4.2. The feasible solution $[\mathrm{FPdim}(R), f_1, f_2, f_3] = [147, 3, 7, 49]$ with $[a, b, c] = [1, 6, 6]$ cannot occur since $\mathrm{rank}(R_{\mathrm{pt}}) = 3$ which is coprime to 2, and there exist only basis elements of two distinct Frobenius-Perron dimensions. Hence $\mathrm{FPdim}(x) = 3$ for any noninvertible $x \in B$ by measuring the Frobenius-Perron dimension of xx^*. The feasible solution $[\mathrm{FPdim}(R), f_1, f_2, f_3] = [315, 3, 7, 45]$ with $[a, b, c] = [2, 3, 12]$ is spurious since this would have a fusion subring of rank 5 with 3 distinct Frobenius-Perron dimensions of basis elements, which does not exist from Section 4. Lastly, the feasible formal codegrees $[\mathrm{FPdim}(R), f_1, f_2, f_3] = [55, 15, 11, 3]$ with unique dimensions $[a, b, c] = [1, 1, 5]$ violate Lemma 2.9 as R_{pt} must act trivially on B in this case. The remaining feasible solutions are listed in Figure 2.

Formal codegrees	Basis FPdims
7, 7, 7, 7, 7, 7, 7	1, 1, 1, 1, 1, 1, 1
39, 13, 13, 13, 13, 3, 3	1, 1, 1, 3, 3, 3, 3
55, 11, 11, 5, 5, 5, 5	1, 1, 1, 1, 1, 5, 5
119, 51, 51, 7, 7, 3, 3	1, 1, 1, 3, 3, 7, 7

FIGURE 2. Feasible rank 7 fusion ring data w. fixed-point-free automorphism of order 2 and integer formal codegrees

The case $\mathrm{FPdim}(R) = 7 = \mathrm{rank}(R)$ must have $R \cong \mathbb{Z}C_7$. For the cases which are not integral groups rings, note that $\mathrm{rank}(R_{\mathrm{pt}})$ is coprime to the number of noninvertible $x \in B$ of any fixed Frobenius-Perron dimension, hence $1 = c_{g,x}^x = c_{x,x^*}^{g^{-1}} = c_{x,x^*}^g$ for any invertible $g \in B$ by the cylic invariance of the fusion rules (Equation (1)) and the fact that xx^* is self-dual.

Assume $\mathrm{FPdim}(R) = 39$ and let $x, y \in B$ be distinct such that $x \neq y^*$ and $\mathrm{FPdim}(x) = \mathrm{FPdim}(y) = 3$. Ordering the noninvertible basis elements x, x^*, y, y^*, the fusion rule for x is

$$
(42)\qquad \left[\begin{array}{c|cc|cc|cc}
0 & 0 & 0 & 0 & 1 & 0 & 0 \\ \hline
0 & 0 & 0 & 0 & 1 & 0 & 0 \\
0 & 0 & 0 & 0 & 1 & 0 & 0 \\ \hline
1 & 1 & 1 & a & a & 1-a & 1-a \\
0 & 0 & 0 & b & a & 3-n & c \\ \hline
0 & 0 & 0 & c & 1-a & d & 2+a-c-d \\
0 & 0 & 0 & 3-n & 1-a & 2a+b+c-d-1 & d
\end{array}\right]
$$

where $a \in \{0,1\}$, $b,c,d \in \{0,1,2,3\}$, and $n := a+b+c$, using only the cyclic symmetry of the fusion rules and the Frobenius-Perron dimension constraints. Commutativity of the above with its transpose determines $a = 0$, and $[b,c,d] \in \{[1,0,0],[2,0,1]\}$ up to a relabeling $y \leftrightarrow y^*$. The case $[b,c,d] = [1,0,0]$ is erroneous since this implies $c_{y,y}^{y} = c_{y,y^*}^{y^*} = c_{y,y^*}^{y} = 1$ forcing $c_{y,y}^{y^*} = -1$ by Frobenius-Perron dimension comparison. The fusion rules for x and y are then determined as

$$
(43)\qquad \left[\begin{array}{c|cc|cc|cc}
0 & 0 & 0 & 0 & 1 & 0 & 0 \\ \hline
0 & 0 & 0 & 0 & 1 & 0 & 0 \\
0 & 0 & 0 & 0 & 1 & 0 & 0 \\ \hline
1 & 1 & 1 & 0 & 0 & 1 & 1 \\
0 & 0 & 0 & 2 & 0 & 1 & 0 \\ \hline
0 & 0 & 0 & 0 & 1 & 1 & 1 \\
0 & 0 & 0 & 1 & 1 & 0 & 1
\end{array}\right]
\qquad
\left[\begin{array}{c|cc|cc|cc}
0 & 0 & 0 & 0 & 0 & 0 & 1 \\ \hline
0 & 0 & 0 & 0 & 0 & 0 & 1 \\
0 & 0 & 0 & 0 & 0 & 0 & 1 \\ \hline
0 & 0 & 0 & 1 & 1 & 0 & 1 \\
0 & 0 & 0 & 1 & 0 & 1 & 1 \\ \hline
1 & 1 & 1 & 1 & 1 & 0 & 0 \\
0 & 0 & 0 & 0 & 1 & 2 & 0
\end{array}\right]
$$

These are the fusion rules of the character ring of the unique nonabelian group of order 39 up to isomorphism, $C_{13} \rtimes C_3$.

The only unknown fusion coefficient for a fusion ring under our current assumptions with $\mathrm{FPdim}(R) = 55$ is $c_{x,x}^{x} = c_{x,x^*}^{x^*}$ since $c_{x,x}^{x^*} = 5 - c_{x,x}^{x}$ and $c_{x,x^*}^{x} = 4 - c_{x,x}^{x}$. But the fusion rule for x must commute with its transpose which forces $c_{x,x}^{x} = 2$, hence the fusion rule for noninvertible x in R with $\mathrm{FPdim}(R) = 55$ is

$$
(44)\qquad \left[\begin{array}{c|cc|cc|cc}
0 & 0 & 0 & 0 & 0 & 0 & 1 \\ \hline
0 & 0 & 0 & 0 & 0 & 0 & 1 \\
0 & 0 & 0 & 0 & 0 & 0 & 1 \\ \hline
0 & 0 & 0 & 0 & 0 & 0 & 1 \\
0 & 0 & 0 & 0 & 0 & 0 & 1 \\ \hline
1 & 1 & 1 & 1 & 1 & 2 & 2 \\
0 & 0 & 0 & 0 & 0 & 3 & 2
\end{array}\right].
$$

These are the fusion rules of the character ring of the unique nonabelian group of order 55 up to isomorphism, $C_{11} \rtimes C_5$.

In the case $\mathrm{FPdim}(R) = 119$, the basis elements of Frobenius-Perron dimensions 1 and 3 form a fusion subring isomorphic to $R_{C_7 \rtimes C_3}$. Indeed the only nonnegative integer solutions to $3^2 = n_1(1) + n_2(3) + n_3(7)$ where $n_1 \in \{0,3\}$ have $n_3 = 0$. If $x, y \in B$ have $\mathrm{FPdim}(x) = 3$ and $\mathrm{FPdim}(y) = 7$, then the only unknown fusion coefficient for x is $c_{x,y}^{y} = c_{x,y^*}^{y^*}$ as by dimension comparison, $c_{x,y}^{y^*} = c_{x,y^*}^{y} = 3 - c_{x,y}^{y}$. Meanwhile, measuring the dimension of yy^* and noting $c_{y,y^*}^{x} = c_{y,y^*}^{x^*}$ and $c_{y,y^*}^{y} = c_{y,y^*}^{y^*}$, the only solution for $7^2 = 3 + 2c_{y,y^*}^{x}(3) + 2c_{y,y^*}^{y}(7)$ is $c_{y,y^*}^{x} = 3$ and $c_{y,y^*}^{y} = 2$ which determines the fusion of x since $3 = c_{y,y^*}^{x} = c_{x^*,y^*}^{y^*} = c_{x,y}^{y}$. By dimension comparison, $c_{y,y}^{y^*} = 5$ and all other fusion coefficients for y are zero except the ones involving invertible basis elements, giving the following fusion for x and y.

$$\left[\begin{array}{c|cc|cc|cc} 0&0&0&0&1&0&0\\ \hline 0&0&0&0&1&0&0\\ 0&0&0&0&1&0&0\\ \hline 1&1&1&1&1&0&0\\ 0&0&0&2&1&0&0\\ \hline 0&0&0&0&0&3&0\\ 0&0&0&0&0&0&3 \end{array}\right] \qquad \left[\begin{array}{c|cc|cc|cc} 0&0&0&0&0&0&1\\ \hline 0&0&0&0&0&0&1\\ 0&0&0&0&0&0&1\\ \hline 0&0&0&0&0&0&3\\ 0&0&0&0&0&0&3\\ \hline 1&1&1&3&3&2&2\\ 0&0&0&0&0&5&2 \end{array}\right] \tag{45}$$

The fusion in (45) along with their duals fail to be associative, hence there is no fusion ring of this type.

Rank	Ring	Prime	Orbits	FPdim	Basis FPdims
1	–	–	–	–	–
2	–	–	–	–	–
3	$\mathbb{Z}C_3$	2	2	3	$1,1,1$
4	$\mathbb{Z}C_2^2$	3	2	4	$1,1,1,1$
5	$\mathbb{Z}C_5$	2	3	5	$1,1,1,1,1$
	$R_{C_7\rtimes C_3}$	2	3	21	$1,1,1,3,3$
6	–	–	–	–	–
7	$\mathbb{Z}C_7$	2, 3	4, 3	7	$1,1,1,1,1,1,1$
	S_2	3	3	16	$1,1,1,1,2,2,2$
	$R_{C_{13}\rtimes C_3}$	2	4	39	$1,1,1,3,3,3,3$
	S_4	3	3	52	$1,1,1,1,4,4,4$
	$R_{C_{11}\rtimes C_5}$	2	4	55	$1,1,1,1,1,5,5$
8	$\mathbb{Z}C_2^3$	7	2	8	$1,1,1,1,1,1,1,1$

FIGURE 3. Fusion rings R w. fixed-point-free automorphisms of prime order, integer formal codegrees, and $\operatorname{rank}(R) < 9$

6. Fusion categories

The primary interest in fusion rings is their realization as Grothendieck rings of *fusion categories*, which have profound applications both representation theory and mathematical physics. A fusion category is equivalent to a fusion ring along with solutions to coherence conditions on the associative relations of the ring, typically referred to as *pentagon equations* [**11**, Section 2.1]. When the fusion coefficients of a fusion ring are larger than 1 or the rank is large, the complexity of these equations becomes untenable and one is forced to develop other tools to determine whether such a ring is *categorifiable*, i.e. realizable as the Grothendieck ring of a fusion category. For example, all categorifications of the integral group rings $\mathbb{Z}G$ for a finite group G are of the form Vec_G^ω for some $\omega \in H^3(G, \mathbb{C}^\times)$, but if there exists a unique noninvertible basis element, this question is unanswered after two decades of research (e.g. [**15**][**20**][**24**][**34**][**33**] and references within). In the remainder of this section we will refer the reader to the standard textbook [**11**] for definitions and basic results in this subject.

6.1. The categorifications of small rank. Here we will determine which of the fusion rings in Figure 3 are categorifiable. This is equivalent to classifying all fusion categories with fixed-point-free fusion automorphisms of prime order and rank less than 9, since the formal codegrees of the Grothendieck ring of such a category must be integers by Theorem 2.3(1) and [**30**, Corollary 2.15]. If a fusion ring R in Figure 3 is an integral group ring, its categorifications have been discussed *ad nauseam* and are the categories of finite-dimensional G-graded vector spaces Vec_G^ω twisted by a 3-cocycle ω. If a fusion ring $R\neq R_{\mathrm{pt}}$ in Figure 3 has Frobenius-Perron dimension pq for distinct primes pq, then there is a unique categorification by the representation category of the unique nonabelian finite group of order pq [**12**, Theorem 6.3]. The only fusion rings in Figure 3 not of these forms are S_2 and S_4.

We will demonstrate that neither S_2 or S_4 are categorifiable by a fusion category $\mathcal{C}$ in the usual manner of showing their *double* $\mathcal{Z}(\mathcal{C})$, a modular fusion category in the sense of [**11**, Section 8.14], cannot exist. The modular fusion category $\mathcal{Z}(\mathcal{C})$ is related to the fusion category $\mathcal{C}$ by the restriction tensor functor $F:\mathcal{Z}(\mathcal{C})\to\mathcal{C}$ and its adjoint $I:\mathcal{C}\to\mathcal{Z}(\mathcal{C})$ [**11**, Section 9.2]. The formal codegrees are of particular importance in such arguments as they control the decomposition of the induction of the tensor unit $I(\mathbb{1}_{\mathcal{C}})\in\mathcal{Z}(\mathcal{C})$ into simple summands [**30**, Theorem 2.13]. Note that to use this result, a *spherical* structure is required. But such a structure is innate for any fusion category with $\mathrm{FPdim}(\mathcal{C})\in\mathbb{Z}$ [**11**, Corollary 9.6.6] because such fusion categories are *pseudounitary* [**11**, Definition 9.4.4]. In this case we have no reason to discuss *categorical* dimensions as they may be replaced with Frobenius-Perron dimensions without any loss of generality.

PROPOSITION 6.1. *There is no categorification of the fusion ring S_2.*

PROOF. Let $\mathcal{C}$ be a categorification of S_2, which must be pseudounitary and moreover spherical [**11**, Corollary 9.6.6], whose simple objects we label

$$g_0:=\mathbb{1}_{\mathcal{C}},g_1,g_2,g_3,X_1,X_2,X_3 \tag{46}$$

with X_j non-invertible and the invertible simple objects forming the group C_2^2. Recall that $\mathrm{FPdim}(X_j)=2$ for $j=1,2,3$. The formal codegrees of S_2 are $16,16,16,16,4,4,4$, hence the simple summands of $I(g_0)$ have dimensions $1,1,1,1,4,4,4$ by [**30**, Theorem 2.13] and all have trivial ribbon twist [**33**, Proposition 3.11]. Using [**11**, Proposition 9.2.2], we compute that

$$F(I(g_j))=7g_j+3\sum_{k\neq j}g_k. \tag{47}$$

Let $Z\subset I(g_0)$ satisfy $\mathrm{FPdim}(Z)=4$. The adjoint relation of F and I gives

$$\dim_{\mathbb{C}}\mathrm{Hom}(I(g_0),Z)=\dim_{\mathbb{C}}\mathrm{Hom}(g_0,F(Z))=1 \tag{48}$$

by [**30**, Theorem 2.13] since S_2 is a commutative fusion ring. Label the three simple summands $Z_1,Z_2,Z_3\subset I(g_0)$ with $\mathrm{FPdim}(Z_j)=4$ for $1\leq j\leq 3$. Note that if $F(Z_j)=g_0+3g_k$ for some $k\neq 0$, then

$$\dim_{\mathbb{C}}(Z_j,I(g_k))=\dim_{\mathbb{C}}(F(Z_j),g_k)=3, \tag{49}$$

thus $\dim_{\mathbb{C}}\mathrm{Hom}(I(g_j),I(g_j))\geq 9$, violating Equation (47). Hence either $2Z_k\oplus Z_\ell\subset I(g_j)$ or $Z_1\oplus Z_2\oplus Z_3\subset I(g_j)$ for some $k\neq\ell\in\{1,2,3\}$ so that Equation (47) is satisfied. But since $\theta_{Z_j} = 1$ for $j \in \{1, 2, 3\}$, in either case by [**30**, Theorem 2.5],

$$0 = \mathrm{Tr}(\theta_{I(g_j)}) = 12 + \sum_k \theta_{Y_k} \mathrm{FPdim}(Y_k) \tag{50}$$

where Y_k are the unknown simple summands of $I(g_j)$ and θ_{Y_k} their ribbon twists which are roots of unity. This would subsequently imply $12 \leq \sum_k |\mathrm{FPdim}(Y_k)| = 4$ by the triangle inequality as $\mathrm{FPdim}(I(g_j)) = 16$. So no categorification of S_2 can exist. □

NOTE 6.2. One could also note that S_2 is a C_2^2-graded extension of $\mathbb{Z}C_2^2$. Thus any categorification of S_2 is subject to the extension theory of [**13**, Theorem 1.3]. The choice to not pursue this option was based on the inevitable need for the induction-restriction functors in the proof of Proposition 6.3, and the length of the argument not significantly decreasing either way.

A description of integral fusion categories of Frobenius-Perron dimension pq^2 for distinct primes p and q was given in [**21**]. One slice of this description are categories which are *group-theoretical*, i.e. those whose doubles coincide with $\mathcal{Z}(\mathrm{Vec}_G^\omega)$ for a finite group G and 3-cocycle ω. It is still a nontrivial task to determine how such a given category might be realized by group-theoretical data, but for the fusion ring S_4 with Frobenius-Perron dimension $52 = 2^2 \cdot 13$, the options are limited.

PROPOSITION 6.3. *There is no categorification of the fusion ring S_4.*

PROOF. Let $\mathcal{C}$ be a categorification of S_4, which must be pseudounitary and moreover spherical [**11**, Corollary 9.6.6]. Even more, $\mathcal{C}$ must be group-theoretical by [**21**, Theorem 1.1], i.e. the double $\mathcal{Z}(\mathcal{C})$ must be braided equivalent to $\mathcal{Z}(\mathrm{Vec}_G^\omega)$ for some finite group G of order 52 and 3-cocycle $\omega \in H^3(G, \mathbb{C}^\times)$. We will first determine there is only one possible such finite group G.

The formal codegrees of S_4 are $52, 13, 13, 13, 4, 4, 4$, hence the simple summands of the induction $I(\mathbb{1}_\mathcal{C})$ have dimensions $1, 4, 4, 4, 13, 13, 13$ by [**30**, Theorem 2.13]. As the forgetful functor $F : \mathcal{Z}(\mathcal{C}) \to \mathcal{C}$ is a tensor functor, if $X \in \mathcal{O}(\mathcal{Z}(\mathcal{C})_{\mathrm{pt}})$, then $\mathbb{1}_\mathcal{C} \cong F(X)^{\otimes 2} \cong F(X^{\otimes 2})$. Thus $X^{\otimes 2} \subset I(\mathbb{1}_\mathcal{C})$ and $X^{\otimes 2}$ is invertible. The unit $\mathbb{1}_{\mathcal{Z}(\mathcal{C})}$ is the only such simple object. Moreover all invertible objects in the double of $\mathcal{C}$ have order at most 2. Now $\mathcal{Z}(\mathrm{Vec}_G^\omega)$ contains a fusion subcategory equivalent to $\mathrm{Rep}(G)$ (refer to [**26**], for example), so the only possibility for a finite group G of order 52 such that $\mathcal{Z}(\mathcal{C}) \simeq \mathcal{Z}(\mathrm{Vec}_G^\omega)$ is a braided equivalence is the dihedral group D_{26} as for $G \not\cong D_{26}$ with $|G| = 52$, G has 1-dimensional representations of order 4 or greater. Analysis of these twisted doubles is possible by inspection, using the methods and formulas found in [**18**], for example. As 13 is a formal codegree of $\mathcal{C}$, there must exist a simple summand $X \subset I(\mathbb{1}_\mathcal{C})$ with $\mathrm{FPdim}(X) = 4$. None of the simple objects of $\mathcal{Z}(\mathrm{Vec}_{D_{26}}^\omega)$ have dimension 4, so no such fusion category $\mathcal{C}$ exists. □

6.2. Fusion automorphisms and tensor autoequivalences. Recall that a *tensor autoequivalence* $F : \mathcal{C} \to \mathcal{C}$ of a fusion category $\mathcal{C}$ is an exact and faithful $\mathbb{C}$-linear functor along with functorial isomorphisms $J_{X,Y} : F(X) \otimes F(Y) \to F(X \otimes Y)$ for all $X, Y \in \mathcal{C}$ such that $F(\mathbb{1}_\mathcal{C}) \cong \mathbb{1}_\mathcal{C}$ and the pair F, J satifies coherence conditions [**11**, Definition 2.4.1]. Needless to say, the disparity between automorphisms of Grothendieck rings and tensor autoequivalences is stark. As this exposition is primarily about the former, we only illustrate this disparity in one example for a fusion ring of small rank from Figure 3 which is not an integral group ring, since these autoequivalences are discussed in detail in [**11**, Section 2.6].

The key observation is that if there exists a fixed-point-free fusion automorphism ϕ of the Grothendieck ring of a fusion category $\mathcal{C}$ of prime order $p \in \mathbb{Z}_{\geq 2}$ which lifts to a tensor autoequivalence of $\mathcal{C}$, this is equivalent to a *categorical action* of C_p on $\mathcal{C}$ [**11**, Definition 4.15.1] and one may consider the C_p-*equivariantization* of $\mathcal{C}$, denoted $\mathcal{C}^{C_p}$ [**11**, Definition 2.7.2]. The simple objects of the fusion category $\mathcal{C}^{C_p}$ are the ϕ-orbits $\tilde{X}$ of the simple objects $X \in \mathcal{C}$ with $\mathrm{FPdim}(\tilde{X}) = p\mathrm{FPdim}(X)$ and invertible objects g for $g \in C_p$ [**11**, Proposition 4.15.9].

PROPOSITION 6.4. *The duality fusion automorphism of* $R_{C_7 \rtimes C_3}$ *does not lift to a tensor autoequivalence of* $\mathrm{Rep}(C_7 \rtimes C_3)$.

PROOF. Assume to the contrary that duality is a tensor autoequivalence of $\mathrm{Rep}(C_7 \rtimes C_3)$. The simple objects of $\mathcal{C} := \mathrm{Rep}(C_7 \rtimes C_3)^{C_2}$, which must be pseudounitary and moreover spherical [**11**, Corollary 9.6.6], are e, g, X, Y where $\mathrm{FPdim}(X) = 2$ and $\mathrm{FPdim}(Y) = 6$. Moreover e, g, X form a fusion subcategory of $\mathcal{C}$ with $\mathrm{Rep}(S_3)$ fusion rules. The fusion rules of $\mathcal{C}$ are then

$$\begin{bmatrix} 1 & 0 & 0 & 0 \\ 0 & 1 & 0 & 0 \\ 0 & 0 & 1 & 0 \\ 0 & 0 & 0 & 1 \end{bmatrix} \quad \begin{bmatrix} 0 & 1 & 0 & 0 \\ 1 & 0 & 0 & 0 \\ 0 & 0 & 1 & 0 \\ 0 & 0 & 0 & 1 \end{bmatrix} \tag{51}$$

$$\begin{bmatrix} 0 & 0 & 1 & 0 \\ 0 & 0 & 1 & 0 \\ 1 & 1 & 1 & 0 \\ 0 & 0 & 0 & 2 \end{bmatrix} \quad \begin{bmatrix} 0 & 0 & 0 & 1 \\ 0 & 0 & 0 & 1 \\ 0 & 0 & 0 & 2 \\ 1 & 1 & 2 & 5 \end{bmatrix} \tag{52}$$

with formal codegrees $42, 7, 3, 2$. Therefore the simple summands $\mathbb{1}_{\mathcal{Z}(\mathcal{C})}$, Z_1, Z_2, Z_3 of $I(e)$ have dimensions 1, 6, 14 and 21, respectively. We compute

$$F(I(e)) = 4e \oplus 2g \oplus 3X \oplus 5Y. \tag{53}$$

As $\mathrm{FPdim}(Y) = 6$ all the summands Y appearing in Equation (53) must be summands of $F(Z_2)$ and $F(Z_3)$. This forces the decompositions $F(Z_1) = e \oplus g \oplus 2X$, $F(Z_2) = e \oplus g \oplus 2Y$, and $F(Z_3) = e \oplus X \oplus 3Y$. Similarly

$$F(I(g) - Z_1 - Z_2) = 2g \oplus X \oplus 3Y. \tag{54}$$

Therefore $I(g)$ has two other nonisomorphic simple summands Z_4, Z_5 whose dimensions must be among the collection 1, 3, 7, 21 as their squares must divide $\mathrm{FPdim}(\mathcal{Z}(\mathcal{C})) = 2^2 \cdot 3^2 \cdot 7^2$ [**11**, Proposition 8.14.6] and they must be odd since they contain g in their forgetful image with multiplicity 1. The only possibility is that $F(Z_4) = g$ and $F(Z_5) = g \oplus X \oplus 3Y$ without loss of generality. Therefore $\mathcal{Z}(\mathcal{C})_{\mathrm{pt}}$ is rank 2. Now note that $\mathcal{C}$ is group-theoretical since its Frobenius-Perron dimension is $2 \cdot 3 \cdot 7$ [**14**, Theorem 9.2], i.e. $\mathcal{Z}(\mathcal{C})$ is braided equivalent to $\mathcal{Z}(\mathrm{Vec}_G^\omega)$ for a finite group G of order 42 which can have at most 2 isomorphism classes of 1-dimensional representations, as these correspond to distinct invertible objects in $\mathrm{Rep}(G) \subset \mathcal{Z}(\mathrm{Vec}_G^\omega)$. Therefore $G \cong D_{21}$. Analysis of these twisted doubles is possible by inspection, using the methods and formulas found in [**18**], for example. We find $\mathrm{FPdim}(X) = 2$ for all simple noninvertible $X \in \mathcal{Z}(\mathrm{Vec}_{D_{21}}^\omega)$ over all $\omega \in H^3(D_{21}, \mathbb{C}^\times)$, and so we conclude no such category $\mathcal{C}$ exists. $\square$

NOTE 6.5. It was shown that there is no *braided* tensor autoequivalence of $\mathrm{Rep}(C_7 \rtimes C_3)$ which restricts to the duality fusion automorphism in Proposition 2.6 and Theorem 2.1 of [**9**].

References

[1] Max A. Alekseyev, Winfried Bruns, Sebastien Palcoux, and Fedor V. Petrov, *Classification of modular data of integral modular fusion categories up to rank 11*, arXiv:, Feb. 2023.

[2] Stephen Bigelow, Emily Peters, Scott Morrison, and Noah Snyder, *Constructing the extended Haagerup planar algebra*, Acta Math. **209** (2012), no. 1, 29–82, DOI 10.1007/s11511-012-0081-7. MR2979509

[3] Harvey I. Blau, *Table algebras*, European J. Combin. **30** (2009), no. 6, 1426–1455, DOI 10.1016/j.ejc.2008.11.008. MR2535395

[4] Harvey I. Blau and Paul-Hermann Zieschang, *Sylow theory for table algebras, fusion rule algebras, and hypergroups*, J. Algebra **273** (2004), no. 2, 551–570, DOI 10.1016/j.jalgebra.2003.09.041. MR2037711

[5] Paul Bruillard, Siu-Hung Ng, Eric C. Rowell, and Zhenghan Wang, *Rank-finiteness for modular categories*, J. Amer. Math. Soc. **29** (2016), no. 3, 857–881, DOI 10.1090/jams/842. MR3486174

[6] Paul Bruillard and Eric C. Rowell, *Modular categories, integrality and Egyptian fractions*, Proc. Amer. Math. Soc. **140** (2012), no. 4, 1141–1150, DOI 10.1090/S0002-9939-2011-11476-X. MR2869100

[7] W. Burnside, *Theory of groups of finite order*, Dover Publications, Inc., New York, 1955. 2d ed. MR69818

[8] Agustina Czenky and Julia Plavnik, *On odd-dimensional modular tensor categories*, Algebra Number Theory **16** (2022), no. 8, 1919–1939, DOI 10.2140/ant.2022.16.1919. MR4516198

[9] Alexei Davydov, *Unphysical diagonal modular invariants*, J. Algebra **446** (2016), 1–18, DOI 10.1016/j.jalgebra.2015.09.007. MR3421083

[10] Jingcheng Dong and Li Dai, *On braided almost square-free fusion categories with odd dimension*, Comm. Algebra **47** (2019), no. 12, 5063–5071, DOI 10.1080/00927872.2019.1612413. MR4019323

[11] Pavel Etingof, Shlomo Gelaki, Dmitri Nikshych, and Victor Ostrik, *Tensor categories*, Mathematical Surveys and Monographs, vol. 205, American Mathematical Society, Providence, RI, 2015, DOI 10.1090/surv/205. MR3242743

[12] Pavel Etingof, Shlomo Gelaki, and Viktor Ostrik, *Classification of fusion categories of dimension pq*, Int. Math. Res. Not. **57** (2004), 3041–3056, DOI 10.1155/S1073792804131206. MR2098028

[13] Pavel Etingof, Dmitri Nikshych, and Victor Ostrik, *Fusion categories and homotopy theory*, Quantum Topol. **1** (2010), no. 3, 209–273, DOI 10.4171/QT/6. With an appendix by Ehud Meir. MR2677836

[14] Pavel Etingof, Dmitri Nikshych, and Victor Ostrik, *Weakly group-theoretical and solvable fusion categories*, Adv. Math. **226** (2011), no. 1, 176–205, DOI 10.1016/j.aim.2010.06.009. MR2735754

[15] David E. Evans and Terry Gannon, *Near-group fusion categories and their doubles*, Adv. Math. **255** (2014), 586–640, DOI 10.1016/j.aim.2013.12.014. MR3167494

[16] Terry Gannon, *The automorphisms of affine fusion rings*, Adv. Math. **165** (2002), no. 2, 165–193, DOI 10.1006/aima.2001.2006. MR1887583

[17] Daniel Gorenstein, *Finite groups*, Harper & Row, Publishers, New York-London, 1968. MR231903

[18] Angus Gruen and Scott Morrison, *Computing modular data for pointed fusion categories*, Indiana Univ. Math. J. **70** (2021), no. 2, 561–593, DOI 10.1512/iumj.2021.70.8309. MR4257620

[19] Seung-moon Hong and Eric Rowell, *On the classification of the Grothendieck rings of non-self-dual modular categories*, J. Algebra **324** (2010), no. 5, 1000–1015, DOI 10.1016/j.jalgebra.2009.11.044. MR2659210

[20] Masaki Izumi, *A Cuntz algebra approach to the classification of near-group categories*, Proceedings of the 2014 Maui and 2015 Qinhuangdao conferences in honour of Vaughan F. R. Jones' 60th birthday, Proc. Centre Math. Appl. Austral. Nat. Univ., vol. 46, Austral. Nat. Univ., Canberra, 2017, pp. 222–343. MR3635673

[21] David Jordan and Eric Larson, *On the classification of certain fusion categories*, J. Noncommut. Geom. **3** (2009), no. 3, 481–499, DOI 10.4171/JNCG/44. MR2511638

[22] W. F. Ke and K. S. Wang, *On the Frobenius groups with kernel of order* 64, Contributions to general algebra, 7 (Vienna, 1990), Hölder-Pichler-Tempsky, Vienna, 1991, pp. 221–233. MR1143086

[23] Edmund Landau, *Über die Klassenzahl der binären quadratischen Formen von negativer Discriminante* (German), Math. Ann. **56** (1903), no. 4, 671–676, DOI 10.1007/BF01444311. MR1511192

[24] Hannah K. Larson, *Pseudo-unitary non-self-dual fusion categories of rank 4*, J. Algebra **415** (2014), 184–213, DOI 10.1016/j.jalgebra.2014.05.032. MR3229513

[25] Helmut Mäurer and Markus Stroppel, *Groups that are almost homogeneous*, Geom. Dedicata **68** (1997), no. 2, 229–243, DOI 10.1023/A:1005090519480. MR1484565

[26] Deepak Naidu, Dmitri Nikshych, and Sarah Witherspoon, *Fusion subcategories of representation categories of twisted quantum doubles of finite groups*, Int. Math. Res. Not. IMRN **22** (2009), 4183–4219, DOI 10.1093/imrn/rnp084. MR2552301

[27] Sonia Natale and Julia Yael Plavnik, *Solvability of a class of braided fusion categories*, Appl. Categ. Structures **22** (2014), no. 1, 229–240, DOI 10.1007/s10485-012-9299-y. MR3163515

[28] Siu-Hung Ng and Peter Schauenburg, *Frobenius-Schur indicators and exponents of spherical categories*, Adv. Math. **211** (2007), no. 1, 34–71, DOI 10.1016/j.aim.2006.07.017. MR2313527

[29] Victor Ostrik, *On formal codegrees of fusion categories*, Math. Res. Lett. **16** (2009), no. 5, 895–901, DOI 10.4310/MRL.2009.v16.n5.a11. MR2576705

[30] Victor Ostrik, *Pivotal fusion categories of rank 3* (English, with English and Russian summaries), Mosc. Math. J. **15** (2015), no. 2, 373–396, 405, DOI 10.17323/1609-4514-2015-15-2-373-396. MR3427429

[31] Peter Rowley, *Finite groups admitting a fixed-point-free automorphism group*, J. Algebra **174** (1995), no. 2, 724–727, DOI 10.1006/jabr.1995.1148. MR1334233

[32] Andrew Schopieray, *Lie theory for fusion categories: a research primer*, Topological phases of matter and quantum computation, Contemp. Math., vol. 747, Amer. Math. Soc., [Providence], RI, [2020] ©2020, pp. 1–26, DOI 10.1090/conm/747/15036. MR4079742

[33] Andrew Schopieray, *Categorification of integral group rings extended by one dimension*, J. Lond. Math. Soc. (2) **108** (2023), no. 4, 1617–1641, DOI 10.1112/jlms.12789. MR4655273

[34] Jacob Siehler, *Near-group categories*, Algebr. Geom. Topol. **3** (2003), 719–775, DOI 10.2140/agt.2003.3.719. MR1997336

[35] Tanzô Takenouchi, *On an indeterminate equation*, Proceedings of the Physico-Mathematical Society of Japan. 3rd Series, **3** (1921), no. 6, 78–92.

[36] John Griggs Thompson, *A proof that finite groups with a fixed-point-free automorphism of prime order are nilpotent*, ProQuest LLC, Ann Arbor, MI, 1959. Thesis (Ph.D.)–The University of Chicago. MR2611480

[37] Andrey V. Vasil'ev and Paul-Hermann Zieschang, *Solvable hypergroups and a generalization of Hall's theorems on finite solvable groups to association schemes*, J. Algebra **594** (2022), 733–750, DOI 10.1016/j.jalgebra.2021.11.026. MR4355908

[38] Yanming Wang and Suling Pang, *Solubility of finite groups admitting a fixed-point-free operator group*, Publ. Math. Debrecen **61** (2002), no. 3-4, 429–437, DOI 10.5486/pmd.2002.2633. MR1943704

Department of Mathematical and Statistical Sciences, Marquette University, Milwaukee, Wisconsin

Email address: andrew.schopieray@marquette.edu

Contemporary Mathematics
Volume **813**, 2025
https://doi.org/10.1090/conm/813/16288

$(3+1)D$ topological orders with only a $\mathbb{Z}_2$-charged particle

Theo Johnson-Freyd

ABSTRACT. There is exactly one bosonic (3+1)-dimensional topological order whose only nontrivial particle is an emergent boson: pure $\mathbb{Z}_2$ gauge theory. There are exactly two (3+1)-dimensional topological orders whose only nontrivial particle is an emergent fermion: pure "spin-$\mathbb{Z}_2$" gauge theory, in which the dynamical field is a spin structure; and an anomalous version thereof. I give three proofs of this classification, varying from hands-on to abstract. Along the way, I provide a detailed study of the braided fusion 2-category $\mathcal{Z}_{(1)}(\Sigma\mathbf{SVec})$ of string and particle operators in pure spin-$\mathbb{Z}_2$ gauge theory.

1. Introduction

1.1. Statement of the main result. A *(3+1)D topological order*, if it has a single local vacuum, is described algebraically by a nondegenerate braided fusion 2-category $\mathcal{B}$ [**KW14, JF22**]. (Topological orders with multiple local vacua are usually rejected by condensed matter theorists because they are unstable: a small deformation to the Hamiltonian will condense the system to one of the local vacua.) Physically, the objects of $\mathcal{B}$ are the (1+1)D (quasi)string excitations in the topological order. Write $\mathbb{1} \in \mathcal{B}$ for the invisible identity string, and $\Omega\mathcal{B} = \mathrm{End}_{\mathcal{B}}(\mathbb{1})$ for the fusion 1-category of its endomorphisms. The objects of $\Omega\mathcal{B}$ are the (0+1)D (quasi)particle excitations in the topological order. It is a nontrivial fact [**JF22**, Theorem 4] that a topological order (with a single local vacuum) is determined just by its strings and particles: a priori there could have been nontrivial information encoded in the 3-category of (2+1)D membrane excitations, but in fact all the membranes arise in a systematic way as networks of strings.

The goal of this paper is to classify (3+1)D topological orders with only one simple particle "e" other than the invisible identity particle 1. Since $\Omega\mathcal{B}$ is automatically symmetric monoidal, e necessarily enjoys a $\mathbb{Z}_2$ fusion rule $e \otimes e \cong 1$: indeed, $\Omega\mathcal{B}$ is isomorphic either to $\mathbf{Rep}(\mathbb{Z}_2)$ or $\mathbf{SVec}$. The particle is called "e" in order to think of it as "electrically charged," but for the gauge group $\mathbb{Z}_2$ rather than the U(1) of Maxwell theory: the world line of an e-particle is a Wilson line for the $\mathbb{Z}_2$ gauge group. The self-braiding $e \otimes e \to e \otimes e$ is $+1$ (times the identity

2020 *Mathematics Subject Classification.* Primary 18M20, 18N10, 81V74, 81T45, 55S10, 55T10.

Research at the Perimeter Institute is supported by the Government of Canada through Industry Canada and by the Province of Ontario through the Ministry of Economic Development and Innovation. The Perimeter Institute is in the Haldimand Tract, land promised to the Six Nations. Dalhousie University is in Mi'kma'ki, the ancestral and unceded territory of the Mi'kmaq.

on $e \otimes e \cong 1$) in $\mathbf{Rep}(\mathbb{Z}_2)$, and -1 in $\mathbf{SVec}$: in other words, e is an *emergent boson* if $\Omega\mathcal{B} \cong \mathbf{Rep}(\mathbb{Z}_2)$ and an *emergent fermion* if $\Omega\mathcal{B} \cong \mathbf{SVec}$. I will prove:

MAIN THEOREM. *There is a unique (up to non-unique isomorphism) (3+1)D topological order with a single nontrivial particle which is an emergent boson (i.e. $\Omega\mathcal{B} \cong \mathbf{Vec}[\mathbb{Z}_2]$). There are exactly two topological orders with a single nontrivial particle which is an emergent fermion ($\Omega\mathcal{B} \cong \mathbf{SVec}$).*

To give names for ease of future reference, I will call these three topological orders "$\mathcal{R}$," "$\mathcal{S}$," and "$\mathcal{T}$." I will give detailed descriptions of all three 2-categories in Section 2. Briefly, the three topological orders are:

- $\mathcal{R}$, the unique topological order in which e is a boson, is plain $\mathbb{Z}_2$ gauge theory. Another name for it is the *3+1D Toric Code*, whose lattice realization was introduced in [**HZW05**] based on the 2+1D version from [**Kit97**], although the continuum gauge-theory model was studied as early as [**Weg71**]; see [**AdR20**] for a survey. A direct analysis of the braided fusion 2-category $\mathcal{R}$ is given in [**KTZ20**]. Algebraically, $\mathcal{R}$ can be described as a Drinfel'd centre in various ways:
$$\mathcal{R} \cong \mathcal{Z}_{(1)}(\Sigma\mathbf{Rep}(\mathbb{Z}_2)) \cong \mathcal{Z}_{(1)}((\Sigma\mathbf{Vec})[\mathbb{Z}_2]).$$
- $\mathcal{S}$ deserves the name "spin-$\mathbb{Z}_2$ gauge theory": instead of a dynamical $\mathbb{Z}_2$ gauge field, its dynamical field is a spin structure η, i.e. a $\mathbb{Z}_2$-valued 1-form solving $d\eta = w_2$, where w_2 is some fixed cocycle representative of the second Stiefel–Whitney class (for instance given by a lattice realization of spacetime as in [**GK16**]). As with plain gauge theory, the action functional is trivial. Algebraically, $\mathcal{S} \cong \mathcal{Z}_{(1)}(\Sigma\mathbf{SVec})$.
- $\mathcal{T}$ is an anomalous version of spin-$\mathbb{Z}_2$ gauge theory studied in [**Tho15**, Section 5] and in [**KLW$^+$20a**, Section III.J.4], and further analyzed in unpublished work of Hao Zheng. It can be constructed in the continuum by positing that its only dynamical field is a spin structure η, but its action functional is $(-1)^{\int \eta w_3} \equiv (-1)^{\int w_2 \mathrm{Sq}^1 \eta}$ (where "$\equiv$" means modulo boundary terms). The non-closedness of η means that this action has a gravitational anomaly equal to $(-1)^{w_2 w_3}$. Algebraically, $\mathcal{S}$ and $\mathcal{T}$ are very similar, with identical fusion rules and almost identical braidings. Indeed, the only difference is that in $\mathcal{S}$, the $\mathbb{Z}_2$ 1-form symmetry generated by the magnetic string is nonanomalous, whereas in $\mathcal{T}$ this symmetry has a nontrivial 't Hooft anomaly.

In the above bullet points and later, I have adopted the following notation, explained more fully in §1.2–1.3. If $\mathcal{A}$ is a fusion n-category, then $\mathcal{Z}_{(1)}(\mathcal{A})$ denotes its *Drinfel'd centre*: roughly speaking, it consists of elements $X \in \mathcal{A}$ together with an isomorphism between left and right multiplication by X. If $\mathcal{B}$ is a braided fusion n-category, then $\mathcal{Z}_{(2)}(\mathcal{B})$ denotes its *Müger centre*: roughly speaking, it consists of elements $X \in \mathcal{B}$ together with an isomorphism between braiding over and under by X. The symbol Σ denotes a *suspension* operation on monoidal Karoubi-complete higher categories developed in [**GJF19a**, **JF22**]: it takes (braided) fusion n-categories to (fusion) $(n+1)$-categories, and is essentially the same as what is called "**Mod**" in [**DN20**].

REMARK 1.1.1. The first sentence of the Main Theorem follows immediately from the classification of 3+1D topological orders without emergent fermions from

[**LKW18**] together with the group cohomology calculation

$$\mathrm{H}^4_{\mathrm{gp}}(\mathbb{Z}_2;\mathbb{C}^\times) = \mathrm{H}^4(\mathbb{Z}_2[1];\mathbb{C}^\times) = 0.$$

(Here and throughout, given an abelian group A, I will write $A[n]$ for the space more typically called $K(A,n)$ or $\mathrm{B}^n A$.) It is worth emphasizing that the arguments in [**LKW18**] and in its sequel [**LW19**] were given before a complete mathematical definition of "fusion 2-category" was first introduced in [**DR18**]. (Indeed, the preprints of [**LKW18, LW19**] appeared on the arXiv in 2017 and early 2018, respectively, whereas [**DR18**] appeared in very late 2018.) Nevertheless, modulo a few small holes filled in [**JF22**], the argument in [**LKW18**] is completely correct, and I will summarize it in Remark 3.0.1.

The sequel [**LW19**], however, contains a small but important mathematical error. Namely, at one point in their argument, the authors of [**LW19**] assume without proof that there is only one (3+1)D topological order $\mathcal{B}$ with $\Omega\mathcal{B} \cong \mathbf{SVec}$; in fact, as stated in the Main Theorem, there are two. The error is quite understandable: as explained in §2.2, the two topological orders with $\mathcal{S}$ and $\mathcal{T}$ are extremely similar, and one must look very closely in order to tell them apart. Furthermore, [**LW19**] never claims itself to be mathematically rigorous, but rather to present a physically reasonable argument. Indeed, [**LW19**] is an attempt to classify "nonanomalous" 3+1D topological orders, without selecting a mathematically precise definition of the term "nonanomalous," and not an attempt to classify all nondegenerate braided fusion 2-categories. Although [**LW19**] did overlook the 2-category $\mathcal{T}$, the Main Theorem provides post hoc support for their classification because, as indicated above, $\mathcal{T}$ arises physically from an anomalous 3+1D system, and probably cannot be realized as a nonanomalous system (compare Remark 1.1.2).

REMARK 1.1.2. Recall that a braided fusion 1-category $\mathcal{C}$ is *slightly degenerate* if its Müger centre $\mathcal{Z}_{(2)}(\mathcal{C})$ is equivalent to $\mathbf{SVec}$. It was a longstanding conjecture, now a theorem of [**JFR23**] (released after the original circulation of this paper), that every slightly degenerate braided fusion 1-category admits a *minimal nondegenerate extension*: an index-2 inclusion into a nondegenerate braided fusion 1-category [**Müg03**]. According to an observation of Dmitri Nikshych reproduced in [**JF22**, Remark V.2], this statement is equivalent to the conjecture that all bosonic 3+1D topological orders are Morita trivial.

Indeed, an even sharper statement (Remark 2.3.9) is available, and serves as one of the motivations of [**DN20**]; I will sketch a proof in §2.3. If $\mathcal{C}$ is slightly degenerate, then $\mathcal{Z}_{(1)}(\Sigma\mathcal{C})$ will be a 3+1D topological order with line operators $\mathbf{SVec}$. Choices of minimal nondegenerate extension of $\mathcal{C}$, if they exist, are in bijection with choices of isomorphism $\mathcal{Z}_{(1)}(\Sigma\mathcal{C}) \cong \mathcal{S}$.

Thus, good control over the topological orders $\mathcal{B}$ with $\Omega\mathcal{B} \cong \mathbf{SVec}$ is quite valuable. In particular, the proof in [**JFR23**] relies on the detailed descriptions from this paper.

For these reasons, the focus of this paper will be on the second sentence of the Main Theorem. I will in fact give three proofs of the second sentence; these proofs occupy Section 3. The three proofs vary from hands-on to abstract, and illustrate different features of the classification problem. Before giving the proofs, Section 2 describes in detail the three braided fusion 2-categories $\mathcal{R}$, $\mathcal{S}$, and $\mathcal{T}$. And before that, I will complete this introductory section with some background on the main definitions.

1.2. Fusion 2-categories. Fusion 2-categories, first axiomatized in [**DR18**], are a special type of monoidal 2-category. Good references on monoidal and braided monoidal 2-categories include [**SP09**, Appendix C] and [**DN20**, §2.2]. All of my "n-categories" are "weak n-categories": composition of ($<n$)-morphisms is associative only up to higher coherence, but n-morphisms form sets and so have strict composition. When $n = 2$, "2-category" is equivalent to "bicategory." I will write "isomorphic" for what many people might term "equivalent": there is no stricter meaningful notion of isomorphism in higher categories.

Fusion higher categories were first axiomatized in [**JF22**], and a very nice treatment is available in [**KZ20**, Section 3]. Much of [**JF22**] was inspired by the earlier informal treatment of fusion n-categories in [**KW14, KWZ15, KWZ17**].

The definition of "fusion 2-category" from [**DR18**] parallels the case of fusion 1-categories:

DEFINITION 1.2.1. A *fusion* 2*-category* $\mathcal{A}$ is a monoidal 2-category such that:

- The underlying 2-category of $\mathcal{A}$ is $\mathbb{C}$-linear finite semisimple.
- The unit object $\mathbb{1} \in \mathcal{A}$ is simple.
- All objects have monoidal duals.

Definition 1.2.1 requires some unpacking, especially concerning the notion of "$\mathbb{C}$-linear finite semisimple 2-category." An n-category $\mathcal{A}$ is $\mathbb{C}$*-linear* if all sets of n-morphisms are $\mathbb{C}$-vector spaces, and all composition maps are multilinear. When $n = 2$, this means that for any objects $X, Y \in \mathcal{A}$, $\hom_{\mathcal{A}}(X, Y)$ should be a $\mathbb{C}$-linear 1-category. A $\mathbb{C}$-linear 2-category $\mathcal{A}$ is *locally finite semisimple* if $\hom_{\mathcal{A}}(X, Y)$ is always a finite semisimple $\mathbb{C}$-linear category: it should be additively and Karoubi complete and completely reducible, with finitely many simple objects.

It is easy to define direct sums, and hence additive-completeness, for n-categories. Let us say that an object is *indecomposable* if it does not decompose as a direct sum. [**DR18**] also defines a notion of "simple" object in a 2-category, which, as in the 1-categorical case, is in general stronger but equivalent to indecomposability if the ambient 2-category is semisimple. As such, I will use the terms "indecomposable" and "simple" interchangeably.

Higher categorical Karoubi completion was developed for 2-categories in [**DR18**] and for general n in [**GJF19a**]. I will not review the details but I will mention a notation: I will write "$X \twoheadrightarrow Y$" when there is a *condensation* of X onto Y; condensation is the n-categorical version of split surjection.

DEFINITION 1.2.2. A 2-category $\mathcal{A}$ is *finite semisimple* over $\mathbb{C}$ if it is locally finite semisimple, all 1-morphisms have adjoints, it is additive and Karoubi complete, and it has only finitely many isomorphism classes of indecomposable objects.

The following fact is used implicitly in [**DR18**], and proved in n-categorical generality in [**JF22**, Lemma IV.1] and in [**KZ20**, Proposition 3.11]:

LEMMA 1.2.3. *Let $\mathcal{A}$ be a finite semisimple* 2*-category over $\mathbb{C}$, and $X, Y \in \mathcal{A}$ a pair of indecomposable objects. Then any nonzero* 1*-morphism $X \to Y$ extends to a condensation $X \twoheadrightarrow Y$.* □

The composition of condensations $X \twoheadrightarrow Y \twoheadrightarrow Z$ is a condensation $X \twoheadrightarrow Z$, and condensations are nonzero. This implies the *Categorical Schur's Lemma* of [**DR18**], which says that "related by a nonzero morphism" is an equivalence relation on the set of indecomposable objects in $\mathcal{A}$. Following [**DR18**, Definition 1.2.22], I will say that two indecomposable objects related by a nonzero morphism are in the same *component*. More generally:

DEFINITION 1.2.4. The *component* of an indecomposable object $X \in \mathcal{A}$ is the full subcategory of $\mathcal{A}$ on the objects Y, not necessarily indecomposable, that admit a condensation $X \twoheadrightarrow Y$.

For example, assuming that $\mathcal{A}$ is fusion so that the unit object $\mathbb{1} \in \mathcal{A}$ is indecomposable, it makes sense to talk about its *identity component*. The components of $\mathcal{A}$ are precisely the indecomposable direct summands of $\mathcal{A}$ itself [**DR18**, Remark 1.2.23]. The set of components is called $\pi_0\mathcal{A}$. I will say that a fusion 2-category $\mathcal{A}$ is *connected* if $\pi_0\mathcal{A} = \{\mathbb{1}\}$, i.e. if the identity component is all of $\mathcal{A}$.

REMARK 1.2.5. Definition 1.2.2 is best only over an algebraically closed field of characteristic 0. If the characteristic is not zero, subtleties can arise from inseparable fusion 1-categories; these do not exist in characteristic zero by [**DSPS20**, Theorem 13]. If the field is not algebraically closed, then the Karoubi-completion requirement will typically be incompatible with the requirement that there be only finitely many isomorphism classes of simple objects, because Karoubi-completion produces simple objects for each Morita equivalence class of simple algebra. Rather, the correct finiteness condition for "finite semisimple 2-category" is to ask that the set $\pi_0\mathcal{A}$ of components, and not the set of equivalence classes of indecomposables, should be finite, as in [**Déc21**].

A typical example of a finite semisimple 2-category is the category of finite semisimple modules for a fusion 1-category $\mathcal{F}$. I will call this 2-category $\Sigma\mathcal{F}$. This follows the notation of [**GJF19a**]. Specifically, suppose $\mathcal{C}$ is an additive and Karoubi complete $\mathbb{C}$-linear n-category equipped with a choice of object $\mathbb{1} \in \mathcal{C}$; then $\Omega\mathcal{C} := \mathrm{End}_{\mathcal{C}}(\mathbb{1})$ is a monoidal additive and Karoubi complete $\mathbb{C}$-linear $(n{-}1)$-category. The adjoint to Ω is Σ, which takes in a monoidal $(n{-}1)$-category, deloops it to a one-object n-category, and Karoubi completes. The unit $\mathrm{id} \to \Omega\Sigma$ of the adjunction $\Sigma \dashv \Omega$, when evaluated on a monoidal Karoubi-complete $(n{-}1)$-category $\mathcal{A}$, is an equivalence $\mathcal{A} \xrightarrow{\sim} \Omega\Sigma\mathcal{A}$. The counit $\Sigma\Omega \to \mathrm{id}$, when evaluated on a pointed n-category $\mathcal{C}$, is fully faithful, and its essential image is the full subcategory of $\mathcal{C}$ on those objects $Y \in \mathcal{C}$ which receive a condensation $\mathbb{1} \twoheadrightarrow Y$. From this it follows that Σ and Ω restrict to an equivalence:

$$\{\text{connected semisimple } n\text{-categories with indecomposable unit}\}$$
$$\simeq \{\text{fusion } (n{-}1)\text{-categories}\}.$$

The Ω and Σ constructions furthermore have good monoidality relations. For instance, if $\mathcal{A}$ is braided monoidal, then $\Sigma\mathcal{A}$ is monoidal; if $\mathcal{A}$ is symmetric monoidal, then so is $\Sigma\mathcal{A}$. Of particular interest is the $(n{+}1)$-category $\Sigma^n\mathbf{Vec}$, and its subgroupoid of invertible objects (and invertible morphisms) $(\Sigma^n\mathbf{Vec})^\times$. The fusion n-category $\Sigma^{n-1}\mathbf{Vec}$ is called simply "$n\mathbf{Vec}$" in [**KZ20**], and in §3.3 I will write $\underline{\mathcal{W}}^n$ for its invertible subgroupoid $\underline{\mathcal{W}}^n := (n\mathbf{Vec})^\times = (\Sigma^{n-1}\mathbf{Vec})^\times$.

1.3. (3+1)D topological orders. The following definition, mentioned in the introduction, is formalized in [**JF22**], based on ideas from [**KW14**]:

DEFINITION 1.3.1. A *(3+1)D topological order with a single local vacuum* is a braided fusion 2-category $\mathcal{B}$ which is *nondegenerate* in the sense of having trivial Müger centre.

Physically, the objects of $\mathcal{B}$ are the stringlike excitations in the topological order. The worldsheet of a string is a 2-dimensional surface, and so strings are also called *surface operators*; the 2-dimensionality is why they are objects of a 2-category. Strings can fuse, and the fusion is braided because in (3+1)D there are two transversal dimensions in which strings can move around. This is why the strings in any (3+1)D topological order form a braided multifusion 2-category $\mathcal{B}$.

The "fusion" condition, demanding that the unit object $\mathbb{1} \in \mathcal{B}$ be indecomposable, is equivalent to the requirement that the local vacuum be unique. In general, if there were multiple local vacua, then there would be vertex (i.e. 0-dimensional) operators that project a state onto a single vacuum; conversely, the set of local vacua is the spectrum of the commutative algebra $\Omega^2\mathcal{B}$ of vertex operators. This algebra being $\mathbb{C}$ is equivalent to the unit object being indecomposable.

The final condition in Definition 1.3.1 to explain is the nondegeneracy of the braiding. As in the case of fusion 1-categories, any fusion 2-category $\mathcal{A}$ has a *Drinfel'd centre* $\mathcal{Z}_{(1)}(\mathcal{A})$, which measures the failure of objects in $\mathcal{A}$ to be detectable by commuting them with other objects. Its objects are objects $X \in \mathcal{A}$ equipped with *centrality data*: a natural-in-Y isomorphism $X \otimes Y \cong Y \otimes X$ plus further data imposing that this isomorphism is multiplicative in Y. The shortest definition is to define $\mathcal{Z}_{(1)}(\mathcal{A})$ as the 2-category of endomorphisms of $\mathcal{A}$-as-an-$\mathcal{A}$-bimodule. By the same token, any braided fusion 2-category $\mathcal{B}$ has a *Müger*, aka *sylleptic*, *centre* $\mathcal{Z}_{(2)}(\mathcal{B})$. It measures the failure of objects in $\mathcal{B}$ to be detectable by braiding with other objects. Explicitly, an object of $\mathcal{Z}_{(2)}(\mathcal{B})$ is an object $X \in \mathcal{B}$ and a natural-in-Y trivialization of the full braiding $\beta_{Y,X} \circ \beta_{X,Y} : X \otimes Y \to Y \otimes X \to X \otimes Y$; this trivialization must furthermore satisfy some braided monoidality requirements. The cleanest definition is to define $\mathcal{Z}_{(2)}(\mathcal{B})$ as the 2-category of endomorphisms of $\mathcal{B}$ as a "braided module" over itself [**DN20**].

The nondegeneracy requirement in Definition 1.3.1 is that $\mathcal{Z}_{(2)}(\mathcal{B})$ should be the trivial: simply the fusion 2-category $2\mathbf{Vec}_{\mathbb{C}} = \Sigma\mathbf{Vec}_{\mathbb{C}}$. This axiom was first stated in [**KW14**], based on ideas in [**Lev13**], and was first given a complete mathematical formulation in [**JF22**]. Physically, the axiom encodes the principle of *remote detectability*: it says that a nontrivial excitation should be detectable by braiding with other excitations.

REMARK 1.3.2. A priori, to specify a (3+1)D topological order one would also need to specify the 3-category of membrane excitations. That this data is not required is a nontrivial theorem about fusion n-categories with trivial Drinfel'd centre: they are always connected [**JF22**, Theorem 4]. To see that it is nontrivial, note that it fails if the "fusion" condition is dropped: a (3+1)D topological order with multiple local vacua is not determined by its braided multifusion 2-category of strings; one needs to include information about the membranes, aka *domain walls*, that separate the different vacua, and the excitations supported by those membranes (see [**DG20**] for a physics discussion).

How might one detect a particle excitation, i.e. an object of the symmetric fusion 1-category $\Omega\mathcal{B}$? The easiest way would be to braid it around a string $X \in \mathcal{B}$. Indeed, $\Omega\mathcal{B}$ is symmetric monoidal, and so particles certainly cannot detect each other, but the movie of a particle moving around a string would evaluate to a number (or matrix), and "detection" occurs when this number is not 1. Conversely, the easiest way to detect a string is by braiding it against particles.

In the special case when the simple particles in $\Omega\mathcal{B}$ form an abelian group A (which occurs when the fusion of any two simple particles is again simple), this leads to the expectation that the indecomposable strings should form the Pontryagin dual group $\hat{A} = \hom(A, \mathrm{U}(1))$. A more precise expectation recognizes that the identity component $\Sigma\Omega\mathcal{B} \subset \mathcal{B}$ is symmetric monoidal (since $\Omega\mathcal{B}$ is) and so should be quotiented out, giving the expectation:

$$\pi_0\mathcal{B} \cong \hat{A}. \tag{1}$$

In the general case with nonabelian particles, Tannakian duality extracts from the symmetric fusion category $\Omega\mathcal{B}$ a finite group G (isomorphic to $\hat{A}$ in the abelian case), and the reader might expect:

$$\pi_0\mathcal{B} \cong G/G := \{\text{conjugacy classes in } G\}. \tag{2}$$

Expectations (1)–(2) are stated fairly explicitly in [**LKW18**, **LW19**], and form a main ingredient in their arguments. There are, however, reasons to hesitate. Particles in a (3+1)D topological order cannot detect each other, but a particle could be detected by a membrane operator rather than a string. Strings can moreover detect each other: indeed, there are typically nontrivial strings in the identity component of a (3+1)D topological order, and those strings are *not* detected by particles but only by other strings. Thus the pairing between the set of strings and the set of particles is not *a priori* perfect.

In spite of these reasons to hesitate, it turns out that expectations (1)–(2) do in fact hold. I do not know a proof of this in general that does not require the full classification [**LKW18**], proved in [**JF22**, Theorem 8 and Corollary V.5] without assuming (1)–(2).

When $\Omega\mathcal{B} = \mathbf{Rep}(\mathbb{Z}_2)$ or $\mathbf{SVec}$, as is of interest in this paper, $G = \hat{A} = \mathbb{Z}_2$. The strings in the non-identity component are called *magnetically charged*. It turns out that in all three cases $\mathcal{R}, \mathcal{S}, \mathcal{T}$, there exists a magnetic string m satisfying the fusion rules $m^2 \cong \mathbb{1}$. This is not automatic: for instance, as explained in §2.1, in $\mathcal{R}$, there is a noninvertible simple magnetic string m', and so it is certainly not true that a random choice of simple magnetic string will solve such a simple fusion rule as $m^2 \cong \mathbb{1}$. For comparison, as explained in §2.2, in $\mathcal{S}$ and $\mathcal{T}$ there are two non-isomorphic indecomposable magnetic strings, each squaring to the vacuum, and no canonical way to pick one. Given a fusion 2-category $\mathcal{C}$, it is an interesting question to ask whether the map $\{\text{indecomposable objects of } \mathcal{C}\} \to \pi_0\mathcal{C}$ splits in any nice way. The explicit calculations of show that for $\mathcal{C} = \mathcal{R}, \mathcal{S}, \mathcal{T}$, this map does split, but not canonically.

2. The three examples

This section describes in detail the three topological orders $\mathcal{R}$, $\mathcal{S}$, and $\mathcal{T}$. Conditional to the expectations that $\pi_0 = \mathbb{Z}_2$ in all cases and that in all cases there exists an invertible magnetic string, the calculations in this section constitute a proof of the Main Theorem. Indeed, the first proof in Section 3 will complete the argument by showing directly that if $\mathcal{B}$ is a 3+1D topological order with $\Omega\mathcal{B} \cong \mathbf{SVec}$, then $\pi_0\mathcal{B} = \mathbb{Z}_2$ and there is an invertible magnetic string. (I was unable to find a proof of the corresponding statement in the "emergent boson" case that did not end up using the full proof of the classification from [**LKW18**].)

2.1. The unique topological order in which e is a boson. I will now describe in some detail the unique 3+1D topological order $\mathcal{R}$ with line operators $\Omega\mathcal{R} \cong \mathbf{Vec}[\mathbb{Z}_2] \cong \mathbf{Rep}(\mathbb{Z}_2)$. As mentioned already in Remark 1.1.1, the main result of [**LKW18**] implies that $\mathcal{R}$ is necessarily the 2-category of operators in plain $\mathbb{Z}_2$ Dijkgraaf–Witten theory, also called the 3+1D Toric Code. The braided fusion categories for general 3+1D Dijkgraaf–Witten theories are studied in [**KTZ19**], and the special case of the 3+1D Toric Code is further studied in [**KTZ20**]; my description is consistent with theirs.

The identity component of $\mathcal{R}$ is $\Sigma\Omega\mathcal{R} = \Sigma\mathbf{Rep}(\mathbb{Z}_2)$. This 2-category is described for example in [**DR18**, Example 1.4.14]. It contains two indecomposable objects, corresponding to the two Morita equivalence classes of simple algebra objects in $\mathbf{Rep}(\mathbb{Z}_2)$. One of these objects is the vacuum, corresponding to the trivial algebra $\mathbb{C} \in \mathbf{Rep}(\mathbb{Z}_2)$. I will call it "$\mathbb{1}$" to distinguish it from the vacuum line "1".

The other object doesn't have a standard name, but corresponds to the algebra object $\mathcal{O}(\mathbb{Z}_2) = \mathbb{C}[x]/(x^2 = 1) \in \mathbf{Rep}(\mathbb{Z}_2)$, where the generator x is nontrivially charged under the $\mathbb{Z}_2$-action. Note that $\mathcal{O}(\mathbb{Z}_2)$ is isomorphic to $\mathbb{C} \oplus \mathbb{C}$ if the $\mathbb{Z}_2$-symmetry is ignored, but it is simple as an algebra object in $\mathbf{Rep}(\mathbb{Z}_2)$: it does not contain any proper $\mathbb{Z}_2$-equivariant ideals. The corresponding quasistring was first described in [**EN17**] under the name "Cheshire string"; for that reason, I will call it plain "c." I will write $e \in \Omega\mathcal{B} = \mathbf{Rep}(\mathbb{Z}_2)$ for the sign representation. The fusion rules are, of course, $e^2 \cong 1$ and

$$c^2 \cong c \oplus c. \tag{3}$$

Indeed, equation (3) corresponds to an isomorphism of $\mathbb{Z}_2$-equivariant algebras

$$\mathcal{O}(\mathbb{Z}_2) \otimes \mathcal{O}(\mathbb{Z}_2) = \mathbb{C}[x, y]/(x^2 = y^2 = 1) \cong \mathcal{O}(\mathbb{Z}_2) \oplus \mathcal{O}(\mathbb{Z}_2)$$

where the two projections correspond to setting $x = y$ or $x = -y$.

REMARK 2.1.1. The fact that c and $\mathbb{1}$ are in the same component means that there is a non-invertible 1-morphism between them. I say "between them" because each 1-morphism has an adjoint, and in a fusion 2-category left- and right-adjoints for 1-morphisms are always isomorphic. In fact, there is a "unique" 1-morphism between $\mathbb{1}$ and c in the sense that the category $\hom_{\Sigma\mathbf{Rep}(\mathbb{Z}_2)}(\mathbb{1}, c)$ is equivalent to $\mathbf{Vec}$. Note that $\mathrm{End}_{\Sigma\mathbf{Rep}(\mathbb{Z}_2)}(\mathbb{1}) = \Omega\Sigma\mathbf{Rep}(\mathbb{Z}_2) = \mathbf{Rep}(\mathbb{Z}_2)$ by construction, and one can calculate that $\mathrm{End}_{\Sigma\mathbf{Rep}(\mathbb{Z}_2)}(c)$ is also equivalent to $\mathbf{Rep}(\mathbb{Z}_2)$.

REMARK 2.1.2. Any choice of gapped nD boundary condition for any n+1D topological order $\mathcal{B}$ realizes $\mathcal{B}$ as the Drinfel'd centre of the fusion $(n-1)$-category of operators on the boundary (provided both the bulk and the boundary have unique local vacua). Every Dijkgraaf–Witten theory has a canonical *Dirichlet* boundary condition. In 3+1D, the Dirichlet boundary operators form the fusion 2-category $(\Sigma\mathbf{Vec})^\alpha[G]$, where $\alpha \in \mathrm{H}^4(G[1]; \mathbb{C}^\times)$ is physically the Dijkgraaf–Witten action, and mathematically the associator data for $(\Sigma\mathbf{Vec})^\alpha[G]$, which is the 2-category with G many components, each of which is a copy of $\Sigma\mathbf{Vec}$ (hence the parenthesis to distinguish it from $\Sigma(\mathbf{Vec}[G])$). This is why [**KTZ19**] calculates the Drinfeld centre $\mathcal{Z}((\Sigma\mathbf{Vec})^\alpha[G])$. (My "$\Sigma\mathbf{Vec}$" is their "2**Vec**.") In particular, the 2-category under consideration arises as

$$\mathcal{R} \cong \mathcal{Z}_{(1)}((\Sigma\mathbf{Vec})[\mathbb{Z}_2]).$$

When the Dijkgraaf–Witten action is trivialized, there is a second canonical boundary condition, namely the *Neumann* boundary. It is characterized by the property that the bulk 't Hooft operators condense on the boundary, and the Wilson lines do not. Said mathematically, the Neumann boundary operators form connected the fusion 2-category $\Sigma\mathbf{Rep}(G)$. In particular, the 2-category under consideration arises as

$$\mathcal{R} \cong \mathcal{Z}_{(1)}(\Sigma\mathbf{Rep}(\mathbb{Z}_2)).$$

This description has an analog for the 3+1D topological order $\mathcal{S}$; see Remark 2.2.2.

From either description in Remark 2.1.2, it is easy to calculate that $\mathcal{R}$ has a single non-identity component: in the notation of [**DR18**, Remark 1.2.23], $\pi_0\mathcal{R} = \mathbb{Z}_2$. The objects in the nontrivial component are the *magnetically charged* strings. The characterizing feature of magnetic charge is that a magnetic string links nontrivially with the electron e: when an e particle is moved all the way around a magnetic string, the state picks up a phase factor of -1.

It turns out that the magnetic component has the same structure as the identity component. There is an indecomposable object $m \in \mathcal{B}$ which is invertible: in fact, $m^2 \cong \mathbb{1}$. This implies already that $\mathrm{End}_{\mathcal{B}}(m) \cong \mathbf{Rep}(\mathbb{Z}_2)$. The other indecomposable object is $m' := m \otimes c$. It is not invertible, but rather satisfies the fusion rules $(m')^2 \cong c \oplus c$. As with the identity component, there is a nonzero, but non-invertible, 1-morphism connecting m and m': $\hom_{\mathcal{C}}(m, m') \cong \mathbf{Vec}$. This is enough to determine $\mathcal{R}$ as a linear 2-category: it has two components, each equivalent to $\Sigma\mathbf{Rep}(\mathbb{Z}_2)$.

To really understand a braided fusion 2-category requires more than the fusion rules: the braided monoidal structure is data. In general, braided monoidality data can be quite complicated. This is familiar from the fusion 1-category, where the associator data, for example, consists of matrices solving some overdetermined inhomogeneous polynomial equations, modulo certain changes of basis. Also familiar from the 1-category case is the fact that these data simplify dramatically for "grouplike" categories in which all the simple objects are invertible. Indeed, fusion categories $\mathcal{C}$ in which the simple objects have fusion rules following a finite group G are classified by ordinary group cohomology $\mathrm{H}^3_{\mathrm{gp}}(G; \mathbb{C}^\times) = \mathrm{H}^3(G[1]; \mathbb{C}^\times)$; if G is abelian, then the braided fusion categories $\mathcal{C}$ with fusion rules G are classified by $\mathrm{H}^4(G[2]; \mathbb{C}^\times)$. Specifically, these classes are the Postnikov invariants for extensions $\mathbb{C}^\times[2].G[1]$ and $\mathbb{C}^\times[3].G[2]$, respectively. These extensions arise as (the classifying spaces of) the invertible subcategories of $\mathcal{C}$, i.e. the subcategories of invertible objects and invertible 1-morphisms. The full category can be reconstructed from its invertible subcategory by promoting the group $\mathbb{C}^\times$ of morphisms back to the $\mathbb{C}$-algebra $\mathbb{C}$, and then additively and Karoubi-completing. I will refer to this process of promoting $\mathbb{C}^\times \rightsquigarrow \mathbb{C}$ and then additively and Karoubi-completing just as *linearization*.

Exactly the same thing happens in fusion 2-categories. In general braided monoidality data is complicated, but it is merely cohomological when the 2-category arises as a linearization of a homotopy type with $\mathbb{C}^\times$ in top degree. This is the situation for $\mathcal{R}$. Indeed, let $\mathcal{R}^\times$ denote its invertible braided fusion sub-2-category. The objects of $\mathcal{R}^\times$ are just $\mathbb{1}$ and m, and the 1-morphisms are just $\mathrm{Aut}(\mathbb{1}) \cong \mathrm{Aut}(m) \cong \{1, e\}$. The 2-morphisms are just $\mathrm{Aut}(1) \cong \mathrm{Aut}(e) \cong \mathbb{C}^\times$. Said another way, we have:

$$\mathcal{R}^\times = \mathbb{C}^\times[2].\mathbb{Z}_2[1].\mathbb{Z}_2[0]. \tag{4}$$

In fact, ignoring the monoidal structure and treating $\mathcal{B}^\times$ just as a 2-category, these extensions all split. Indeed, the only possible Postnikov extension datum is the connecting map for $\mathbb{C}^\times[2].\mathbb{Z}_2[1]$, which lives in $\mathrm{H}^3(\mathbb{Z}_2[1];\mathbb{C}^\times)$ and encodes the associator (but not braiding) on $\Omega\mathcal{R} = \mathbf{Rep}(\mathbb{Z}_2)$. This associator is trivial. Furthermore:

LEMMA 2.1.3. *$\mathcal{R}$ is the linearization of $\mathcal{R}^\times$.*

PROOF. The Lemma essentially follows from the earlier discussion about $\Sigma\mathbf{Rep}(\mathbb{Z}_2)$. To understand the details, it helps to work in layers. The top two layers of $\mathcal{R}^\times$ are $\mathbb{C}^\times[2].\mathbb{Z}_2[1]$, which linearize to the 1-category $\mathbf{Rep}(\mathbb{Z}_2)$. To produce a 2-category requires further Karoubi completion, and this is how the Cheshire string c is created. The bottom layer $\mathbb{Z}_2[0]$ means that there is a second equivalent component. □

It follows that to describe the braided monoidal structure on $\mathcal{B}$, it suffices to describe the braided monoidal structure on $\mathcal{R}^\times$. This is in turn described by its double classifying space $\mathrm{B}^2\mathcal{R}^\times$, which has the same homotopy groups shifted by 2:

$$\mathrm{B}^2\mathcal{R}^\times = \mathbb{C}^\times[4].\mathbb{Z}_2[3].\mathbb{Z}_2[2]. \tag{5}$$

There are many spaces with these homotopy groups. To find the right one, it helps to first study the quotient extension $\mathbb{Z}_2[3].\mathbb{Z}_2[2] = \{1,e\}[3].\{\mathbb{1},m\}[2]$, which is classified by a class $\beta \in \mathrm{H}^4(\mathbb{Z}_2[2];\mathbb{Z}_2) \cong \mathbb{Z}_2$. In terms of braided category theory, this class records the self-braiding

$$\beta_{m,m} : m \otimes m \to m \otimes m.$$

Specifically, β is trivial or not depending on whether this self-braiding is 1 or e. Using either description from Remark 2.1.2, it is not hard to show that $\beta_{m,m}$ is trivial, and so the sub-extension $\mathbb{Z}_2[3].\mathbb{Z}_2[2]$ is isomorphic to a product $\mathbb{Z}_2[3]\times\mathbb{Z}_2[2]$.

To complete the description (5) of $\mathcal{R}$, it therefore suffices to give a class

$$\alpha \in \mathrm{H}^5(\mathbb{Z}_2[3] \times \mathbb{Z}_2[2];\mathbb{C}^\times) = \mathrm{H}^5(\{1,e\}[3] \times \{\mathbb{1},m\}[2];\mathbb{C}^\times). \tag{6}$$

Using the Künneth and Hurewicz theorems, one finds:

$$\begin{aligned}\mathrm{H}^5(\mathbb{Z}_2[3] \times \mathbb{Z}_2[2];\mathbb{C}^\times) &\cong \underbrace{\mathrm{H}^5(\mathbb{Z}_2[3];\mathbb{C}^\times)}_{\cong\mathbb{Z}_2} \\ &\quad\times \underbrace{\mathrm{H}^3(\mathbb{Z}_2[3];\mathbb{C}^\times)\otimes\mathrm{H}^2(\mathbb{Z}_2[2];\mathbb{C}^\times)}_{\cong\mathbb{Z}_2\otimes\mathbb{Z}_2\cong\mathbb{Z}_2} \times \underbrace{\mathrm{H}^5(\mathbb{Z}_2[2];\mathbb{C}^\times)}_{\cong\mathbb{Z}_2}.\end{aligned} \tag{7}$$

(Hurewicz implies that the "extension" classes from Künneth vanish in this degree.)

It is useful to give names to the generators on the right-hand side of (7). Recall that the $\mathbb{Z}_2$-cohomology ring $\mathrm{H}^\bullet(\mathbb{Z}_2[n];\mathbb{Z}_2)$ is freely generated over the Steenrod algebra by a single degree-n generator. I will write "E" for the degree-3 generator of $\mathrm{H}^\bullet(\{1,e\}[3];\mathbb{Z}_2)$, and "$M$" for the degree-2 generator of $\mathrm{H}^\bullet(\{1,m\}[2];\mathbb{Z}_2)$. Then for example the $\mathbb{Z}_2$-cohomology of $\{1,e\}[3] \times \{1,m\}[2]$ is generated over the Steenrod algebra by E and M. There is a map $x \mapsto (-1)^x$ from $\mathbb{Z}_2$-cohomology to $\mathbb{C}^\times$-cohomology. In general, not every $\mathbb{C}^\times$-cohomology class is in the image of this map: for example, $\mathrm{H}^4(\mathbb{Z}_2[2];\mathbb{C}^\times) \cong \mathbb{Z}_4$, and so any map from $\mathbb{Z}_2$-cohomology cannot by surjective. But it turns out that the map $\mathrm{H}^5(\{1,e\}[3] \times \{1,m\}[2];\mathbb{Z}_2) \to \mathrm{H}^5(\{1,e\}[3] \times \{1,m\}[2];\mathbb{C}^\times)$ is surjective (although not injective). Specifically, the three $\mathbb{Z}_2$s on the right-hand side of (7) are generated by the classes

$$(-1)^{\mathrm{Sq}^2 E}, \quad (-1)^{EM}, \quad \text{and} \quad (-1)^{\mathrm{Sq}^2\mathrm{Sq}^1 M} = (-1)^{M\mathrm{Sq}^1 M}. \tag{8}$$

What product $\alpha(E, M)$ of these classes describes the braiding on $\mathcal{R}^\times$ (and hence on $\mathcal{R}$)? The restriction $\alpha|_{M=0}$ to $\{1, e\}[3]$ describes the braiding on $\Omega\mathcal{R} = \mathbf{Rep}(\mathbb{Z}_2)$. The statement that e is a boson is the statement that this braiding is trivial. Thus $\alpha|_{M=0} = +1$, and the term $(-1)^{\mathrm{Sq}^2 E}$ does not appear in $\alpha(E, M)$. On the other hand, the term $(-1)^{EM}$ must appear in α, since it is what encodes that the e-particle and the m-string detect each other. What about the third term $(-1)^{\mathrm{Sq}^2\mathrm{Sq}^1 M} = (-1)^{M\mathrm{Sq}^1 M}$? The answer is that it doesn't matter:

LEMMA 2.1.4. *With notation as above, the classes $(-1)^{EM}$ and $(-1)^{EM+M\mathrm{Sq}^1 M}$ lead to equivalent braided fusion 2-categories.*

PROOF. A straightforward calculation shows that $\pi_0 \operatorname{Aut}(\mathbb{Z}_2[3] \times \mathbb{Z}_2[2]) \cong \mathbb{Z}_2$. The nontrivial automorphism can be understood in coordinates as

$$(E, M) \mapsto (E', M') = (E + \mathrm{Sq}^1 M, M).$$

In words, this is the automorphism of $\mathbb{Z}_2[3] \times \mathbb{Z}_2[2]$ which restricts to the identity on $\mathbb{Z}_2[3]$ and projects to the identity on $\mathbb{Z}_2[2]$, but uses the nontrivial class in $\mathrm{H}^3(\mathbb{Z}_2[2]; \mathbb{Z}_2)$ to choose a different splitting. Pulling back $(-1)^{EM}$ along this automorphism gives $(-1)^{E'M'} = (-1)^{(E+\mathrm{Sq}^1 M)M}$, and so this automorphism exchanges the two classes. $\square$

REMARK 2.1.5. Lemma 2.1.4 could be understood as "experimental confirmation" of the uniqueness of the 3+1D topological order $\mathcal{R}$ with $\Omega\mathcal{R} = \mathbf{Rep}(\mathbb{Z}_2)$. As explained in Remark 1.1.1, this uniqueness follows directly from the general classification given in [**LKW18**].

Suppose that this general classification was unavailable, but that there was a direct proof that $\pi_0\mathcal{R} = \mathbb{Z}_2$ and that $\mathcal{R}$ contained an invertible magnetic string m. Then $\mathcal{R}$ would arise as an extension of shape "$\Sigma\mathbf{Rep}(\mathbb{Z}_2).\mathbb{Z}_2^F$," where the notation is supposed to indicate that this is not a central extension, but rather $\pi_0\mathcal{R} = \{\mathbb{1}, m\}$ acts on the identity component as the 1-form symmetry $(-1)^F$. After remembering that m does indeed source a 1-form symmetry, one sees that extensions of this shape are the same as extensions of shape $(\mathbb{C}^\times[4] \times \mathbb{Z}_2[3]).\mathbb{Z}_2^f[2]$, where $\mathbb{C}^\times[4] \times \mathbb{Z}_2[3]$ arises as a classifying space of the braided monoidal 2-groupoid $\Sigma\mathbf{Rep}(\mathbb{Z}_2)^\times$. The symmetric monoidal structure on $\mathbf{Rep}(\mathbb{Z}_2)$ makes it natural to think of $\mathbb{C}^\times[4] \times \mathbb{Z}_2[3]$ as the degree-4 term in the spectrum for the cohomology theory $\mathrm{H}^\bullet(-; \mathbb{C}^\times \times \mathbb{Z}_2[-1]) := \mathrm{H}^\bullet(-; \mathbb{C}^\times) \times \mathrm{H}^{\bullet-1}(-; \mathbb{Z}_2)$. Extensions of shape $(\mathbb{C}^\times[4] \times \mathbb{Z}_2[3]).\mathbb{Z}_2^f[2]$ are then classified by a *twisted* cohomology group $\mathrm{H}^5(\mathbb{Z}_2^f[2]; \mathbb{C}^\times \times \mathbb{Z}_2[-1])$, where the twisting encodes that $\mathbb{Z}_2^F$ is supposed to act nontrivially on $\mathbb{C}^\times[4] \times \mathbb{Z}_2[3]$ and is recorded in the notation by the f superscript.

There is an (easy!) Atiyah–Hirzebruch spectral sequence computing this twisted cohomology. The E_2 page has two rows:

$$\begin{Bmatrix} E_2^{\bullet,1} = \mathrm{H}^\bullet(\mathbb{Z}_2[2]; \mathbb{Z}_2) \\ E_2^{\bullet,0} = \mathrm{H}^\bullet(\mathbb{Z}_2[2]; \mathbb{C}^\times) \end{Bmatrix} \Rightarrow \mathrm{H}^\bullet(\mathbb{Z}_2^f[2]; \mathbb{C}^\times \times \mathbb{Z}_2[-1]).$$

As above, write M for the degree-2 generator of $\mathrm{H}^\bullet(\mathbb{Z}_2[2]; \mathbb{Z}_2)$. Then the only differential in this spectral sequence is

$$\mathrm{d}_2 : \mathrm{H}^\bullet(\mathbb{Z}_2[2]; \mathbb{Z}_2) \to \mathrm{H}^{\bullet+2}(\mathbb{Z}_2[2]; \mathbb{C}^\times), \quad X \mapsto (-1)^{XM}.$$

The unique nonzero class $(-1)^{M\mathrm{Sq}^1 M} = (-1)^{\mathrm{Sq}^2\mathrm{Sq}^1 M} \in \mathrm{H}^5(\mathbb{Z}_2[2]; \mathbb{C}^\times)$ is in the image of this differential, and the unique nonzero class $M^2 = \mathrm{Sq}^2 M \in \mathrm{H}^4(\mathbb{Z}_2[2]; \mathbb{Z}_2)$

is not in the kernel of this differential, and so:

$$\mathrm{H}^5(\mathbb{Z}_2^F[2]; \mathbb{C}^\times \times \mathbb{Z}_2[-1]) = 0.$$

This is another way to package the earlier calculations, and establishes directly the uniqueness of $\mathcal{R}$ subject to the existence of an invertible magnetic string.

2.2. The two topological orders in which e is a fermion. I now turn to the case of topological orders $\mathcal{B}$ with $\Omega\mathcal{B} \cong \mathbf{SVec}$. As in §2.1, the first step is to understand the identity component $\Sigma\mathbf{SVec}$. Since $\mathbf{SVec}$ and $\mathbf{Rep}(\mathbb{Z}_2)$ are monoidally equivalent, $\Sigma\mathbf{SVec}$ and $\Sigma\mathbf{Rep}(\mathbb{Z}_2)$ are equivalent as semisimple 2-categories. In particular, $\Sigma\mathbf{SVec}$ contains two simple objects (up to equivalence): the identity object $\mathbb{1}$ and a Cheshire string c.

The first difference between the emergent-boson and emergent-fermion cases shows up in the fusion rules for the Cheshire string. In the bosonic case, the Cheshire string satisfied $c^2 \cong c \oplus c$ (3). When e is a fermion, however, the Cheshire string corresponds to the simple associative superalgebra $\mathrm{Cliff}(1) \in \mathbf{SVec}$. The tensor product of algebra objects depends on the braiding in the ambient category. As superalgebras, there is a well-known isomorphism [**Wal64**]

$$\mathrm{Cliff}(1) \otimes \mathrm{Cliff}(1) = \mathrm{Cliff}(2) \cong \mathrm{Mat}(1|1) \simeq \mathbb{C}.$$

Here $\mathrm{Mat}(1|1)$ means the matrix algebra of endomorphisms of $\mathbb{C}^{1|1} \in \mathbf{SVec}$, and $\simeq$ denotes a super Morita equivalence. As such, in $\Sigma\mathbf{SVec}$ the Cheshire string satisfies the fusion rule:

$$c^2 \cong \mathbb{1}. \tag{9}$$

Since c is invertible, its self-braiding $\beta_{c,c} : c \otimes c \to c \otimes c$ is either the identity or is (the identity times) the fermion e. The latter holds in fact holds:

$$\beta_{c,c} \cong e : c \otimes c \to c \otimes c \tag{10}$$

The self-braiding of the fermion $\beta_{e,e}$ is also nontrivial. In other words, the braided monoidal groupoid $\Sigma\mathbf{SVec}^\times$ corresponds to a space

$$\mathrm{B}^2\Sigma\mathbf{SVec}^\times = \mathbb{C}^\times[4].\mathbb{Z}_2[3].\mathbb{Z}_2[2] = \mathbb{C}^\times[4].\{1, e\}[3].\{\mathbb{1}, c\}[2]$$

with both Postnikov k-invariants nontrivial:

$$\mathrm{Sq}^2 : \{\mathbb{1}, c\} \to \{1, e\}, \qquad (-1)^{\mathrm{Sq}^2} : \{1, e\} \to \mathbb{C}^\times.$$

These Postnikov invariants are stable — just given by Steenrod operations — because $\Sigma\mathbf{SVec}$ is symmetric monoidal and not just braided monoidal. Indeed, as a symmetric monoidal object, $\Sigma\mathbf{SVec}^\times$ corresponds to the "extended supercohomology" $\mathrm{SH}^\bullet$ of [**WG17**]. (My degree conventions are that $\mathrm{SH}^n(-)$ classifies maps into $\mathbb{C}^\times[n].\mathbb{Z}_2[n-1].\mathbb{Z}_2[n-1]$.) A priori to determine a three-level extension like this requires more than just the k-invariants, but in the $\mathrm{SH}^\bullet$ case the two possible extensions with these k-invariants are isomorphic [**GJF19b**, §5.4].

To extend $\Sigma\mathbf{SVec}$ to a topological order requires adding in a magnetic component — for now, I will take for granted the expectation (1) that $\pi_0\mathcal{B} = \mathbb{Z}_2$, so there is a single magnetic component. Choose a simple magnetic string m in this component. By [**JFY21**, Theorem B], any choice m is invertible, and the other simple magnetic string is $m' = m \otimes c$. Since m is invertible, its self-braiding

$\beta_{m,m} : m \otimes m \to m \otimes m$ is given by an invertible particle $x \in \{1, e\}$. The self-braiding on m^2 is then given by x^4. In particular, equation (10) is inconsistent with $m^2 \cong c$, and so:

$$m^2 \cong \mathbb{1}. \tag{11}$$

REMARK 2.2.1. Unlike the case considered in Lemma 2.1.3, $\Sigma\mathbf{SVec}$ is *not* the linearization of its invertible subgroupoid $\Sigma\mathbf{SVec}^\times$. Rather, $\Sigma\mathbf{SVec}$ is already the linearization of just its sub-groupoid $\mathbb{C}^\times[2].\{1, e\}[1]$, since $\mathbf{SVec}$ is the linearization of $\mathbb{C}^\times[1].\{1, e\}[0]$. If one linearized all of $\Sigma\mathbf{SVec}^\times$, one would get a 2-category with two components. Similarly, the full topological order $\mathcal{B}$ is not the linearization of its invertible subgroupoid $\mathcal{B}^\times$, but just of $\mathbb{C}^\times[2].\{1, e\}[1].\{\mathbb{1}, m\}[0]$.

Thus to classify the topological orders $\mathcal{B}$ with $\Omega\mathcal{B} \cong \mathbf{SVec}$, one must simply work out the possible extensions $\mathbb{C}^\times[4].\{1, e\}[3].\{\mathbb{1}, m\}[2]$ consistent with the braiding requirements. The first question to work out is the quotient extension $\{1, e\}[3].\{\mathbb{1}, m\}[2]$, which classifies the self-braiding $\beta_{m,m} : m \otimes m \to m \otimes m$ up to isomorphism. In Proposition 3.1.3, I will prove that the self-braiding of any magnetic string is necessarily trivial, and so $\{1, e\}[3].\{\mathbb{1}, m\}[2]$ is (noncanonically) a product. Then, as in equation (6) and the surrounding discussion, to determine the topological order it suffices to give a class

$$\alpha(E, M) \in \mathrm{H}^5(\{1, e\}[3] \times \{\mathbb{1}, m\}[2]; \mathbb{C}^\times) = \{(\pm 1)^{\mathrm{Sq}^2 E}\} \times \{(\pm 1)^{EM}\} \times \{(\pm 1)^{\mathrm{Sq}^2 \mathrm{Sq}^1 M}\}.$$

The statement that e is a fermion means that $\alpha|_{M=0}$ should be nontrivial, and so the term $(-1)^{\mathrm{Sq}^2 E}$ is present. The term $(-1)^{EM}$ is also present. As in the case of $\mathcal{R}$, the nontrivial coefficient on EM is what enforces remote detectability: it says that an e-particle and an m-string link nontrivially. This leaves two choices for α:

$$\mathcal{S} \quad \leftrightarrow \quad \alpha(E, M) = (-1)^{\mathrm{Sq}^2 E + EM}, \tag{12}$$

$$\mathcal{T} \quad \leftrightarrow \quad \alpha(E, M) = (-1)^{\mathrm{Sq}^2 E + EM + \mathrm{Sq}^2 \mathrm{Sq}^1 M}. \tag{13}$$

It remains to prove that, unlike the case considered in Lemma 2.1.4, these two choices really do lead to inequivalent topological orders. First, one could wonder about the automorphism $(E, M) \mapsto (E', M') = (E + \mathrm{Sq}^1 M, M)$ used in that Lemma, which corresponded to choosing a different splitting of the extension $\{1, e\}.\{\mathbb{1}, m\}$. Pulling back the class in (12) along this automorphism, one calculates:

$$(-1)^{\mathrm{Sq}^2 E' + E'M'} = (-1)^{\mathrm{Sq}^2 (E + \mathrm{Sq}^1 M) + (E + \mathrm{Sq}^1 M) M} = (-1)^{\mathrm{Sq}^2 E + \mathrm{Sq}^2 \mathrm{Sq}^1 M + EM + M \mathrm{Sq}^1 M}$$

by the linearity of Sq^2. But $(-1)^{M \mathrm{Sq}^1 M} = (-1)^{\mathrm{Sq}^2 \mathrm{Sq}^1 M}$ as classes in $\mathrm{H}^5(\mathbb{Z}_2[2]; \mathbb{C}^\times)$, and so one ends up back to $(-1)^{\mathrm{Sq}^2 E + EM}$. The same calculation shows that class in (13) is also fixed by $(E, M) \mapsto (E', M') = (E + \mathrm{Sq}^1 M, M)$.

The other possible worry is whether the choice of magnetic string matters: perhaps $\mathcal{S}$ and $\mathcal{T}$ are exchanged by switching $m \mapsto m' = m \otimes c$. In fact, this does not happen. Indeed, I will show in Proposition 2.3.10 that $\mathcal{S}$ has an automorphism that switches m with m', and so $\mathcal{S}$ and $\mathcal{T}$ cannot be exchanged by such a switch (since the previous paragraph already establishes that $\mathcal{S}$ and $\mathcal{T}$ are not exchanged by any automorphism that fixes m). Thus the two topological orders truly are inequivalent. Moreover, $\mathcal{T}$ also has an automorphism switching m with m'.

REMARK 2.2.2. The topological order $\mathcal{S}$ is achieved physically by replacing the $\mathbb{Z}_2$ gauge field in §2.1 with a spin structure, to produce the TFT that counts spin structures. As explained in Remark 2.1.2, any boundary condition for a topological order realizes that topological order as a Drinfel'd centre. There is no sensible Dirichlet boundary condition for spin-$\mathbb{Z}_2$ gauge theory because there is no "trivial" spin structure and so no sensible meaning to the sentence "the spin-structure trivializes at the boundary." However, there is a Neumann boundary condition for spin-$\mathbb{Z}_2$ gauge theory in which the spin structure is arbitrary on the boundary. This boundary condition corresponds to realizing $\mathcal{S}$ as

$$\mathcal{S} \cong \mathcal{Z}_{(1)}(\Sigma\mathbf{SVec}).$$

The topological order $\mathcal{T}$ is achieved by the anomalous spin-$\mathbb{Z}_2$ gauge theory studied in [**Tho15**, Section 5] and [**KLW**$^+$**20a**, Section III.J.4]. I do not expect that it has a description as a Drinfel'd centre; compare Remarks 1.1.2 and 3.3.9.

REMARK 2.2.3. Suppose you meet a topological order $\mathcal{B}$ in the wild, and check that $\Omega\mathcal{B} \cong \mathbf{SVec}$. How do you tell whether $\mathcal{B} \cong \mathcal{S}$ or $\mathcal{T}$? The answer can be read off from the extension $\alpha(E, M)$ in (12)–13 as follows. First, choose a magnetic string m. Second, choose an isomorphism $m^2 \cong 1$: this is the choice of splitting $\{1, e\}.\{\mathbb{1}, m\} \cong \{1, e\} \times \{\mathbb{1}, m\}$. These two choices together allow you to find inside $\mathcal{B}$ a braided sub-2-category whose only objects are $\{\mathbb{1}, m\}$, whose only 1-morphisms are identities, and whose only 2-morphisms are scalars. The braided monoidal structure on this sub-2-category is then described by the class $\alpha|_{E=0} \in \mathrm{H}^5(\mathbb{Z}_2[2]; \mathbb{C}^\times)$. If this class is trivial, then $\mathcal{B} \cong \mathcal{S}$, whereas if it is nontrivial, then $\mathcal{B} \cong \mathcal{T}$. The paper [**JFR23**], which develops further the calculations in this paper, includes an explicit Klein-bottle-shaped anomaly indicator that measures the class $\alpha|_{E=0}$.

Incidentally, the nontrivial choice $(-1)^{\mathrm{Sq}^2\mathrm{Sq}^1 M} \in \mathrm{H}^5(\mathbb{Z}_2[2]; \mathbb{C}^\times) \cong \mathbb{Z}_2$ is a stable class, meaning that this braided monoidal sub-2-category is in fact symmetric monoidal (in a unique way, since the desuspension map $\mathrm{H}^{k+3}(\mathbb{Z}_2[k]; \mathbb{C}^\times) \to \mathrm{H}^5(\mathbb{Z}_2[2]; \mathbb{C}^\times)$ is an isomorphism for $k \geq 2$).

REMARK 2.2.4. Forgetting the braiding on $\mathcal{S}$ or $\mathcal{T}$, and remembering only the monoidal structure, corresponds to working with the extension

$$\mathbb{C}^\times[3].(\{1, e\}[2] \times \{\mathbb{1}, m\}[1])$$

classified by the desuspension of the class $\alpha(E, M)$ from (12)–13. It is easy to desuspend stable classes like $(-1)^{\mathrm{Sq}^2 E}$ and $(-1)^{\mathrm{Sq}^2\mathrm{Sq}^1 M}$: their formulas stay the same, just with E and M now interpreted as the degree-2 and degree-1 generators of $\mathrm{H}^\bullet(\{1, e\}[2]; \mathbb{Z}_2)$ and $\mathrm{H}^\bullet(\{\mathbb{1}, m\}[1]; \mathbb{Z}_2)$, respectively. But $\mathrm{H}^4(\mathbb{Z}_2[1]; \mathbb{C}^\times)$ is trivial, and indeed $(-1)^{\mathrm{Sq}^2\mathrm{Sq}^1 M} = (-1)^{M^4} = +1$, since M has now degree 1. The class $(-1)^{EM}$ also desuspends to the trivial class. All together, one finds that $\mathcal{S}$ and $\mathcal{T}$ are monoidally, but not braided monoidally, equivalent, and both are furthermore monoidally equivalent to the symmetric monoidal fusion 2-category $(\Sigma\mathbf{SVec})[\mathbb{Z}_2]$.

REMARK 2.2.5. As in Remark 2.1.5, there is a more direct spectral-sequence-based proof of the classification of topological orders $\mathcal{B}$ with $\Omega\mathcal{B} \cong \mathbf{SVec}$, subject only to the assumption that $\pi_0\mathcal{B} = \mathbb{Z}_2$. First, by [**JFY21**, Theorem B], every indecomposable magnetic string is invertible. Thus the question is to classify braided extensions of shape "$(\Sigma\mathbf{SVec}).\mathbb{Z}_2^F$," where the superscript f indicates that the fibre $\Sigma\mathbf{SVec}$ is not central, but rather transforms nontrivially under the 1-form symmetry group $\mathbb{Z}_2^F = \pi_0\mathcal{B}$. Since $(\Sigma\mathbf{SVec})^\times$ classifies supercohomology $\mathrm{SH}^\bullet$, extensions of this shape are classified by the "twisted supercohomology" group $\mathrm{SH}^5(\mathbb{Z}_2^F[2])$.

There is an Atiyah–Hirzebruch spectral sequence computing this twisted cohomology group:

$$\mathrm{H}^i(\mathbb{Z}_2^F[2]; \mathrm{SH}^j(\mathrm{pt})) \Rightarrow \mathrm{SH}^{i+j}(\mathbb{Z}_2^F[2]).$$

Here $\mathrm{SH}^0(\mathrm{pt}) = \mathbb{C}^\times$, $\mathrm{SH}^1(\mathrm{pt}) = \mathbb{Z}_2$, $\mathrm{SH}^2(\mathrm{pt}) = \mathbb{Z}_2$, and all other rows vanish. The twisting does not affect the E_2 page (because a 1-form symmetry cannot act on an ordinary group like $\mathrm{SH}^\bullet(\mathrm{pt})$), but rather appears in the differentials. The d_2 differentials are

$$\mathrm{d}_2 : E_2^{i,2} = \mathrm{H}^i(\mathbb{Z}_2^F[2]; \mathbb{Z}_2) \to E_2^{i+2,1} = \mathrm{H}^{i+2}(\mathbb{Z}_2^F[2]; \mathbb{Z}_2) \qquad X \mapsto \mathrm{Sq}^2 X + MX$$
$$\mathrm{d}_2 : E_2^{i,1} = \mathrm{H}^i(\mathbb{Z}_2^F[2]; \mathbb{Z}_2) \to E_2^{i+2,0} = \mathrm{H}^{i+2}(\mathbb{Z}_2^F[2]; \mathbb{C}^\times) \qquad X \mapsto (-1)^{\mathrm{Sq}^2 X + MX}$$

The E_2 page with nontrivial differentials drawn in looks like:

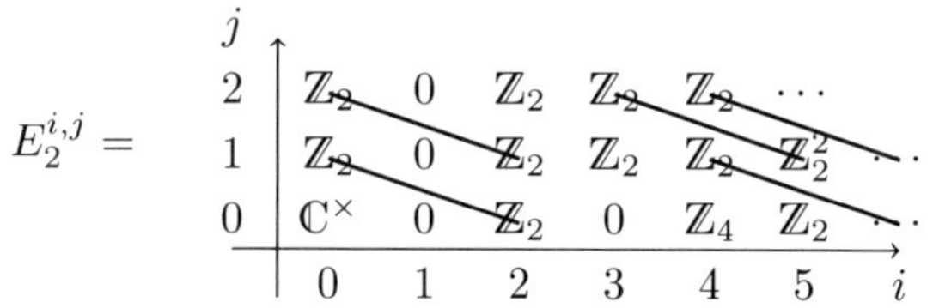

This is a truncation to degrees $j \leq 2$ of the spectral sequence considered in §3.3. In particular, Lemma 3.3.6 implies that the limit of this spectral sequence in total degree 4 is

(14) $$\mathrm{SH}^4(\mathbb{Z}_2^F[1]) \cong \mathbb{Z}_{16}.$$

Indeed, this group is identifiable with the kernel of the map $\mathcal{W} \to S\mathcal{W}$ from the "Witt group" of (Morita equivalence classes of) nondegenerate braided fusion 1-categories to the "super Witt group" of slightly-degenerate braided fusion 1-categories, which is shown to be a $\mathbb{Z}_{16}$ in [**DNO13**, 5.3]. I will give another description of this isomorphism in Remark 2.3.9.

It follows that all the entries in total degree 4 survive to the E_∞ page, and so the d_3 differential vanishes in this range of degrees. Therefore the $\mathbb{Z}_2$ in bidegree $(5,0)$ survives to E_∞, and:

(15) $$\mathrm{SH}^5(\mathbb{Z}_2^F[2]) \cong \mathbb{Z}_2.$$

Thus there are two topological orders $\mathcal{B}$ with $\Omega\mathcal{B} = \mathbf{SVec}$ and $\pi_0\mathcal{B} = \mathbb{Z}_2$, and they are distinguished by a class in ordinary cohomology $\mathrm{H}^5(\mathbb{Z}_2[2]; \mathbb{C}^\times)$, reproducing Remark 2.2.3.

The fact that the $\mathbb{Z}_2$ in bidegree $(3,2)$ supports a nontrivial d_2 differential is "the same" as the fact (11) that $m^2 \cong \mathbb{1}$ and not c. The fact that the $\mathbb{Z}_2$ in bidegree $(4,1)$ supports a nontrivial d_2 differential is "the same" as the fact that the self-braiding $\beta_{m,m}$ is necessarily 1, which I will prove directly in Proposition 3.1.3.

2.3. Automorphisms of $\mathcal{R}$, $\mathcal{S}$, and $\mathcal{T}$. To fully understand a topological order, it is interesting to work out its symmetry group. The following language has become fairly standard in the physics literature, following [**GKSW15**]:

DEFINITION 2.3.1. The *p-form symmetries* of a higher-categorical object $\mathcal{X}$ are $\pi_p \operatorname{Aut}(\mathcal{X})$.

Let $\mathcal{A}$ be the fusion n-category of all operators (including higher-dimensional membrane operators) in an n+1D topological order. It is almost true that $\mathrm{Aut}(\mathcal{A}) = \mathcal{A}^\times$ is the monoidal sub-3-groupoid of invertible operators. The reason this fails is that of course the centre of $\mathcal{A}$ acts trivially. The "remote detectability" axiom for a general topological order requires that this centre is as trivial as possible:

$$\mathcal{Z}_{(1)}(\mathcal{A}) = \Sigma^{n-1}\mathbf{Vec} = n\mathbf{Vec}.$$

The correct general statement asserts that $\mathrm{Aut}(\mathcal{A})$ is the quotient of $\mathcal{A}^\times$ by

$$\mathcal{Z}_{(1)}(\mathcal{A})^\times = (\Sigma^{n-1}\mathbf{Vec})^\times :$$

PROPOSITION 2.3.2. *Let $\mathcal{A}$ be an n+1D topological order in the sense of* [**JF22**], *i.e. a multifusion n-category with trivial Drinfel'd centre. Then there is a fibre sequence*

$$\mathrm{B}\mathcal{A}^\times \to \mathrm{B}\,\mathrm{Aut}(\mathcal{A}) \xrightarrow{\kappa} \mathrm{B}(\Sigma^n\mathbf{Vec})^\times.$$

This proposition should be thought of as a topological (and noninfinitesimal) version of Noether's theorem stating that every (infinitesimal) automorphism is inner. The spaces $(\Sigma^\bullet\mathbf{Vec})^\times$ fit together into a loop spectrum: $\Omega((\Sigma^\bullet\mathbf{Vec})^\times) = (\Sigma^{\bullet-1}\mathbf{Vec})^\times$. Following §3.3, I will call (a cohomological shift of) this spectrum by the name "$\underline{\mathcal{W}}^\bullet$":

$$\underline{\mathcal{W}}^n := (\Sigma^{n-1}\mathbf{Vec})^\times = (n\mathbf{Vec})^\times. \tag{16}$$

Complete details, and an explanation of the name, can be found in §3.3. The extension $\mathcal{A}^\times = (n\mathbf{Vec})^\times.\,\mathrm{Aut}(\mathcal{A})$ has as its Postnikov class $\kappa \in \underline{\mathcal{W}}^{n+2}(\mathrm{B}\,\mathrm{Aut}(\mathcal{A}))$, which should be thought of as the *universal 't Hooft anomaly* for $\mathcal{A}$.

PROOF. Any automorphism $f \in \mathrm{Aut}(\mathcal{A})$ of $\mathcal{A}$ can be encoded as an invertible $\mathcal{A}$-$\mathcal{A}$ bimodule "$\mathcal{A}_f$": it is $\mathcal{A}$ as a left-module, but the right action is twisted by f. To remember f, one remembers the "pointing" $1_\mathcal{A} \in \mathcal{A}_f$. Since $\mathcal{A}$ is a topological order, it is invertible in the Morita (n+2)-category of multifusion n-categories [**JF22**, Theorem 2], and so its monoidal (n+1)-category of (unpointed) bimodules is trivial, i.e. equivalent to $\Sigma^n\mathbf{Vec}$. In other words, the group map $\mathrm{Aut}(\mathcal{A}) \to (\Sigma^n\mathbf{Vec})^\times$ sending $f \mapsto \mathcal{A}_f$, as an unpointed bimodule, has kernel the choice of pointing. But the choice of pointing is exactly an element of $\mathcal{A}^\times$. □

In the case at hand, $n = 3$ and $\mathcal{A} = \Sigma\mathcal{B}$ for $\mathcal{B} \in \{\mathcal{R}, \mathcal{S}, \mathcal{T}\}$ — $\mathcal{A}$ has trivial Drinfel'd centre because $\mathcal{B}$ has trivial Müger centre [**JF22**, Corollary IV.2]. In particular, $\mathrm{Aut}(\mathcal{A}) = \mathrm{Aut}(\mathcal{B})$ and $\pi_i\mathcal{A}^\times = \pi_{i-1}\mathcal{B}^\times$ for $i \geq 1$. Unfolding the fibre sequence from Proposition 2.3.2 and applying the notation from (16) produces a long exact sequence in homotopy groups:

$$\begin{aligned}1 \to \mathbb{C}^\times \to \pi_2\mathcal{B}^\times \to \pi_3\,\mathrm{Aut}(\mathcal{B}) \to \pi_0\underline{\mathcal{W}}^1 \to \pi_1\mathcal{B}^\times \to \pi_2\,\mathrm{Aut}(\mathcal{B}) \to \pi_0\underline{\mathcal{W}}^2 \\ \to \pi_0\mathcal{B}^\times \to \pi_1\,\mathrm{Aut}(\mathcal{B}) \to \pi_0\underline{\mathcal{W}}^3 \to \pi_0(\Sigma\mathcal{B})^\times \to \pi_0\,\mathrm{Aut}(\mathcal{B}) \to \pi_0\underline{\mathcal{W}}^4\end{aligned} \tag{17}$$

Very explicitly, $\pi_p\,\mathrm{Aut}(\mathcal{B})$ are the p-form symmetries of $\mathcal{B}$, whereas $\pi_{p-1}\mathcal{B}^\times$ are the invertible $(p-1)$-morphisms in $\mathcal{B}$. It is not true in general that this sequence is exact on the far right: in the proof of Proposition 2.3.2, not every invertible bimodule needs to be isomorphic to one of the form $\mathcal{A}_f$ for an automorphism f.

A few of the terms in (17) are quick to fill in. Since $\mathcal{B}$ is fusion, $\pi_2\mathcal{B}^\times = \mathbb{C}$ and the first arrow is an isomorphism. As explained in Lemma 3.3.2, $\pi_0\underline{\mathcal{W}}^1$, $\pi_0\underline{\mathcal{W}}^2$, and

$\pi_0\underline{\mathcal{W}}^3$ are all trivial. Inspecting the descriptions of $\mathcal{R}, \mathcal{S}, \mathcal{T}$ in §2.1 and §2.2 then gives:

COROLLARY 2.3.3. *The 2-form symmetry groups of $\mathcal{R}$, $\mathcal{S}$, and $\mathcal{T}$ are*

$$\pi_2 \operatorname{Aut}(\mathcal{R}) \cong \pi_2 \operatorname{Aut}(\mathcal{S}) \cong \pi_2 \operatorname{Aut}(\mathcal{T}) \cong \mathbb{Z}_2 = \{1, e\}.$$

The 1-form symmetry groups are

$$\pi_1 \operatorname{Aut}(\mathcal{R}) \cong \mathbb{Z}_2 = \{\mathbb{1}, m\}, \qquad \pi_1 \operatorname{Aut}(\mathcal{S}) \cong \pi_1 \operatorname{Aut}(\mathcal{T}) \cong \mathbb{Z}_2^2 = \{\mathbb{1}, c, m, m'\},$$

where in the latter case m and m' are the two magnetic strings. Moreover, the extension $\pi_2.\pi_1$ is trivializable in the $\mathcal{R}$-case, whereas in the case of $\mathcal{S}$ and $\mathcal{T}$ it is classified by the nontrivial braiding $\beta_{c,c} = e$. □

REMARK 2.3.4. As is clear from their explicit descriptions in §2.2, the two braided fusion 2-categories $\mathcal{S}$ and $\mathcal{T}$ are almost isomorphic. They are monoidally isomorphic (see Remark 2.2.4), and most of their braiding data agrees: the only difference is a few ±1s in the self-braidings of the magnetic strings. As a result, the automorphism groups $\operatorname{Aut}(\mathcal{S})$ and $\operatorname{Aut}(\mathcal{T})$ are isomorphic, and their actions on $\mathcal{S}$ and $\mathcal{T}$ are "the same." The difference is only in the anomalies: one can summarize Remark 2.2.3 by saying that in $\mathcal{S}$ the 1-form $\mathbb{Z}_2$-symmetry generated by m is nonanomalous, whereas in $\mathcal{T}$ is has a nontrivial 't Hooft anomaly.

This leaves the zero-form symmetries. The triviality of $\pi_0\underline{\mathcal{W}}^3$ gives a left-exact sequence

$$1 \to (\Sigma\mathcal{B})^\times \to \pi_0 \operatorname{Aut}(\mathcal{B}) \to \pi_0\underline{\mathcal{W}}^4.$$

It is worth explaining this sequence physically. The right-most term is nothing but the "quantum Witt group" $\mathcal{W}$ of nondegenerate braided fusion categories studied in [**DMNO13, DNO13**]. The middle term is of course the group of 0-form automorphisms of $\mathcal{B}$. Any 0-form automorphism can be realized as a gapped topological 2+1D $\mathcal{B}$-$\mathcal{B}$ interface aka membrane operator. The objects of $\Sigma\mathcal{B}$, on the other hand, are the membrane operators X that can be built from condensing networks of surface operators, i.e. for which there is a condensation $\mathrm{Vac} \twoheadrightarrow X$, where Vac denotes the trivial (aka invisible or vacuum) 2+1D operator. These are precisely the gapped topological 2+1D membrane operators that admit a gapped topological 1+1D boundary condition. Other invertible 2+1D membrane operators may only admit gapless boundary conditions. The map $\pi_0 \operatorname{Aut}(\mathcal{B}) \to \mathcal{W}$ records the obstruction to gappability of the boundary by recording the equivalence class, modulo gapped 0+1D boundary interfaces, of any conformal boundary condition.

As this discussion illustrates, it will be hard to analyze $\pi_0 \operatorname{Aut}(\mathcal{B})$ by directly computing $\Sigma\mathcal{B}$. Fortunately, these groups are accessible by inspecting the spectral sequences in Remarks 2.1.5 and 2.2.5, which in both cases describe $\mathcal{B}$ as an extension of shape $\Sigma\Omega\mathcal{B}.\mathbb{Z}_2^F$. Indeed, it is a general feature of such cohomological computations that if the choices are parameterized by degree-n cohomology, then the automorphisms (up to higher isomorphism) of a given choice are parameterized by degree-$(n-1)$ cohomology. As such, there are isomorphisms

$$\pi_0 \operatorname{Aut}_{\Omega\mathcal{R}}(\mathcal{R}) \cong \mathrm{H}^4(\mathbb{Z}_2^F[2]; \mathbb{C}^\times \times \mathbb{Z}_2[-1]) \cong \mathbb{Z}_2, \tag{18}$$

$$\pi_0 \operatorname{Aut}_{\Omega\mathcal{S}}(\mathcal{S}) \cong \pi_0 \operatorname{Aut}_{\Omega\mathcal{T}}(\mathcal{T}) \cong \mathrm{SH}^4(\mathbb{Z}_2^F[2]) \cong \mathbb{Z}_{16}. \tag{19}$$

The subscript indicates that these are the groups of automorphisms trivialized on $\Omega\mathcal{R} = \mathbf{Rep}(\mathbb{Z}_2)$ and $\Omega\mathcal{S} = \Omega\mathcal{T} = \mathbf{SVec}$.

I will describe in some detail these groups, focusing on the case of $\mathrm{Aut}_{\Omega\mathcal{S}}(\mathcal{S})$ for definiteness.

It will be convenient to work with a pair of closely-related symmetric monoidal 5-categories, which are built following the general constructions in [**JFS17**, Section 8]. First is the *pointed Morita category* $\mathbf{Alg}_2(\mathbf{2KarCat}_\mathbb{C})$ with:

0-morphisms: Braided monoidal Karoubi-complete $\mathbb{C}$-linear 2-categories.
1-morphisms: Monoidal Karoubi-complete $\mathbb{C}$-linear 2-category bimodules.
2-morphisms: Pointed Karoubi-complete $\mathbb{C}$-linear 2-category bimodules.
3-morphisms: Pointed functors of 2-category bimodules.
4-morphisms: Pointed natural transformations of bimodule functors.
5-morphisms: Pointed modifications of natural transformations.

Details of what it means for a braided fusion 2-category to act on a fusion 2-category may be found in [**DN20**]. By "pointed," I mean that the bimodules come equipped with distinguished objects, which are to be preserved by higher morphisms. The composition of 1- and 2-morphisms is a balanced tensor product. In order to match with order of composition, I will consider a bimodule ${}_\mathcal{A}\mathcal{M}_\mathcal{B}$ to be a morphism $\mathcal{B} \to \mathcal{A}$.

Second is the *unpointed Morita category* $\mathbf{Mor}_2(\mathbf{2KarCat}_\mathbb{C})$ with:

0-morphisms: Braided monoidal Karoubi-complete $\mathbb{C}$-linear 2-categories.
1-morphisms: Monoidal Karoubi-complete $\mathbb{C}$-linear 2-category bimodules.
2-morphisms: Karoubi-complete $\mathbb{C}$-linear 2-category bimodules.
3-morphisms: Functors of 2-category bimodules.
4-morphisms: Natural transformations of bimodule functors.
5-morphisms: Modifications of natural transformations.

Given a braided monoidal Karoubi-complete $\mathbb{C}$-linear 2-category $\mathcal{B}$, I will write $[\mathcal{B}]$ for the corresponding object in $\mathbf{Mor}_2(\mathbf{2KarCat}_\mathbb{C})$. There is obviously a forgetful functor $\mathbf{Alg}_2(\mathbf{2KarCat}_\mathbb{C}) \to \mathbf{Mor}_2(\mathbf{2KarCat}_\mathbb{C})$ which forgets the pointing data.

Although seemingly very similar, these categories end up having quite different flavour:

PROPOSITION 2.3.5. *Two braided monoidal Karoubi-complete $\mathbb{C}$-linear 2-categories are equivalent in $\mathbf{Alg}_2(\mathbf{2KarCat}_\mathbb{C})$ if and only if they are braided-monoidally equivalent; automorphism groups computed in $\mathbf{Alg}_2(\mathbf{2KarCat}_\mathbb{C})$ agree with automorphism groups in the braided-monoidal sense. Every object (resp. 1-morphism) in $\mathbf{Alg}_2(\mathbf{2KarCat}_\mathbb{C})$ is 2- (resp. 1-) dualizable, but the only 5-dualizable object in $\mathbf{Alg}_2(\mathbf{2KarCat}_\mathbb{C})$ is the identity object $2\mathbf{Vec}$.*

Equivalences in $\mathbf{Mor}_2(\mathbf{2KarCat}_\mathbb{C})$ are Morita equivalences. An object $[\mathcal{B}] \in \mathbf{Mor}_2(\mathbf{2KarCat}_\mathbb{C})$ is 5-dualizable if and only if $\mathcal{B}$ is fusion, and $[\mathcal{B}]$ is invertible if and only if $\mathcal{B}$ is a topological order.

PROOF. The first sentence follows from [**Sch14**, Proposition 3.2.45]. The second sentence is essentially to [**GS18**, Theorem 6.1]; see also [**JF20**, Theorem 7.3]. The third sentence is essentially the definition of "Morita equivalence." The last sentence is [**JF22**, Theorems 1 and 2 and Corollary IV.2]; a similar result in a slightly different context is [**BJSS20**, Theorem 1.1]. □

Now fix an isomorphism $N : \mathcal{S} \overset{\sim}{\to} \mathcal{Z}_{(1)}(\Sigma\mathbf{SVec})$; as explained in Remark 2.2.2, such an isomorphism arises by realizing $\mathcal{S}$ as spin-$\mathbb{Z}_2$ gauge theory and choosing the Neumann boundary condition, hence the name N for the isomorphism. This

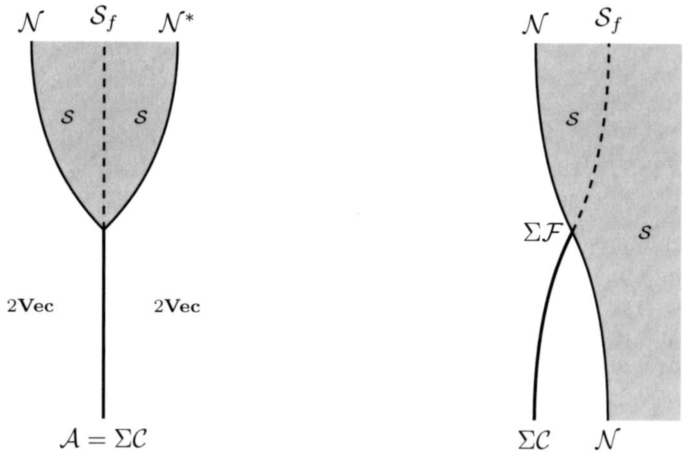

FIGURE 1. Left: a graphical depiction of the definition (20) of the 2+1D topological order $\mathcal{A}$ as a "slab" built from the Neumann boundary condition $\mathcal{N} : \mathcal{S} \to \mathbf{2Vec}$ and the automorphism $f \in \mathrm{Aut}(\mathcal{S})$. Right: the same figure, rotated a bit, recovers the boundary condition $\mathcal{N} \circ \mathcal{S}_f$ from a Morita equivalence $\mathcal{F} : \mathcal{C} \otimes \mathbf{SVec} \simeq \mathbf{SVec}$.

is equivalent to choosing a way for $\mathcal{S}$ to act (from the right, say) on the fusion 2-category $\Sigma\mathbf{SVec}$. In other words, it realizes $\mathcal{N} := \Sigma\mathbf{SVec}_{\mathcal{S}}$ as a 1-morphism $\mathcal{N} : \mathcal{S} \to \mathbf{2Vec}$ in $\mathbf{Alg}_2(\mathbf{2KarCat}_{\mathbb{C}})$. Furthermore, the corresponding 1-morphism $[\mathcal{N}] : [\mathcal{S}] \to [\mathbf{2Vec}]$ in $\mathbf{Mor}_2(\mathbf{2KarCat}_{\mathbb{C}})$ is an equivalence because the original map N was an isomorphism. $\mathcal{Z}_{(1)}(\Sigma\mathbf{SVec})$ also acts on $\Sigma\mathbf{SVec}$ from the left; this produces the dual 1-morphism $\mathcal{N}^* : \mathbf{2Vec} \to \mathcal{S}$ in $\mathbf{Alg}_2(\mathbf{2KarCat}_{\mathbb{C}})$ and its image $[\mathcal{N}^*] = [\mathcal{N}]^{-1} : [\mathbf{2Vec}] \to [\mathcal{S}]$ in $\mathbf{Mor}_2(\mathbf{2KarCat}_{\mathbb{C}})$.

For any $f \in \mathrm{Aut}_{\Omega\mathcal{S}}(\mathcal{S})$, write $\mathcal{S}_f$ for the corresponding bimodule $\mathcal{S}$-$\mathcal{S}$ bimodule: as a right $\mathcal{S}$-module, $\mathcal{S}_f = \mathcal{S}$, whereas as a left $\mathcal{S}$-module, the action is twisted by f. Then $N \circ f$ is another isomorphism of braided 2-categories $\mathcal{S} \xrightarrow{\sim} \mathcal{Z}_{(1)}(\Sigma\mathbf{SVec})$, i.e. another boundary condition. It corresponds to the 1-morphism $\mathcal{N} \circ \mathcal{S}_f = \mathcal{N} \otimes_{\mathcal{S}} \mathcal{S}_f : \mathcal{S} \to \mathbf{2Vec}$ in $\mathbf{Alg}_2(\mathbf{2KarCat}_{\mathbb{C}})$. Physically, $\mathcal{S}_f$ is an invertible membrane operator in the 3+1D "bulk" $\mathcal{S}$, and $\mathcal{N} \circ \mathcal{S}_f$ is the boundary condition built by bringing this membrane to the 2+1D boundary $\mathcal{N}$.

Consider now a "sandwich" or "slab," drawn in the left-hand part of Figure 1, in which the interior is the 3+1D topological order $\mathcal{S}$ with left-hand boundary $\mathcal{N} \circ \mathcal{S}_f$ and right-hand boundary the original Neumann right-boundary $\mathcal{N}^*$. In terms of the 5-category $\mathbf{Alg}_2(\mathbf{2KarCat}_{\mathbb{C}})$, this slab is precisely the composition

$$\mathcal{A} := \mathcal{N} \circ \mathcal{S}_f \circ \mathcal{N}^* = \Sigma\mathbf{SVec} \underset{\mathcal{S}}{\otimes} \mathcal{S}_f \underset{\mathcal{S}}{\otimes} \Sigma\mathbf{SVec} \tag{20}$$

Note that the image $[\mathcal{A}] : [\mathbf{2Vec}] \to [\mathbf{2Vec}]$ in $\mathbf{Mor}_2(\mathbf{2KarCat}_{\mathbb{C}})$ is invertible, since it is the composition of $\mathbf{Mor}$-invertible 1-morphisms $[\mathcal{N}] \circ [\mathcal{S}_f] \circ [\mathcal{N}]^{-1}$. Thus, $\mathcal{A}$ is a 2+1D topological order in the sense of [**JF22**]: it is a multifusion 2-category with trivial centre.

As with any "sandwich" or "slab" construction, the k-dimensional operators in $\mathcal{A}$ are realized as triples consisting of: a $(k+1)$-dimensional operator in the bulk $\mathcal{S}$, a way for this operator to end on the left-boundary, and a way for this operator to end on the right boundary. For example, to give a vertex operator in $\mathcal{A}$ requires giving a line in $\mathcal{S}$ which can end on both boundaries. But the worldline of the fermion e cannot end on the Neumann boundary — it does not "condense" on the boundary — and so the only vertex operators in $\mathcal{A}$ are ends of the vacuum line. Thus $\mathcal{A}$ is in fact fusion. By [**JF22**, Theorem 5], the lines in $\mathcal{A}$ form a nondegenerate braided fusion 1-category $\mathcal{C} := \Omega\mathcal{A}$, and $\mathcal{A}$ is recovered as its suspension $\mathcal{A} \cong \Sigma\mathcal{C}$.

Thus to understand $\mathcal{A}$ it suffices to understand its category $\mathcal{C}$ of line operators. Every line operator is realized as a surface operator in $\mathcal{S}$ together with ways for that surface to end on the boundaries. The trivial surface operator $\mathbb{1} \in \mathcal{S}$ provides two lines to $\mathcal{A}$: the vacuum identity line 1 and a fermion e. In other words, $\mathcal{C} \supset \mathbf{SVec}$. The Cheshire string $c \in \mathcal{S}$ does not condense on either boundary, and may be ignored. More precisely, the Cheshire string can end in the bulk in a noninvertible way, and this ending may be moved to the boundary; the resulting object in $\mathcal{C}$ is $1 \oplus e$.

REMARK 2.3.6. Each boundary condition $\mathcal{N}, \mathcal{N}^*$ provides an inclusion $\mathbf{SVec} \subset \mathcal{C}$, since those boundary conditions are each ways that $\mathcal{S}$ can act on $\Sigma\mathbf{SVec}$ — the copies of $\mathbf{SVec}$ arise as the left- and right- endpoints of the vacuum string in $\mathcal{S}$. Because the lines can be lifted through the bulk, these two subcategories $\mathbf{SVec} \subset \mathcal{C}$ are in fact equal as subcategories. If $\mathbf{SVec}$ admitted 0-form automorphisms, then the two inclusions $\mathbf{SVec} \hookrightarrow \mathcal{C}$ might differ by an automorphism of $\mathbf{SVec}$, but this cannot happen since $\pi_0 \operatorname{Aut}(\mathbf{SVec})$ is trivial. However, $\mathbf{SVec}$ does has 1-form automorphisms, which could show up if one were to consider an isomorphism $f \cong f'$ of elements in $\operatorname{Aut}(\mathcal{S})$. The cleanest statement is to require f to live in the kernel of $\operatorname{Aut}(\mathcal{S}) \to \operatorname{Aut}(\Omega\mathcal{S}) = \operatorname{Aut}(\mathbf{SVec})$.

This leaves strings from the magnetic component of $\mathcal{S}$. These strings are characterized by having nontrivial linking with the fermion e. It follows that any line in $\mathcal{A}$ produced from their endings will also link nontrivially with e. In other words:

PROPOSITION 2.3.7. *The nondegenerate braided fusion category $\mathcal{C} = \Omega\mathcal{A}$ is a* minimal modular extension (MME) *of* $\mathbf{SVec} = \{1, e\}$*: the relative Müger centre of the inclusion* $\mathbf{SVec} \subset \mathcal{C}$ *is just* $\mathbf{SVec}$. □

It is a famous fact, due to [**Kit06, DNO13**], that $\mathbf{SVec}$ has exactly 16 minimal modular extensions, namely the modular tensor categories $\operatorname{Spin}(k)_1$ (which depend only on k mod 16); see [**LKW17**, §4.4] for details.

The construction $f \mapsto \mathcal{C}$ can be reversed. Suppose $\mathcal{C}$ is a minimal modular extension of $\mathbf{SVec}$, and consider the braided fusion category $\mathcal{C} \otimes \mathbf{SVec}$. Since the braiding in $\mathcal{C}$ is nondegenerate, $\mathcal{Z}_{(1)}(\Sigma\mathcal{C}) \cong \Sigma\mathbf{Vec}$ is trivial, and so there is a canonical isomorphism of 3+1D topological orders

$$\begin{aligned} \mathcal{Z}_{(1)}(\Sigma(\mathcal{C} \otimes \mathbf{SVec})) &\cong \mathcal{Z}_{(1)}(\Sigma\mathcal{C} \otimes \Sigma\mathbf{SVec}) \\ &\cong \mathcal{Z}_{(1)}(\Sigma\mathcal{C}) \otimes \mathcal{Z}_{(1)}(\Sigma\mathbf{SVec}) \cong \mathcal{Z}_{(1)}(\Sigma\mathbf{SVec}). \end{aligned} \tag{21}$$

Physically, this isomorphism corresponds to realizing $\mathcal{C} \otimes \mathbf{SVec}$ as a boundary condition for $\mathcal{S} = \mathcal{Z}_{(1)}(\Sigma\mathbf{SVec})$ simply by taking the Neumann boundary $\mathcal{N}$ and stacking on a noninteracting copy of $\mathcal{A} = \Sigma\mathcal{C}$. That it gives a true boundary condition is precisely the statement that the composition $\mathcal{A} \circ \mathcal{N} : \mathcal{S} \to \Sigma\mathbf{Vec}$ in $\mathbf{Alg}_2(\mathbf{2KarCat}_{\mathbb{C}})$ lives over an invertible morphism $[\mathcal{A}] \circ [\mathcal{N}]$ in $\mathbf{Mor}_2(\mathbf{2KarCat}_{\mathbb{C}})$.

Since $\mathcal{C}$ is a minimal modular extension of $\mathbf{SVec}$, it comes equipped with a distinguished subcategory $\mathcal{C} \supset \mathbf{SVec}$, and so $\mathcal{C} \otimes \mathbf{SVec} \supset \mathbf{SVec} \otimes \mathbf{SVec}$. Condense the diagonal in $\mathbf{SVec} \otimes \mathbf{SVec}$, i.e. condense the composite boson $e \otimes e \in \mathcal{C} \otimes \mathbf{SVec}$. The result is yet another copy of $\mathbf{SVec}$, and the condensation procedure produces a Morita equivalence of braided fusion 1-categories

$$\mathcal{F} : \mathcal{C} \otimes \mathbf{SVec} \overset{\sim}{\to} \mathbf{SVec}. \tag{22}$$

The fusion category $\mathcal{F}$ is a copy of $\mathcal{C}$, and its $\mathcal{C}$-action is the standard one.

Since Morita equivalences induce isomorphisms of centres, (22) leads to an isomorphism

$$\mathcal{Z}_{(1)}(\Sigma(\mathcal{C} \otimes \mathbf{SVec})) \cong \mathcal{Z}_{(1)}(\Sigma\mathbf{SVec}). \tag{23}$$

Composing the isomorphisms (21) and (23) gives an automorphism $f \in \mathrm{Aut}(\mathcal{S})$. By its construction, this automorphism trivializes on the identity component $\Sigma\mathbf{SVec}$ of $\mathcal{S}$.

In the 5-categorical language, (22) corresponds to a 2-morphism

$$\Sigma\mathcal{F} : \mathcal{A} \circ \mathcal{N} \Rightarrow \mathcal{N} \circ \mathcal{S}_f \tag{24}$$

of 1-morphisms $\mathcal{S} \to 2\mathbf{Vec}$ in $\mathbf{Alg}_2(2\mathbf{KarCat}_{\mathbb{C}})$. That $\mathcal{F}$ was a Morita equivalence translates to the statement that $[\Sigma\mathcal{F}] : [\mathcal{N}] \Rightarrow [\mathcal{A}] \circ [\mathcal{N}]$ is invertible in $\mathbf{Mor}_2(2\mathbf{KarCat}_{\mathbb{C}})$.

PROPOSITION 2.3.8. *Theses constructions $f \mapsto \mathcal{C}$ and $\mathcal{C} \mapsto f$ are inverse to each other.*

PROOF. The 2-morphism $\Sigma\mathcal{F}$ from (24) is nothing but the "mate" of the isomorphism (20), i.e. its image under the equivalence

$$\hom(\mathcal{A} \Rightarrow \mathcal{N} \circ \mathcal{S}_f \circ \mathcal{N}^*) \cong \hom(\mathcal{A} \circ \mathcal{N} \Rightarrow \mathcal{N} \circ \mathcal{S}_f).$$

The only thing to confirm is that, given $\mathcal{C}$, the corresponding 2-morphism $\Sigma\mathcal{F} \in \hom(\mathcal{A}\circ\mathcal{N} \Rightarrow \mathcal{N}\circ\mathcal{S}_f)$ has an isomorphism as its mate. Since $\Sigma\mathcal{F}$ lives over a Morita equivalence between Morita invertible 1-morphisms, its mate is certainly a Morita equivalence. By Proposition 2.3.7, it is a Morita equivalence between (suspensions of) minimal modular extensions of $\mathbf{SVec}$. Unpacking definitions shows that the mate of $\Sigma\mathcal{F}$ is simply $\Sigma\mathcal{F}$ again, but thought of as a bimodule in a different way. On the other hand, the construction of $\mathcal{F}$ showed that, as a $\mathcal{C}$-module, it was the standard one. This implies that the 2-morphism $\mathcal{A} = \Sigma\mathcal{C} \Rightarrow \mathcal{N} \circ \mathcal{S}_f \circ \mathcal{N}^*$ in $\mathbf{Alg}_2(2\mathbf{KarCat}_{\mathbb{C}})$ is the "modulation" of a homomorphism $\mathcal{A} \to \mathcal{N} \circ \mathcal{S}_f \circ \mathcal{N}^*$ of fusion 2-categories, or equivalently a homomorphism $\mathcal{C} = \Omega\mathcal{A} \to \Omega(\mathcal{N} \circ \mathcal{S}_f \circ \mathcal{N}^*)$ of braided fusion 2-categories. But a homomorphism of minimal modular extensions of $\mathbf{SVec}$ is automatically an isomorphism. □

REMARK 2.3.9. The same construction in fact proves the following much more general statement, first conjectured in [**KLW$^+$20b**, Theorem^{ph} 3.37] (see also [**KLW$^+$20a**, Propositions 33 and 34]). Let $\mathcal{C}$ be a braided fusion 1-category with Müger centre $\mathcal{E} = \mathcal{Z}_{(2)}(\mathcal{C})$. Then the groupoid of minimal modular extensions of $\mathcal{C}$ is naturally isomorphic to the groupoid of isomorphisms $\mathcal{Z}_{(1)}(\Sigma\mathcal{C}) \cong \mathcal{Z}_{(1)}(\Sigma\mathcal{E})$ whose restriction to $\Sigma\mathcal{E}$ is the identity. (Given the higher-categorical nature of the problem, there is data involved in identifying the restriction to $\Sigma\mathcal{E}$ with the identity isomorphism. The necessity of this data is indicated in Remark 2.3.6.) In particular, if $\mathcal{C} = \mathcal{E}$ and if $\mathcal{E}$ has no 0-form automorphisms (as is the case for $\mathcal{E} = \mathbf{SVec}$), then $\pi_0 \operatorname{Aut}(\mathcal{Z}_{(1)}(\Sigma\mathcal{E}))$ is isomorphic to the group of minimal modular extensions of $\mathcal{E}$ from [**LKW17**].

For example, for any finite group G, the minimal modular extensions of $\mathbf{Rep}(G)$ are classified by $\mathrm{H}^3(\mathrm{B}G; \mathbb{C}^\times)$ [**LKW17**, Theorem 4.22]. Taking $G = \mathbb{Z}_2$ confirms equation (18). For $\mathcal{E} = \mathbf{SVec}$, this confirms (the $\mathcal{S}$ part of) equation (19), and also explains the relationship, alluded to in Remark 2.2.5, between $\mathrm{SH}^4(\mathbb{Z}_2^F[2]) \cong \mathbb{Z}_{16}$ and the kernel of the map $\mathcal{W} \to \mathcal{SW}$ from the bosonic to fermionic Witt groups.

The last remaining question is to understand "computationally" how an automorphism $f \in \operatorname{Aut}_{\Omega\mathcal{S}}(\mathcal{S})$ determines the corresponding minimal modular extension $\mathcal{C}$ of $\mathbf{SVec}$. As mentioned above, two of the simple objects of $\mathcal{C}$ — the vacuum 1 and the fermion e — arise as ways that the invisible string $\mathbb{1} \in \mathcal{S}$ can end, and the other objects come from the magnetic strings m and m'. In the Neumann boundary condition $\mathcal{N}$, one of these — m, say — condenses, whereas the $m' \mapsto c$, the Cheshire string in $\Sigma\mathbf{SVec}$. Suppose that m also condenses in $\mathcal{N} \circ \mathcal{S}_f$. Then the magnetic particles in $\mathcal{C}$, i.e. the simple objects that braid nontrivially with e, are invertible, and there are two of them, differing by a copy of e. This follows simply because $m \in \mathcal{S}$ can end in an invertible way on both boundary conditions. On the other hand, suppose that m' condenses in new $\mathcal{N} \circ \mathcal{S}_f$ boundary condition, and $m \mapsto c$. The Cheshire string c can end, but only in an noninvertible way, and that ending is fixed under fusing with e. As a result, $\mathcal{C}$ contains a noninvertible simple object fixed under fusing with e.

In other words, the automorphism f exchanges which string in $\mathcal{S}$ condenses if and only if the corresponding braided fusion category $\mathcal{C}$ is one of the eight Ising-type categories among the 16 minimal modular extensions of $\mathbf{SVec}$. The existence of such Ising-type categories implies:

PROPOSITION 2.3.10. *$\mathcal{S}$ (and hence $\mathcal{T}$) admits an automorphism which exchanges the magnetic strings m and m'.* □

3. Three proofs of the Main Theorem

In this section I will prove the Main Theorem. As explained in the introduction, I will focus on the second sentence. The reason is that I do now know any proof of the first sentence that does not simply repeat the argument from [**LKW18**]. As for the second sentence, I will give three proofs. The first, in §3.1 is extremely hands-on and direct: I will complete the proof started in §2.2 by showing directly that any topological order $\mathcal{B}$ with $\Omega\mathcal{B} \cong \mathbf{SVec}$ has only two components $\pi_0\mathcal{B} \cong \mathbb{Z}_2$, thus realizing the expectation (1). The problem with hands-on proofs is that they tend to use special features of the problem, and it is often not clear how to apply their ideas to other cases. In the second proof, in §3.2, is more generalizable: by using a super version of the classification from [**LKW18**], the "base change" $\mathcal{B} \otimes \Sigma\mathbf{SVec}$ of any topological $\mathcal{B}$ with $\Omega\mathcal{B} \cong \mathbf{SVec}$ is known, and the Main Theorem reduces to a "Galois descent" problem for the "Galois extension" $\Sigma\mathbf{Vec} \to \Sigma\mathbf{SVec}$. The last proof, in §3.3, is the most abstract: I will directly apply the classification statement of [**JF22**, Corollary V.5].

REMARK 3.0.1. For completeness, I will very briefly review the argument from [**LKW18**], specialized to the case in the first sentence of the Main Theorem. Suppose that $\mathcal{B}$ is a 3+1D topological order with $\Omega\mathcal{B} = \mathbf{Rep}(\mathbb{Z}_2)$. Then the emergent boson e is condensible. More precisely, the algebra object $\mathcal{O}(\mathbb{Z}_2) \in \mathbf{Rep}(\mathbb{Z}_2)$ is commutative, and this provides a (commutative) algebra structure to the Cheshire string $c \in \Sigma\mathbf{Rep}(\mathbb{Z}_2)$, which is moreover a condensation monad in the sense of [**GJF19a**]. The image of this condensation monad is a new 3+1D topological order $\mathcal{B}'$ equipped with a Morita equivalence to $\mathcal{B}$; physically, the 3+1D topological orders $\mathcal{B}$ and $\mathcal{B}'$ are separated by a gapped topological 2+1D interface. But since e condenses, $\mathcal{B}'$ has no line operators, and so is trivial by [**JF22**, Remark IV.3], and the interface is a boundary condition for $\mathcal{B}$. Thus (by Remark 2.1.2) $\mathcal{B} = \mathcal{Z}_{(1)}(\mathcal{A})$, where $\mathcal{A}$ is the fusion 2-category of surfaces on this boundary condition. Finally, the whole procedure started with the choice of commutative algebra object $\mathcal{O}(\mathbb{Z}_2) \in \mathbf{Rep}(\mathbb{Z}_2)$, and its automorphism group is $\mathbb{Z}_2$ (as a 0-form group, with no higher-form symmetries). This means that the boundary condition has 0-form automorphism group $\mathbb{Z}_2$, implemented by invertible surface operators, and it has no higher-dimensional operators, and so $\mathcal{A} = (\Sigma\mathbf{Vec})^{\omega}[\mathbb{Z}_2]$ for some class $\omega \in \mathrm{H}^4(\mathbb{Z}_2[1]; \mathbb{C}^\times) = 0$. (By [**JFY21**], there are no non-invertible indecomposable surface operators on the boundary.)

3.1. First proof of the Main Theorem: direct analysis of $\pi_0\mathcal{B}$. I will now give a hands-on direct proof of (the second sentence of) the Main Theorem, building on the calculations in §2.2. Throughout this subsection, I assume that $\mathcal{B}$ is a (possibly-degenerate) braided fusion 2-category with $\Omega\mathcal{B} \cong \mathbf{SVec}$. To start off, note that:

LEMMA 3.1.1 ([**JFY21**, Theorem B]). *The indecomposable objects of $\mathcal{B}$ are all invertible. Each component of $\mathcal{B}$ has exactly two objects, related by tensoring with the Cheshire string $c \in \Sigma\Omega\mathcal{B}$. The components also form a group $\pi_0\mathcal{B} = \{$indecomposable objects$\}/\{\mathbb{1}, c\}$.* □

Thus it suffices to work just with the invertible subgroupoid $\mathcal{B}^\times \subset \mathcal{B}$, and I will simply say "string" when I mean an object of $\mathcal{B}^\times$, leaving the word "indecomposable" implicit.

Since the fermion e squares to the vacuum, the linking of e with any string x is given just by a sign ± 1. Furthermore, the function

$$\text{link with } e : \{\text{strings}\} \to \{\pm 1\}$$

is a homomorphism. Since e links trivially with the Cheshire string, this homomorphism factors through the group of components $\pi_0\mathcal{B}$. Call x *magnetically charged* if e links nontrivially with x, and *magnetically neutral* if they link trivially.

The invertibility of a string x furthermore implies that the self-braiding $\beta_{x,x} : x \otimes x \to x \otimes x$ is either the identity or it is (the identity times) the fermion e. This is analogous to asking whether an invertible particle is a boson or a fermion. Due to this, I will say say that x is a *bosonic string* if $\beta_{x,x} \cong \mathrm{id}_{x\otimes x}$, and a *fermionic string* if $\beta_{x,x} \cong e \otimes \mathrm{id}_{x\otimes x}$. Note that this distinction records some but not all of the self-statistics of x. For example, as explained in Remark 2.2.3, in both topological orders $\mathcal{S}$ and $\mathcal{T}$ the magnetic string m is "bosonic," but the full braiding data is different in the two cases.

The following lemma should remind the reader of the 2+1D Toric Code and its fermionic cousin $\mathrm{Spin}(8)_1$:

LEMMA 3.1.2. *If x is magnetically neutral, then one of x and $x \otimes c$ is bosonic and the other is fermionic. If x is magnetically charged, then x and $x \otimes c$ are either both bosonic or both fermionic.*

PROOF. As mentioned already (10), a superalgebra calculation shows that the Cheshire string itself is fermionic: $\beta_{c,c} \cong e \otimes \mathrm{id}_{c,c}$. The behaviour of the identity component $\Sigma\Omega\mathcal{B}$ under endofunctors is fully determined by the behaviour of the fermion e. In particular, the full braiding map $\beta_{-,x} \circ \beta_{x,-} : \Sigma\Omega\mathcal{B} \to \Sigma\Omega\mathcal{B}$ is the identity if x is magnetically neutral, and acts as the 1-form symmetry $(-1)^F$ if x is magnetically charged. It is a fun superalgebra calculation that the automoprhism $(-1)^F$ of Cliff(1), when converted into a bimodule, is isomorphic to the parity-reversal of the identity bimodule. In other words, the natural automorphism $(-1)^F$ of $\Sigma\mathbf{SVec}$, when evaluated at the Cheshire string c, gives $e \otimes \mathrm{id}_c$:

$$\beta_{c,x} \circ \beta_{x,c} \cong \begin{cases} \mathrm{id}_{x,c}, & \text{if } x \text{ is magnetically neutral,} \\ e \otimes \mathrm{id}_{x,c}, & \text{if } x \text{ is magnetically charged.} \end{cases} \tag{25}$$

Finally, if x and y are any invertible objects in a braided monoidal n-category, then

$$\beta_{x\otimes y, y \otimes x} \cong \beta_{x,x} \otimes \beta_{y,y} \otimes (\beta_{x,y} \circ \beta_{y,x}) \tag{26}$$

up to identity maps. (The isomorphisms in (25)–(26) are not quite canonical in general, but the statement only requires knowing β up to isomorphism.) □

But in fact a much stronger statement holds:

PROPOSITION 3.1.3. *All magnetically charged strings in $\mathcal{B}$ are bosonic.*

PROOF. Let x be a magnetic string. It selects a homomorphism $\mathbb{Z} \to \{\text{strings}\}$. Consider the full braided sub-2-groupoid $\mathcal{G}$ of $\mathcal{B}^\times$ tensor-generated by the strings $\mathbb{1}, c, x$ and of course the 1-morphisms $1, e$ and 2-morphisms $\mathbb{C}^\times$. This groupoid lives over the cyclic subgroup of $\{\text{strings}\}$ generated by x; pulling it back along the homomorphism $\mathbb{Z} \to \{\text{strings}\}$ gives a braided 2-groupoid of shape

$$\mathcal{G} = (\mathbb{C}^\times[2].\{1,e\}[1].\{\mathbb{1},c\}[0]).\mathbb{Z}^f[0],$$

or in other words a space

$$\mathrm{B}^2\mathcal{G} = (\mathbb{C}^\times[4].\{1,e\}[3].\{\mathbb{1},c\}[2]).\mathbb{Z}^f[2].$$

The supscript f reminds that the base $\mathbb{Z}$ acts by nontrivial 1-form symmetries of the fibre $(\Sigma\mathbf{SVec})^\times = \mathbb{C}^\times[2].\{1,e\}[1].\{\mathbb{1},c\}[0]$.

As in Remark 2.2.5, extensions $\mathcal{G}$ of this shape can be classified by computing a twisted supercohomology group $\mathrm{SH}^5(\mathbb{Z}^f[2])$, which can be computed by a spectral sequence.

$$\mathrm{H}^i(\mathbb{Z}[2]; \mathrm{SH}^j(\mathrm{pt})) \Rightarrow \mathrm{SH}^{i+j}(\mathbb{Z}^f[2]).$$

Note that the Eilenberg–Mac Lane space $\mathbb{Z}[2] = K(\mathbb{Z},2)$ is nothing but $\mathbb{CP}^\infty$, which has a single cell in each even degree. Thus the E_2 page is very easy to write down. With the nonzero d_2 differentials drawn in, it reads:

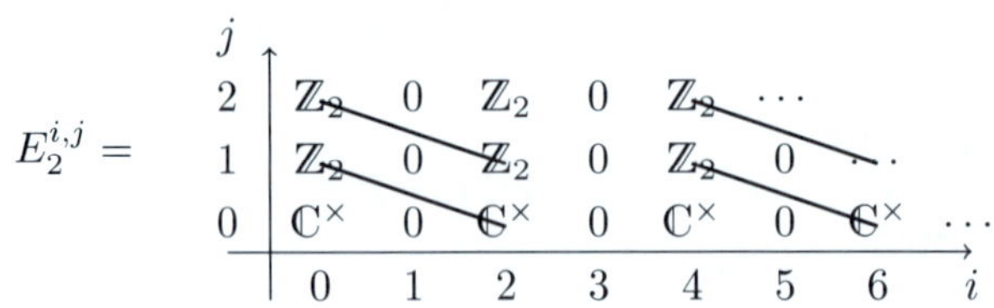

The self-braiding $\beta_{x,x}$ corresponds to the $\mathbb{Z}_2$ in bidegree $(4,1)$. The fact that that $\mathbb{Z}_2$ supports a nonzero differential is equivalent to the statement in the Proposition.

In fact, this calculation shows $\mathrm{SH}^5(\mathbb{Z}^f[2])$ is trivial, meaning that there is a unique isomorphism class of extensions of shape $(\mathbb{C}^\times[4].\{1,e\}[3].\{\mathbb{1},c\}[2]).\mathbb{Z}^f[2]$. □

REMARK 3.1.4. It is worth emphasizing that spectral sequence calculations like in the above proof are nothing but a clean way of working with braiding data, etc., in higher monoidal groupoids.

COROLLARY 3.1.5. *Suppose x is a bosonic magnetically neutral string, and y is a magnetically charged string. Then x and y link trivially: the full braiding $\beta_{y,x}\circ\beta_{x,y}$ is isomorphic to $\mathrm{id}_{x\otimes y}$.*

PROOF. Otherwise, equation (26) would say that one of the magnetically-charged strings y and $x\otimes y$ is fermionic and the other is bosonic. But this contradicts Proposition 3.1.3. □

PROOF OF THE MAIN THEOREM. Suppose $\mathcal{B}$ is a 3+1D topological order with $\Omega\mathcal{B}\cong\mathbf{SVec}$. Then $\mathcal{B}$ contains a magnetic string m, as otherwise the electron e would be undetectable. If $\pi_0\mathcal{B}$ is not just $\mathbb{Z}_2$, then $\mathcal{B}$ contains a nontrivial magnetically neutral bosonic string x. Because it is magnetically neutral, it is not detected by linking with the fermion e (or the Cheshire string c). By Corollary 3.1.5, it is not detected by linking with any magnetically charged string. Furthermore, if it were detected by linking with a magnetically neutral string y, then it would be detected by linking with the magnetically charged string $y\otimes m$, again contradicting Corollary 3.1.5. So x is undetectable, violating the remote detectability axiom of topological orders: x would lift to a nontrivial element in the Müger centre $\mathcal{Z}_{(2)}(\mathcal{B})$.

Thus $\pi_0\mathcal{B}=\{\mathbb{1},m\}=\mathbb{Z}_2^F$. The calculations from §2.2 then imply that $\mathcal{B}\cong\mathcal{S}$ or $\mathcal{T}$. □

3.2. Second proof of the Main Theorem: categorified Galois descent. For my second proof of the Main Theorem, I will rely on the analysis of $\mathrm{Aut}(\mathcal{S})$ from §2.3, but I will not otherwise require the calculations from §2.2.

The strategy is the following. Given a 3+1D topological order $\mathcal{B}$, consider the super topological order $\mathbf{S}\mathcal{B}:=\mathcal{B}\otimes\Sigma\mathbf{SVec}$ given by base-changing along the "field extension" $\mathbf{Vec}\to\mathbf{SVec}$; I will call such base-change the *superification* of $\mathcal{B}$. By definition, a 3+1D *super topological order* is a nondegenerate braided fusion 2-*supercategory*, meaning an 2-category enriched over $\mathbf{SVec}$; "nondegenerate" means that its enriched Müger centre is $\Sigma\mathbf{SVec}$ (as a 2-supercategory).

The Tannakian formalism of super fibre functors from [**Del02**] owes its strength to the fact that $\mathbf{Vec}\to\mathbf{SVec}$ is a *categorified Galois extension* [**JF17**]. By this I mean in particular that $\mathbf{Vec}$ arises as the fixed points of the 1-form Galois group $\mathbb{Z}_2^F[1]=\mathrm{Aut}(\mathbf{SVec})=:\mathrm{Gal}(\mathbf{SVec}/\mathbf{Vec})$ acting on $\mathbf{SVec}$. Furthermore, *Galois descent* says that the algebra of bosonic higher categories is "the same" as the algebra of $\mathbb{Z}_2^F[1]$-equivariant higher supercategories.

For example, any bosonic topological order $\mathcal{B}$ can be recovered as the fixed points of the $\mathbb{Z}_2^F[1]$-action on $\mathbf{S}\mathcal{B}$ induced from the base change $\mathbf{S}\mathcal{B}=\mathcal{B}\otimes\Sigma\mathbf{SVec}$. Thus, to classify bosonic topological orders with some property, it suffices to classify supertopological orders with that property, together with their possible Galois actions. This is a useful and highly generalizable strategy, and is the reason that I give this second proof of the Main Theorem.

Now suppose that $\mathcal{B}$ has a unique nontrivial particle, and that it is an emergent fermion. Since I want to work over the "base field" $\mathbf{SVec}$, I will not use the same name for $\Omega\mathcal{B}$. Rather, I will call it $\Omega\mathcal{B} \cong \mathbf{Rep}(\mathbb{Z}_2, \epsilon)$, where ϵ is the nontrivial element in $\mathbb{Z}_2$, and the notation follows the Tannakian formalism of [**Del02**]: given a finite group G and a central order-2 element $\epsilon \in Z(G)$, $\mathbf{Rep}(G,\epsilon)$ denotes the (bosonic!) symmetric fusion 1-category of super representations of G on which ϵ acts by $(-1)^F$. Then $\Omega\mathbf{S}\mathcal{B} \cong \mathbf{SVec} \otimes \mathbf{Rep}(\mathbb{Z}_2, \epsilon)$, and it is not hard to show in general that $\mathbf{SRep}(G,\epsilon) := \mathbf{SVec} \otimes \mathbf{Rep}(G,\epsilon) \cong \mathbf{SRep}(G)$ is independent of the choice ϵ. Indeed, the choice $\epsilon \in Z(G) = \pi_1 \operatorname{Aut}_{\mathbf{SVec}}(\mathbf{SRep}(G))$ is nothing but a Galois descent datum to recover $\mathbf{Rep}(G,\epsilon)$ from its superification $\mathbf{SRep}(G)$.

The classification argument from [**LKW18**] reviewed in Remark 3.0.1 applies just as well in the super case as in the bosonic case:

THEOREM 3.2.1 ([**LKW18**], [**JF22**, Corollary V.4]). *Suppose $\mathbf{S}\mathcal{B}$ is a 3+1D super topological order with $\Omega\mathbf{S}\mathcal{B} \cong \mathbf{SRep}(G)$. Then $\mathbf{S}\mathcal{B} \cong \mathcal{Z}_{(1)}(\mathbf{SVec}^\omega[G])$ for some supercohomology class $\omega \in \mathrm{SH}^4(G[1])$. (Here $\mathcal{Z}_{(1)}$ denotes the super Drinfel'd centre.)* □

But $\mathrm{SH}^4(\mathbb{Z}_2[1]) = 0$. This can be computed for example by an Atiyah–Hirzebruch spectral sequence $\mathrm{H}^i(\mathbb{Z}_2[1]; \mathrm{SH}^j(\mathrm{pt})) \Rightarrow \mathrm{SH}^{i+j}(\mathbb{Z}_2[1])$, whose d_2 differentials are $\mathrm{Sq}^2 : \mathrm{SH}^2(\mathrm{pt}) = \mathbb{Z}_2 \to \mathrm{SH}^1(\mathrm{pt}) = \mathbb{Z}_2$ and $(-1)^{\mathrm{Sq}^2} : \mathrm{SH}^1(\mathrm{pt}) = \mathbb{Z}_2 \to \mathrm{SH}^0(\mathrm{pt}) = \mathbb{C}^\times$. The E_2 page with d_2 differentials drawn in reads:

$$E_2^{i,j} = \begin{array}{c|cccccc} j & & & & & & \\ 2 & \mathbb{Z}_2 & \mathbb{Z}_2 & \cancel{\mathbb{Z}_2} & \cancel{\mathbb{Z}_2} & \dots & \\ 1 & \mathbb{Z}_2 & \mathbb{Z}_2 & \mathbb{Z}_2 & \cancel{\mathbb{Z}_2} & \cancel{\mathbb{Z}_2} & \dots \\ 0 & \mathbb{C}^\times & \mathbb{Z}_2 & 0 & \mathbb{Z}_2 & 0 & \cancel{\mathbb{Z}_2} \quad \dots \\ \hline & 0 & 1 & 2 & 3 & 4 & 5 \quad i \end{array}$$

Thus $\mathbf{S}\mathcal{B}$ is independent of $\mathcal{B}$, provided only that $\Omega\mathcal{B} \cong \mathbf{Rep}(\mathbb{Z}_2)$ or $\mathbf{Rep}(\mathbb{Z}_2, \epsilon)$. In particular:

$$\mathbf{S}\mathcal{B} \cong \mathbf{S}\mathcal{R} \cong \mathbf{S}\mathcal{S} \cong \dots.$$

In particular, to classify the choices for $\mathcal{B}$, it suffices to pick one of these "understood" topological orders, for instance $\mathcal{S}$, and work out the automorphism group and possible Galois descent data on $\mathbf{S}\mathcal{S}$.

REMARK 3.2.2. Superification preserves the set of components. Thus $\pi_0\mathcal{B} \cong \pi_0\mathbf{S}\mathcal{B} \cong \pi_0\mathbf{S}\mathcal{S} \cong \pi_0\mathcal{S} \cong \mathbb{Z}_2$, and so one could finish the proof by citing the calculations of §2.2. I will instead analyze the Galois descent data directly.

Since $\mathbf{S}\mathcal{S}$ is a super topological order, its $\mathbf{SVec}$-linear automorphism group $\operatorname{Aut}_{\mathbf{SVec}}(\mathbf{S}\mathcal{S})$ can be analyzed as in §2.3. In particular, there is a super analogue of the "topological Noether's theorem" Proposition 2.3.2: there is a fibre sequence

$$\mathrm{B}\mathbf{S}\mathcal{S}^\times \to \mathrm{B}\operatorname{Aut}_{\mathbf{SVec}}(\mathbf{S}\mathcal{S}) \to \mathrm{B}(\Sigma^3\mathbf{SVec})^\times. \tag{27}$$

As in §3.3, set $\underline{\mathrm{S}\mathcal{W}}^n := (\Sigma^{n-1}\mathbf{SVec})^\times$. Then (27) unfolds to a long exact sequence:

(28)

$$1 \to \mathbb{C}^\times \to \pi_2(\mathbf{S}\mathcal{S}^\times) \to \pi_3 \operatorname{Aut}_{\mathbf{SVec}}(\mathbf{S}\mathcal{S}) \to \pi_0\underline{\mathrm{S}\mathcal{W}}^1 \to \pi_1(\mathbf{S}\mathcal{S}^\times) \to \pi_2 \operatorname{Aut}_{\mathbf{SVec}}(\mathbf{S}\mathcal{S})$$
$$\to \pi_0\underline{\mathrm{S}\mathcal{W}}^2 \to \pi_0(\mathbf{S}\mathcal{S}^\times) \to \pi_1 \operatorname{Aut}_{\mathbf{SVec}}(\mathbf{S}\mathcal{S}) \to \pi_0\underline{\mathrm{S}\mathcal{W}}^3 \to \dots$$

The first arrow $\mathbb{C}^\times \to \pi_2 \mathbf{S}\mathcal{B}^\times$ is an isomorphism. To describe the other maps, I will need some notation. As above, write $1, e \in \Omega\mathcal{S} = \mathbf{Rep}(G, \epsilon)$ for the simple objects before superification. These are different from the simple objects $\pi_0 \underline{\mathcal{SW}}^1$ in the "base field" **SVec**, which I will call $1 := \mathbb{C}^{1|0}$ and $f := \mathbb{C}^{0|1}$. With this notation in hand, the arrow $\pi_0(\mathbf{SVec}^\times) = \pi_0\underline{\mathcal{SW}}^1 \to \pi_1(\mathbf{S}\mathcal{S}^\times) = \pi_0(\mathbf{SRep}(\mathbb{Z}_2, \epsilon)^\times)$ is the inclusion

$$\{1, f\} \to \{1, f, e, fe\} \cong \mathbb{Z}_2^2.$$

Similarly, it will be helpful to distinguish the identity-component strings $\mathbb{1}, c \in \mathcal{S}$ from the "base field" strings $\mathbb{1} = \mathrm{Cliff}(0)$ and $a := \mathrm{Cliff}(1)$. In other words, $\pi_0\underline{\mathcal{SW}}^2 = \pi_0(\Sigma\mathbf{SVec}^\times) = \{1, a\} \cong \mathbb{Z}_2$.

It is a slightly nontrivial task to work out $\pi_0(\mathbf{S}\mathcal{S}^\times)$. The reason is that to compute $\mathbf{S}\mathcal{S} = \mathcal{S} \otimes \Sigma\mathbf{SVec}$, one first forms a "naive tensor product" whose objects are just ordered pairs, but then one must Karoubi-complete, and this latter step produces new indecomposable objects, some of which might be invertible. In the case at hand, $\pi_0\mathbf{S}\mathcal{S} = \mathbb{Z}_2 = \{1, m\}$, and the question is just to understand the identity component $\Sigma\Omega\mathbf{S}\mathcal{S} = \Sigma(\mathbf{SRep}(\mathbb{Z}_2, \epsilon))$.

LEMMA 3.2.3. *There are six indecomposable objects in the identity component* $\Sigma(\mathbf{SRep}(\mathbb{Z}_2, \epsilon))$: $\mathbb{1}, a, c, ac$, *and two that are not invertible.*

PROOF. After forgetting the braiding and supercategory structure, there is a monoidal equivalence $\mathbf{SRep}(\mathbb{Z}_2, \epsilon) \cong \mathbf{Vec}[\mathbb{Z}_2^2]$. Thus the simple objects in $\Sigma(\mathbf{SRep}(\mathbb{Z}_2, \epsilon))$ are indexed by simple algebra objects in $\mathbf{Vec}[\mathbb{Z}_2^2]$. It is a general fact for any finite group G that simple algebra objects in $\mathbf{Vec}[G]$ are indexed by pairs (K, α) where $K \subset G$ is a subgroup and $\alpha \in \mathrm{H}^2_{\mathrm{gp}}(K; \mathbb{C}^\times)$ (see e.g. [**EGNO15**, Example 7.4.10]). Specifically, writing $\mathbb{C}g$ for the simple object of $\mathbf{Vec}[G]$ corresponding to $g \in G$, the algebra corresponding to (K, α) is $\bigoplus_{k \in K} \mathbb{C}k$ made into a twisted group algebra via (any cocycle representative of) α.

The group $G = \mathbb{Z}_2^2$ has five subgroups, and one of them (namely G itself) has two possible choices for α. It then remains to compute the fusion rules for these algebras, which depends on the braidings. The identity algebra is $\mathbb{1} \leftrightarrow (K = \{1\}, \alpha = \mathrm{triv})$. Two of the nontrivial elements of $G = \mathbb{Z}_2^2$, namely f and e, are fermions in the sense that $\beta_{f,f} = \beta_{e,e} = -1$, and so the algebras $a = 1 \oplus f$ and $c = 1 \oplus e$, corresponding to two of the $\mathbb{Z}_2$ subgroups of $\mathbb{Z}_2^2$, are invertible. Their tensor product is then also invertible: it is one of the algebra structures on $1 \oplus e \oplus f \oplus fe$. On the other hand, $1 \oplus fe$, with its unique algebra structure, is not invertible, since fe is a boson. Finally, the other algebra structure on $1 \oplus e \oplus f \oplus fe$ is isomorphic to the tensor product of the noninvertible algebra $1 \oplus fe$ with either of the invertible algebras a and c, and so is not invertible. □

Finally, as I will explain in Lemma 3.3.3, $\pi_0\underline{\mathcal{SW}}^3$ is trivial. Therefore, the long exact sequence (28) implies that $\pi_3 \mathrm{Aut}_{\mathbf{SVec}}(\mathbf{S}\mathcal{S})$ vanishes and

(29) $$\pi_2 \mathrm{Aut}_{\mathbf{SVec}}(\mathbf{S}\mathcal{S}) \cong \mathbb{Z}_2 = \{1, e\}, \quad \pi_1 \mathrm{Aut}_{\mathbf{SVec}}(\mathbf{S}\mathcal{S}) \cong \mathbb{Z}_2^2 = \{\mathbb{1}, c, m, m'\}.$$

Of course, π_3 had to vanish because $\mathbf{S}\mathcal{S}$, being merely a 2-categorical object, has only 2- and lower-form symmetries. As in the previous sections, $m' := m \otimes c$. The extension $\pi_2[2].\pi_1[1]$ is the same as for $\mathrm{Aut}(\mathcal{S})$: $\beta_{c,c} = e$. In other words:

LEMMA 3.2.4. *The canonical map* $\mathrm{Aut}(\mathcal{S}) \to \mathrm{Aut}_{\mathbf{SVec}}(\mathbf{S}\mathcal{S})$ *is an isomorphism on 1- and 2-form symmetries.* □

Equation (29) describes the "**SVec**-linear" 1-form and 2-form symmetries of $\mathbf{S}\mathcal{S}$. (There are also 0-form symmetries that I did not describe.) But Galois descent data is by definition nonlinear: in addition to $\mathrm{Aut}_{\mathbf{SVec}}(\mathbf{S}\mathcal{S})$, we also need to know $\mathrm{Aut}_{\mathbf{Vec}}(\mathbf{S}\mathcal{S})$. The trick is that any automorphism restricts to an automorphism of the centre $\Sigma\mathbf{SVec}$, and so there is a fibre sequence

$$\begin{aligned}(30)\qquad &\mathrm{B}\,\mathrm{Aut}_{\mathbf{SVec}}(\mathbf{S}\mathcal{S}) \to \mathrm{B}\,\mathrm{Aut}_{\mathbf{Vec}}(\mathbf{S}\mathcal{S})\\ &\qquad \to \mathrm{B}\,\mathrm{Aut}_{\mathbf{Vec}}(\mathbf{SVec}) = \mathrm{B}\,\mathrm{Gal}(\mathbf{SVec}/\mathbf{Vec}) = \mathbb{Z}_2^F[2].\end{aligned}$$

Galois descent data, of course, consists of a choice of splitting of the extension (30). But the existence of the bosonic form $\mathcal{S}$ means that this map splits, and moreover Lemma 3.2.4 means that it splits as a product (and not just a semidirect product). In other words:

COROLLARY 3.2.5. *The natural map* $\mathrm{Gal}(\mathbf{SVec}/\mathbf{Vec}) \times \mathrm{Aut}(\mathcal{S}) \to \mathrm{Aut}_{\mathbf{Vec}}(\mathbf{S}\mathcal{S})$ *is an isomorphism on 1- and 2-form symmetries.* □

Since there are no nontrivial 0-form symmetries of **SVec**, this implies:

COROLLARY 3.2.6. *The possible Galois descent data are precisely the orbits of the* $\pi_0\,\mathrm{Aut}_{\mathbf{Vec}}(\mathbf{S}\mathcal{S})$*-action on*

$$\pi_0\,\mathrm{maps}\big(\mathbb{Z}_2^F[2],\ \underbrace{\{1,e\}[3].\{\mathbb{1},c,m,m'\}[2]}_{\textit{connective cover of }\mathrm{B}\,\mathrm{Aut}(\mathbf{S}\mathcal{S})}\ \big). \quad \square$$

PROOF OF THE MAIN THEOREM. A map $\mathbb{Z}_2^F[2] \to \{1,e\}[3].\{\mathbb{1},c,m,m'\}[2]$ consists of: a homomorphism $\mathbb{Z}_2^F \to \{\mathbb{1},c,m,m'\}$, together with a trivialization of the corresponding class in $\mathrm{H}^4(\mathbb{Z}_2^F[2];\{1,e\}) \cong \mathbb{Z}_2$. If the nontrivial element of $\mathbb{Z}_2^F$ maps to $x \in \{\mathbb{1},c,m,m'\}$, then the class that needs to be trivialized is $\beta_{x,x} \in \{1,e\}$, thought of as an element in $\mathrm{H}^4(\mathbb{Z}_2^F[2];\{1,e\}) \cong \{1,e\}$. For $x = c$, there is no choice of trivialization (10), but there is for $x \in \{1,m,m'\}$, in which case there are $\mathrm{H}^3(\mathbb{Z}_2^F[2];\{1,e\}) \cong \mathbb{Z}_2$ many choices for the trivialization.

This seems to provide $\{1,m,m'\} \times \mathrm{H}^3(\mathbb{Z}_2^F[2];\{1,e\}) = 3\times 2 = 6$ options for the Galois descent data. But note that the data must be considered up to the action of $\pi_0\,\mathrm{Aut}_{\mathbf{Vec}}(\mathbf{S}\mathcal{S}) \supset \pi_0\,\mathrm{Aut}(\mathcal{S}) \cong \mathbb{Z}_{16}$. As explained in Proposition 2.3.10, the odd elements in this $\mathbb{Z}_{16}$ exchange the two magnetic strings m and m'. Moreover, the different choices in $\mathrm{H}^3(\mathbb{Z}_2^F[2];\{1,e\})$ are exchanged if the map $\mathbb{Z}_2^F \to \{\mathbb{1},c,m,m'\}$ selects one of the magnetic strings. (This is essentially the same as Lemma 2.1.4.)

At the end of the day, there are three choices for the Galois descent data:

- The trivial map $\mathbb{Z}_2^F \to \{\mathbb{1},c,m,m'\}$, and the trivial trivialization in $\mathrm{H}^3(\mathbb{Z}_2^F[2];\{1,e\})$, recovers the topological order $\mathcal{S}$.
- The two nontrivial maps $\mathbb{Z}_2^F \to \{\mathbb{1},c,m,m'\}$, and the two choices of trivialization, are all isomorphic. The corresponding bosonic topological $\mathcal{B}$ order is the fixed-points of the $\mathbb{Z}_2^F[1]$-action on $\mathbf{S}\mathcal{S}$. This action assigns a nontrivial charge to the central fermion f, and, since it uses the magnetic string, it also assigns a nontrivial charge to the non-central fermion e. The only unscreened particle is the boson ef, and so $\Omega\mathcal{B} \cong \mathbf{Rep}(\mathbb{Z}_2)$ rather than $\mathbf{Rep}(\mathbb{Z}_2,\epsilon)$. In other words, $\mathcal{B} \cong \mathcal{R}$.
- The last option is the trivial map $\mathbb{Z}_2^F \to \{\mathbb{1},c,m,m'\}$, but with the nontrivial trivialization in $\mathrm{H}^3(\mathbb{Z}_2^F[2];\{1,e\})$. It screens the invisible fermion f and the boson ef, leaving the fermion e unscreened, and so the fixed topological order $\mathcal{B}$ has $\Omega\mathcal{B} \cong \mathbf{Rep}(\mathbb{Z}_2,\epsilon)$ rather than $\mathbf{Rep}(\mathbb{Z}_2)$.

This truly is a different topological order $\mathcal{T}$ than the fixed points $\mathcal{S}$ for the "trivial" Galois descent data. (Of course, $\mathcal{T}$ must be producible from some Galois descent data, and $\mathcal{T} \not\cong \mathcal{S}$ by the discussion after (12), (13).) To see the difference in terms of Galois descent data, note first that the 2-form symmetry (of $\mathbf{S}\mathcal{S}$) sourced by e has a nontrivial 't Hooft anomaly $(-1)^{\mathrm{Sq}^2 E} \in \mathrm{H}^5(\mathbb{Z}_2[3]; \mathbb{C}^\times)$, exactly because e is a fermion rather than a boson. (The notation is as in §2.2.) This anomaly restricts along the nontrivial map $\mathrm{Sq}^1 \in \mathrm{H}^3(\mathbb{Z}_2^F[2]; \mathbb{Z}_2) = \mathrm{maps}(\mathbb{Z}_2[2], \mathbb{Z}_2[3])$ to $(-1)^{\mathrm{Sq}^2\mathrm{Sq}^1 F} \neq 1 \in \mathrm{H}^5(\mathbb{Z}_2^F[2]; \mathbb{C}^\times)$. In other words, the Galois action that selects $\mathcal{T}$ is anomalous.

This nontrivial anomaly of the Galois action manifests directly in $\mathcal{T}$ as the nontrivial anomaly of the magnetic string m described in Remarks 2.2.3 and 2.3.4. □

REMARK 3.2.7. Anomalies of Galois actions are familiar to physicists. Indeed, time-reversing symmetries are $\mathbb{C}$-antilinear: $\mathbb{Z}_2^T = \mathrm{Gal}(\mathbb{C}/\mathbb{R})$, and anomalies of $\mathbb{Z}_2^T$-symmetries are one of the basic ingredients in topological insulators (see e.g. [**WS13**]).

Mathematicians are also familiar with anomalies of Galois actions. For example, for any finite-dimensional central simple algebra A over a field $\mathbb{K}$, there is a finite-degree field extension $\mathbb{K} \to \mathbb{L}$ and an isomorphism $A \otimes_{\mathbb{K}} \mathbb{L} \cong \mathrm{Mat}(n, \mathbb{L})$, the algebra of $n \times n$ matrices over $\mathbb{L}$. Since $\mathrm{Aut}_{\mathbb{L}}(\mathrm{Mat}(n, \mathbb{L})) = \mathrm{PGL}(n, \mathbb{L})$, A is determined by a Galois descent datum in the twisted cohomology group $\mathrm{H}^1(\mathrm{BGal}(\mathbb{L}/\mathbb{K}); \mathrm{PGL}(n, \mathbb{L}))$. The *anomaly* of this Galois symmetry is the class of A in the Brauer group $\mathrm{Br}(\mathbb{K})$. Explicitly, there is a "universal anomaly" in $\mathrm{H}^2(\mathrm{BPGL}(n, \mathbb{L}); \mathbb{L}^\times)$ classifying the extension $\mathrm{GL}(n, \mathbb{L}) = \mathbb{L}^\times.\mathrm{PGL}(n, \mathbb{L})$. Restricting this universal anomaly along the Galois descent datum produces a class in $\mathrm{H}^2(\mathrm{BGal}(\mathbb{L}/\mathbb{K}); \mathbb{L}^\times)$, and $\mathrm{Br}(\mathbb{K})$ is the limit of these groups as $\mathbb{L}$ approaches the separable closure of $\mathbb{K}$.

3.3. Third proof of the Main Theorem: a long exact sequence. The last proof that I will give of the Main Theorem relies on the classification result [**JF22**, Corollary V.5]. This result extends (and, as mentioned in Remark 1.1.1, corrects) the classifications announced in [**LKW18, LW19**], the first of which I reviewed in Remark 3.0.1.

I will state this classification as Theorem 3.3.8. The statement will require a bit of notation. Much of what I will say below reproduces material from [**JF22**], and in particular Remark V.2 therein.

As mentioned at the end of §1.2, the symmetric fusion category **Vec** has an infinite sequence of suspensions $\Sigma^\bullet\mathbf{Vec}$ constructed in [**GJF19a**]. Another name for the fusion n-category $\Sigma^{n-1}\mathbf{Vec}$ is "$n\mathbf{Vec}$" [**KZ20**]; by convention, $0\mathbf{Vec} = \Sigma^{-1}\mathbf{Vec} = \Omega\mathbf{Vec} = \mathbb{C}$. These categories fit together as a categorical version of a loop spectrum: $\Omega\Sigma^\bullet\mathbf{Vec} = \Sigma^{\bullet-1}\mathbf{Vec}$. In particular, define $\underline{\mathcal{W}}^n := (n\mathbf{Vec})^\times = (\Sigma^{n-1}\mathbf{Vec})^\times$. Then the spaces $\underline{\mathcal{W}}^\bullet$ fit together into a loop spectrum in the usual sense. I will use the same name for the corresponding generalized cohomology theory: for any space X, I will write

$$\underline{\mathcal{W}}^\bullet(X) := \pi_0 \,\mathrm{maps}(X, \underline{\mathcal{W}}^\bullet).$$

Here and throughout, I will use cohomological degree conventions. The *homotopy groups* of a spectrum use the opposite conventions:

$$\pi_n \underline{\mathcal{W}} := \underline{\mathcal{W}}^{-n}(\mathrm{pt}) = \pi_0(\underline{\mathcal{W}}^{-n}).$$

REMARK 3.3.1. Note that $\underline{\mathcal{W}}^0 = \mathbb{C}^\times$ (with the discrete topology), and so $\underline{\mathcal{W}}^\bullet$ is coconnective: its positive-homotopical-degree homotopy groups (the negative cohomology of a point) vanish. This allows for many simplifications. For example, coconnectivity implies that all Atiyah–Hirzebruch spectral sequences for $\underline{\mathcal{W}}$-cohomology converge strongly [**Boa99**]. It also implies that Borel-equivariant cohomology, i.e. the cohomology of the classifying spaces, is genuinely equivariant. Finally, coconnectivity allows all the subtleties of Brown's representability theorem to be ignored: when working with non-coconnective spectra, there can be nontrivial maps which induce the zero map on cohomology theories, but these do not occur for coconnective spectra. As such, I will freely use the terms "generalized cohomology theory" and "spectrum" interchangeably.

Every object of $\Sigma^n\mathbf{Vec}$ is $\Sigma\mathcal{A}$ for some fusion $(n-1)$-category $\mathcal{A}$, defined up to Morita equivalence of fusion $(n-1)$-categories. (With "fusion" replaced by "multifusion," this is essentially true by construction; see e.g. [**KZ20**, Theorem 3.21]. That one can always select $\mathcal{A}$ to be fusion requires that $\mathbb{C}$ is algebraically closed: one needs the fact that the commutative algebra $\Omega^{n-1}\mathcal{A}$ admits a map to $\mathbb{C}$.) An object in $\Sigma^n\mathbf{Vec}$ is invertible if and only if the corresponding fusion $(n-1)$-category $\mathcal{A}$ has trivial Drinfel'd centre [**JF22**, Theorem 2], in which case it is $\mathcal{A} = \Sigma\mathcal{B}$ for a nondegenerate braided fusion n-category $\mathcal{B} = \Omega\mathcal{A}$ [**JF22**, Theorem 4]. In other words:

$$\begin{aligned}\pi_0\underline{\mathcal{W}}^n &= \frac{\{\text{invertible fusion } (n-2)\text{-categories}\}}{\text{Morita equivalence}} \\ &= \frac{\{\text{nondeg braided fusion } (n-3)\text{-categories}\}}{\text{Morita equivalence}}.\end{aligned}$$

In particular, $\pi_0\underline{\mathcal{W}}^4 = \mathcal{W}$ is the "bosonic Witt group" of nondegenerate braided fusion 1-categories studied in [**DMNO13**]. This is why I chose the letter "$\mathcal{W}$" for the spectrum.

LEMMA 3.3.2. *The low-degree homotopy groups of* $\underline{\mathcal{W}}$ *are:*

(0) $\underline{\mathcal{W}}^0(\mathrm{pt}) = \mathbb{C}^\times$.
(1) $\underline{\mathcal{W}}^1(\mathrm{pt}) = 0$, *because every* $\otimes$*-invertible vector space is one-dimensional.*
(2) $\underline{\mathcal{W}}^2(\mathrm{pt}) = 0$, *because every central simple algebra over* $\mathbb{C}$ *is Morita-trivial.*
(3) $\underline{\mathcal{W}}^3(\mathrm{pt}) = 0$, *because a braided fusion* 0*-category is simply a commutative algebra over* $\mathbb{C}$*, and the only one with trivial centre is* $\mathbb{C}$ *itself.*
(4) $\underline{\mathcal{W}}^4(\mathrm{pt}) = \mathcal{W}$ *is the bosonic Witt group of* [**DMNO13**]. □

There is also a "super" version $\underline{S\mathcal{W}}$ of the spectrum $\underline{\mathcal{W}}$:

$$\underline{S\mathcal{W}}^\bullet := (\Sigma^{n-1}\mathbf{SVec})^\times.$$

Its homotopy groups measure the super topological orders up to Morita equivalence. In particular:

LEMMA 3.3.3. (0) $\underline{\mathcal{SW}}^0(\mathrm{pt}) = \mathbb{C}^\times$.
(1) $\underline{\mathcal{SW}}^1(\mathrm{pt}) = \pi_0\mathbf{SVec}_{\mathbb{C}}^\times = \{\mathbb{C}^{1|0}, \mathbb{C}^{0|1}\} \cong \mathbb{Z}_2$. *Every $\otimes$-invertible super vector space is one-dimensional, but in the super world, that one dimension can be either bosonic or fermionic.*
(2) $\underline{\mathcal{SW}}^2(\mathrm{pt}) = \pi_0\mathbf{SAlg}_{\mathbb{C}}^\times = \{\mathbb{C}, \mathrm{Cliff}(1)\} \cong \mathbb{Z}_2$. *Every central simple superalgebra over $\mathbb{C}$ is isomorphic either to a super matrix algebra (Morita trivial) or to an odd Clifford algebra (Morita-equivalent to* Cliff(1)*).*
(3) $\underline{\mathcal{SW}}^3(\mathrm{pt}) = 0$. *The bosonic proof applies without change.*
(4) $\underline{\mathcal{SW}}^4(\mathrm{pt}) = \mathcal{SW}$ *is the "fermionic Witt group" of* [**DNO13**]. □

Furthermore, Theorem 3.2.1 implies:

LEMMA 3.3.4. *Every 3+1D super topological order is Morita trivial, i.e.* $\underline{\mathcal{SW}}^5(\mathrm{pt}) = 0$. □

REMARK 3.3.5. The extended supercohomology $\mathrm{SH}^\bullet$ used in Remark 2.2.5 is the connective cover of $\underline{\mathcal{SW}}^\bullet$ in which one only uses the first four homotopy groups. In other words, for any space X there is a map $\mathrm{SH}^\bullet(X) \to \underline{\mathcal{SW}}^\bullet(X)$ which is an isomorphism in degrees $\bullet < 4$ and an injection in degree $\bullet = 4$.

Since $\underline{\mathcal{SW}}$ was built from $\mathbf{SVec}$, it inherits an action by the (purely 1-form) categorified Galois group $\mathrm{Gal}(\mathbf{SVec}/\mathbf{Vec}) = \mathrm{Aut}_{\mathbf{Vec}}(\mathbf{SVec}) = \mathbb{Z}_2^F[1]$. Categorified Galois descent then says that $\underline{\mathcal{W}}^\bullet$ is nothing but the fixed-point spectrum of the action of $\mathbb{Z}_2^F[1]$ on $\underline{\mathcal{SW}}^\bullet$. In particular, the $\underline{\mathcal{W}}^\bullet$-cohomology of a point is precisely the *twisted* $\underline{\mathcal{SW}}^\bullet$-cohomology of the classifying space $\mathbb{Z}_2^F[2] = \mathrm{pt}/\mathbb{Z}_2^F[1]$ of $\mathrm{Gal}(\mathbf{SVec}/\mathbf{Vec})$. Consistent with Remark 2.2.5, I will write simply $\underline{\mathcal{SW}}^\bullet(\mathbb{Z}_2^F[2])$ for this twisted cohomology:

$$\underline{\mathcal{W}}^\bullet(\mathrm{pt}) = \underline{\mathcal{SW}}^\bullet(\mathbb{Z}_2^F[2]). \tag{31}$$

More generally, given a space X with a $\mathbb{Z}_2^F[1]$-action, I will write $X/\mathbb{Z}_2^F[1]$ for the corresponding homotopy quotient, and $\underline{\mathcal{SW}}^\bullet(X/\mathbb{Z}_2^F[1])$ for its twisted cohomology.

There is an Atiyah–Hirzebruch spectral sequence computing the right-hand side $\underline{\mathcal{SW}}^\bullet(\mathbb{Z}_2^F[2])$ of (31):

$$\mathrm{H}^i(\mathbb{Z}_2^F[2]; \underline{\mathcal{SW}}^j(\mathrm{pt})) \Rightarrow \underline{\mathcal{SW}}^{i+j}(\mathbb{Z}_2^F[2]). \tag{32}$$

Note that $\mathbb{Z}_2^F[1]$ acts trivially on the 0-form groups $\mathrm{SH}^j(\mathrm{pt})$. If the cohomology were untwisted, then the d_2 differentials would be simply the Postnikov k-invariants for $\underline{\mathcal{SW}}$, which in the bottom degrees are:

$$(-1)^{\mathrm{Sq}^2} : \underline{\mathcal{SW}}^1(\mathrm{pt}) \to \underline{\mathcal{SW}}^0(\mathrm{pt}), \qquad \mathrm{Sq}^2 : \underline{\mathcal{SW}}^2(\mathrm{pt}) \to \underline{\mathcal{SW}}^1(\mathrm{pt}).$$

The twisting, however, modifies these. Write $T \in \mathrm{H}^2(\mathbb{Z}_2^F[2]; \mathbb{Z}_2)$ for the degree-2 generator; the whole cohomology ring $\mathrm{H}^\bullet(\mathbb{Z}_2[2]; \mathbb{Z}_2)$ is freely generated by T under the Steenrod-algebra action. Then the twisted d_2 differentials for the spectral sequence (32) are:

$$\mathrm{d}_2 = (-1)^{T+\mathrm{Sq}^2} : \underline{\mathcal{SW}}^1(\mathrm{pt}) \to \underline{\mathcal{SW}}^0(\mathrm{pt}), \qquad \mathrm{d}_2 = T + \mathrm{Sq}^2 : \underline{\mathcal{SW}}^2(\mathrm{pt}) \to \underline{\mathcal{SW}}^1(\mathrm{pt}).$$

Thus the E_2 page with nonzero d_2 differentials drawn in reads:

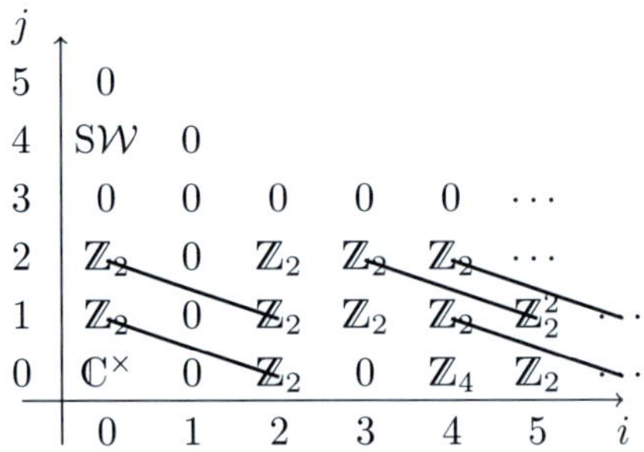

The differentials supported in bidegrees $(i,j) = (0,1)$ and $(0,2)$ are forced by Lemma 3.3.2, which says that in the limit there can be no cohomology of total degree $i + j \in \{1, 2, 3\}$. Furthermore, in degree $i + j = 4$ the limit of this spectral sequence is the bosonic Witt group $\mathcal{W}$, and the restriction to bidegree $(i, j) = (0, 4)$ is the canonical map $\mathcal{W} \to \mathrm{S}\mathcal{W}$. This map is known to have kernel of order 16 [**DNO13**, Propositions 5.13 and 5.14], and so:

LEMMA 3.3.6. *All of the* $\mathbb{Z}_2$*s and* $\mathbb{Z}_4$ *in total degree* 4 *survive to the* E_∞ *page.* □

REMARK 3.3.7. The d_4 differential automatically vanishes. The only other potential differential in this range of degrees is $\mathrm{d}_5 : \mathrm{S}\mathcal{W} = E_5^{0,4} \to \mathbb{Z}_2 = E_5^{5,0}$. Either it vanishes, in which case $\underline{\mathcal{W}}^5(\mathrm{pt}) = \mathbb{Z}_2$ and $\mathcal{W} \to \mathrm{S}\mathcal{W}$ is a surjection, or it does not vanish, in which case $\underline{\mathcal{W}}^5(\mathrm{pt}) = 0$ and $\mathrm{coker}(\mathcal{W} \to \mathrm{S}\mathcal{W}) \cong \mathbb{Z}_2$.

I do not know which case occurs, but the good money is on the former. This is because, as pointed out by Dmitri Nikshych (see [**JF22**, Remark V.2]), the surjectivity of the map $\mathcal{W} \to \mathrm{S}\mathcal{W}$ is equivalent to the conjecture that every slightly-degenerate braided fusion 1-category admits a minimal modular extension, and most experts believe that conjecture.

I will need one more notion in order to give the complete classification from [**JF22**]. For any space X, there is a canonical map $\pi : X \to \mathrm{pt}$, and so for any cohomology theory E a restriction map $\pi^* : E^\bullet(\mathrm{pt}) \to E^\bullet(X)$. The *reduced* $E^\bullet$-cohomology of X, denoted $\widetilde{E}^\bullet(X)$, is by definition the homotopy cokernel of this map: a cocycle for a class in $\widetilde{E}^n(X)$ is a pair (α, β), where α is a cocycle for a class in $\widetilde{E}^{n+1}(\mathrm{pt})$, and β is a degree-n cochain on X solving $\mathrm{d}\beta = \pi^*\alpha$. Fix a point $x \in X$. This choice provides a restriction $E^\bullet(X) \to E^\bullet(\{x\})$, and the composition $E^\bullet(\mathrm{pt}) \to E^\bullet(X) \to E^\bullet(\{x\})$ is the identity. As a result, $E^\bullet(\mathrm{pt})$ is a direct summand of $E^\bullet(X)$, and is simply the other direct summand:

$$\widetilde{E}^\bullet(X) = E^\bullet(X) \ominus E^\bullet(\mathrm{pt}). \tag{33}$$

Now suppose that $E^\bullet$ is a twisted $\mathcal{G}$-equivariant cohomology theory, for instance $E^\bullet = \underline{\mathrm{S}\mathcal{W}}^\bullet$ and $\mathcal{G} = \mathbb{Z}_2^F[1]$. Then there is still a *reduced* twisted-equivariant $E^\bullet$-cohomology of any $\mathcal{G}$-space X. As above, let $E^\bullet(X/\mathcal{G})$ denote the unreduced twisted $\mathcal{G}$-equivariant cohomology of X. The reduced cohomology $\widetilde{E}^\bullet(X/\mathcal{G})$ is defined to be the homotopy cokernel of the map $\pi^* : E^\bullet(\mathrm{pt}/\mathcal{G}) \to E^\bullet(X/\mathcal{G})$ that restricts along the canonical map $X/\mathcal{G} \to \mathrm{pt}/\mathcal{G}$ (i.e. the canonical G-equivariant map $X \to \mathrm{pt}$).

If X contains a $\mathcal{G}$-fixed point $x \in X$ (equivalently, if the map $X/\mathcal{G} \to \mathrm{pt}/\mathcal{G}$ splits), then the discussion leading to equation (33) describes $\widetilde{E}^\bullet(X/\mathcal{G})$ as a direct summand of $E^\bullet(X/\mathcal{G})$. But in general, there are no such fixed points. Indeed, $\pi^* : E^\bullet(\mathrm{pt}/\mathcal{G}) \to E^\bullet(X/\mathcal{G})$ can fail to be an injection. Rather, in place of equation (33), there is a long exact sequence:

(34)
$$\cdots \to E^\bullet(\mathrm{pt}/\mathcal{G}) \to E^\bullet(X/\mathcal{G}) \to \widetilde{E}^\bullet(X/\mathcal{G}) \to E^{\bullet+1}(\mathrm{pt}/\mathcal{G}) \to E^{\bullet+1}(X/\mathcal{G}) \to \dots$$

With these concepts in place, I can now state the promised generalization [**JF22**, Corollary V.5] of the classification described in Remark 3.0.1:

THEOREM 3.3.8. *Let G be a finite group. An action of $\mathbb{Z}_2^F[1]$ on the classifying space $\mathrm{B}G$ is the same data as a central element $\epsilon \in Z(G)$ such that $\epsilon^2 = 1$; the action is nontrivial if $\epsilon \neq 1 \in G$, in which case BG contains no $\mathbb{Z}_2^F[1]$-fixed points. Let $\mathbf{Rep}(G,\epsilon)$ denote the corresponding (bosonic) symmetric fusion category: it consists of superrepresentations of G on which ϵ acts by $(-1)^f$. Then the 3+1D topological orders $\mathcal{B}$ with line operators $\Omega\mathcal{B} \cong \mathbf{Rep}(G,\epsilon)$ are classified by $\widetilde{\underline{\mathcal{SW}}}^4(\mathrm{B}G/\mathbb{Z}_2^F[1])$, the reduced $\mathbb{Z}_2^F[1]$-twisted-equivariant degree-4 $\underline{\mathcal{SW}}$-cohomology of $\mathrm{B}G$.* □

Note that, by [**Del02**], every symmetric fusion category is equivalent to some $\mathbf{Rep}(G,\epsilon)$. The proof of Theorem 3.3.8 essentially amounts to a careful study of how categorified Galois descent interacts with Theorem 3.2.1, which described the classification of super topological orders. When $\epsilon = 1$, one can choose a $\mathbb{Z}_2^F[1]$-fixed point in $\mathrm{B}G$, and $\underline{\mathcal{SW}}^\bullet(\mathrm{B}G/X) = \underline{\mathcal{W}}^\bullet(\mathrm{B}G)$, and so Theorem 3.3.8 reduces to the bosonic classification described in Remark 3.0.1.

PROOF OF THE MAIN THEOREM. I will apply Theorem 3.3.8 in the case when $\mathbf{Rep}(G,\epsilon) = \mathbf{Rep}(\mathbb{Z}_2,\epsilon) = \mathbf{SVec}$; I will write simply "want" for the desired group $\widetilde{\underline{\mathcal{SW}}}^4(\mathrm{B}\mathbb{Z}_2/\mathbb{Z}_2^F[1])$.

The action of $\mathbb{Z}_2^F[1]$ on $\mathrm{B}G = \mathrm{B}\mathbb{Z}_2 = \mathbb{Z}_2[1]$ corresponding to the nontrivial ϵ is the canonical one "by multiplication," and so the quotient is $\mathrm{B}G/\mathbb{Z}_2^F[1] = \mathrm{pt}$. Set $E^\bullet = \underline{\mathcal{SW}}^\bullet$ and $X = \mathrm{B}G$ and $\mathcal{G} = \mathbb{Z}_2^F[1]$. Then the entries $E^\bullet(\mathrm{pt}/\mathcal{G})$ and $E^\bullet(X/\mathcal{G})$ in the long exact sequence (34) are

$$E^\bullet(\mathrm{pt}/\mathcal{G}) = \underline{\mathcal{SW}}^\bullet(\mathbb{Z}_2^F[2]) = \underline{\mathcal{W}}^\bullet(\mathrm{pt}), \qquad E^\bullet(X/\mathcal{G}) = \underline{\mathcal{SW}}^\bullet(\mathrm{pt}).$$

Thus the long exact sequence reads:

$$\cdots \to \underline{\mathcal{W}}^4(\mathrm{pt}) \to \underline{\mathcal{SW}}^4(\mathrm{pt}) \to \text{want} \to \underline{\mathcal{W}}^5(\mathrm{pt}) \to \underline{\mathcal{SW}}^5(\mathrm{pt}) \to \dots$$

Filling in the known values from Lemmas 3.3.2, 3.3.3, and 3.3.4 gives:

$$\cdots \to \mathcal{W} \to \mathcal{SW} \to \text{want} \to \underline{\mathcal{W}}^5(\mathrm{pt}) \to 0 \to \dots$$

Finally, Remark 3.3.7 says that either $\mathrm{coker}(\mathcal{W} \to \mathcal{SW}) = \mathbb{Z}_2$ and $\underline{\mathcal{W}}^5 = 0$, or $\mathrm{coker}(\mathcal{W} \to \mathcal{SW}) = 0$ and $\underline{\mathcal{W}}^5 = \mathbb{Z}_2$. In either case, the desired group "want" is isomorphic to $\mathbb{Z}_2$. □

REMARK 3.3.9. This proof makes clear that if $\pi_0\underline{\mathcal{W}}^5 = \pi_0(5\mathbf{Vec})^\times$ is nontrivial, then $\mathcal{T}$ represents the nontrivial class; contrapositively, if $\mathcal{T}$ is a Drinfel'd centre, then so is every bosonic 3+1D topological order. Translating through Remarks 1.1.2 and 3.3.7, one finds that the minimal modularity conjecture for slightly-degenerate fusion 1-categories is equivalent to the statement that there exists a Morita-nontrivial 3+1D topological order.

Actually, that fact that there is at most one nontrivial class in $\pi_0\underline{\mathcal{W}}^5$, and that if it exists it is represented by $\mathcal{T}$, follows directly from the construction in

[**LW19**] together with the Main Theorem from this paper. Indeed, [**LW19**] gives an algorithm to condense any bosonic 3+1D topological order $\mathcal{B}$ either directly to the vacuum (in which case $\mathcal{B}$ was Morita trivial) or to a topological order with a single emergent fermion and nothing else. Since, by the Main Theorem, there are only two topological orders with this particle content, one of which ($\mathcal{S} \cong \mathcal{Z}_{(1)}(\Sigma\mathbf{SVec})$) is Morita-trivial, one finds that if a topological order $\mathcal{B}$ is not Morita trivial, then it must be Morita equivalent to $\mathcal{T}$.

Acknowledgments

I thank Davide Gaiotto, Tian Lan, Dmitri Nykshych, David Reutter, Chong Wang, Matthew Yu, and Hao Zheng for conversations related to this project.

References

[AdR20] M. F. Araujo de Resende, *A pedagogical overview on 2D and 3D toric codes and the origin of their topological orders*, Rev. Math. Phys. **32** (2020), no. 2, 2030002, 32, DOI 10.1142/S0129055X20300022. MR4073072

[BJSS20] Adrien Brochier, David Jordan, Pavel Safronov, and Noah Snyder, *Invertible braided tensor categories*, Algebr. Geom. Topol. **21** (2021), no. 4, 2107–2140, DOI 10.2140/agt.2021.21.2107. MR4302495

[Boa99] J. Michael Boardman, *Conditionally convergent spectral sequences*, Homotopy invariant algebraic structures (Baltimore, MD, 1998), Contemp. Math., vol. 239, Amer. Math. Soc., Providence, RI, 1999, pp. 49–84, DOI 10.1090/conm/239/03597. MR1718076

[Déc21] Thibault D. Décoppet, *Compact semisimple 2-categories*, Trans. Amer. Math. Soc. **376** (2023), no. 12, 8309–8336, DOI 10.1090/tran/9044. MR4669298

[Del02] P. Deligne, *Catégories tensorielles* (French, with English summary), Mosc. Math. J. **2** (2002), no. 2, 227–248, DOI 10.17323/1609-4514-2002-2-2-227-248. Dedicated to Yuri I. Manin on the occasion of his 65th birthday. MR1944506

[DG20] Diego Delmastro and Jaume Gomis, *Domain walls in 4d $\mathcal{N} = 1$ SYM*, J. High Energy Phys. **3** (2021), Paper No. 259, 52, DOI 10.1007/jhep03(2021)259. MR4260969

[DMNO13] Alexei Davydov, Michael Müger, Dmitri Nikshych, and Victor Ostrik, *The Witt group of non-degenerate braided fusion categories*, J. Reine Angew. Math. **677** (2013), 135–177, DOI 10.1515/crelle.2012.014. MR3039775

[DN20] Alexei Davydov and Dmitri Nikshych, *Braided Picard groups and graded extensions of braided tensor categories*, Selecta Math. (N.S.) **27** (2021), no. 4, Paper No. 65, 87, DOI 10.1007/s00029-021-00670-1. MR4281262

[DNO13] Alexei Davydov, Dmitri Nikshych, and Victor Ostrik, *On the structure of the Witt group of braided fusion categories*, Selecta Math. (N.S.) **19** (2013), no. 1, 237–269, DOI 10.1007/s00029-012-0093-3. MR3022755

[DR18] Christopher L. Douglas and David J. Reutter, *Fusion 2-categories and a state-sum invariant for 4-manifolds*, `arXiv:1812.11933`, 2018.

[DSPS20] Christopher L. Douglas, Christopher Schommer-Pries, and Noah Snyder, *Dualizable tensor categories*, Mem. Amer. Math. Soc. **268** (2020), no. 1308, vii+88, DOI 10.1090/memo/1308. MR4254952

[EGNO15] Pavel Etingof, Shlomo Gelaki, Dmitri Nikshych, and Victor Ostrik, *Tensor categories*, Mathematical Surveys and Monographs, vol. 205, American Mathematical Society, Providence, RI, 2015, DOI 10.1090/surv/205. MR3242743

[EN17] Dominic V. Else and Chetan Nayak, *Cheshire charge in (3+1)-dimensional topological phases*, Phys. Rev. B, **96**(4):045136, 2017.

[GJF19a] Davide Gaiotto and Theo Johnson-Freyd. *Condensations in higher categories*, `arXiv:1905.09566` 2019.

[GJF19b] Davide Gaiotto and Theo Johnson-Freyd, *Symmetry protected topological phases and generalized cohomology*, J. High Energy Phys. **5** (2019), 007, 34, DOI 10.1007/jhep05(2019)007. MR3978827

[GK16] Davide Gaiotto and Anton Kapustin, *Spin TQFTs and fermionic phases of matter*, Internat. J. Modern Phys. A, **31**(28n29):1645044, 2016.

[GKSW15] Davide Gaiotto, Anton Kapustin, Nathan Seiberg, and Brian Willett, *Generalized global symmetries*, J. High Energy Phys. **2** (2015), 172, front matter+61, DOI 10.1007/JHEP02(2015)172. MR3321281

[GS18] Owen Gwilliam and Claudia Scheimbauer, *Duals and adjoints in higher Morita categories*, `arXiv:1804.10924`, 2018.

[HZW05] Alioscia Hamma, Paolo Zanardi, and Xiao-Gang Wen, *String and membrane condensation on 3d lattices*, Phys. Rev. B, **72**(3):035307, 2005.

[JF17] Theo Johnson-Freyd, *Spin, statistics, orientations, unitarity*, Algebr. Geom. Topol. **17** (2017), no. 2, 917–956, DOI 10.2140/agt.2017.17.917. MR3623677

[JF20] Theo Johnson-Freyd, *Heisenberg-picture quantum field theory*, Representation theory, mathematical physics, and integrable systems, Progr. Math., vol. 340, Birkhäuser/Springer, Cham, [2021] ©2021, pp. 371–409, DOI 10.1007/978-3-030-78148-4_13. MR4391354

[JF22] Theo Johnson-Freyd, *On the classification of topological orders*, Comm. Math. Phys. **393** (2022), no. 2, 989–1033, DOI 10.1007/s00220-022-04380-3. MR4444089

[JFR23] Theo Johnson-Freyd and David Reutter, *Minimal nondegenerate extensions*, J. Amer. Math. Soc. **37** (2024), no. 1, 81–150, DOI 10.1090/jams/1023. MR4654609

[JFS17] Theo Johnson-Freyd and Claudia Scheimbauer, *(Op)lax natural transformations, twisted quantum field theories, and "even higher" Morita categories*, Adv. Math. **307** (2017), 147–223, DOI 10.1016/j.aim.2016.11.014. MR3590516

[JFY21] Theo Johnson-Freyd and Matthew Yu, *Fusion 2-categories with no line operators are grouplike*, Bull. Aust. Math. Soc. **104** (2021), no. 3, 434–442, DOI 10.1017/S0004972721000095. MR4338473

[Kit97] Alexei Kitaev, *Quantum error correction with imperfect gates*, Quantum Communication, Computing, and Measurement (Osamu Hirota, A.S. Holevo, and C.M. Caves, editors) pages 181–188. Springer US.

[Kit06] Alexei Kitaev, *Anyons in an exactly solved model and beyond*, Ann. Physics **321** (2006), no. 1, 2–111, DOI 10.1016/j.aop.2005.10.005. MR2200691

[KLW+20a] Liang Kong, Tian Lan, Xiao-Gang Wen, Zhi-Hao Zhang, and Hao Zheng. *Algebraic higher symmetries and categorical symmetry—a holographic entanglement view of symmetry*, `arXiv:2005.14178`, 2020.

[KLW+20b] Liang Kong, Tian Lan, Xiao-Gang Wen, Zhi-Hao Zhang, and Hao Zheng, *Classification of topological phases with finite internal symmetries in all dimensions*, J. High Energy Phys. **9** (2020), 093, 48, DOI 10.1007/jhep09(2020)093. MR4203166

[KTZ19] Liang Kong, Yin Tian, and Shan Zhou, *The center of monoidal 2-categories in* 3 + 1D *Dijkgraaf-Witten theory*, Adv. Math. **360** (2020), 106928, 25, DOI 10.1016/j.aim.2019.106928. MR4037491

[KTZ20] Liang Kong, Yin Tian, and Zhi-Hao Zhang, *Defects in the 3-dimensional toric code model form a braided fusion 2-category*, `arXiv:2009.06564`, 2020.

[KW14] Liang Kong and Xiao-Gang Wen, *Braided fusion categories, gravitational anomalies, and the mathematical framework for topological orders in any dimensions*, `arXiv:1405.5858`, 2014.

[KWZ15] Liang Kong, Xiao-Gang Wen, and Hao Zheng, *Boundary-bulk relation for topological orders as the functor mapping higher categories to their centers*, `arXiv:1502.01690`, 2015.

[KWZ17] Liang Kong, Xiao-Gang Wen, and Hao Zheng, *Boundary-bulk relation in topological orders*, Nuclear Phys. B, **922** (2017), no. 9, 62–76.

[KZ20] Liang Kong and Hao Zheng, *Categories of quantum liquids I*, J. High Energy Phys. **8** (2022), Paper No. 70, 43, DOI 10.1007/jhep08(2022)070. MR4467269

[Lev13] Michael Levin, *Protected edge modes without symmetry*, Phys. Rev. X, **3**:021009, May 2013.

[LKW17] Tian Lan, Liang Kong, and Xiao-Gang Wen, *Modular extensions of unitary braided fusion categories and* 2 + 1D *topological/SPT orders with symmetries*, Comm. Math. Phys. **351** (2017), no. 2, 709–739, DOI 10.1007/s00220-016-2748-y. MR3613518

[LKW18] Tian Lan, Liang Kong, and Xiao-Gang Wen, *Classification of* $(3+1)$D *bosonic topological orders: The case when pointlike excitations are all bosons*, Phys. Rev. X, **8**:021074, Jun 2018.

[LW19] Tian Lan and Xiao-Gang Wen, *Classification of* $3+1$D *bosonic topological orders (ii): The case when some pointlike excitations are fermions*, *Phys. Rev. X*, **9**:021005, Apr 2019.

[Müg03] Michael Müger, *On the structure of modular categories*, Proc. London Math. Soc. (3) **87** (2003), no. 2, 291–308, DOI 10.1112/S0024611503014187. MR1990929

[Sch14] Claudia Scheimbauer, *Factorization homology as a fully extended topological field theory*. PhD thesis, ETH Zürich, 2014. `https://www.research-collection.ethz.ch/handle/20.500.11850/154981`.

[SP09] Christopher John Schommer-Pries, *The classification of two-dimensional extended topological field theories*, ProQuest LLC, Ann Arbor, MI, 2009. Thesis (Ph.D.)–University of California, Berkeley. MR2713992

[Tho15] Ryan Thorngren, *Framed Wilson operators, fermionic strings, and gravitational anomaly in 4d*, J. High Energy Phys. **2** (2015), 152, front matter+14, DOI 10.1007/JHEP02(2015)152. MR3321301

[Wal64] C. T. C. Wall, *Graded Brauer groups*, J. Reine Angew. Math. **213** (1963/64), 187–199, DOI 10.1515/crll.1964.213.187. MR167498

[Weg71] Franz J. Wegner, *Duality in generalized Ising models and phase transitions without local order parameters*, J. Mathematical Phys. **12** (1971), 2259–2272, DOI 10.1063/1.1665530. MR289087

[WG17] Qing-Rui Wang and Zheng-Cheng Gu, *Towards a complete classification of fermionic symmetry protected topological phases in 3d and a general group supercohomology theory*, `arXiv:1703.10937`, 2017.

[WS13] Chong Wang and T. Senthil, *Boson topological insulators: A window into highly entangled quantum phases*, Phys. Rev. B, **87**(23):235122, 2013.

Perimeter Institute for Theoretical Physics, Waterloo, Ontario; and Department of Mathematics, Dalhousie University, Halifax, Nova Scotia

Email address: `theojf@pitp.ca; theojf@dal.ca`

Contemporary Mathematics
Volume **813**, 2025
https://doi.org/10.1090/conm/813/16289

Grothendieck-Verdier duality in categories of bimodules and weak module functors

Jürgen Fuchs, Gregor Schaumann, Christoph Schweigert, and Simon Wood

Abstract. Various monoidal categories, including suitable representation categories of vertex operator algebras, admit natural Grothendieck-Verdier duality structures. We recall that such a Grothendieck-Verdier category comes with two tensor products which should be related by distributors obeying pentagon identities. We discuss in which circumstances these distributors are isomorphisms. This is achieved by taking the perspective of module categories over monoidal categories, using in particular the natural weak module functor structure of internal Homs and internal coHoms. As an illustration, we exhibit these concepts concretely in the case of categories of bimodules over associative algebras.

1. Introduction

Dualities play a pivotal role in various applications of monoidal categories. A familiar example of a duality is a rigid structure. For many applications, for instance in representation theory and in linear logic, it is, however, necessary to consider more general notions of duality. For example, if a tensor product is not exact, then the category does not admit a rigid duality.

A framework that generalizes rigidity and in which structures familiar from rigidity, such as internal Homs, persist, is furnished by a $*$-autonomous structure, also known as a *Grothendieck-Verdier duality*. Categories endowed with a Grothendieck-Verdier duality have a wide range of applications, from linear logic to the representation theory of vertex operator algebras. In this paper we present some important aspects of Grothendieck-Verdier dualities, concentrating on such which are related to internal Hom and coHom functors. Our point of view is the one of module categories. In particular, we will regard a monoidal category as a module category over itself. Specifically, a crucial ingredient is the structure of internal Homs and coHoms as weak module functors. This allows us to introduce distributors, which play a role similar to the associator for the tensor product.

2020 *Mathematics Subject Classification.* Primary 18M10, 16B50, 81R10.

We thank Chelsea Walton for helpful comments. The first author was supported by VR under project no. 2022-02931. The third author was supported by the Deutsche Forschungsgemeinschaft (DFG, German Research Foundation) under SCHW1162/6-1 and under Germany's Excellence Strategy - EXC 2121 "Quantum Universe" - 390833306. The fourth author was supported by the Engineering and Physical Sciences Research Council (EPSRC) EP/V053787/1 and by the Alexander von Humboldt Foundation.

Before discussing these concepts in appropriate generality, we illustrate some of them in the, arguably, simplest relevant situation – finite-dimensional bimodules over finite-dimensional algebras. After presenting the general theory, we return to this example and present explicit formulas for the distributors in categories of bimodules.

2. Bimodules

To start with, let us consider some very basic mathematical objects which are familiar already to undergraduates. We fix an algebraically closed field $\Bbbk$.

Let A be a (unital associative) $\Bbbk$-algebra. The category of A-bimodules can be endowed with a monoidal structure given by the tensor product $\otimes_A$ over A, which has A as its monoidal unit. The tensor product $\otimes_A$ is defined by the coequalizer diagram

$$B_1 \otimes_\Bbbk A \otimes_\Bbbk B_2 \xrightarrow{\varphi} B_1 \otimes_\Bbbk B_2 \longrightarrow B_1 \otimes_A B_2 \,. \tag{2.1}$$

Here the first arrow is the linear map

$$\varphi = \rho^{B_1} \otimes_\Bbbk \mathrm{id}_{B_2} - \mathrm{id}_{B_1} \otimes_\Bbbk \lambda^{B_2}, \tag{2.2}$$

where ρ^{B_1} is the right A-action on B_1 and λ^{B_2} the left A-action on B_2. Here, as well as largely below, we suppress the associator α of a monoidal category, and likewise we will suppress the unitors l and r.

The tensor product $\otimes_A$ is right exact. Since our primary interest in this note is in duality structures, it is convenient to impose suitable finiteness conditions. Specifically, unless stated otherwise, in the sequel all $\Bbbk$-algebras are assumed to be finite-dimensional, and all modules and bimodules to be finite-dimensional as $\Bbbk$-vector spaces. Then in particular the tensor product is right exact and equips the category A-bimod of A-bimodules with a monoidal structure, and we can make contact to the theory of coalgebras and comodules.

When A is commutative, then every left A-module is canonically a right module and even a bimodule. Accordingly we may then also consider the category A-mod of left A-modules, which is a monoidal subcategory of the category A-bimod of A-bimodules, with A as a left module as the monoidal unit. One motivation to consider commutative algebras in this elementary study comes from the representation theory of vertex operator algebras. Indeed, commutative algebras can be seen as particular examples of vertex operator algebras [**FB**, Sect. 1.3], and finite-dimensional algebras as particular examples of C_2-cofinite vertex operator algebras. Vertex operator algebras that are commutative algebras are known to arise as the result of BRST cohomology or quantum Hamiltonian reduction, see e.g. Section 2.2 of [**LZ**] for an infinite-dimensional example. BRST cohomology can also result in finite-dimensional commutative algebras.[1]

A fundamental operation in linear algebra is to take the linear dual of a vector space. The category $\mathrm{vect}_\Bbbk$ of finite-dimensional vector spaces is *rigid*, that is, every vector space V has both a left dual ${}^\vee V$ and a right dual $V^\vee$ (both of which are isomorphic to $\mathrm{Hom}_\Bbbk(V, \Bbbk)$ since $\mathrm{vect}_\Bbbk$ is symmetric) endowed with evaluation and

[1] We thank Sven Möller for pointing out an (unpublished) example of such a commutative algebra.

coevaluation maps, given by

$$\begin{aligned} \mathrm{ev}\colon V^\vee \otimes_{\Bbbk} V &\longrightarrow \Bbbk\,, & \mathrm{coev}\colon \Bbbk &\longrightarrow V \otimes_{\Bbbk} V^\vee, \\ f \otimes_{\Bbbk} v &\longmapsto f(v)\,, & \xi &\longmapsto \xi \textstyle\sum_i v_i \otimes_{\Bbbk} v^i \end{aligned} \tag{2.3}$$

for the right dual, and analogously for the left dual, where (v_i) is a basis of V and (v^i) the dual basis of $V^\vee$. The notion of rigidity pervades the theory of Hopf algebras and quantum topology. Indeed, a (quasi-)Hopf algebra is essentially an algebra H endowed with additional structure that is precisely such that the category H-mod of finite-dimensional H-modules is rigid monoidal.

Taking the double dual gives a monoidal endofunctor $(-)^{\vee\vee}$ of the monoidal category $\mathrm{vect}_{\Bbbk}$. The double dual $V^{\vee\vee}$ of a finite-dimensional vector space is canonically identified with V. More precisely, the category $\mathrm{vect}_{\Bbbk}$ is pivotal, i.e. there is a monoidal isomorphism $\mathrm{id}_{\mathrm{vect}_{\Bbbk}} \to (-)^{\vee\vee}$ between the identity functor and the double dual. For an arbitrary rigid monoidal category $\mathcal{C}$ such a monoidal isomorphism may or may not exist; and if it exists, then there might be several of them. The choice of a specific one is then called a pivotal structure on $\mathcal{C}$. For $\mathcal{C} = H$-mod the category of finite-dimensional modules over a Hopf algebra H, a pivotal structure amounts to a finding a group-like element $g \in H$ such that the square of the antipode S of H is given by $S^2(h) = ghg^{-1}$.

It is tempting to regard rigidity as the generic paradigm for a duality structure on monoidal categories. However, already long ago more general duality structures have been encountered (compare e.g. Remark 4.8 below). Such generalized dualities occur in important situations, in particular for representation categories of vertex operator algebras to which the HLZ-theory of tensor products [**HLZ**] applies. The purpose of the present section is to illustrate these structures in the simplest possible case: for bimodules over a $\Bbbk$-algebra.

Indeed it is easy to see that in general the category A-bimod cannot be rigid, even if the algebra A is finite-dimensional and only finite-dimensional bimodules are considered. In an abelian rigid monoidal category, the functors $X \otimes -$ and $- \otimes X$ of tensoring from the left or from the right with an object X are exact [**EGNO**, Prop. 4.2.1]. This property does not hold for bimodules, in general. A simple instructive counter example is the following: For any field $\Bbbk$, the algebra of dual numbers, i.e. the two-dimensional quotient

$$A_2 := \Bbbk[x] \,/\, \langle x^2 \rangle \tag{2.4}$$

of the polynomial algebra $\Bbbk[x]$, is a commutative associative unital algebra. A canonical basis of A_2 is given by the classes $[1] = 1 \bmod x^2$ and $[x] = x \bmod x^2$; to unburden notation, we drop the class symbols and just write 1 and x for the elements of this basis. The algebra A_2 has, up to isomorphism, a single simple module S, which is one-dimensional with generator s; the element x of A_2 acts on it by zero. The only other indecomposable finite-dimensional A_2-module, up to isomorphism, is the free module $P = A_2$ of rank 1, with basis elements 1 and x; the element $x \in A_2$ acts as $1 \mapsto x$ and $x \mapsto x^2 = 0$.

The following argument shows that the simple module S is not flat. The injective morphism $\iota\colon S \to P$ is given by $s \mapsto x$. One immediately verifies that $S \otimes_{A_2} S \cong S$ and, since P is the monoidal unit, $S \otimes_{A_2} P \cong S$. The structure map of the coequalizer $S \otimes_{\Bbbk} P \to S \otimes_{A_2} P$ maps $s \otimes_{\Bbbk} x \mapsto 0$. It follows that tensoring with S maps the monomorphism ι to the zero morphism: thus indeed the module S is not flat. As a consequence the category A_2-mod is not rigid.

Still, even in the non-rigid case we do have dual vector spaces at our disposal. We account for their presence by introducing, for any pair A and A' of algebras, the contravariant functor

$$G_{A,A'}: \quad A\text{-}A'\text{-bimod} \longrightarrow A'\text{-}A\text{-bimod} \tag{2.5}$$

that sends a bimodule $B \in A\text{-}A'\text{-bimod}$ to its linear dual $B^* = \mathrm{Hom}_{\Bbbk}(B, \Bbbk)$ endowed with the canonical action

$$(a'.\beta.a)(b) := \beta(a.b.a') \tag{2.6}$$

for $a \in A$, $a' \in A'$, $b \in B$ and $\beta \in B^*$. Obviously this functor squares to the identity, $G_{A',A} \circ G_{A,A'} = \mathrm{id}$. In particular, for $A' = A$ we get this way an A-bimodule A^*; this furnishes a distinguished object in the monoidal category A-bimod.

The distinguished bimodule A^* is, in general, not isomorphic to the monoidal unit A of A-bimod, as is already illustrated by the simple example

$$B_3 := \Bbbk[x, y] \,/\, \langle x^2, y^2, xy \rangle \tag{2.7}$$

of a three-dimensional commutative algebra. A canonical basis of B_3 is given by the classes $[1]$, $[x]$ and $[y]$, which again we abbreviate just as 1, x and y. The element 1 is the unit of B_3, while x and y satisfy $xx = yy = xy = 0$. The algebra B_3 has a unique (up to isomorphism) simple (bi)module S, which is one-dimensional. There is a short non-split exact sequence

$$0 \longrightarrow S \oplus S \longrightarrow B_3 \longrightarrow S\,, \tag{2.8}$$

for B_3, from which by dualizing we obtain a short exact sequence

$$0 \longrightarrow S \longrightarrow B_3^* \longrightarrow S \oplus S \longrightarrow 0 \tag{2.9}$$

for B_3^*. In particular, the modules B_3 and B_3^* have different socle and hence are not isomorphic. (As a consequence, B_3 does not admit the structure of a Frobenius algebra; it is in fact the simplest example of an algebra with this property.)

These observations raise the question: What is the appropriate *categorical* duality structure (which cannot be rigidity) on the monoidal category A-bimod that accomodates the distinguished bimodule B_3^* and the duality (2.5) inherited from the duality on vector spaces?

Before discussing the appropriate duality structure in detail, let us recall a few further properties that the monoidal category A-bimod of bimodules over a finite-dimensional algebra inherits from $\mathrm{vect}_{\Bbbk}$. These will eventually find their explanation in the framework of the general duality structures.

We first note the somewhat less well-known fact that for a finite-dimensional algebra A the abelian category A-bimod admits a second monoidal product. To see this, note that the vector space A^* has a natural structure of a coalgebra. Further, any right A-module $(M, \rho\colon M \otimes_{\Bbbk} A \to M)$ becomes a right A^*-comodule as follows: Select a basis (a_i) of A, which yields a dual basis (a^i) of A^*. Then the prescription

$$\tilde\rho(m) := \sum_i m.a_i \otimes_{\Bbbk} a^i \tag{2.10}$$

defines a right A^*-coaction. A similar construction turns a left A-module into a left A^*-comodule. One then defines a tensor product $\otimes^A$ of comodules as an equalizer

$$B_1 \otimes^A B_2 \longrightarrow B_1 \otimes_{\Bbbk} B_2 \xrightarrow{\varphi} B_1 \otimes_{\Bbbk} A^* \otimes_{\Bbbk} B_2 \tag{2.11}$$

with φ the linear map

$$\varphi = \tilde{\rho}^{B_1} \otimes_{\Bbbk} \mathrm{id}_{B_2} - \mathrm{id}_{B_1} \otimes_{\Bbbk} \tilde{\lambda}^{B_2}. \tag{2.12}$$

Since this tensor product is defined as a limit, it is, as usual for coalgebras, left exact. Its monoidal unit is the bimodule A^*.

We thus have two tensor products on A-bimod; the tensor product $\otimes^A$ is left exact, while $\otimes_A$ is right exact. This suggests to study their left and right adjoints, respectively. Doing so leads to the notions of internal Homs and internal coHoms; we introduce these for general linear monoidal categories:

DEFINITION 2.1. Let $\mathcal{C}$ be a $\Bbbk$-linear monoidal category with tensor product $\otimes$, and let $Y, Z \in \mathcal{C}$. The *internal Hom* $\underline{\mathrm{Hom}}^{\mathrm{r}}(Y,Z)$ is an object representing the functor $\mathcal{C}^{\mathrm{opp}} \to \mathrm{vect}_{\Bbbk}$ that is given by $X \mapsto \mathrm{Hom}(X \otimes Y, Z)$, i.e. we have functorial isomorphisms

$$\mathrm{Hom}(X \otimes Y, Z) \cong \mathrm{Hom}(X, \underline{\mathrm{Hom}}^{\mathrm{r}}(Y,Z)) . \tag{2.13}$$

Similarly a second internal Hom $\underline{\mathrm{Hom}}^{\mathrm{l}}$ is obtained by keeping the second tensor factor [**BoD**, Remarks 2.1(3)], i.e.

$$\mathrm{Hom}(X \otimes Y, Z) \cong \mathrm{Hom}(Y, \underline{\mathrm{Hom}}^{\mathrm{l}}(X,Z)) . \tag{2.14}$$

The category $\mathcal{C}$ is called *right closed* if all internal Homs $\underline{\mathrm{Hom}}^{\mathrm{r}}(Y,Z)$ exist, and it is called *left closed* if all internal Homs $\underline{\mathrm{Hom}}^{\mathrm{l}}(Y,Z)$ exist.

A necessary condition for the existence of the internal Homs is that the tensor product $\otimes$ is right exact. The analogous notion for the left exact tensor product $\otimes^A$ on A-bimod is the left adjoint of $\otimes^A$. Again this can be generalized:

DEFINITION 2.2. Let $\mathcal{C}$ be a $\Bbbk$-linear monoidal category with tensor product $\otimes$, and let $Y, Z \in \mathcal{C}$. Then the *internal coHom* is the object characterized by functorial isomorphisms

$$\mathrm{Hom}(X, Y \otimes Z) \cong \mathrm{Hom}(\underline{\mathrm{coHom}}^{\mathrm{r}}(Z,X), Y) . \tag{2.15}$$

Similarly there is a second internal coHom $\underline{\mathrm{coHom}}^{\mathrm{l}}$.

For the case of the monoidal category of A-bimodules considered here, internal Homs and coHoms do exist. Using the canonical pivotal structure of finite-dimensional vector spaces, they can be expressed in terms of the tensor products $\otimes^A$ and $\otimes_A$, respectively:

$$\underline{\mathrm{Hom}}^{\mathrm{r}}(X,Y) = Y \otimes^A X^* \quad \text{and} \quad \underline{\mathrm{coHom}}^{\mathrm{r}}(X,Y) = Y \otimes_A X^*, \tag{2.16}$$

as well as

$$\underline{\mathrm{Hom}}^{\mathrm{l}}(X,Y) = X^* \otimes^A Y \quad \text{and} \quad \underline{\mathrm{coHom}}^{\mathrm{l}}(X,Y) = X^* \otimes_A Y. \tag{2.17}$$

To show that the category A-bimod does admit internal Homs, we provide them explicitly. In fact, for any triple of algebras A, B and C and bimodules ${}_CM_B$, ${}_BN_A$ and ${}_CX_A$ there is the adjunction

$$\begin{aligned} \mathrm{Hom}_{C,A}({}_CM_B \otimes_B {}_BN_A, {}_CX_A) &\cong \mathrm{Hom}_{C,B}({}_CM_B, \mathrm{Hom}_A(N_A, X_A)) \\ &\cong \mathrm{Hom}_{C,B}({}_CM_B, X_A \otimes^A (N_A)^*) , \end{aligned} \tag{2.18}$$

where we indicate only one of the algebra actions on a bimodule in cases when it is more relevant than the other, e.g. N_A is the bimodule ${}_BN_A$ seen as a right A-module, and where in the second expression, $\mathrm{Hom}_A(N_A, X_A)$ is regarded as (C, B)-bimodule via the left module structures of N_A and X_A. We can thus read off that

$$\underline{\mathrm{Hom}}^{\mathrm{r}}(N, X) = \mathrm{Hom}_A(N_A, X_A)\,. \tag{2.19}$$

Similarly we find

$$\underline{\mathrm{Hom}}^{\mathrm{l}}(N, X) = \mathrm{Hom}_A({}_AN, {}_AX)\,. \tag{2.20}$$

Note that (2.19) exhibits the internal Hom as a sub-bimodule

$$\mathrm{Hom}_A(N_A, X_A) \subset \mathrm{Hom}_{\Bbbk}(N, X) \tag{2.21}$$

of the bimodule obtained by equating the right A-actions on X and N, and analogously for (2.20). In contrast, internal coHoms are defined as quotients. For finite-dimensional bimodules this fits with the description (2.16) and (2.17) of internal Homs and coHoms in terms of the two tensor products $\otimes^A$ and $\otimes_A$. Also note that according to (2.19) and (2.20), left and right internal Homs are, in general, non-isomorphic.

Finally we specialize (2.18) to the case that $X = A^*$. Restricting further to the special case that M and N are A-bimodules we then get

$$\begin{aligned}
\mathrm{Hom}_{A,A}(M \otimes_A N, G({}_AA_A)) &\overset{(2.18)}{\cong} \mathrm{Hom}_{A,A}(M, \mathrm{Hom}_A(N_A, (A^*)_A)) \\
&= \mathrm{Hom}_{A,A}(M, \mathrm{Hom}_{\Bbbk,A}(N, \mathrm{Hom}_{\Bbbk}(A, \Bbbk))) \\
&\overset{(2.18)}{\cong} \mathrm{Hom}_{A,A}(M, \mathrm{Hom}_{\Bbbk,\Bbbk}(N \otimes_A A, \Bbbk)) \\
&\cong \mathrm{Hom}_{A,A}(M, G(N))\,.
\end{aligned} \tag{2.22}$$

REMARK 2.3. Recall that to any A-bimodule M one can associate its center

$$Z(M) := \{b \in M \mid a.m = m.a \text{ for all } a \in A\} \tag{2.23}$$

as a subobject and its trace

$$\mathrm{tr}\, M := M/\langle a.m - m.a\rangle \tag{2.24}$$

as a quotient. By comparison with the A-bimodule structure on the vector space $\mathrm{Hom}_{\Bbbk}(N, X)$ defined above, we see that internal Homs are centers and internal coHoms are traces. Also note that for the definition of the center and trace we do not need to require that the bimodule M is finite-dimensional. By considering bimodules of the form $M \otimes_{\Bbbk} N$, this implies that the two tensor products $\otimes_A$ and $\otimes^A$ still exist for infinite-dimensional bimodules. Similarly, the contravariant functor (2.5) that sends a bimodule to its linear dual and the adjunction (2.18) are still present in the infinite-dimensional case.

3. Grothendieck-Verdier duality

3.1. First definitions. The observations in the preceding section motivate the following

DEFINITION 3.1. Let $\mathcal{C} = (\mathcal{C}, \otimes, 1, \alpha, l, r)$ be a monoidal category.

(1) A *dualizing object* of $\mathcal{C}$ is an object $K \in \mathcal{C}$ such that for every $y \in \mathcal{C}$ the functor $x \mapsto \mathrm{Hom}(x \otimes y, K)$ is representable by some object $Gy \in \mathcal{C}$ and the so defined contravariant functor $G\colon \mathcal{C} \to \mathcal{C}$ is an anti-equivalence.

We thus have isomorphisms

$$\varpi_{x,y}: \quad \mathrm{Hom}(x\otimes y, K) \xrightarrow{\cong} \mathrm{Hom}(x, Gy) \tag{3.1}$$

for $x, y \in \mathcal{C}$. G is called the *duality functor with respect to K*.

(2) A *Grothendieck-Verdier structure* on $\mathcal{C}$ is the choice of a dualizing object $K \in \mathcal{C}$. A *Grothendieck-Verdier category*, or *GV-category*, for short, is a monoidal category together with a choice of a Grothendieck-Verdier structure.

While the functor G depends on the choice of K, we suppress it in the notation.[2] This duality structure has been introduced in [**Ba1**] under the name $*$-autonomous category, compare also [**Ba2**, **Ba3**, **Ba4**]. The term Grothendieck-Verdier category dates back to [**BoD**]. GV-categories appear for instance in the study of quadratic algebras and operads [**Ma**] and also play an important role in linear logic, see e.g. [**Se**].

The result (2.22) in Section 2 tells us that A-bimod is a GV-category with dualizing object A^*, thereby giving a categorical meaning to this distinguished bimodule. This should not come as a surprise: it is known [**FSS**, Rem. 3.17] that this fact can be formulated in a Morita invariant manner. The dualizing object is in fact *structure*; if K is a dualizing object, then any object of the form $K\otimes D$ with D an invertible object of $\mathcal{C}$ is dualizing as well [**BoD**, Prop. 1.3]. It should, however, be appreciated that the dualizing object itself need not be invertible. GV-categories with non-invertible dualizing object indeed arise as representation categories of vertex operator algebras [**GRW**]. Below, we also give an example involving bimodules.

Definition 3.2. An *r-category* is a monoidal category for which the monoidal unit 1 is a dualizing object.

Clearly, every rigid monoidal category is an r-category. But there also exist non-rigid r-categories; for an example, see e.g. [**BoD**, Ex. 1.9].

In the case of categories A-bimod of bimodules, certain invertible bimodules can be obtained by twisting either the left or right action of A by an algebra automorphism ψ of A. For the isomorphism class of the bimodule A_ψ obtained in this way, ψ only matters up to inner automorphisms. For instance, the group of outer automorphisms of the three-dimensional algebra $B_3 = \Bbbk[x,y]/\langle x^2, y^2, xy\rangle$ considered in (2.7) is isomorphic to GL(2), acting on the two generators in the obvious way. This gives a simple example of a GV-category admitting different GV-structures.

We also have already noticed that in this case the vector space dual B_3^* is not isomorphic to B_3 as a bimodule, so B_3 is not a Frobenius algebra. The following consideration shows that B_3^* is not even invertible: B_3 has, up to isomorphism, a single simple module S, which is one-dimensional and on which the elements x and y of B_3 act as zero. A direct calculation shows that $B_3^* \otimes_A B_3^* \cong S^{\oplus 4}$. Thus if B_3^* were invertible, then $S^{\oplus 4}$ would be too, i.e. there would be a bimodule X such that $S^{\oplus 4} \otimes_A X \cong A$. But this is impossible, simply because A is indecomposable as a module over itself. In fact, one can show that the regular bimodule B_3 is, up to isomorphism, the only invertible B_3-bimodule.

[2] It is worth noting that this (standard) definition of a GV-category differs from the one in Definition 2.3 of [**HasL**], a paper which also discusses the relation to linearly distributive categories.

3.2. Internal Homs. The existence of a GV-structure on a monoidal category implies much structure familiar from category theory. In particular, recall from Definition 2.1 that, for $\mathcal{C}$ a linear monoidal category and $y, z \in \mathcal{C}$, the internal Hom $\underline{\mathrm{Hom}}^{\mathrm{r}}(y, z) \in \mathcal{C}$ is an object representing the functor $x \mapsto \mathrm{Hom}(x \otimes y, z)$, as expressed by the functorial isomorphisms (2.13). Since $\mathcal{C}$ is not assumed to be braided, there is also a separate variant $\underline{\mathrm{Hom}}^{\mathrm{l}}$ of the internal Hom in which the second tensor factor is kept, with functorial isomorphisms (2.14).

A GV-category admits internal Homs. In fact, from the defining properties of the functor G it follows that

$$\underline{\mathrm{Hom}}^{\mathrm{r}}(x, z) \cong G(x \otimes G^{-1}z) \qquad \text{and} \qquad \underline{\mathrm{Hom}}^{\mathrm{l}}(x, z) \cong G^{-1}(Gz \otimes x)\,. \tag{3.2}$$

In particular we have

$$\underline{\mathrm{Hom}}^{\mathrm{r}}(x, K) \cong G(x \otimes G^{-1}(K)) \cong G(x \otimes 1) \cong G(x)\,, \tag{3.3}$$

i.e. there is an isomorphism

$$\underline{\mathrm{Hom}}^{\mathrm{r}}(-, K) \cong G \tag{3.4}$$

of functors. This explicitly shows how the duality functor G is determined in terms of the dualizing object K.

There are canonical evaluation morphisms

$$\underline{\mathrm{ev}}^{\mathrm{r}}_{x,y} \colon \underline{\mathrm{Hom}}^{\mathrm{r}}(x, y) \otimes x \to y \qquad \text{and} \qquad \underline{\mathrm{ev}}^{\mathrm{l}}_{x,y} \colon x \otimes \underline{\mathrm{Hom}}^{\mathrm{l}}(x, y) \to y \tag{3.5}$$

for the internal Homs, given by the images of the identity morphism in the spaces $\mathrm{End}(\underline{\mathrm{Hom}}^{\mathrm{r}}(x, y))$ and $\mathrm{End}(\underline{\mathrm{Hom}}^{\mathrm{l}}(x, y))$ under the defining adjunctions (2.13) and (2.14), respectively. When combined with (3.4), this immediately gives left and right evaluation morphisms for the tensor product. In contrast, compatible co-evaluations do not exist, in general; otherwise the tensor product on any abelian GV-category would be exact. Given the evaluations, associative multiplication morphisms

$$\begin{aligned} &\underline{\mu}^{\mathrm{r}}_{x,y,z} \colon && \underline{\mathrm{Hom}}^{\mathrm{r}}(y, z) \otimes \underline{\mathrm{Hom}}^{\mathrm{r}}(x, y) \longrightarrow \underline{\mathrm{Hom}}^{\mathrm{r}}(x, z) \\ \text{and} \quad &\underline{\mu}^{\mathrm{l}}_{x,y,z} \colon && \underline{\mathrm{Hom}}^{\mathrm{l}}(x, y) \otimes \underline{\mathrm{Hom}}^{\mathrm{l}}(y, z) \longrightarrow \underline{\mathrm{Hom}}^{\mathrm{l}}(x, z) \end{aligned} \tag{3.6}$$

are obtained by standard arguments [**EGNO**, Sect. 7.9]. In particular, $\underline{\mathrm{Hom}}^{\mathrm{l}}(z, z)$ and $\underline{\mathrm{Hom}}^{\mathrm{r}}(z, z)$ have the structure of (unital) associative algebras.

3.3. A second tensor product. It is natural to introduce on a GV-category a second monoidal structure by

$$x \bullet y := G^{-1}(Gy \otimes Gx)\,. \tag{3.7}$$

The two tensor products are, in general, different, and hence in general $G(x \otimes y)$ and $G(y) \otimes G(x)$ are not isomorphic, in contrast to the case of rigid categories. The double dual G^2, however, is monoidal [**BoD**, Prop. 5.2]. It describes how cyclic invariance of invariant tensors of Grothendieck-Verdier categories is violated [**BoD**, Rem. 2.1(2)]:

$$\begin{aligned} \mathrm{Hom}(G^2(y) \otimes x, K) &\cong \mathrm{Hom}(G^2(y), Gx) \\ &\cong \mathrm{Hom}(x, Gy) \cong \mathrm{Hom}(x \otimes y, K)\,. \end{aligned} \tag{3.8}$$

There is also another variant of the second monoidal structure, given by

$$x \bullet' y := G(G^{-1}y \otimes G^{-1}x)\,. \tag{3.9}$$

The monoidal structure on the functor G^2 provides a canonical identification of $\bullet$ and $\bullet'$; accordingly we will henceforth identify $\bullet$ and $\bullet'$.

REMARK 3.3. If $\mathcal{C}$ is an r-category, then the monoidal products $\otimes$ and $\bullet$ are connected by canonical functorial morphisms

$$\varphi_{x,y}: \quad x\otimes y \longrightarrow x\bullet y\,. \tag{3.10}$$

The morphisms $\varphi_{x,y}$ are functorial in x and y and compatible with the associativity constraints for the two tensor products. In general they are not isomorphisms.

REMARK 3.4. As follows from (2.16), in the categories of bimodules considered in Section 2, the two tensor products $\otimes$ and $\bullet$ are realized as the $\otimes_A$- and $\otimes^A$-tensor products of bimodules, respectively. These have counterparts in the representation theory of vertex operator algebras. There are two commonly used constructions for the tensor product of modules over a vertex operator algebra. One of them, developed in the physics literature [**Na**, **GK**], considers, for a given pair of modules M and N, two actions of the vertex operator algebra on the vector space $M\otimes_{\mathbb{C}} N$ and performs a coequalizer construction resembling the one in the definition of $\otimes_A$. The other one [**HLZ**], which is e.g. briefly summarized in Section 2.2 of [**ALSW**], starts instead with two actions on the vector space $\mathrm{Hom}_{\mathbb{C}}(M\otimes_{\mathbb{C}} N,\mathbb{C})$ and restricts to the subspace on which they coincide, similarly as in the equalizer construction of $\otimes^A$, and afterwards takes the dual of this subspace. We refer to [**KR**] for a more detailed account of the two approaches. We expect that in favorable circumstances the two constructions can be related through a Grothendieck-Verdier structure in a similar way as the tensor products $\otimes_A$ and $\otimes^A$ of bimodules; we intend to come back to this issue elsewhere.

Recall from Definition 2.2 that, for $(\mathcal{C},\otimes,1)$ a linear monoidal category and $x,z\in\mathcal{C}$, the internal coHom $\underline{\mathrm{coHom}}^{\mathrm{r}}(z,x)\in\mathcal{C}$ is the object characterized by the functorial isomorphisms (2.15). As seen in (3.2), for any GV-category internal Homs for $\otimes$ exist, and so do internal coHoms for $\bullet$. In terms of the two tensor products, they can be written as

$$\underline{\mathrm{coHom}}^{\mathrm{r}}(x,y) = y\otimes G^{-1}x \quad \text{and} \quad \underline{\mathrm{Hom}}^{\mathrm{r}}(x,y) = y\bullet Gx\,, \tag{3.11}$$

and as

$$\underline{\mathrm{coHom}}^{\mathrm{l}}(x,y) = Gx\otimes y \quad \text{and} \quad \underline{\mathrm{Hom}}^{\mathrm{l}}(x,y) = G^{-1}x\bullet y\,, \tag{3.12}$$

generalizing the expressions (2.16) and (2.17).

It is an instructive exercise to derive the concrete formulas we gave for the monoidal category A-bimod for the tensor product $\otimes^A$ and the internal (co)Hom from these general statements. Also, in analogy to (3.6) it follows immediately that for any object $c\in\mathcal{C}$ the internal coHom $\underline{\mathrm{coHom}}(c,c)$ is a counital coassociative coalgebra for the tensor product $\bullet$. By duality, this tensor product comes with coevaluations.

As usual, the trivialization of double duals is interesting:

DEFINITION 3.5. A *pivotal* GV-category is a GV-category together with natural isomorphisms

$$\psi_{x,y}: \quad \mathrm{Hom}(x\otimes y,K) \xrightarrow{\cong} \mathrm{Hom}(y\otimes x,K) \tag{3.13}$$

such that

$$\psi_{x,y}\circ\psi_{y,x} = \mathrm{id} \quad \text{and} \quad \psi_{x\otimes y,z}\circ\psi_{y\otimes z,x}\circ\psi_{z\otimes x,y} = \mathrm{id}\,, \tag{3.14}$$

or, more explicitly when not suppressing the associatior,

$$\psi_{x\otimes y,z} \circ \alpha_{x,y,z} \circ \psi_{y\otimes z,x} \circ \alpha_{y,z,x} \circ \psi_{z\otimes x,y} \circ \alpha_{z,x,y} = \mathrm{id}\,. \tag{3.15}$$

We have

PROPOSITION 3.6 ([**BoD**, Prop. 6.7]). *Pivotal structures on $\mathcal{C}$ are in bijection to monoidal isomorphisms of functors $\pi\colon \mathrm{id}{\to}G^2$ whose component $\pi_K\colon K \xrightarrow{\cong} G^2(K)$ is given by the canonical isomorphism $K \xrightarrow{\cong} G1 = G^2G^{-1}1 \xrightarrow{\cong} G^2K$.*

There also exists a notion of a *ribbon* GV-category [**BoD**, Sect. 8]. It has been shown that ribbon GV-categories lead to *ansular functors* [**MW**], and non-degeneracy conditions on the braiding have been identified which guarantee that they lead to modular functors [**BrW**]. It should also be appreciated that the representation category of a vertex operator algebra to which the HLZ tensor product theory applies has a natural structure of a ribbon GV-category [**ALSW**]. (This fits with the facts that a vertex operator algebra and its gradewise dual are not necessarily isomorphic as modules, and that the tensor product of vertex operator algebra modules need not be exact.)

3.4. A symmetric formulation of the Eilenberg-Watts theorem. To connect the abstract considerations of Sections 3.1 – 3.3 with the particular case of A-bimod that was studied in Section 2, let us focus on the case of a GV-category that is in addition abelian.

LEMMA 3.7. *Let $\mathcal{C}$ be an abelian category that has the structure of a GV-category with biadditive tensor product bifunctor $\otimes$. Then $\otimes$ is right exact and the second tensor product $\bullet$ is left exact.*

This statement applies in particular to the abelian category A-bimod of finite-dimensional bimodules over finite-dimensional algebras which we considered in Section 2. In this case we observe, as a by-product, the following application of dualities: the GV-structure on A-bimod allows one to rewrite the classical Eilenberg-Watts theorem in a form that treats left exact and right exact functors in a completely symmetric manner.

The classical Eilenberg-Watts theorem describes right exact functors between categories of modules: Let mod-A and mod-A' be the categories of finite-dimensional right modules over finite-dimensional algebras A and A'. Then any right exact functor

$$F: \quad \text{mod-}A \longrightarrow \text{mod-}A' \tag{3.16}$$

that commutes with direct sums obeys

$$F(-) \cong - \otimes_A F(A)\,. \tag{3.17}$$

Here it should be appreciated that $F(A)$ is by definition an A'-right module, while it is also a left A-module by applying F to the endomorphisms of A, as a right module over itself, given by the action of F on left multiplication. A similar formula is valid for a functor $F'\colon A\text{-mod} \to A'\text{-mod}$ of categories of left modules:

$$F'(-) \cong F(A) \otimes_A -\,. \tag{3.18}$$

There is also a version of the Eilenberg-Watts theorem for left exact functors, in which the role of the tensor product functor in (3.17) is taken over by a Hom functor [**Iv**, Thms. 2.6 & 2.7]: If (and only if) H is left exact, then

$$H \cong \mathrm{Hom}_A(M,-) \quad \text{with} \quad M = (H({}_A A^*_A))^* . \tag{3.19}$$

(In particular, H is representable.)

Now consider the particular instance

$$G_{A,\Bbbk}: \quad A\text{-mod} \longrightarrow \text{mod-}A \tag{3.20}$$

of the duality functor given in (2.5) and the analogous functor $G_{\Bbbk,A}$ mapping right modules to left modules. Suppose that

$$H: \quad A\text{-mod} \to A'\text{-mod} \tag{3.21}$$

is a left exact functor of left modules. Then

$$F := G_{A',\Bbbk} \circ H \circ G_{\Bbbk,A}: \quad \text{mod-}A \longrightarrow \text{mod-}A' \tag{3.22}$$

is a right exact functor of right modules. The Eilenberg-Watts equivalence (3.17) for this functor F reads

$$G_{A',\Bbbk} \circ H \circ G_{\Bbbk,A}(-) \cong - \otimes_A \big(G_{A',\Bbbk}\, H\, G_{\Bbbk,A}(A)\big) . \tag{3.23}$$

Now for $M \in A$-mod set $\widetilde{M} := G_{A,\Bbbk}(M) \in$ mod-A. Applying the isomorphism (3.23) to $\widetilde{M}$ and then $G_{\Bbbk,A'}$ to the so obtained isomorphism gives

$$\begin{aligned} H(M) &= G_{\Bbbk,A'}\big(\widetilde{M} \otimes_A G_{A',\Bbbk} H G_{\Bbbk,A}(A)\big) \\ &= G_{\Bbbk,A'}\big(G_{A,\Bbbk}(M) \otimes_A G_{A',\Bbbk}\, H(A^*)\big) = H(A^*) \otimes^A M . \end{aligned} \tag{3.24}$$

(A corresponding result for comodules is shown in [**BW**, 12.1].) This description is beautiful, even for very classical algebra, and promising. In short, with the help of the duality functor, *left exact* functors can be expressed in terms of the second tensor product $\otimes^A$ in completely the same way as in the classical Eilenberg-Watts description (3.17) of *right exact* functors: apply the functor to the monoidal unit – this time A^* – and then take the tensor product – this time $\otimes^A$.

For $\varphi\colon A \longrightarrow A'$ a homomorphism of associative $\Bbbk$-algebras, the pullback functor Res_φ: mod-$A' \longrightarrow$ mod-A assigns to any right A'-module $(M', \rho_{M'})$ the right A-module $(M', \rho_{M'} \circ (\varphi \otimes_\Bbbk \mathrm{id}_{M'}))$ with A-action given by pulling back the A'-action along φ. (The notation Res indicates that if φ is injective, then Res_φ is the restriction of the A'-action to a subalgebra A.) For easier distinction we write $\varphi^*(M')$ for the image of M' under φ^* even though it coincides with M' as a vector space. On morphism spaces φ^* is then the first inclusion in the sequence

$$\mathrm{Hom}_{A'}(M',N') \subset \mathrm{Hom}_A(\varphi^*(M'),\varphi^*(N')) \subset \mathrm{Hom}_\Bbbk(M',N') . \tag{3.25}$$

Thus the functor Res_φ is faithful (albeit not necessarily full or essentially surjective) and exact. The symmetric formulation of the Eilenberg-Watts theorem obtained above yields two expressions

$$M' \otimes_{A'} \mathrm{Res}_\varphi(A') = \mathrm{Res}_\varphi(M') = M' \otimes^{A'} \mathrm{Res}_\varphi(A'^*) \tag{3.26}$$

for the pullback functor. Denoting the left and right adjoint of Res_φ, which in the subalgebra case are the induction and coinduction functors, by Ind_φ and coInd_φ, respectively, the adjunction (2.18) thus gives

COROLLARY 3.8. *The left and right adjoints of the pullback functor* Res_φ *can be written symmetrically as*

$$\mathrm{Ind}_\varphi(M) = M \otimes_A \mathrm{Res}_\varphi(A'^*)^* \quad \textit{and} \quad \mathrm{coInd}_\varphi(M) = M \otimes^A \mathrm{Res}_\varphi(A')^* . \tag{3.27}$$

4. Distributors

In a monoidal category with tensor product $\otimes$, different bracketings of multiple tensor products are related through the associator $\alpha = \alpha^{\otimes}$. (A pentagon identity for α ensures coherence.) In a GV-category there is another associator $\alpha^{\bullet}$ for multiple $\bullet$-tensor products, whose components are directly obtained from those of α as images under the duality functor G. As a consequence, they are isomorphisms and also $\alpha^{\bullet}$ satisfies a pentagon equation. In addition there are similar coherent families of morphisms that relate multiple products involving both $\bullet$ and $\otimes$. These are best understood when viewed from the perspective of module categories, involving in particular the notion of a weak module functor. Recall that a module category is a categorification of the notion of a module over a ring, see e.g. [**EGNO**, Def. 7.1.1],

DEFINITION 4.1. Let $\mathcal{M}$ and $\mathcal{N}$ be left module categories over a monoidal category $\mathcal{C}$. A *lax module functor* from $\mathcal{M}$ to $\mathcal{N}$ is a functor $G\colon \mathcal{M} \to \mathcal{N}$ endowed with a natural family of morphisms $g_{c,m}\colon c \rhd G(m) \to G(c \rhd m)$ (called the lax *module constraint* of G) such that the appropriate pentagon diagram is fulfilled. Analogously, for an *oplax module functor* $F\colon \mathcal{M} \to \mathcal{N}$ there is a corresponding coherent family of morphisms $F(c \rhd m) \to c \rhd F(m)$.

A lax or oplax module functor is also called a *weak module functor*.

A *strong module functor* is a lax (or oplax) module functor for which the lax (or oplax) module constraints are isomorphisms.

The notion of internal Homs and coHoms generalizes from monoidal categories to module categories in a straightforward way. For $(\mathcal{M}, \rhd)$ a left module category over a monoidal category $(\mathcal{C}, \otimes)$, one defines internal Homs, similarly as $\underline{\mathrm{Hom}}^{\mathrm{r}}$ in Definition 2.1 for $\mathcal{C}$ itself, as objects representing the functors $c \mapsto \mathrm{Hom}_{\mathcal{M}}(c \rhd m, n)$, giving rise to coherent isomorphisms

$$\mathrm{Hom}_{\mathcal{M}}(c \rhd m, n) \cong \mathrm{Hom}_{\mathcal{C}}(c, \underline{\mathrm{Hom}}^{\mathrm{r}}(m, n)) \tag{4.1}$$

for $c \in \mathcal{C}$ and $m, n \in \mathcal{M}$. Denote, for any object m in a left $(\mathcal{C}, \otimes)$-module category $\mathcal{M}$, by $Q^{\rhd}_m\colon {}_{\mathcal{C}}\mathcal{C} \to \mathcal{M}$, with ${}_{\mathcal{C}}\mathcal{C}$ standing for the category $\mathcal{C}$ regarded as a left module over itself, the strong $(\mathcal{C}, \otimes)$-module functor obtained by acting on m, i.e.

$$Q^{\rhd}_m(c) := c \rhd m \tag{4.2}$$

for $c \in \mathcal{C}$. Then the *internal Hom* $\underline{\mathrm{Hom}}^{\mathrm{r}}(m, -)$ is the right adjoint of the module functor $Q^{\rhd}_m$. Indeed, all all module functors $\mathcal{C} \to \mathcal{M}$ are of this type:

LEMMA 4.2. *Let $\mathcal{C}$ be a monoidal category, $\mathcal{M}$ a left $\mathcal{C}$-module, and $F\colon \mathcal{C} \to \mathcal{M}$ a strong module functor. Then there exists an object $m \in \mathcal{M}$ with a module natural isomorphism $F \xrightarrow{\cong} Q^{\rhd}_m$. The object m is unique up to unique isomorphism.*

PROOF. We set $m := F(1)$. From the module constraint f of F we obtain a natural isomorphism $f_{c,1}\colon F(c) = F(c \otimes 1) \xrightarrow{\cong} c \rhd F(1) = c \rhd m$ for $c \in \mathcal{C}$. From the pentagon axiom of the module constraint it follows that this family of isomorphisms constitutes a module natural isomorphism $F \xrightarrow{\cong} Q^{\rhd}_m$. By a Yoneda argument, m is unique up to unique isomorphism. □

If $(\mathcal{C}, \otimes)$ is a GV-category, then the regular left module ${}_{\mathcal{C}}\mathcal{C}$ does admit internal Homs (which are given by the formulas (3.2)). In contrast, for an arbitrary module category over a GV-category, the existence of such right adjoints is, of course, not guaranteed. Accordingly we give

DEFINITION 4.3. A *left GV-module category* over a GV-category $\mathcal{C}$ is a left module category $(\mathcal{M}, \triangleright)$ over $(\mathcal{C}, \otimes)$ such that both the action functor $Q^{\triangleright}_m : {}_{\mathcal{C}}\mathcal{C} \to \mathcal{M}$ defined by (4.2) and the functor $c \triangleright - : \mathcal{M} \to \mathcal{M}$ admit a right adjoint, for every $m \in \mathcal{M}$ and every $c \in \mathcal{C}$, respectively.

Note that the GV-structure of $\mathcal{C}$ is in fact not used in this definition. The separate terminology 'GV-module category' is chosen to indicate that an additional condition is imposed.

Since any left adjoint functor preserves all colimits, for $\mathcal{M}$ a GV-module category over a GV-category $\mathcal{C}$ the functor $Q^{\triangleright}_m$ is right exact for every $m \in \mathcal{M}$. By definition, for any $c \in \mathcal{C}$ the endofunctor $c \triangleright -$ has a right adjoint; denote this right adjoint by $R_c : \mathcal{M} \to \mathcal{M}$. We have

PROPOSITION 4.4. *Let $(\mathcal{M}, \triangleright)$ be a left GV-module category over a GV-category $(\mathcal{C}, \otimes)$. Then the bifunctor $\blacktriangleright : \mathcal{C} \times \mathcal{M} \to \mathcal{M}$ given by*

$$c \blacktriangleright m := R_{Gc}(m) \tag{4.3}$$

is left exact in each variable and defines a left module category structure over $(\mathcal{C}, \bullet)$.

PROOF. In the defining adjunction $\mathrm{Hom}_{\mathcal{M}}(c \triangleright n, m) \cong \mathrm{Hom}_{\mathcal{M}}(n, R_c(m))$ of R_c, the left hand side is manifestly left exact in $m \in \mathcal{M}$, and thus so is $\blacktriangleright$. Left exactness in $c \in \mathcal{C}$ follows by right exactness of $\triangleright$ from the isomorphism

$$\mathrm{Hom}_{\mathcal{M}}(m, c \blacktriangleright n) = \mathrm{Hom}_{\mathcal{M}}(m, R_{Gc}(n)) \cong \mathrm{Hom}_{\mathcal{M}}(Gc \triangleright m, n)\,. \tag{4.4}$$

Further, for $m, n \in \mathcal{M}$ and $b, c \in \mathcal{C}$ there are natural isomorphisms

$$\begin{aligned} \mathrm{Hom}_{\mathcal{M}}(m, b \blacktriangleright (c \blacktriangleright n)) &\cong \mathrm{Hom}_{\mathcal{M}}(Gb \triangleright m, c \blacktriangleright n) \\ &\cong \mathrm{Hom}_{\mathcal{M}}(Gc \triangleright (Gb \triangleright m), n) \\ &\cong \mathrm{Hom}_{\mathcal{M}}((Gc \otimes Gb) \triangleright m), n) \\ &\cong \mathrm{Hom}_{\mathcal{M}}(m, (b \bullet c) \blacktriangleright n)\,. \end{aligned} \tag{4.5}$$

From these isomorphisms the module constraints for $\blacktriangleright$ follow by the Yoneda lemma. Moreover, the Yoneda embedding transports the pentagon identity for the module constraint of $\triangleright$ to the one for the module constraint of $\blacktriangleright$. □

Setting

$$Q^{\blacktriangleright}_m(c) := c \blacktriangleright m \tag{4.6}$$

we thus obtain for any left GV-module category $(\mathcal{M}, \triangleright)$ over a GV-category $(\mathcal{C}, \otimes)$ also a $(\mathcal{C}, \bullet)$-module functor $Q^{\blacktriangleright}_m : {}^{\bullet}_{\mathcal{C}}\mathcal{C} \to \mathcal{M}$. Since $Q^{\blacktriangleright}_m$ is defined via the right adjoint of the $(\mathcal{M}, \triangleright)$-module functor $Q^{\triangleright}_m$, it has a left adjoint. Indeed we have natural isomorphisms

$$\begin{aligned} \mathrm{Hom}_{\mathcal{M}}(n, Q^{\blacktriangleright}_m(c)) = \mathrm{Hom}_{\mathcal{M}}(n, c \blacktriangleright m) &\cong \mathrm{Hom}_{\mathcal{M}}(Gc \triangleright n, m) \\ &\cong \mathrm{Hom}_{\mathcal{C}}(Gc, \underline{\mathrm{Hom}}^{\mathrm{r}}(n, m)) \cong \mathrm{Hom}_{\mathcal{C}}(G^{-1}(\underline{\mathrm{Hom}}^{\mathrm{r}}(n, m)), c)\,. \end{aligned} \tag{4.7}$$

We call the left adjoint of the $(\mathcal{C}, \bullet)$-module functor $Q^{\blacktriangleright}_m$ the *internal coHom* and denote it by $\underline{\mathrm{coHom}}^{\mathrm{r}}(m, -)$. By definition, there are thus coherent isomorphisms

$$\mathrm{Hom}_{\mathcal{M}}(n, c \blacktriangleright m) \cong \mathrm{Hom}_{\mathcal{C}}(\underline{\mathrm{coHom}}^{\mathrm{r}}(m, n), c) \tag{4.8}$$

for all $c \in \mathcal{C}$ and all $m, n \in \mathcal{M}$ and, by (4.7),

$$\underline{\mathrm{coHom}}^{\mathrm{r}}(m,n) \cong G^{-1}(\underline{\mathrm{Hom}}^{\mathrm{r}}(n,m)) \tag{4.9}$$

for all $m, n \in \mathcal{M}$.

It should be appreciated that the adjoint of a strong module functor is, in general, only a weak module functor. To understand this, it is convenient to consider linear categories and work with module profunctors. Following [**Sh**, Def. 2.1] we give

DEFINITION 4.5. Let $\mathcal{M}$ and $\mathcal{N}$ be left modules over a $\Bbbk$-linear monoidal category $\mathcal{C}$. A *$\mathcal{C}$-module profunctor* from $\mathcal{M}$ to $\mathcal{N}$ is a bilinear functor $H\colon \mathcal{M}^{\mathrm{opp}} \times \mathcal{N} \to \mathrm{vect}_\Bbbk$ together with a family

$$\theta_{m,n,c}: \quad H(m,n) \longrightarrow H(c \triangleright m, c \triangleright n) \tag{4.10}$$

of morphisms that is natural in $m \in \mathcal{M}$ and in $n \in \mathcal{N}$, is dinatural in $c \in \mathcal{C}$, and is coherent with respect to the monoidal structure and the unit of $\mathcal{C}$, i.e. satisfies $\theta_{m,n,c\otimes d} = \theta_{d\triangleright m, d\triangleright n, c} \circ \theta_{m,n,d}$ and $\theta_{m,n,1} = \mathrm{id}_{H(m,n)}$ for all $m \in \mathcal{M}$, $n \in \mathcal{N}$ and $c, d \in \mathcal{C}$. (Here, analogously as we do for monoidal categories, we suppress the mixed associators and unitors of a module category.)

A *morphism* $\varphi\colon H_1 \to H_2$ *of module profunctors* between $\mathcal{C}$-module profunctors H_1 and H_2 is a natural transformation that commutes with the respective (di)natural transformations.

The aspect of module profunctors that is relevant to us is

LEMMA 4.6 ([**Sh**, Lemma 2.3]). *Let $\mathcal{M}$ and $\mathcal{N}$ be left modules over a $\Bbbk$-linear monoidal category $\mathcal{C}$, and let $F\colon \mathcal{M} \to \mathcal{N}$ and $G\colon \mathcal{N} \to \mathcal{M}$ be $\Bbbk$-linear functors. There is a canonical bijection between oplax module functor structures on F and $\mathcal{C}$-module profunctor structures on*

$$\mathrm{Hom}_{\mathcal{N}}(F(-),-): \quad \mathcal{M}^{\mathrm{opp}} \times \mathcal{N} \longrightarrow \mathrm{vect}_\Bbbk\,, \tag{4.11}$$

as well as a canonical bijection between lax module functor structures on G and $\mathcal{C}$-module profunctor structures on

$$\mathrm{Hom}_{\mathcal{M}}(-,G(-)): \quad \mathcal{M}^{\mathrm{opp}} \times \mathcal{N} \longrightarrow \mathrm{vect}_\Bbbk\,. \tag{4.12}$$

This allows one to transport lax to oplax module functor structures and vice versa:

COROLLARY 4.7 ([**Sh**, Lemmas 2.4 & 2.5]). *Let $\mathcal{M}$ and $\mathcal{N}$ be left modules over a $\Bbbk$-linear monoidal category $\mathcal{C}$, and let $F\colon \mathcal{M} \to \mathcal{N}$ be a linear functor with right adjoint $G\colon \mathcal{N} \to \mathcal{M}$. There is a canonical bijection between oplax $\mathcal{C}$-module functor structures on F and lax $\mathcal{C}$-module structures on G, such that the adjunction φ with components*

$$\varphi_{m,n}: \quad \mathrm{Hom}(F(m),n) \longrightarrow \mathrm{Hom}(m,G(n)) \tag{4.13}$$

is an isomorphism of $\mathcal{C}$-module profunctors. If F is even a strong module functor, then G has a unique structure of a lax $\mathcal{C}$-module functor such that the unit and counit of the adjunction φ are module natural transformations.

It follows in particular, by considering the module functor $F = Q_m^{\triangleright}$, that for every object m in a left $\mathcal{C}$-module category $\mathcal{M}$ the internal Hom $\underline{\mathrm{Hom}}^{\mathrm{r}}(m,-)$ is a lax module functor. Hence there are natural coherent morphisms

$${}^{\mathrm{r}}\delta^m_{c,n}: \quad c \otimes \underline{\mathrm{Hom}}^{\mathrm{r}}(m,n) \longrightarrow \underline{\mathrm{Hom}}^{\mathrm{r}}(m, c \triangleright n) \tag{4.14}$$

for $c \in \mathcal{C}$ and $m, n \in \mathcal{M}$.

REMARK 4.8. The observation that internal Homs are not necessarily strong module functors is, of course, not new, e.g. it is used in the definition of rigidity in [**GR**, Def. 9.1.2]. Conditions for being strong have been considered in the literature. For instance, for finitely generated projective modules the internal Hom is a strong module functor [**Bo**, p. 269]. For internal coHoms, which naturally appear in the study of categories of comodules over a coalgebra, a similar statement, namely that one has strong module functors for quasi-finite and injective comodules, has been obtained in [**Ta**, Prop. 1.14]. See also Lemma 5.5 for equivalent conditions for internal Homs and coHoms in categories of bimodules to be strong module functors.

The particular case of interest to us is that $\mathcal{C}$ is a GV-category and that $\mathcal{M} = {}_{\mathcal{C}}\mathcal{C}$ is the regular left $\mathcal{C}$-module, for which according to (3.11) we have $\underline{\mathrm{Hom}}^{\mathrm{r}}(x,y) = y \bullet G(x)$. In this case we obtain

LEMMA 4.9. *Let $\mathcal{C}$ be a $\Bbbk$-linear GV-category. Then there is a family*

$$ {}^{\mathrm{r}}\delta^{x}_{c,y} : \; c \otimes \underline{\mathrm{Hom}}^{\mathrm{r}}(x,y) = c \otimes (y \bullet G(x)) \longrightarrow (c \otimes y) \bullet G(x) = \underline{\mathrm{Hom}}^{\mathrm{r}}(x, c \otimes y) \tag{4.15} $$

of morphisms, for $c, x, y \in \mathcal{C}$, which endows the internal Hom $\underline{\mathrm{Hom}}^{\mathrm{r}}(x,y) = y \bullet G(x)$ with a lax module functor structure for $\mathcal{C}$ as the regular left $\mathcal{C}$-module.

While $\underline{\mathrm{Hom}}^{\mathrm{r}}(x,-) = - \bullet G(x)$ is, in general, only a weak module functor for the right exact tensor product $\otimes$, it is a strong module functor for the left exact tensor product $\bullet$. Indeed, the module functor

$$ I_x := \underline{\mathrm{Hom}}^{\mathrm{r}}(x,-) : \quad {}^{\bullet}_{\mathcal{C}}\mathcal{C} \rightarrow {}^{\bullet}_{\mathcal{C}}\mathcal{C} , \tag{4.16} $$

with ${}^{\bullet}_{\mathcal{C}}\mathcal{C}$ the regular $(\mathcal{C}, \bullet)$-left module, is strong because the associator $\alpha^{\bullet}$ provides us with isomorphisms

$$ I_y(c \bullet x) = (c \bullet x) \bullet G(y) \xrightarrow{\cong} c \bullet (x \bullet G(y)) = c \bullet I_y(x) . \tag{4.17} $$

Invoking Lemma 4.6, this implies that the left adjoint $Q_y = - \otimes y \colon {}^{\bullet}_{\mathcal{C}}\mathcal{C} \rightarrow {}^{\bullet}_{\mathcal{C}}\mathcal{C}$ of I_y is an oplax module functor. Accordingly there are coherent morphisms

$$ \begin{aligned} {}^{\mathrm{l}}\delta^{x}_{y,c} : \quad & (G^{-1}(x) \bullet y) \otimes c = Q_c(G^{-1}(x) \bullet y) \\ & \longrightarrow G^{-1}(x) \bullet Q_c(y) = G^{-1}(x) \bullet (y \otimes c) . \end{aligned} \tag{4.18} $$

Borrowing terminology from the theory of linearly distributive categories (see e.g. [**CS, Pa, HasL**]), we give

DEFINITION 4.10. The natural transformations ${}^{\mathrm{l}}\delta$ and ${}^{\mathrm{r}}\delta$ introduced in (4.15) and (4.18) are called the left and right *distributors* of the GV-category $\mathcal{C}$, respectively.

In fact, as stated in the literature [**Pa, HasL**], GV-categories are the same as linearly distributive categories with negation. The interpretation of distributors in terms of GV-structures has, however, (to the best of our knowledge) not been established explicitly. In our approach, the definition of distributors via weak module functors implies

PROPOSITION 4.11. *The distributors ${}^{\mathrm{l}}\delta$ and ${}^{\mathrm{r}}\delta$ satisfy all the compatibility conditions with the unitors for $1^{\otimes}$ and $1^{\bullet}$ (four mixed triangle identities) and with the associators $\alpha^{\otimes}$ and $\alpha^{\bullet}$ (four mixed pentagon identities) that the distributors in a linearly distributive category have to obey.*

PROOF. Since by definition a weak module functor satisfies an appropriate pentagon diagram, the two paths in the pentagon diagram that is built from the associator $\alpha^{\otimes}$ and the weak module structure (4.15), which induces the left distributor, commute:

$$(4.19)\qquad \begin{aligned} &(c\otimes d)\otimes \underline{\mathrm{Hom}}^{\mathrm{r}}(x,y) = (c\otimes d)\otimes (y\bullet G(x)) \\ &\qquad \rightrightarrows (c\otimes (d\otimes y))\bullet G(x) = \underline{\mathrm{Hom}}^{\mathrm{r}}(x, c\otimes (d\otimes y)) \,. \end{aligned}$$

Further, for all $x_1, x_2 \in \mathcal{C}$ there is a canonical isomorphism

$$(4.20)\qquad L_{x_2}\circ L_{x_1} \cong L_{x_1\otimes x_2}$$

of strong module functors. This isomorphism induces an isomorphism of the respective right adjoint weak module functors, and thus translates into a commuting diagram

$$(4.21)\qquad c\otimes ((y\bullet G(x_2))\bullet G(x_1)) \rightrightarrows (c\otimes y)\bullet (G(x_2)\bullet G(x_1)) \,.$$

Two further pentagon diagrams, which are mirror images of (4.19) and (4.21) and involve the right distributor commute by the analogous arguments for the strong module functors $L^{\bullet}_{x}$. Inspection shows that these pentagons are the same as those obeyed by the distributors in a linearly distributive category, as given e.g. in [**CS**, Sect. 2.1.3]. The triangle diagrams involving the unitors (see e.g. [**CS**, Sect. 2.1.2]) are immediate. □

It can also be shown that the two additional pentagon identities that are valid in a linearly distributive category, each of which involves both ${}^{\mathrm{l}}\delta$ and ${}^{\mathrm{r}}\delta$, are fulfilled as well. However, the proof of this statement we know of is less conceptual and considerably more indirect; we refrain from presenting it here. Also note that, as a consequence of their construction via *weak* module functors, the distributors ${}^{\mathrm{r}}\delta$ and ${}^{\mathrm{l}}\delta$ are, in general, not isomorphisms. They *are* isomorphisms if and only if the category is rigid, see Lemma 5.5.

5. Subcategories of rigid objects

Next we characterize, for right closed monoidal categories $\mathcal{C}$, those objects $x\in\mathcal{C}$ for which $\underline{\mathrm{Hom}}^{\mathrm{r}}(x,-)$ is a *strong* module functor, i.e. for which the coherence morphisms ${}^{\mathrm{l}}\delta^{x}_{y,z}$ in (4.15) are *iso*morphisms for all objects $y,z\in\mathcal{C}$. We start with the standard observation (compare e.g. [**EGNO**, Prop. 2.10.8]) that the monoidal equivalence of $\mathcal{C}$ and $\mathrm{End}_{\mathcal{C}}({}_{\mathcal{C}}\mathcal{C})$ directly implies

LEMMA 5.1. *Let $(\mathcal{C},\otimes)$ be a monoidal category and let $x\in\mathcal{C}$. An object $x^{\vee}\in\mathcal{C}$ is a right dual of x if and only if the functor $-\otimes x^{\vee}$ is right adjoint to the functor $-\otimes x$ as a module functor.*

Note that it does not suffice to merely require that $-\otimes x^{\vee}$ is right adjoint to $-\otimes x$ as a linear functor (a counter example is given in [**HalZ**]). A statement analogous to Lemma 5.1 holds for left duals.

PROPOSITION 5.2. *Let $(\mathcal{C},\otimes)$ be a right closed monoidal category. For $x\in\mathcal{C}$ the lax module functor $\underline{\mathrm{Hom}}^{\mathrm{r}}(x,-)\colon \mathcal{C}\to\mathcal{C}$ is a strong module functor if and only if x has a right dual object $x^{\vee}$. Moreover, if this is the case, then $x^{\vee}\cong\underline{\mathrm{Hom}}^{\mathrm{r}}(x,1)$ as objects, and*

$$(5.1)\qquad \underline{\mathrm{Hom}}^{\mathrm{r}}(x,-)\cong -\otimes x^{\vee}$$

as module functors.

PROOF. Assume that the object x has a right dual $x^{\vee}$. Then the left $\mathcal{C}$-module functor $Q_x\colon \mathcal{C} \to \mathcal{C}$ with $Q_x(y) = y \otimes x$ has as a right adjoint the left $\mathcal{C}$-module functor $Q_{x^{\vee}}\colon \mathcal{C} \to \mathcal{C}$, in such a way that the unit and counit of the adjunction $\mathrm{Hom}_{\mathcal{C}}(Q_x(y), z) \cong \mathrm{Hom}_{\mathcal{C}}(y, Q_{x^{\vee}}(z))$ consist of module natural transformations. Thus it follows from Corollary 4.7 that $\underline{\mathrm{Hom}}^{\mathrm{r}}(x, -) = - \otimes x^{\vee}$ as lax module functors. This, in turn, implies that $\underline{\mathrm{Hom}}^{\mathrm{r}}(x, -)$ is in fact a strong module functor, with the associator of $\mathcal{C}$ as module constraint.

Conversely, assume that $\underline{\mathrm{Hom}}^{\mathrm{r}}(x, -)\colon \mathcal{C} \to \mathcal{C}$ is a strong module functor. By Lemma 4.2 it then follows that there is a natural isomorphism

$$\underline{\mathrm{Hom}}^{\mathrm{r}}(x, -) \cong - \otimes \underline{\mathrm{Hom}}^{\mathrm{r}}(x, 1) \tag{5.2}$$

of module functors. Thus $- \otimes \underline{\mathrm{Hom}}^{\mathrm{r}}(x, 1)$ is right adjoint to $- \otimes x$ as a module functor, and so by Lemma 5.1 the object $\underline{\mathrm{Hom}}^{\mathrm{r}}(x, 1) \in \mathcal{C}$ is a right dual of x. □

Now the class of objects of $(\mathcal{C}, \otimes)$ that admit a right dual is closed under the monoidal product $\otimes$. Thus we have

COROLLARY 5.3. *The full subcategory on those objects $y \in \mathcal{C}$ for which $\underline{\mathrm{Hom}}^{\mathrm{r}}(y, -)$ is a strong module functor is a unital monoidal subcategory of $\mathcal{C}$.*

Note that, as for instance the category B_3-bimod of bimodules over the algebra (2.7) illustrates, for a GV-category the dualizing object K need not be contained in this subcategory. Also, clearly, there are analogues for left closed monoidal categories and for internal coHoms:

LEMMA 5.4. *Let $(\mathcal{C}, \otimes)$ be a monoidal category.*

(1) *Assume that the monoidal category $(\mathcal{C}, \otimes^{\mathrm{opp}})$ obtained by reversing the monoidal structure is left closed. Then for every $x \in \mathcal{C}$ the left internal Hom $\underline{\mathrm{Hom}}^{\mathrm{l}}(x, -)$ is a lax right module functor, with module constraint $\underline{\mathrm{Hom}}^{\mathrm{l}}(x, y) \otimes z \to \underline{\mathrm{Hom}}^{\mathrm{l}}(x, y \otimes z)$.*

   Moreover, $\underline{\mathrm{Hom}}^{\mathrm{l}}(x, -)$ is strong if and only if x has a left dual.

(2) *Assume that the opposite category $(\mathcal{C}^{\mathrm{opp}}, \otimes)$ is left closed. A right internal coHom of $(\mathcal{C}, \otimes)$ is a right internal Hom of $(\mathcal{C}^{\mathrm{opp}}, \otimes)$. As a consequence, $\underline{\mathrm{coHom}}^{\mathrm{r}}(x, -)$ is an oplax left module functor with module constraint $\underline{\mathrm{coHom}}^{\mathrm{r}}(x, y \otimes z) \to y \otimes \underline{\mathrm{coHom}}^{\mathrm{r}}(x, z)$.*

   Moreover, $\underline{\mathrm{coHom}}^{\mathrm{r}}(x, -)$ is strong if and only if x has a left dual.

(3) *Assume that the monoidal category $(\mathcal{C}^{\mathrm{opp}}, \otimes^{\mathrm{opp}})$ is left closed. Then the left internal coHom $\underline{\mathrm{coHom}}^{\mathrm{l}}$ is an oplax right module functor with module constraint $\underline{\mathrm{coHom}}^{\mathrm{l}}(x, y \otimes z) \to \underline{\mathrm{coHom}}^{\mathrm{l}}(x, y) \otimes z$.*

   Moreover, $\underline{\mathrm{coHom}}^{\mathrm{l}}(x, -)$ is strong if and only if x has a right dual.

As an illustration, for categories of bimodules we get

LEMMA 5.5. *Let A be a finite-dimensional $\Bbbk$-algebra, and let ${}_A M_A \in A$-bimod be a finite-dimensional A-bimodule.*

The following statements are equivalent:

(i) *$\underline{\mathrm{Hom}}^{\mathrm{r}}(M, -)$ is a strong module functor.*
(ii) *M has an $\otimes_A$-right dual.*
(iii) *M_A is projective as a right A-module.*
(iv) *${}_A(M^*)$ is injective as a left A-module.*

(v) *For all* $X, Y \in A$-bimod *the distributor* ${}^{\mathrm{r}}\delta^{M}_{X,Y}\colon X \otimes_A (Y \otimes^A M^*) \longrightarrow (X \otimes_A Y) \otimes^A M^*$ *is an isomorphism.*

Likewise, the following statements are equivalent:

(i′) $\underline{\mathrm{Hom}}^{\mathrm{l}}(M,-)$ *is a strong module functor.*
(ii′) M *has an* $\otimes_A$*-left dual.*
(iii′) ${}_AM$ *is projective as a left* A*-module.*
(iv′) $(M^*)_A$ *is injective as a right* A*-module.*
(v′) *For all* $X, Y \in A$-bimod *the distributor* ${}^{\mathrm{l}}\delta^{M}_{X,Y}\colon (M^* \otimes^A X) \otimes_A Y \longrightarrow M^* \otimes^A (X \otimes_A Y)$ *is an isomorphism.*

Proof. (ii) follows from (i) with Proposition 5.2. The equivalence (ii) $\Longleftrightarrow$ (iii) is standard (compare e.g. [**nLab**]). Since $(-)^*$ is an antiequivalence, (iv) is equivalent to (iii). Finally, equivalence of (v) and (i) follows again from Proposition 5.2. □

We can also determine the duals explicitly. For instance, the $\otimes_A$-right dual of ${}_AM_A$ is

$$G^{-1}(M \otimes K) = (M \otimes_A A^*)^* = \mathrm{Hom}_A(M_A, A_A)\,, \tag{5.3}$$

in accordance with (2.19).

6. Distributors for bimodules

Let us now describe in detail the distributors for the case of categories of finite-dimensional bimodules over finite-dimensional $\Bbbk$-algebras considered in Section 2. Recall that we assume all algebras A, B, etc. as well as all bimodules to be finite-dimensional.

Making use of equalizer inclusions and coequalizer surjections, for any triple X, Y, Z of (A, A)-bimodules we will define four (A, A)-bimodule homomorphisms

$$\begin{aligned} &\partial^{\mathrm{l}}_{X,Y,Z}\,,\ \widetilde{\partial}^{\mathrm{l}}_{X,Y,Z}: \quad X \otimes_A (Y \otimes^A Z) \longrightarrow (X \otimes_A Y) \otimes^A Z \\ \text{and}\quad &\partial^{\mathrm{r}}_{X,Y,Z}\,,\ \widetilde{\partial}^{\mathrm{r}}_{X,Y,Z}: \quad (X \otimes^A Y) \otimes_A Z \longrightarrow X \otimes^A (Y \otimes_A Z) \end{aligned} \tag{6.1}$$

which, up to vector space dualities, are natural candidates for the distributors introduced in Definition 4.10. A priori, these are all different, but we will show that in fact $\widetilde{\partial}^{\mathrm{l}}_{X,Y,Z} = \partial^{\mathrm{l}}_{X,Y,Z}$ and $\widetilde{\partial}^{\mathrm{r}}_{X,Y,Z} = \partial^{\mathrm{r}}_{X,Y,Z}$.

Let us first present the definition of $\partial^{\mathrm{l}}_{X,Y,Z}$ in some detail. Consider the diagram

$$\begin{array}{ccc} X \otimes_{\Bbbk} (Y \otimes_{\Bbbk} Z) & \xrightarrow[\cong]{\alpha_{\mathrm{vect}}^{-1}} & (X \otimes_{\Bbbk} Y) \otimes_{\Bbbk} Z \\ {\scriptstyle X \otimes_{\Bbbk} \imath_{Y,Z}} \uparrow & \nearrow & \uparrow {\scriptstyle \imath^{\mathrm{l}}_{X,Y,Z}} \\ X \otimes_{\Bbbk} (Y \otimes^A Z) & & (X \otimes_{\Bbbk} Y) \otimes^A Z \end{array} \tag{6.2}$$

where we explicitly display the associator α_{vect} of the category of finite-dimensional vector spaces, and where

$$\begin{aligned} &\imath_{Y,Z}: \quad Y \otimes^A Z \hookrightarrow Y \otimes_{\Bbbk} Z \\ \text{and}\quad &\imath^{\mathrm{l}}_{X,Y,Z}: \quad (X \otimes_{\Bbbk} Y) \otimes^A Z \hookrightarrow (X \otimes_{\Bbbk} Y) \otimes_{\Bbbk} Z \end{aligned} \tag{6.3}$$

are the equalizer inclusions of $Y \otimes^A Z$ into $Y \otimes_{\Bbbk} Z$ and of $(X \otimes_{\Bbbk} Y) \otimes^A Z$ into $(X \otimes_{\Bbbk} Y) \otimes_{\Bbbk} Z$, respectively. Note that since the tensor product of vector spaces is exact, the map $X \otimes_{\Bbbk} \imath_{Y,Z}$ is injective. Explicitly we have

$$\begin{aligned} \alpha_{\text{vect}}^{-1} \circ (X \otimes_{\Bbbk} \imath_{Y,Z}) : \quad & X \otimes_{\Bbbk} (Y \otimes^A Z) \longrightarrow (X \otimes_{\Bbbk} Y) \otimes_{\Bbbk} Z \,, \\ & \sum_i x \otimes (y_i \otimes z_i) \mapsto \sum_i (x \otimes y_i) \otimes z_i \,, \end{aligned} \tag{6.4}$$

where $x \in X$ and we write a generic element of $Y \otimes^A Z$ as a finite sum $\sum_i y_i \otimes_{\Bbbk} z_i$ with $y_i \in Y$ and $z_i \in Z$, obeying $\sum_i y_i.a \otimes_{\Bbbk} z_i = \sum_i y_i \otimes_{\Bbbk} a.z_i$ for all $a \in A$. The map (6.4) is cobalanced, i.e. $\sum_i (x \otimes y_i) \otimes a.z_i = \sum_i (x \otimes y_i.a) \otimes z_i = \sum_i (x \otimes y_i).a \otimes z_i$ for all $a \in A$; hence by the universal property of equalizers there exists a unique homomorphism $\gamma^1_{X,Y,Z} \colon X \otimes_{\Bbbk} (Y \otimes^A Z) \to (X \otimes_{\Bbbk} Y) \otimes^A Z$ such that the diagram

$$\begin{array}{ccc} X \otimes_{\Bbbk} (Y \otimes_{\Bbbk} Z) & \xrightarrow[\cong]{\ \alpha_{\text{vect}}^{-1}\ } & (X \otimes_{\Bbbk} Y) \otimes_{\Bbbk} Z \\ {\scriptstyle X \otimes_{\Bbbk} \imath_{Y,Z}} \uparrow & & \uparrow {\scriptstyle \imath^1_{X,Y,Z}} \\ X \otimes_{\Bbbk} (Y \otimes^A Z) & \xrightarrow{\ \gamma^1_{X,Y,Z}\ } & (X \otimes_{\Bbbk} Y) \otimes^A Z \end{array} \tag{6.5}$$

commutes. Since $\imath^1_{X,Y,Z}$ and $X \otimes_{\Bbbk} \imath_{Y,Z}$ are injective and α_{vect} is an isomorphism, and since the images of $\imath_{Y,Z}$ and of (6.4) coincide, $\gamma^1_{X,Y,Z}$ is an isomorphism. Next consider the diagram

$$\begin{array}{ccc} X \otimes_{\Bbbk} (Y \otimes^A Z) & \xrightarrow[\cong]{\ \gamma^1_{X,Y,Z}\ } & (X \otimes_{\Bbbk} Y) \otimes^A Z \\ {\scriptstyle \pi^1_{X,Y,Z}} \downarrow & \searrow & \downarrow {\scriptstyle \pi_{X,Y} \otimes^A Z} \\ X \otimes_A (Y \otimes^A Z) & & (X \otimes_A Y) \otimes^A Z \end{array} \tag{6.6}$$

where

$$\begin{aligned} \pi_{X,Y} : \quad & X \otimes_{\Bbbk} Y \twoheadrightarrow X \otimes_A Y \\ \text{and} \quad \pi^1_{X,Y,Z} : \quad & X \otimes_{\Bbbk} (Y \otimes^A Z) \twoheadrightarrow X \otimes_A (Y \otimes^A Z) \end{aligned} \tag{6.7}$$

are the coequalizer surjections from $X \otimes_{\Bbbk} Y$ onto $X \otimes_A Y$ and from $X \otimes_{\Bbbk} (Y \otimes^A Z)$ onto $X \otimes_A (Y \otimes^A Z)$, respectively. Note that since the functor $- \otimes^A Z$ need not be right exact, $\pi_{X,Y} \otimes^A Z$ need not be surjective. Explicitly we have

$$\begin{aligned} (\pi_{X,Y} \otimes^A Z) \circ \gamma^1_{X,Y,Z} : \quad & X \otimes_{\Bbbk} (Y \otimes^A Z) \longrightarrow (X \otimes_A Y) \otimes^A Z \,, \\ & \sum_i x \otimes_{\Bbbk} (y_i \otimes_{\Bbbk} z_i) \longmapsto \sum_i [x \otimes_{\Bbbk} y_i] \otimes_{\Bbbk} z_i \,, \end{aligned} \tag{6.8}$$

where $[x \otimes_{\Bbbk} y]$ is the element of $X \otimes_A Y$ represented by $x \otimes_{\Bbbk} y$ with $x \in X$ and $y \in Y$. The map (6.8) is balanced, i.e. $\sum_i [x.a \otimes_{\Bbbk} y_i] \otimes_{\Bbbk} z_i = \sum_i [x \otimes_{\Bbbk} a.y_i] \otimes_{\Bbbk} z_i$ for all $a \in A$, By the universal property of the coequalizer defining the $\otimes_A$-tensor product, there is thus a unique map $\partial^{\mathrm{l}}_{X,Y,Z}\colon X \otimes_A (Y \otimes^A Z) \longrightarrow (X \otimes_A Y) \otimes^A Z$ such that the diagram

$$
\begin{array}{ccc}
X \otimes_{\Bbbk} (Y \otimes^A Z) & \xrightarrow[\cong]{\gamma^{\mathrm{l}}_{X,Y,Z}} & (X \otimes_{\Bbbk} Y) \otimes^A Z \\
{\scriptstyle \pi^{\mathrm{l}}_{X,Y,Z}}\downarrow & & \downarrow{\scriptstyle \pi_{X,Y} \otimes^A Z} \\
X \otimes_A (Y \otimes^A Z) & \xrightarrow{\partial^{l}_{X,Y,Z}} & (X \otimes_A Y) \otimes^A Z
\end{array}
\tag{6.9}
$$

commutes. Explicitly we have

$$
\partial^{\mathrm{l}}_{X,Y,Z}: \quad \sum_i [x \otimes_{\Bbbk} (y_i \otimes_{\Bbbk} z_i)] \longmapsto \sum_i [x \otimes_{\Bbbk} y_i] \otimes_{\Bbbk} z_i \,. \tag{6.10}
$$

The other three maps in (6.1) are constructed similarly, leading to

DEFINITION 6.1. Let X, Y and Z be (A,A)-bimodules. The homomorphism $\partial^{\mathrm{l}}_{X,Y,Z}\colon X \otimes_A (Y \otimes^A Z) \longrightarrow (X \otimes_A Y) \otimes^A Z$ is the one that is determined by the commutativity of the two squares (6.5) and (6.9).

The homomorphisms $\widetilde{\partial}^{\mathrm{l}}_{X,Y,Z}\colon X \otimes_A (Y \otimes^A Z) \longrightarrow (X \otimes_A Y) \otimes^A Z$ and $\partial^{\mathrm{r}}_{X,Y,Z}$, $\widetilde{\partial}^{\mathrm{r}}_{X,Y,Z}\colon (X \otimes^A Y) \otimes_A Z \longrightarrow X \otimes^A (Y \otimes_A Z)$ are the ones that are determined by the commutativity of the following pairs of squares, respectively (using self-explanatory notation similar to the one in (6.3) and (6.7)):

- $\widetilde{\partial}^{l}_{X,Y,Z}$:

$$
\begin{array}{ccc}
X \otimes_{\Bbbk} (Y \otimes_{\Bbbk} Z) & \xrightarrow[\cong]{\alpha_{\mathrm{vect}}^{-1}} & (X \otimes_{\Bbbk} Y) \otimes_{\Bbbk} Z \\
{\scriptstyle \widetilde{\pi}^{\mathrm{l}}_{X,Y,Z}}\downarrow & & \downarrow{\scriptstyle \pi_{X,Y} \otimes_{\Bbbk} Z} \\
X \otimes_A (Y \otimes_{\Bbbk} Z) & \xrightarrow[\cong]{\widetilde{\gamma}^{\mathrm{l}}_{X,Y,Z}} & (X \otimes_A Y) \otimes_{\Bbbk} Z
\end{array}
\quad \text{and} \quad
\begin{array}{ccc}
X \otimes_A (Y \otimes_{\Bbbk} Z) & \xrightarrow[\cong]{\widetilde{\gamma}^{\mathrm{l}}_{X,Y,Z}} & (X \otimes_A Y) \otimes_{\Bbbk} Z \\
{\scriptstyle X \otimes_A \imath_{Y,Z}}\uparrow & & \uparrow{\scriptstyle \widetilde{\imath}^{\mathrm{l}}_{X,Y,Z}} \\
X \otimes_A (Y \otimes^A Z) & \xrightarrow{\widetilde{\partial}^{\mathrm{l}}_{X,Y,Z}} & (X \otimes_A Y) \otimes^A Z
\end{array}
\tag{6.11}
$$

- $\partial^{\mathrm{r}}_{X,Y,Z}$:

$$
\begin{array}{ccc}
(X \otimes_{\Bbbk} Y) \otimes_{\Bbbk} Z & \xrightarrow[\cong]{\alpha_{\mathrm{vect}}} & X \otimes_{\Bbbk} (Y \otimes_{\Bbbk} Z) \\
{\scriptstyle \imath_{X,Y} \otimes_{\Bbbk} Z}\uparrow & & \uparrow{\scriptstyle \imath^{\mathrm{r}}_{X,Y,Z}} \\
(X \otimes^A Y) \otimes_{\Bbbk} Z & \xrightarrow[\cong]{\gamma^{\mathrm{r}}_{X,Y,Z}} & X \otimes^A (Y \otimes_{\Bbbk} Z)
\end{array}
\quad \text{and} \quad
\begin{array}{ccc}
(X \otimes^A Y) \otimes_{\Bbbk} Z & \xrightarrow[\cong]{\gamma^{\mathrm{r}}_{X,Y,Z}} & X \otimes^A (Y \otimes_{\Bbbk} Z) \\
{\scriptstyle \pi^{\mathrm{r}}_{X,Y,Z}}\downarrow & & \downarrow{\scriptstyle X \otimes^A \pi_{Y,Z}} \\
(X \otimes^A Y) \otimes_A Z & \xrightarrow{\partial^{\mathrm{r}}_{X,Y,Z}} & X \otimes^A (Y \otimes_A Z)
\end{array}
\tag{6.12}
$$

- $\widetilde{\partial}^{\mathrm{r}}_{X,Y,Z}$:

(6.13)
$$
\begin{array}{ccc}
(X \otimes_{\Bbbk} Y) \otimes_{\Bbbk} Z & \xrightarrow[\cong]{\alpha_{\mathrm{vect}}} & X \otimes_{\Bbbk} (Y \otimes_{\Bbbk} Z) \\
{\scriptstyle \widetilde{\pi}^{\mathrm{r}}_{X,Y,Z}} \big\downarrow & & \big\downarrow {\scriptstyle X \otimes_{\Bbbk} \pi_{Y,Z}} \\
(X \otimes_{\Bbbk} Y) \otimes_A Z & \overset{\widetilde{\gamma}^{\mathrm{r}}_{X,Y,Z}}{\underset{\cong}{\dashrightarrow}} & X \otimes_{\Bbbk} (Y \otimes_A Z)
\end{array}
$$

and

$$
\begin{array}{ccc}
(X \otimes_{\Bbbk} Y) \otimes_A Z & \xrightarrow[\cong]{\widetilde{\gamma}^{\mathrm{r}}_{X,Y,Z}} & X \otimes_{\Bbbk} (Y \otimes_A Z) \\
{\scriptstyle \imath_{X,Y} \otimes_A Z} \big\uparrow & & \big\uparrow {\scriptstyle \widetilde{\imath}^{\mathrm{r}}_{X,Y,Z}} \\
(X \otimes^A Y) \otimes_A Z & \overset{\widetilde{\partial}^{\mathrm{r}}_{X,Y,Z}}{\dashrightarrow} & X \otimes^A (Y \otimes_A Z)
\end{array}
$$

The so defined homomorphisms indeed provide us with the distributors:

PROPOSITION 6.2. *Let X, Y and Z be (A,A)-bimodules and let*

(6.14)
$$
\begin{aligned}
{}^{\mathrm{r}}\delta^{Z}_{X,Y} : &\quad X \otimes_A \underline{\mathrm{Hom}}^{\mathrm{r}}(Z,Y) \longrightarrow \underline{\mathrm{Hom}}^{\mathrm{r}}(Z, X \otimes_A Y) \\
\text{and} \quad {}^{\mathrm{l}}\delta^{Z}_{X,Y} : &\quad \underline{\mathrm{Hom}}^{\mathrm{l}}(Z,X) \otimes_A Y \longrightarrow \underline{\mathrm{Hom}}^{\mathrm{l}}(Z, X \otimes_A Y)
\end{aligned}
$$

be the respective lax module functor structures on the functors $\underline{\mathrm{Hom}}^{\mathrm{r}}(Z,-)$ *and* $\underline{\mathrm{Hom}}^{\mathrm{l}}(Z,-)$*, as defined in* (4.15) *and* (4.18)*. Then we have*

(6.15)
$$
\partial^{\mathrm{r}}_{X,Y,Z} = {}^{\mathrm{l}}\delta^{X^*}_{Y,Z} = \widetilde{\partial}^{\mathrm{r}}_{X,Y,Z} \qquad \text{and} \qquad \partial^{\mathrm{l}}_{X,Y,Z} = {}^{\mathrm{r}}\delta^{Z^*}_{X,Y} = \widetilde{\partial}^{\mathrm{l}}_{X,Y,Z} \,.
$$

Further, ∂^{r} and ∂^{l} are explicitly given by

(6.16)
$$
\begin{aligned}
&\partial^{\mathrm{r}}_{X,Y,Z}\big(\sum_i [(x_i \otimes_{\Bbbk} y_i) \otimes_{\Bbbk} z]\big) = \sum_i x_i \otimes [y \otimes_{\Bbbk} z] \\
\text{and} \quad &\partial^{\mathrm{l}}_{X,Y,Z}\big(\sum_j [x \otimes_{\Bbbk} (y_j \otimes_{\Bbbk} z_j)]\big) = \sum_j [x \otimes_{\Bbbk} y_j] \otimes_{\Bbbk} z_j \,.
\end{aligned}
$$

PROOF. We show that $\partial^{\mathrm{l}}_{X,Y,Z}$, $\widetilde{\partial}^{\mathrm{l}}_{X,Y,Z}$ and ${}^{\mathrm{l}}\delta^{Z^*}_{X,Y}$ are all given by the second formula in (6.16). We have already obtained that formula for $\partial^{\mathrm{l}}_{X,Y,Z}$ in (6.10). The same result is found for $\widetilde{\partial}^{\mathrm{l}}_{X,Y,Z}$ when performing an analogous calculation as the one leading to (6.10) instead for the diagrams (6.11). To derive the formula also for ${}^{\mathrm{l}}\delta^{Z^*}_{X,Y}$ we invoke Lemma 4.6 and Corollary 4.7 to compute the module functor structure from the associator $\alpha^{\otimes}$ of the $\otimes_A$-tensor product. Consider the adjoint functors $Q_{Z^*}(-) = - \otimes_A Z^*$ and $I_{Z^*}(-) = \underline{\mathrm{Hom}}^{\mathrm{r}}(Z^*,-)$. The associator $\alpha^{\otimes}$ provides a strong module functor structure on Q_{Z^*} and can be used to construct a transformation

(6.17)
$$
\begin{aligned}
\theta^{\otimes}_{U,Y,X} : \quad \mathrm{Hom}(Q_{Z^*}(U), Y) &\longrightarrow \mathrm{Hom}(Q_{Z^*}(X \otimes_A U), X \otimes_A Y) \,, \\
\psi &\mapsto (\mathrm{id}_C \otimes_A \psi) \circ (\alpha^{\otimes}_{X,U,Z^*})^{-1}
\end{aligned}
$$

that is natural in U and Y as well as dinatural in X. The internal Hom adjunction $\phi_{V,W} \colon \mathrm{Hom}(Q_{Z^*}(V), W) \xrightarrow{\cong} \mathrm{Hom}(V, I_{Z^*}(W))$ can then be used to construct a (di)natural transformation

(6.18)
$$
\theta^{\delta}_{U,Y,X} : \quad \mathrm{Hom}(U, I_{Z^*}(Y)) \longrightarrow \mathrm{Hom}(X \otimes_A U, I_{Z^*}(X \otimes_A Y))
$$

by setting $\theta^{\delta}_{U,Y,X} := \phi_{X\otimes_A U, X\otimes_A Y} \circ \theta^{\otimes}_{U,Y,X} \circ \phi^{-1}_{U,Y}$. That is, $\theta^{\delta}_{U,Y,X}$ is exactly such that the diagram

$$\begin{array}{ccc} \mathrm{Hom}(Q_{Z^*}(U), Y) & \xrightarrow{\theta^{\otimes}_{U,Y,X}} & \mathrm{Hom}(Q_{Z^*}(X \otimes_A U), X \otimes_A Y) \\ \downarrow{\scriptstyle \phi_{U,V}} & & \downarrow{\scriptstyle \phi_{X\otimes_A U, X\otimes_A V}} \\ \mathrm{Hom}(U, I_{Z^*}(Y)) & \xrightarrow{\theta^{\delta}_{U,Y,X}} & \mathrm{Hom}(X \otimes_A U, I_{Z^*}(X \otimes_A Y)) \end{array} \tag{6.19}$$

commutes. The transported lax module structure on I_{Z^*} is then

$$ {}^{\mathrm{l}}\delta^{Z^*}_{X,Y} = \theta^{\delta}_{I_{Z^*}(Y),Y,X}(\mathrm{id}_{I_{Z^*}(Y)}) \,. \tag{6.20}$$

Evaluating this map explicitly on generic elements in X, $Y \otimes^A Z$ and Z^*, respectively, finally leads again to the expression given for $\partial^{\mathrm{l}}_{X,Y,Z}$ in (6.16).

The formulas for $\partial^{\mathrm{r}}_{X,Y,Z}$, $\widetilde{\partial}^{\mathrm{r}}_{X,Y,Z}$ and ${}^{\mathrm{r}}\delta^{Z^*}_{XY}$ follow by analogous arguments. □

We also note the following properties of the distributors, which follow independently of Lemma 5.5:

PROPOSITION 6.3. *Let X be an (A,A)-bimodule.*

(1) *If X is left $\otimes_A$-flat, then $\partial^{\mathrm{l}}_{X,Y,Z}$ is injective for all $Y, Z \in A$*-bimod.
(2) *If X is right $\otimes^A$-flat, then $\partial^{\mathrm{l}}_{Y,Z,X}$ is surjective for all $Y, Z \in A$*-bimod.
(3) *If X is right $\otimes_A$-flat, then $\partial^{\mathrm{r}}_{Y,Z,X}$ is injective for all $Y, Z \in A$*-bimod.
(4) *If X is left $\otimes^A$-flat, then $\partial^{\mathrm{r}}_{X,Y,Z}$ is surjective for all $Y, Z \in A$*-bimod.

PROOF. Part (1): Consider the second of the diagrams (6.11). The composite $\widetilde{\gamma}^{\mathrm{l}}_{X,Y,Z} \circ (X \otimes_A \imath_{Y,Z})$ is cobalanced, and $\partial^{\mathrm{l}}_{X,Y,Z}$ is the unique homomorphism determined by the universal property of the equalizer. Thus the kernel of $\partial^{\mathrm{l}}_{X,Y,Z}$ equals the kernel of $X \otimes_A \imath_{Y,Z}$. Now note that this map is obtained by applying the functor $X \otimes_A -$ to the inclusion in the equalizer definition of $\otimes^A$. Hence the kernel of $X \otimes_A \imath_{Y,Z}$ is determined by torsion, and hence $\partial^{l}_{X,Y,Z}$ is injective if X is left $\otimes_A$-flat.

Part (2): Recall that in the diagram (6.9) the composite $(\pi_{X,Y} \otimes^A Z) \circ \gamma^{\mathrm{l}}_{X,Y,Z}$ is balanced, and that $\partial^{\mathrm{l}}_{X,Y,Z}$ is the unique homomorphism determined by the universal property of the coequalizer. Thus the image of $\pi_{X,Y} \otimes^A Z$ equals the image of $\partial^{\mathrm{l}}_{X,Y,Z}$. Note further that $\pi_{X,Y} \otimes^A Z$ is obtained by applying the functor $-\otimes^A$ to the surjection in the coequalizer definition of $\otimes_A$. Hence the image of $\partial^{\mathrm{l}}_{X,Y,Z}$ is determined by cotorsion, and thus $\partial^{\mathrm{l}}_{X,Y,Z}$ is surjective if Z is right $\otimes^A$-flat.

The statements (3) and (4) are shown analogously. □

It is worth pointing out that the failure of the distributors to be isomorphisms can concern both their kernel and their image. We mention two concrete examples: First, the three-dimensional algebra $B_3 = \Bbbk[x,y]/\langle x^2, y^2, xy\rangle$ considered in (2.7) has two non-isomorphic three-dimensional indecomposable modules, namely the projective $P = B_3 = \mathrm{span}_{\Bbbk}\{1, x, y\}$ and the injective $I = P^* = \mathrm{span}_{\Bbbk}\{1^*, x^*, y^*\}$. We find that

$$
\begin{aligned}
\ker(\partial^{\mathrm{l}}_{I,P,P}) = \mathrm{span}_{\Bbbk}\{\, & [x^* \otimes_{\Bbbk} (x \otimes_{\Bbbk} x)] - [y^* \otimes_{\Bbbk} (y \otimes_{\Bbbk} x)]\,, \\
& [x^* \otimes_{\Bbbk} (x \otimes_{\Bbbk} y)] - [y^* \otimes_{\Bbbk} (y \otimes_{\Bbbk} y)]\,, [x^* \otimes_{\Bbbk} (y \otimes_{\Bbbk} x)]\,, \\
& [x^* \otimes_{\Bbbk} (y \otimes_{\Bbbk} y)]\,, [y^* \otimes_{\Bbbk} (x \otimes_{\Bbbk} x)]\,, [y^* \otimes_{\Bbbk} (x \otimes_{\Bbbk} y)] \,\}\,, \\
\mathrm{im}(\partial^{\mathrm{l}}_{I,P,P}) = \mathrm{span}_{\Bbbk}\{\, & [1^* \otimes_{\Bbbk} 1] \otimes_{\Bbbk} x, [1^* \otimes_{\Bbbk} 1] \otimes_{\Bbbk} y \,\}\,,
\end{aligned}
\tag{6.21}
$$

i.e. $\partial^{\mathrm{l}}_{I,P,P}$ is neither injective nor surjective. Thus I is not $\otimes_A$-flat, while P is not $\otimes^A$-flat. Second, for the algebra (2.4) of dual numbers, which up to isomorphism has two indecomposable modules $S = \mathrm{span}_{\Bbbk}\{s\}$ and $P = A_2 = \mathrm{span}_{\Bbbk}\{1, x\}$, we find an example for which the distributor vanishes: $\partial^{\mathrm{l}}_{S,P,S} = 0$. So S is neither $\otimes_A$-flat nor $\otimes^A$-flat.

Both the concrete examples and the general results studied in this note reveal that GV-dualities are natural structures and that they deserve further study, in particular in the realm of quantum algebra and quantum topology.

References

[ALSW] R. Allen, S. Lentner, C. Schweigert, and S. Wood, *Duality structures for module categories of vertex operator algebras and the Feigin Fuchs boson*, `arXiv:2107.05718 [math.QA]`, 2021.

[Ba1] M. Barr, *∗-autonomous categories*, Lecture Notes in Mathematics, vol. 752, Springer, Berlin, 1979. With an appendix by Po Hsiang Chu. MR550878

[Ba2] M. Barr, *Nonsymmetric ∗-autonomous categories*, Theoret. Comput. Sci. **139** (1995), no. 1-2, 115–130, DOI 10.1016/0304-3975(94)00089-2. MR1320237

[Ba3] M. Barr, *∗-autonomous categories, revisited*, J. Pure Appl. Algebra **111** (1996), no. 1-3, 1–20, DOI 10.1016/0022-4049(95)00040-2. MR1394341

[Ba4] M. Barr, *∗-autonomous categories: once more around the track*, Theory Appl. Categ. **6** (1999), 5–24. The Lambek Festschrift. MR1732459

[Bo] N. Bourbaki, *Algebra. I. Chapters 1–3*, Elements of Mathematics (Berlin), Springer-Verlag, Berlin, 1989. Translated from the French; Reprint of the 1974 edition. MR979982

[BoD] M. Boyarchenko and V. Drinfeld, *A duality formalism in the spirit of Grothendieck and Verdier*, Quantum Topol. **4** (2013), no. 4, 447–489, DOI 10.4171/QT/45. MR3134025

[BrW] A. Brochier and L. Woike, *A classification of modular functors via factorization homology*, `arXiv:2212.11259 [math.QA]`, 2022.

[BW] T. Brzezinski and R. Wisbauer, *Corings and comodules*, London Mathematical Society Lecture Note Series, vol. 309, Cambridge University Press, Cambridge, 2003, DOI 10.1017/CBO9780511546495. MR2012570

[CS] J. R. B. Cockett and R. A. G. Seely, *Weakly distributive categories*, J. Pure Appl. Algebra **114** (1997), no. 2, 133–173, DOI 10.1016/0022-4049(95)00160-3. MR1426767

[EGNO] P. Etingof, S. Gelaki, D. Nikshych, and V. Ostrik, *Tensor categories*, Mathematical Surveys and Monographs, vol. 205, American Mathematical Society, Providence, RI, 2015, DOI 10.1090/surv/205. MR3242743

[FB] E. Frenkel and D. Ben-Zvi, *Vertex algebras and algebraic curves*, 2nd ed., Mathematical Surveys and Monographs, vol. 88, American Mathematical Society, Providence, RI, 2004, DOI 10.1090/surv/088. MR2082709

[FSS] J. Fuchs, G. Schaumann, and C. Schweigert, *Eilenberg-Watts calculus for finite categories and a bimodule Radford S^4 theorem*, Trans. Amer. Math. Soc. **373** (2020), no. 1, 1–40, DOI 10.1090/tran/7838. MR4042867

[GK] M. R. Gaberdiel and H. G. Kausch, *Indecomposable fusion products*, Nuclear Phys. B **477** (1996), no. 1, 293–318, DOI 10.1016/0550-3213(96)00364-1. MR1413262

[GRW] M. R. Gaberdiel, I. Runkel, and S. Wood, *Logarithmic bulk and boundary conformal field theory and the full centre construction*, Conformal field theories and tensor categories, Math. Lect. Peking Univ., Springer, Heidelberg, 2014, pp. 93–168. MR3585367

[GR] D. Gaitsgory and N. Rozenblyum, *A study in derived algebraic geometry. Vol. I. Correspondences and duality*, Mathematical Surveys and Monographs, vol. 221, American Mathematical Society, Providence, RI, 2017, DOI 10.1090/surv/221.1. MR3701352

[HalZ] S. Halbig and T. Zorman, *Duality in monoidal categories*, `arXiv:2301.03545 [math.CT]`, 2023.

[HasL] M. Hasegawa and J.-S. P. Lemay, *Linear distributivity with negation, star-autonomy, and Hopf monads*, Theory Appl. Categ. **33** (2018), Paper No. 37, 1145–1157. MR3909240

[HLZ] Y.-Z. Huang, J. Lepowsky, and L. Zhang, *Logarithmic tensor category theory for generalized modules for a conformal vertex algebra, I: introduction and strongly graded algebras and their generalized modules*, Conformal field theories and tensor categories, Math. Lect. Peking Univ., Springer, Heidelberg, 2014, pp. 169–248. MR3585368

[Iv] S. O. Ivanov, *Nakayama functors and Eilenberg-Watts theorems* (Russian, with English and Russian summaries), Zap. Nauchn. Sem. S.-Peterburg. Otdel. Mat. Inst. Steklov. (POMI) **388** (2011), no. Voprosy Teorii Predstavleniĭ Algebr i Grupp. 21, 179–188, 311, DOI 10.1007/s10958-012-0831-2; English transl., J. Math. Sci. (N.Y.) **183** (2012), no. 5, 675–680. MR2822520

[KR] S. Kanade and D. Ridout, *NGK and HLZ: fusion for physicists and mathematicians*, Affine, vertex and W-algebras, Springer INdAM Ser., vol. 37, Springer, Cham, 2019, pp. 135–181, DOI 10.1007/978-3-030-32906-8_7. MR4290628

[LZ] B. H. Lian and G. J. Zuckerman, *New perspectives on the BRST-algebraic structure of string theory*, Comm. Math. Phys. **154** (1993), no. 3, 613–646. MR1224094

[Ma] Yu. I. Manin, *Grothendieck-Verdier duality patterns in quantum algebra* (Russian, with English summary), Izv. Ross. Akad. Nauk Ser. Mat. **81** (2017), no. 4, 158–166, DOI 10.4213/im8620; English transl., Izv. Math. **81** (2017), no. 4, 818–826. MR3682786

[MW] L. Müller and L. Woike, *Dimensional reduction, extended topological field theories and orbifoldization*, Bull. Lond. Math. Soc. **53** (2021), no. 2, 392–403, DOI 10.1112/blms.12427. MR4239182

[Na] W. Nahm, *Quasi-rational fusion products*, Internat. J. Modern Phys. B **8** (1994), no. 25-26, 3693–3702, DOI 10.1142/S0217979294001597. Perspectives on solvable models. MR1305167

[nLab] nLab, *Dualizable module*, `https://ncatlab.org/nlab/show/dualizable+module`

[Pa] C. Pastro, *Note on star-autonomous comonads*, Theory Appl. Categ. **26** (2012), No. 7, 194–203, DOI 10.1007/s13347-012-0089-0. MR2948487

[Se] R. A. G. Seely, *Linear logic, $*$-autonomous categories and cofree coalgebras*, Categories in computer science and logic (Boulder, CO, 1987), Contemp. Math., vol. 92, Amer. Math. Soc., Providence, RI, 1989, pp. 371–382, DOI 10.1090/conm/092/1003210. MR1003210

[Sh] K. Shimizu, *Relative Serre functor for comodule algebras*, J. Algebra **634** (2023), 237–305, DOI 10.1016/j.jalgebra.2023.07.015. MR4624433

[Ta] M. Takeuchi, *Morita theorems for categories of comodules*, J. Fac. Sci. Univ. Tokyo Sect. IA Math. **24** (1977), no. 3, 629–644. MR472967

Teoretisk fysik, Karlstads Universitet, Universitetsgatan 21, S – 651 88 Karlstad, Sweden

Email address: `juerfuch@kau.se`

Mathematische Physik, Institut für Mathematik, Universität Würzburg, Emil-Fischer-Strasse 31, D – 97 074 Würzburg, Germany

Email address: `gregor.schaumann@mathematik.uni-wuerzburg.de`

Fachbereich Mathematik, Universität Hamburg, Bereich Algebra und Zahlentheorie, Bundesstrasse 55, D – 20 146 Hamburg, Germany

Email address: `christoph.schweigert@uni-hamburg.de`

School of Mathematics, Cardiff University, Abacws, Senghennydd Road, Cardiff, CF24 4AG, United Kingdom; and Fachbereich Mathematik, Universität Hamburg, Bereich Algebra und Zahlentheorie, Bundesstrasse 55, D – 20 146 Hamburg, Germany

Email address: `woodsi@cardiff.ac.uk`

Contemporary Mathematics
Volume **813**, 2025
https://doi.org/10.1090/conm/813/16290

Action of subgroups of the mapping class group on Heisenberg homologies

Christian Blanchet, Martin Palmer, and Awais Shaukat

ABSTRACT. In previous work we constructed twisted representations of mapping class groups of surfaces, depending on a choice of representation V of the Heisenberg group $\mathcal{H}$. For certain V we were able to untwist these mapping class group representations. Here, we study the restrictions of our twisted representations to different subgroups of the mapping class group. In particular, we prove that these representations may be untwisted on the Torelli group for any given representation V of $\mathcal{H}$. When V is the Schrödinger representation, we also construct untwisted representations of subgroups defined as kernels of crossed homomorphisms studied by Earle and Morita.

Introduction

In recent work [**BPS21**], we constructed a twisted action of the mapping class group of any compact, connected, oriented surface Σ with one boundary component on the homology of configuration spaces with local coefficients determined by a representation V of the discrete Heisenberg group $\mathcal{H} = \mathcal{H}(\Sigma)$. The details of this construction are recalled briefly in §1. For specific representations V of $\mathcal{H}$ we were able to untwist and obtain genuine, untwisted linear representations of the mapping class group (for the linearisation $\mathcal{H}\oplus\mathbb{Z}$ of the affine translation action of $\mathcal{H}$ on itself) or linear representations of central extensions of the mapping class group (for the Schrödinger representation of $\mathcal{H}$).

Our goal here is to complete the study of this action on Heisenberg homology. In §2 we identify the kernel of the action of the mapping class group on the Heisenberg group as the Chillingworth subgroup (Proposition 2.6); hence we obtain a linear representation of this subgroup for any representation V of $\mathcal{H}$ (Theorem 2.12). We also identify the *projective kernel* of this action (i.e. the subgroup of elements that act by inner automorphisms) with the Torelli group (Proposition 2.7) and use this fact in §3 to obtain untwisted linear representations of the Torelli group for any V (Theorem 3.3). In the special case where V is the Schrödinger representation – where in [**BPS21**] we obtained an action of the stably universal central extension

2020 *Mathematics Subject Classification.* Primary 57K20, 55R80, 55N25, 20C12, 19C09.

The first and third authors are thankful for the support of the Abdus Salam School of Mathematical Sciences.

The second author was partially supported by a grant of the Romanian Ministry of Education and Research, CNCS - UEFISCDI, project number PN-III-P4-ID-PCE-2020-2798, within PNCDI III.

of the mapping class group – we show in §4 that, restricting to a so-called *Earle-Morita subgroup* defined in Definition 2.1, we obtain linear representations without passing to any extension (Theorem 4.14).

Using the local system given by the Schrödinger representation at an odd root of 1, De Renzi and Martel [**DM22**] produced a homological model for TQFT representations derived from quantum $sl(2)$. Our results shed light on a few points in their paper. First, Proposition 2.6 identifies a certain subgroup of the mapping class group denoted by $\mathscr{M}_g^{\mathbb{H}}$ in [**DM22**, Proposition 2.21] with the Chillingworth subgroup. Second, the construction of [**DM22**, §6.2] depends on the identification (Proposition 2.7) of the projective kernel of the action of the mapping class group on $\mathcal{H}$ with the Torelli group.

1. Twisted representations of the mapping class group

1.1. A review of Heisenberg homology. Let $\Sigma = \Sigma_{g,1}$, for $g \geq 1$, be a compact, connected, oriented surface of genus g with one boundary component. For $n \geq 2$, the unordered configuration space of n points in $\Sigma_{g,1}$ is

$$\mathcal{C}_n(\Sigma_{g,1}) = \{\{c_1, \ldots, c_n\} \subset \Sigma_{g,1} \mid c_i \neq c_j \text{ for } i \neq j\}.$$

The surface braid group is then defined as $\mathbb{B}_n(\Sigma) = \pi_1(\mathcal{C}_n(\Sigma), *)$. A presentation for this group was first obtained by G. P. Scott [**Sco70**] and subsequently revisited by González-Meneses [**GM01**] and Bellingeri [**Bel04**]. We fix a collection of based loops, $\alpha_1, \ldots, \alpha_g, \beta_1, \ldots, \beta_g$, as depicted in Figure 1. The base point $*_1$ belongs to the base configuration $*$. We will use the same notation α_r, β_s for the corresponding braids where only the first point is moving.

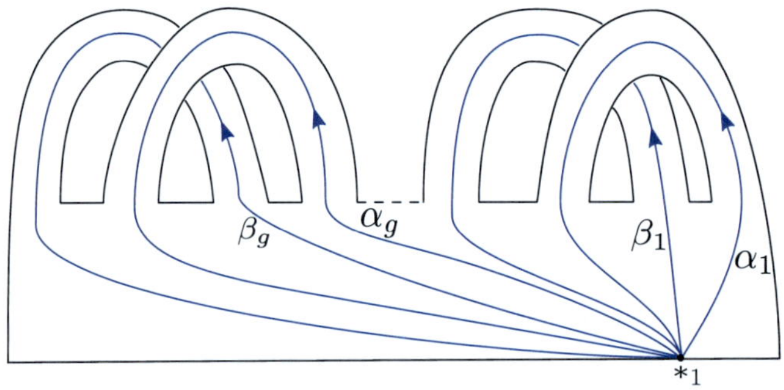

FIGURE 1. Model for Σ.

The braid group $\mathbb{B}_n(\Sigma)$ has generators $\alpha_1, \ldots, \alpha_g, \beta_1, \ldots, \beta_g$, together with the classical generators $\sigma_1, \ldots, \sigma_{n-1}$, and relations:

$$(1.1)\quad \begin{cases} \textbf{(BR1)}\ [\sigma_i, \sigma_j] = 1 & \text{for } |i-j| \geq 2, \\ \textbf{(BR2)}\ \sigma_i\sigma_j\sigma_i = \sigma_j\sigma_i\sigma_j & \text{for } |i-j| = 1, \\ \textbf{(CR1)}\ [\zeta, \sigma_i] = 1 & \text{for } i > 1 \text{ and all } \zeta \text{ among the } \alpha_r, \beta_s, \\ \textbf{(CR2)}\ [\zeta, \sigma_1\zeta\sigma_1] = 1 & \text{for all } \zeta \text{ among the } \alpha_r, \beta_s, \\ \textbf{(CR3)}\ [\zeta, \sigma_1^{-1}\eta\sigma_1] = 1 & \text{for all } \zeta \neq \eta \text{ among the } \alpha_r, \beta_s, \text{ with} \\ & \{\zeta, \eta\} \neq \{\alpha_r, \beta_r\}, \\ \textbf{(SCR)}\ \sigma_1\beta_r\sigma_1\alpha_r\sigma_1 = \alpha_r\sigma_1\beta_r & \text{for all } r. \end{cases}$$

Composition of loops is written from right to left.

We will use the notation $x.y$ for the standard intersection form on $H_1(\Sigma;\mathbb{Z})$. The *Heisenberg group* $\mathcal{H}(\Sigma)$ is the central extension of the homology group $H_1(\Sigma;\mathbb{Z})$ defined using the 2-cocycle $(x,y)\mapsto x.y$. As a set, it is equal to $\mathbb{Z}\times H_1(\Sigma;\mathbb{Z})$, and the operation is given by

$$(k,x)(l,y)=(k+l+x.y,x+y). \tag{1.2}$$

We will often denote the Heisenberg group simply by $\mathcal{H}=\mathcal{H}(\Sigma)$ when the surface Σ under consideration is clear. We will use the notation a_r, b_s, for the homology classes of α_r, β_s. From the presentation (1.1) we deduced the following in [**BPS21**, §1].

PROPOSITION 1.1. *For each $g\geq 1$ and $n\geq 2$, the quotient of the braid group $\mathbb{B}_n(\Sigma)$ by the subgroup $[\sigma_1,\mathbb{B}_n(\Sigma)]^N$ normally generated by the commutators $[\sigma_1,x]$, $x\in\mathbb{B}_n(\Sigma)$, is isomorphic to the Heisenberg group $\mathcal{H}(\Sigma)$. An isomorphism*

$$\mathbb{B}_n(\Sigma)/[\sigma_1,\mathbb{B}_n(\Sigma)]^N\cong\mathcal{H}(\Sigma) \tag{1.3}$$

is represented by the surjective homomorphism

$$\phi\colon\mathbb{B}_n(\Sigma)\longrightarrow\mathcal{H}(\Sigma)$$

sending each σ_i to $u=(1,0)$, α_r to $\tilde{a}_r=(0,a_r)$ and β_s to $\tilde{b}_s=(0,b_s)$.

From the homomorphism ϕ we obtain a regular covering $\widetilde{\mathcal{C}}_n(\Sigma)$ of the configuration space $\mathcal{C}_n(\Sigma)$. The homology of this covering space is the homology of $\mathcal{C}_n(\Sigma)$ with local coefficients defined by ϕ, which we call *Heisenberg homology* and denote by $H_*(\mathcal{C}_n(\Sigma),\mathbb{Z}[\mathcal{H}])$. It is equipped with a right $\mathbb{Z}[\mathcal{H}]$-module structure defined by deck transformations.

Let us denote by $\mathcal{S}_*(\widetilde{\mathcal{C}}_n(\Sigma))$ the singular chain complex of the Heisenberg covering $\widetilde{\mathcal{C}}_n(\Sigma)$; this is a complex of right $\mathbb{Z}[\mathcal{H}]$-modules. Given a (left) representation $\rho\colon\mathcal{H}\to GL(V)$, the corresponding twisted homology is that of the complex

$$\mathcal{S}_*(\mathcal{C}_n(\Sigma);V):=\mathcal{S}_*(\widetilde{\mathcal{C}}_n(\Sigma))\otimes_{\mathbb{Z}[\mathcal{H}]}V \tag{1.4}$$

This will be called the *Heisenberg homology* of surface configurations with coefficients in V.

We also consider the Borel-Moore homology

$$H_*^{BM}(\mathcal{C}_n(\Sigma);V)=\varprojlim_T H_*(\mathcal{C}_n(\Sigma),\mathcal{C}_n(\Sigma)\setminus T;V), \tag{1.5}$$

where the inverse limit is taken over all compact subsets $T\subset\mathcal{C}_n(\Sigma)$. We denote by $\mathcal{C}_n(\Sigma,\partial^-(\Sigma))$ the closed subspace of configurations containing at least one point in a fixed closed interval $\partial^-(\Sigma)\subset\partial\Sigma$. The relative Borel-Moore homology is defined similarly as

$$\begin{aligned}&H_*^{BM}(\mathcal{C}_n(\Sigma),\mathcal{C}_n(\Sigma,\partial^-(\Sigma));V)\\&\qquad=\varprojlim_T H_*\big(\mathcal{C}_n(\Sigma),\mathcal{C}_n(\Sigma,\partial^-(\Sigma))\cup(\mathcal{C}_n(\Sigma)\setminus T);V\big).\end{aligned} \tag{1.6}$$

The following theorem computes the relative Borel-Moore homology as a module.

THEOREM 1.2 ([**BPS21**, §2]). *Let V be any representation of the discrete Heisenberg group $\mathcal{H}(\Sigma)$. Then, for $n\geq 2$, there is an isomorphism of modules*

$$H_n^{BM}(\mathcal{C}_n(\Sigma),\mathcal{C}_n(\Sigma,\partial^-(\Sigma));V)\cong\bigoplus_{k\in\mathcal{K}}V.$$

Furthermore, this is the only non-vanishing] module, in $H_*^{BM}(\mathcal{C}_n(\Sigma),\mathcal{C}_n(\Sigma,\partial^-(\Sigma));V)$.

There is also an explicit geometric description of the indexing set $\mathcal{K}$ of the direct sum decomposition of Theorem 1.2, which is explained in [**BPS21**, §2]; however, it will not be essential for the present paper.

1.2. An action of the mapping class group on the Heisenberg group. The *mapping class group* of Σ, denoted by $\mathfrak{M}(\Sigma)$, is the group of orientation-preserving diffeomorphisms of Σ fixing the boundary pointwise, modulo isotopies relative to the boundary. The isotopy class of a diffeomorphism f is denoted by $[f]$. An orientation-preserving self-diffeomorphism $f\colon \Sigma \to \Sigma$ fixing the boundary pointwise gives a homeomorphism $\mathcal{C}_n(f)\colon \mathcal{C}_n(\Sigma) \to \mathcal{C}_n(\Sigma)$, defined by $\{x_1, x_2, \ldots, x_n\} \mapsto \{f(x_1), f(x_2), \ldots, f(x_n)\}$. If we ensure that the basepoint configuration of $\mathcal{C}_n(\Sigma)$ is contained in $\partial\Sigma$, then it is fixed by $\mathcal{C}_n(f)$ and this in turn induces an automorphism $f_{\mathbb{B}_n(\Sigma)} = \pi_1(\mathcal{C}_n(f))\colon \mathbb{B}_n(\Sigma) \to \mathbb{B}_n(\Sigma)$, which depends only on the isotopy class $[f]$ of f. In [**BPS21**, §3] we proved:

PROPOSITION 1.3. *There exists a unique automorphism $f_{\mathcal{H}}\colon \mathcal{H} \to \mathcal{H}$ such that the following square commutes:*

$$\begin{array}{ccc} \mathbb{B}_n(\Sigma) & \xrightarrow{f_{\mathbb{B}_n(\Sigma)}} & \mathbb{B}_n(\Sigma) \\ \phi\downarrow & & \downarrow\phi \\ \mathcal{H} & \xrightarrow{f_{\mathcal{H}}} & \mathcal{H} \end{array} \tag{1.7}$$

Thus, there is an action of $\mathfrak{M}(\Sigma)$ on the Heisenberg group $\mathcal{H}$ given by

$$\Psi\colon f \mapsto f_{\mathcal{H}}\colon \mathfrak{M}(\Sigma) \longrightarrow \mathrm{Aut}(\mathcal{H}). \tag{1.8}$$

1.3. Twisted representations of the mapping class group. In [**BPS21**] we explained how to use the action (1.8) to obtain twisted representations of $\mathfrak{M}(\Sigma)$ for each representation V of $\mathcal{H}$ over a ring R and each integer $n \geq 2$. We recall this briefly here.

For a (left) representation $\rho\colon \mathcal{H} \to \mathrm{Aut}_R(V)$ and an automorphism $\tau \in \mathrm{Aut}(\mathcal{H})$, the τ-twisted representation $\rho \circ \tau$ is denoted by ${}_\tau V$. Also, for any representation V of $\mathcal{H}$, we denote the induced local system

$$\mathbb{Z}[\widetilde{\mathcal{C}}_n(\Sigma)] \otimes_{\mathbb{Z}[\mathcal{H}]} V$$

on the configuration space $\mathcal{C}_n(\Sigma)$ simply by V, by abuse of notation. We then write

$$\mathcal{V}_n(V) = H_n^{BM}(\mathcal{C}_n(\Sigma), \mathcal{C}_n(\Sigma, \partial^-(\Sigma)); V) \tag{1.9}$$

for the relative Borel-Moore homology with coefficients in this local system. In this notation, [**BPS21**, §4.1] explains that each mapping class $f \in \mathfrak{M}(\Sigma)$ induces an automorphism of $\mathcal{C}_n(\Sigma)$ covered by an isomorphism

$${}_{\tau \circ f_{\mathcal{H}}}V \longrightarrow {}_\tau V$$

of local systems for each $\tau \in \mathrm{Aut}(\mathcal{H})$. Taking relative Borel-Moore homology, we therefore obtain isomorphisms

$$\mathcal{V}_n\bigl({}_{\tau \circ f_{\mathcal{H}}}V\bigr) \longrightarrow \mathcal{V}_n\bigl({}_\tau V\bigr) \tag{1.10}$$

of R-modules. This may be described succinctly as a representation of the *action groupoid* associated to the action (1.8) of $\mathfrak{M}(\Sigma)$ on $\mathcal{H}$.

DEFINITION 1.4. For a group G with a left action $a\colon G \to \mathrm{Sym}(X)$ on a set X, the *action groupoid* $\mathrm{Ac}(G \curvearrowright X)$ is the groupoid whose set of objects is $a(G)$, whose set of morphisms $\sigma \to \tau$ is the subset $a^{-1}(\tau^{-1}\sigma) \subseteq G$ and whose composition is given by multiplication in G.

THEOREM 1.5 ([**BPS21**, Theorem A(b)]). *Associated to any representation V of $\mathcal{H}$ over R and any integer $n \geq 2$, there is a functor*

$$\mathrm{Ac}(\mathfrak{M}(\Sigma) \curvearrowright \mathcal{H}) \longrightarrow \mathrm{Mod}_R \tag{1.11}$$

sending each object $\tau\colon \mathcal{H} \to \mathcal{H}$ to the R-module $\mathcal{V}_n({}_\tau V)$ and sending each morphism $f\colon \tau \circ f_{\mathcal{H}} \to \tau$ to the R-linear isomorphism (1.10).

REMARK 1.6. The basic strategy to upgrade this to an *untwisted* representation is to try to construct coefficient isomorphisms $V \cong {}_{f_{\mathcal{H}}}V$ for each f. Given this, one may then pre-compose (1.10) with the induced isomorphism of twisted homology groups to obtain *automorphisms* of $\mathcal{V}_n(V)$. We explained how to do this in [**BPS21**, §4.2] when V is the linearisation $L = \mathcal{H} \oplus \mathbb{Z}$ of the (affine) translation action of $\mathcal{H}$ on itself. We also explained in [**BPS21**, §5] how to do this – after passing to a certain central extension of $\mathfrak{M}(\Sigma)$ – when V is the Schrödinger representation of $\mathcal{H}$ (this is recalled briefly in §4.1). The goal of the present paper is to explain how to untwist on the Torelli group $\mathfrak{T}(\Sigma) \subset \mathfrak{M}(\Sigma)$ for *any* representation V of $\mathcal{H}$, as well as how to untwist on Earle-Morita subgroups of $\mathfrak{M}(\Sigma)$ when V is the Schrödinger representation of $\mathcal{H}$ (without passing to any central extension).

2. Action on the Heisenberg group

The goal of this section is to study the action (1.8) of the mapping class group $\mathfrak{M}(\Sigma)$ on the Heisenberg group $\mathcal{H} = \mathcal{H}(\Sigma)$, as well as the crossed homomorphism naturally associated to this action.

2.1. Automorphisms of the Heisenberg group.

As a first step, we recall a semi-direct product decomposition of an index-2 subgroup of the automorphism group $\mathrm{Aut}(\mathcal{H})$.

Since the element $u = (1,0)$ of $\mathcal{H}$ generates its centre, which is infinite cyclic, any automorphism of $\mathcal{H}$ must send it either to itself or its inverse. We denote the group of automorphisms of $\mathcal{H}$ that fix u by $\mathrm{Aut}^+(\mathcal{H})$. The structure of this group was studied in [**BPS21**, §3]. There is a natural homomorphism $\mathcal{L}\colon \mathrm{Aut}^+(\mathcal{H}) \to Sp(H)$, where we write $H = H_1(\Sigma;\mathbb{Z})$, and a split short exact sequence

$$1 \longrightarrow H^1(\Sigma;\mathbb{Z}) \xrightarrow{\ j\ } \mathrm{Aut}^+(\mathcal{H}) \xrightarrow{\ \mathcal{L}\ } Sp(H) \longrightarrow 1 \tag{2.1}$$

where $j(c) = [(k,x) \mapsto (k + c(x), x)]$. The splitting gives a decomposition $\mathrm{Aut}^+(\mathcal{H}) \cong Sp(H) \ltimes H^1(\Sigma;\mathbb{Z})$, where the semi-direct product structure on the right-hand side is induced by the natural action of $Sp(H)$ on $\mathrm{Hom}(H,\mathbb{Z}) \cong H^1(\Sigma;\mathbb{Z})$. The projection onto the right-hand factor of this decomposition is a function $(-)^\diamond\colon \mathrm{Aut}^+(\mathcal{H}) \to H^1(\Sigma;\mathbb{Z}) \cong \mathrm{Hom}(H,\mathbb{Z})$ (which is not a group homomorphism) given by the assignment $\varphi \mapsto \varphi^\diamond = \mathrm{pr}_1(\varphi(0,-))$.

For a mapping class $f \in \mathfrak{M}(\Sigma)$, the map $f_{\mathcal{H}}$ of (1.7) is represented as follows:

$$f_{\mathcal{H}}\colon (k,x) \mapsto (k + \delta_f(x), f_*(x)), \tag{2.2}$$

where $\delta_f = (f_{\mathcal{H}})^\diamond \in H^1(\Sigma;\mathbb{Z}) \cong \mathrm{Hom}(H,\mathbb{Z})$. We proved in [**BPS21**, §3] that the map $\delta\colon \mathfrak{M}(\Sigma) \to H^1(\Sigma;\mathbb{Z})$ given by $f \mapsto \delta_f$ is a crossed homomorphism, meaning that

$$\delta_{g\circ f}(x) = \delta_f(x) + f^*(\delta_g)(x).$$

We also identified this crossed homomorphism, showing that it coincides with the combinatorially-defined crossed homomorphism $\mathfrak{d}$ constructed by Morita [**Mor89a**], who proved that it represents a generator of $H^1(\mathfrak{M}(\Sigma); H^1(\Sigma;\mathbb{Z}))$, which is infinite cyclic. In fact, a different crossed homomorphism ψ had been constructed somewhat earlier by Earle [**Ear78**], and turned out, in light of [**Mor89a**], also to represent a generator of $H^1(\mathfrak{M}(\Sigma); H^1(\Sigma;\mathbb{Z}))$. The precise relationship between the crossed homomorphisms $\mathfrak{d}$ and ψ was elucidated in [**Kun09**]. In §2.3 below, we discuss the relationship of $\mathfrak{d} = \delta$ to the *Trapp representation* [**Tra92**] (see Proposition 2.10).

DEFINITION 2.1. The *Earle-Morita subgroup* $\mathrm{Mor}(\Sigma) \subseteq \mathfrak{M}(\Sigma)$ is defined to be the kernel of $\mathfrak{d}$.

REMARK 2.2. The Earle-Morita subgroup is not a normal subgroup of $\mathfrak{M}(\Sigma)$, despite being a kernel; this is because it is a kernel of a *crossed* homomorphism. We remark also that there are many Earle-Morita subgroups, since the definition of $\mathfrak{d}$ (or, equivalently, the definition of δ) depends on a choice. The definition of δ above uses the splitting of the short exact sequence (2.1). Note that it depends on the choice of isomorphism (1.3) in Proposition 1.1, identifying the relevant quotient of the surface braid group with (an explicit model for) the Heisenberg group.

2.2. The Chillingworth subgroup. Recall that the *Torelli subgroup* $\mathfrak{T}(\Sigma) \subseteq \mathfrak{M}(\Sigma)$ consists of those elements of the mapping class group whose natural action on $H_1(\Sigma;\mathbb{Z})$ is trivial. The restriction of the crossed homomorphism $\delta\colon f \mapsto \delta_f$ to the Torelli group is a homomorphism. We will first describe this homomorphism in relation with the action of the Torelli group on homotopy classes of vector fields. Recall that the set $\Xi(\Sigma)$ of homotopy classes of non-vanishing vector fields on Σ supports a natural simply transitive action of $H^1(\Sigma;\mathbb{Z})$ (in other words, an affine structure over $\mathbb{Z}$ with associated $\mathbb{Z}$-module $H^1(\Sigma;\mathbb{Z})$), and the action of $\mathfrak{M}(\Sigma)$ is compatible with this action. It follows that the Torelli group acts by translation on $\Xi(\Sigma)$, which defines a homomorphism $e\colon \mathfrak{T}(\Sigma) \to H^1(\Sigma;\mathbb{Z})$. A formula for $e(f)([\gamma])$, where γ is a regular curve, is given by the variation of the winding number. For convenience we recall some details about the winding number below.

Fix a Riemannian metric on Σ. A non-vanishing vector field X gives a trivialisation of the unit tangent bundle $T_1(\Sigma) \cong \Sigma \times S^1$. The winding number $\omega_X(\gamma)$ of a regular oriented curve γ is the degree of the second component of the unit tangent vector. It can be computed as follows. Assuming that γ is transverse to X except at a finite set $\gamma \pitchfork X$ of points, where it looks locally as in Figure 2, then

$$\omega_X(\gamma) = \sum_{p\in\gamma\pitchfork X} \mathrm{sgn}(p),$$

where $\mathrm{sgn}(p)$ is defined in Figure 2. Notice that only the points p where the tangent vector of γ is *parallel* to X count towards this sum; those that point in the opposite direction to X do not.

DEFINITION 2.3. The *Chillingworth homomorphism* $e\colon \mathfrak{T}(\Sigma) \to H^1(\Sigma;\mathbb{Z})$, studied in [**Chi72, Joh80**], is defined by

$$e(f)([\gamma]) = \omega_X(f\circ\gamma) - \omega_X(\gamma). \tag{2.3}$$

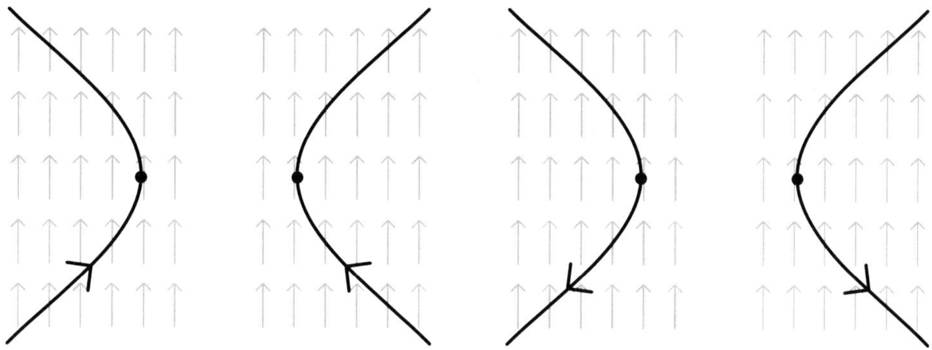

$\mathrm{sgn}(p) = 1$ $\mathrm{sgn}(p) = -1$ $\mathrm{sgn}(p) = 0$ $\mathrm{sgn}(p) = 0$

FIGURE 2. The sign of a point on γ that is tangent to X.

Its kernel is the *Chillingworth subgroup* $\mathrm{Chill}(\Sigma)$. We note that e does not depend on the choice of non-vanishing vector field X, but extends to a crossed homomorphism $e_X \colon \mathfrak{M}(\Sigma) \to H^1(\Sigma;\mathbb{Z})$ that does, as we shall discuss in §2.3 (see Definition 2.9).

REMARK 2.4. Since we have $\delta = \mathfrak{d}$ (as recalled in §2.1 above) and $e = \delta$ on the Torelli group (Lemma 2.5 below), we have:

$$\mathrm{Chill}(\Sigma) = \ker(e) = \mathfrak{T}(\Sigma) \cap \ker(\mathfrak{d}) = \mathfrak{T}(\Sigma) \cap \mathrm{Mor}(\Sigma).$$

Equivalently, we may say that $\mathrm{Chill}(\Sigma)$ is the intersection of the kernels of $\mathfrak{s}$ and $\mathfrak{d}$, in other words it is the kernel of $(\mathfrak{s},\mathfrak{d}) \colon \mathfrak{M}(\Sigma) \to Sp(H) \ltimes H$. Notice in particular that, although the Earle-Morita subgroup $\mathrm{Mor}(\Sigma)$ depends on a non-canonical choice (see Remark 2.2), its intersection with the Torelli group does not.

The following lemma is Proposition 3.7 of [**Bre02**]. The proof there uses a result of [**Mor93**]. We give an independent proof below.

LEMMA 2.5. *The homomorphisms δ and e coincide on the Torelli group and have image $\delta(\mathfrak{T}(\Sigma)) = 2.H^1(\Sigma;\mathbb{Z})$.*

From formula (2.2) it follows that the kernel of the action $\Psi \colon \mathfrak{M}(\Sigma) \to \mathrm{Aut}^+(\mathcal{H})$ is contained in the Torelli group; we may therefore identify this kernel as a corollary of Lemma 2.5.

PROPOSITION 2.6. *For any genus $g \geq 1$, we have* $\ker(\Psi) = \mathrm{Chill}(\Sigma)$.

PROOF. From formula (2.2) we see that $\ker(\Psi) = \mathfrak{T}(\Sigma) \cap \ker(\delta)$; by Lemma 2.5 this is equal to $\ker(e) = \mathrm{Chill}(\Sigma)$. □

Denote by $\mathrm{Inn}(\mathcal{H})$ the group of *inner* automorphisms of the Heisenberg group $\mathcal{H}$. From Lemma 2.5, we may also identify the *projective kernel* of the action Ψ, namely the subgroup $\Psi^{-1}(\mathrm{Inn}(\mathcal{H}))$ that acts by inner automorphisms.

PROPOSITION 2.7. *For any genus $g \geq 1$, we have* $\Psi^{-1}(\mathrm{Inn}(\mathcal{H})) = \mathfrak{T}(\Sigma)$.

PROOF. Conjugation in the Heisenberg group $\mathcal{H}$ is given by the formula

$$(l,x)(k,y)(-l,-x) = (l,y)(k,x)(-l,-x) = (k+2x.y, y) \tag{2.4}$$

First, if $\Psi(f) = f_{\mathcal{H}}$ is an inner automorphism, then its induced action on H must be trivial. This means that f lies in the Torelli group. Conversely, if $f \in \mathfrak{T}(\Sigma)$, we have from Lemma 2.5 that δ_f is in $2.H^1(\Sigma;\mathbb{Z})$. Using Poincaré duality, we obtain $x \in H$ such that $\delta_f(y) = 2x.y$ for every $y \in H$. Comparing formulas (2.2) and (2.4), we deduce that $f_{\mathcal{H}}$ is inner. □

We will use the following basic lemma about crossed homomorphisms in the proof of Lemma 2.5.

LEMMA 2.8. *Let G be a group acting on an abelian group K and denote by $N \subseteq G$ the kernel of this action. Suppose that $S \subseteq N$ normally generates N in G. If two crossed homomorphisms $\theta_1, \theta_2 \colon G \to K$ agree on S, then they agree on N.*

PROOF. The assumption that S normally generates N in G means that

$$T = \{gsg^{-1} \mid s \in S, g \in G\} \subseteq N$$

generates N. It will therefore suffice to show that θ_1 and θ_2 agree on T. Let $s \in S$ and $g \in G$. We know by hypothesis that $\theta_1(s) = \theta_2(s)$, and we need to show that $\theta_1(g^{-1}sg) = \theta_2(g^{-1}sg)$. First, observe that, for $i = 1, 2$, we have

$$\theta_i(g) + g.\theta_i(g^{-1}) = \theta_i(g^{-1}g) = \theta_i(1) = 0.$$

Using this, and the fact that $s \in N$, so it acts trivially on K, we deduce that

$$\begin{aligned}\theta_i(g^{-1}sg) &= \theta_i(g) + g.\theta_i(s) + gs.\theta_i(g^{-1})\\ &= \theta_i(g) + g.\theta_i(s) + g.\theta_i(g^{-1})\\ &= g.\theta_i(s).\end{aligned}$$

Thus $\theta_1(g^{-1}sg) = g.\theta_1(s) = g.\theta_2(s) = \theta_2(g^{-1}sg)$, as required. □

PROOF OF LEMMA 2.5. The Torelli group is generated by genus-one bounding pair diffeomorphisms [**Joh79**, Theorem 2], and this generating set is a single conjugacy class in the full mapping class group. It follows that the Torelli group is normally generated by a(ny) single genus-one bounding pair diffeomorphism f. We may therefore apply Lemma 2.8 to the setting where $G = \mathfrak{M}(\Sigma)$, $K = H^1(\Sigma;\mathbb{Z})$, $N = \mathfrak{T}(\Sigma)$ and $S = \{f\}$. Both δ and e extend to crossed homomorphisms defined on the full mapping class group (for e such an extension is given by the Trapp representation, as recalled in §2.3 below). To prove that δ and e coincide on $\mathfrak{T}(\Sigma)$, it is therefore sufficient to show that δ and e agree on the single element f. Moreover, to show that $\delta(\mathfrak{T}(\Sigma)) \subseteq 2.H^1(\Sigma;\mathbb{Z})$ it is enough to show that this common value $\delta_f = e(f)$ lies in $2.H^1(\Sigma;\mathbb{Z})$. Specifically, we will take this element to be

$$f = BP(\gamma,\delta) = T_\gamma.T_\delta^{-1},$$

the genus one bounding pair diffeomorphism depicted in Figure 3, and we will show that the elements $e(f)$ and δ_f of $H^1(\Sigma;\mathbb{Z}) \cong \mathrm{Hom}(H,\mathbb{Z})$ are both equal to the homomorphism $H = H_1(\Sigma;\mathbb{Z}) \to \mathbb{Z}$ given by

$$(2.5) \qquad a_1 \mapsto 2 \quad , \quad a_i \mapsto 0 \text{ for } i \geq 2 \quad \text{and} \quad b_i \mapsto 0 \text{ for } i \geq 1.$$

As explained just above, this calculation will show that $\delta = e$ on $\mathfrak{T}(\Sigma)$ and that $\delta(\mathfrak{T}(\Sigma)) \subseteq 2.H^1(\Sigma;\mathbb{Z})$. The opposite inclusion will also follow, since $\delta_f = (2.5)$ is a free generator of $2.H^1(\Sigma;\mathbb{Z})$ and the other $2g - 1$ elements of the evident free generating set may be realised similarly as $\delta_{f'}$ for analogous elements f'.

It therefore remains to calculate that $\delta_f = e(f) = (2.5)$.

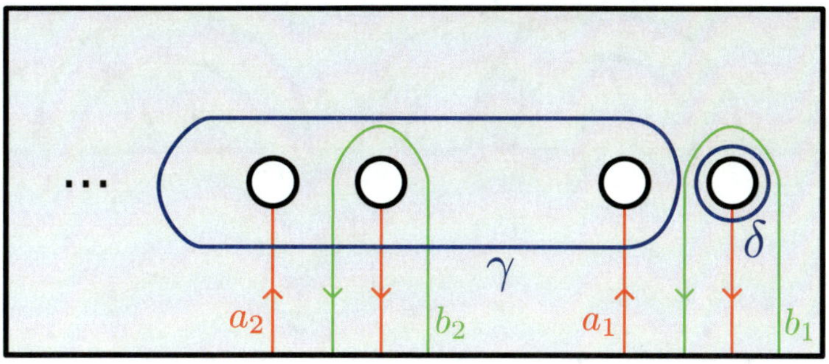

FIGURE 3. The surface Σ is obtained by identifying the $2g$ interior boundary components (four of which are depicted above) in g pairs by reflections. The bounding pair map from the proof of Lemma 2.5 is $BP(\gamma, \delta) = T_\gamma . T_\delta^{-1}$, for the blue curves γ and δ. The red and green arcs form a symplectic basis for the first homology of Σ relative to the bottom edge $\partial^-(\Sigma)$.

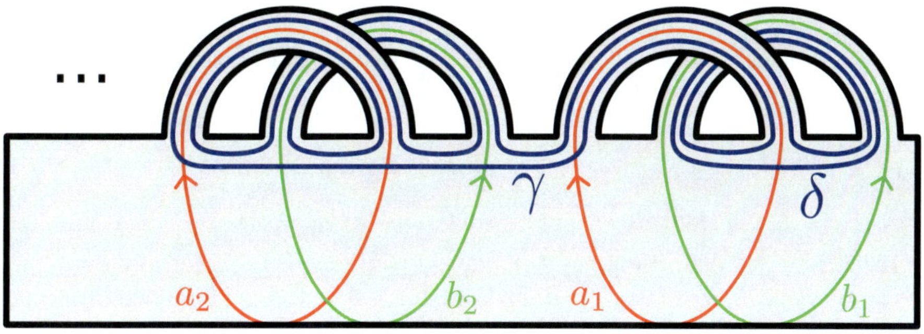

FIGURE 4. An alternative model for the surface Σ, the bounding pair (γ, δ) and the symplectic basis for the first homology of Σ relative to the bottom edge $\partial^-(\Sigma)$.

We first calculate δ_f from the automorphism $f_{\mathcal{H}}$. We may directly read off from Figure 3 the effect of $f_{\mathcal{H}}$ on the elements $\tilde{a}_i$ and $\tilde{b}_i$ of $\mathcal{H}$. It clearly acts trivially except possibly on the three elements $\tilde{a}_2 = (0, a_2)$, $\tilde{b}_2 = (0, b_2)$ and $\tilde{a}_1 = (0, a_1)$, since the others may be realised disjointly from $\gamma \cup \delta$, and:

$$\begin{aligned}
\tilde{a}_1 &\mapsto [\tilde{a}_2, \tilde{b}_2].\tilde{a}_1 = u^2 \tilde{a}_1 = (2, a_1) \\
\tilde{a}_2 &\mapsto [\tilde{a}_2, \tilde{b}_2].\tilde{a}_1 \tilde{b}_1 \tilde{a}_1^{-1}.\tilde{a}_2.\tilde{a}_1 \tilde{b}_1^{-1} \tilde{a}_1^{-1}.[\tilde{a}_2, \tilde{b}_2]^{-1} = \tilde{a}_2 \\
\tilde{b}_2 &\mapsto [\tilde{a}_2, \tilde{b}_2].\tilde{a}_1 \tilde{b}_1 \tilde{a}_1^{-1}.\tilde{b}_2.\tilde{a}_1 \tilde{b}_1^{-1} \tilde{a}_1^{-1}.[\tilde{a}_2, \tilde{b}_2]^{-1} = \tilde{b}_2.
\end{aligned}$$

This gives the calculation $\delta_f = (2.5)$.

To calculate $e(f)$, we use the alternative model for the surface Σ, the bounding pair (γ, δ) and the symplectic basis a_i, b_i for H depicted in Figure 4. This model for Σ has the advantage of having an obvious non-vanishing vector field X, which simply points *upwards* according to the standard framing of the page.

Using this vector field X and comparing to Figure 2, we observe that the winding numbers of the symplectic generators a_i and b_i (more precisely, their smooth,

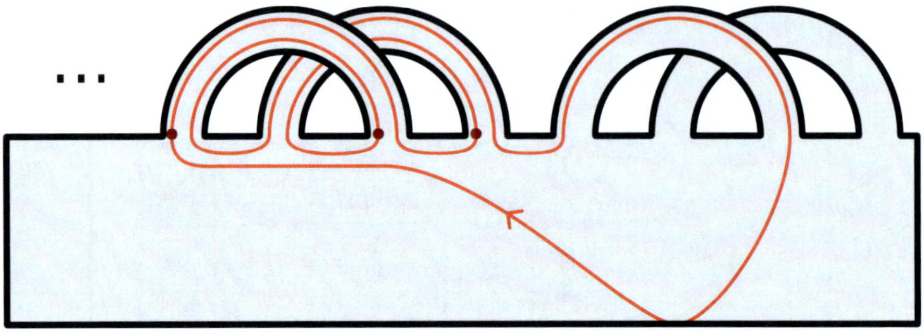

FIGURE 5. The curve $f \circ a_1$ for $f = T_\gamma . T_\delta^{-1}$. The three points where its tangent vector points vertically upwards are marked with dark red points: the left-most one is negative according to Figure 2, and the other two are positive. [At first sight it may look like there are two more, but these are not allowed since they do not fit either of the local models of Figure 2. We therefore perturb the curve slightly to get rid of these two tangencies with the vector field X. Alternatively, we may perturb it differently, to turn each of these disallowed tangencies into a pair of two allowed tangencies with opposite signs, which will therefore cancel in the expression for $\omega_X(f \circ a_1)$.]

closed representatives pictured in Figure 4) are given by

$$\omega_X(a_i) = -1 \qquad \text{and} \qquad \omega_X(b_i) = +1.$$

We recall that, by definition, $e(f)(c) = \omega_X(f \circ \bar{c}) - \omega_X(\bar{c}) \in \mathbb{Z}$ for any $c = [\bar{c}] \in H$. We clearly have $f \circ \bar{c} = \bar{c}$ for $\bar{c} = a_i$ or b_i with $i \geq 3$ or for $\bar{c} = b_1$, since these curves may be represented disjointly from $\gamma \cup \delta$. Hence $e(f)([\bar{c}]) = 0$ for these $\bar{c}$.

The curve $f \circ a_1$ is depicted in Figure 5.

There are precisely three points on this curve where its tangent vector is equal to the vector field X, i.e., where its tangent vector is pointing vertically upwards: two are positive and one is negative (compare the local models in Figure 2), hence

$$\begin{aligned} e(f)(a_1) &= \omega_X(f \circ a_1) - \omega_X(a_1) \\ &= (2-1) - (-1) \\ &= 2. \end{aligned}$$

Now let $\bar{c}$ be either a_2 or b_2. In this case the effect of f is simply to conjugate $\bar{c}$ by γ, so we have that

$$\begin{aligned} \omega_X(f \circ \bar{c}) &= \omega_X(\gamma) + \omega_X(\bar{c}) - \omega_X(\gamma) \\ &= \omega_X(\bar{c}), \end{aligned}$$

since positive/negative tangencies with X for γ are negative/positive tangencies with X for γ^{-1} respectively, and so $e(f)([\bar{c}]) = \omega_X(f \circ \bar{c}) - \omega_X(\bar{c}) = 0$. Thus we have shown that $e(f) \colon H \to \mathbb{Z}$ is also given by (2.5). □

2.3. The Trapp representation. We next recall the *Trapp representation* [**Tra92**], and show that our representation (1.8) of $\mathfrak{M}(\Sigma)$ on $\mathcal{H}$ may be identified with it, up to "coboundaries", when the genus g of Σ is at least 2. This provides an alternative proof of Proposition 2.6 (except when $g = 1$), since the kernel of the Trapp representation is equal to the Chillingworth subgroup Chill(Σ) when $g \geq 2$ [**Tra92**, Corollary 2.7].

DEFINITION 2.9. Write $H = H_1(\Sigma;\mathbb{Z})$ and $H^* = \mathrm{Hom}(H,\mathbb{Z}) \cong H^1(\Sigma;\mathbb{Z})$. The representation of Trapp [**Tra92**] is defined as a homomorphism

$$\Phi_X : \mathfrak{M}(\Sigma) \longrightarrow Sp(H) \ltimes H^* \subset GL_{2g+1}(\mathbb{Z}) \tag{2.6}$$

lifting the standard symplectic action $\mathfrak{s}\colon \mathfrak{M}(\Sigma) \to Sp(H)$. Having fixed this choice of symplectic action, the homomorphism (2.6) corresponds to a choice of crossed homomorphism

$$e_X : \mathfrak{M}(\Sigma) \longrightarrow H^*. \tag{2.7}$$

This crossed homomorphism is given by the variation of the winding number with respect to a fixed non-vanishing vector field X on Σ, as already discussed in §2.2; see the formula (2.3).

We therefore have two homomorphisms

$$\Psi = (1.8) \text{ and } \Phi_X = (2.6)\colon \mathfrak{M}(\Sigma) \longrightarrow Sp(H) \ltimes H^*$$

corresponding to crossed homomorphisms δ and $e_X : \mathfrak{M}(\Sigma) \to H^*$. We have proven above (Lemma 2.5) that these crossed homomorphisms are equal when restricted to the Torelli group. We now strengthen this to show that δ and e_X agree, modulo coboundaries, on the whole mapping class group.

PROPOSITION 2.10. *For $g \geq 2$, the crossed homomorphisms δ and e_X represent the same cohomology class in $H^1(\mathfrak{M}(\Sigma);H^*) \cong \mathbb{Z}$. In other words, they are equal modulo principal crossed homomorphisms, i.e. coboundaries.*

PROOF. We will use the homomorphism

$$H^1(\mathfrak{M}(\Sigma);H^*) \longrightarrow \mathrm{Hom}(\mathfrak{T}(\Sigma),H^*) \tag{2.8}$$

given by restricting a crossed homomorphism $\mathfrak{M}(\Sigma) \to H^*$ to the Torelli group. This is well-defined since principal crossed homomorphisms (coboundaries) are trivial on the Torelli group. The right-hand side of (2.8) is rather large: by a theorem of Johnson [**Joh85**], the abelianisation of $\mathfrak{T}(\Sigma)$ is isomorphic to $\wedge^3 H \oplus$ (torsion), so $\mathrm{Hom}(\mathfrak{T}(\Sigma),H^*) \cong \mathrm{Hom}(\wedge^3 H,H^*)$, which is free abelian of rank $2g\binom{2g}{3}$. However, it has the advantage that it is easy to detect when its elements are equal, since it is just a group of homomorphisms (rather than crossed homomorphisms modulo principal ones). On the other hand, the left-hand side of (2.8) is much smaller. Indeed, Morita proved in [**Mor89a**, Proposition 6.4] that the group $H^1(\mathfrak{M}(\Sigma);H^*)$ is infinite cyclic. (In fact, it is generated by $[\mathfrak{d}]$, which we showed in [**BPS21**] is equal to $[\delta]$, but we will not need this.) In Lemma 2.5 we have proven that δ and e_X coincide, and are non-trivial, on the Torelli subgroup. Since $\mathrm{Hom}(\mathfrak{T}(\Sigma),H^*)$ is torsion-free, the homomorphism (2.8) is injective and the result follows. □

REMARK 2.11. In summary, we have considered three crossed homomorphisms

$$\delta, \mathfrak{d}, e_X : \mathfrak{M}(\Sigma) \longrightarrow H^* \cong H^1(\Sigma;\mathbb{Z}),$$

where δ is the crossed homomorphism corresponding to the action (1.8) of the mapping class group on the Heisenberg group, $\mathfrak{d}$ is Morita's crossed homomorphism (whose precise relationship to Earle's crossed homomorphism ψ is described in [**Kun09**]) and e_X is Chillingworth's crossed homomorphism, depending on a choice of non-vanishing vector field X on Σ. We showed in [**BPS21**] that $\delta = \mathfrak{d}$ on $\mathfrak{M}(\Sigma)$. In this section, we have shown (Lemma 2.5) that $\delta = e_X$ when restricted to $\mathfrak{T}(\Sigma)$ and, moreover, that $\delta = e_X$ on $\mathfrak{M}(\Sigma)$ modulo coboundaries (Proposition 2.10). We note, however, that only the weaker statement of Lemma 2.5 was needed to deduce (Propositions 2.6 and 2.7) that the kernel of Ψ is $\mathrm{Chill}(\Sigma)$ and the projective kernel of Ψ is $\mathfrak{T}(\Sigma)$.

2.4. Restricting to the Chillingworth subgroup. Using Proposition 2.6, we deduce that the twisted representations of $\mathfrak{M}(\Sigma)$ constructed in Theorem 1.5 are in fact untwisted when restricted to $\mathrm{Chill}(\Sigma) \subset \mathfrak{M}(\Sigma)$.

THEOREM 2.12. *Associated to any representation V of $\mathcal{H}$ over R and any integer $n \geq 2$, there is a representation*

$$\mathrm{Chill}(\Sigma) \longrightarrow \mathrm{Aut}_R\big(H_n^{BM}(\mathcal{C}_n(\Sigma), \mathcal{C}_n(\Sigma, \partial^-(\Sigma)); V)\big), \tag{2.9}$$

which is a restriction of (1.11) *to a single object of the action groupoid.*

PROOF. This follows from the construction described in §1.3, with each element $f \in \mathrm{Chill}(\Sigma)$ acting by the automorphism (1.10) (setting $\tau = \mathrm{id}$), since we know from Proposition 2.6 that $f_{\mathcal{H}} = \Psi(f) = \mathrm{id}$ for each $f \in \mathrm{Chill}(\Sigma)$. □

3. Representations of the Torelli group

We now restrict to the Torelli group $\mathfrak{T}(\Sigma) \subseteq \mathfrak{M}(\Sigma)$. By Proposition 2.7, the Torelli group acts by inner automorphisms under Ψ, so we have a homomorphism

$$\Psi \colon \mathfrak{T}(\Sigma) \longrightarrow \mathrm{Inn}(\mathcal{H}) \subset \mathrm{Aut}(\mathcal{H}). \tag{3.1}$$

We may therefore pull back the $\mathbb{Z}$-central extension

$$1 \to \mathbb{Z} \cong \mathcal{Z}(\mathcal{H}) \longrightarrow \mathcal{H} \longrightarrow \mathrm{Inn}(\mathcal{H}) \to 1 \tag{3.2}$$

along (3.1) to obtain a $\mathbb{Z}$-central extension

$$1 \to \mathbb{Z} \longrightarrow \widetilde{\mathfrak{T}}(\Sigma) \longrightarrow \mathfrak{T}(\Sigma) \to 1 \tag{3.3}$$

and a homomorphism

$$\widetilde{\Psi} \colon \widetilde{\mathfrak{T}}(\Sigma) \longrightarrow \mathcal{H} \tag{3.4}$$

lifting (3.1).

REMARK 3.1. The inner automorphism group $\mathrm{Inn}(\mathcal{H})$ naturally identifies with the first homology group $H = H_1(\Sigma; \mathbb{Z})$. Under this identification, (3.1) becomes the crossed homomorphism $\delta \colon \mathfrak{T}(\Sigma) \to H^*$ (which is a *homomorphism* on the Torelli group) composed with the Poincaré duality isomorphism $H^* \cong H$. A 2-cocycle representing the extension (3.2) is the intersection form . on H. A 2-cocycle representing (3.3) is therefore given by $(f, f') \mapsto \delta(f)^\sharp . \delta(f')^\sharp$, where $(\ )^\sharp$ denotes Poincaré duality. As mentioned above, we showed in [**BPS21**] that $\delta = \mathfrak{d}$, so this 2-cocycle may also be written as $(f, f') \mapsto \mathfrak{d}(f)^\sharp . \mathfrak{d}(f')^\sharp$.

Recall from §1.1 that, for any representation V of $\mathcal{H}$, the Heisenberg homology module $H_n^{BM}(\mathcal{C}_n(\Sigma), \mathcal{C}_n(\Sigma, \partial^-(\Sigma)); V)$ is obtained from the singular chain complex $\mathcal{S}_*(\widetilde{\mathcal{C}}_n(\Sigma))$ of the Heisenberg covering $\widetilde{\mathcal{C}}_n(\Sigma)$ by:

- tensoring over $\mathbb{Z}[\mathcal{H}]$ with V (1.4);
- taking the quotient of this complex by the subcomplex corresponding to the subspace $\mathcal{C}_n(\Sigma, \partial^-(\Sigma)) \cup (\mathcal{C}_n(\Sigma) \setminus T)$, for each compact $T \subset \mathcal{C}_n(\Sigma)$;
- passing to homology and then taking the inverse limit over T (1.6).

The isomorphism (1.10) (for $\tau = (f_{\mathcal{H}})^{-1}$) is induced by the natural twisted action of $f \in \mathfrak{M}(\Sigma)$ on $\mathcal{S}_*(\widetilde{\mathcal{C}}_n(\Sigma))$, which is of the form

$$\mathcal{S}_*(\widetilde{\mathcal{C}}_n(\Sigma)) \longrightarrow \mathcal{S}_*(\widetilde{\mathcal{C}}_n(\Sigma))_{f_{\mathcal{H}}}. \tag{3.5}$$

Now, for an element $h \in \mathcal{H}$, let us denote by $c_h = h{-}h^{-1}$ the corresponding inner automorphism $c_h \in \mathrm{Inn}(\mathcal{H})$. One may verify that the isomorphism

$$- \cdot h \colon \mathcal{S}_*(\widetilde{\mathcal{C}}_n(\Sigma))_{c_h} \longrightarrow \mathcal{S}_*(\widetilde{\mathcal{C}}_n(\Sigma)) \tag{3.6}$$

of singular chain complexes given by the right-action of h is $\mathbb{Z}[\mathcal{H}]$-linear. For each $\widetilde{f} \in \widetilde{\mathfrak{T}}(\Sigma)$, we may then take the composition

$$\mathcal{S}_*(\widetilde{\mathcal{C}}_n(\Sigma)) \xrightarrow{(3.5)} \mathcal{S}_*(\widetilde{\mathcal{C}}_n(\Sigma))_{f_{\mathcal{H}}} \xrightarrow{(3.6)} \mathcal{S}_*(\widetilde{\mathcal{C}}_n(\Sigma)), \tag{3.7}$$

where f denotes the projection of $\widetilde{f}$ to $\mathfrak{T}(\Sigma)$ and we set $h = \widetilde{\Psi}(\widetilde{f}) \in \mathcal{H}$. The fact that $f_{\mathcal{H}} = c_h$ follows from the fact that (3.4) is a lift of (3.1). This defines an untwisted, $\mathbb{Z}[\mathcal{H}]$-linear action of $\widetilde{\mathfrak{T}}(\Sigma)$ on the singular chain complex $\mathcal{S}_*(\widetilde{\mathcal{C}}_n(\Sigma))$. By the construction recalled above, this in turn induces an untwisted, R-linear action of $\widetilde{\mathfrak{T}}(\Sigma)$ on Heisenberg homology:

$$\widetilde{\mathfrak{T}}(\Sigma) \longrightarrow \mathrm{Aut}_R\big(H_n^{BM}(\mathcal{C}_n(\Sigma), \mathcal{C}_n(\Sigma, \partial^-(\Sigma)); V)\big). \tag{3.8}$$

To complete the construction, we show that:

LEMMA 3.2. *The central extension $\widetilde{\mathfrak{T}}(\Sigma)$ of $\mathfrak{T}(\Sigma)$ is trivial, i.e. it is isomorphic to the product $\mathfrak{T}(\Sigma) \times \mathbb{Z}$.*

THEOREM 3.3. *Associated to any representation V of $\mathcal{H}$ over R and any integer $n \geq 2$, there is a well-defined representation of the Torelli group*

$$\mathfrak{T}(\Sigma) \longrightarrow \mathrm{Aut}_R\big(H_n^{BM}(\mathcal{C}_n(\Sigma), \mathcal{C}_n(\Sigma, \partial^-(\Sigma)); V)\big) \tag{3.9}$$

that lifts a projective action of $\mathfrak{T}(\Sigma)$ on this homology module.

PROOF. Let us abbreviate $\mathcal{V}_n(V) = H_n^{BM}(\mathcal{C}_n(\Sigma), \mathcal{C}_n(\Sigma, \partial^-(\Sigma)); V)$. The group homomorphism (3.8) must send the subgroup $\mathbb{Z} \subset \widetilde{\mathfrak{T}}(\Sigma)$ (the kernel of the central extension of $\mathfrak{T}(\Sigma)$) to the centre of $\mathrm{Aut}_R(\mathcal{V}_n(V))$, so it descends to

$$\mathfrak{T}(\Sigma) \longrightarrow \mathrm{PAut}_R(\mathcal{V}_n(V)), \tag{3.10}$$

where the projective automorphism group $\mathrm{PAut}_R(A)$ of an R-module A is the quotient of $\mathrm{Aut}_R(A)$ by its centre. Note that the centre of $\mathrm{Aut}_R(A)$ is equal to $\{- \cdot \lambda \mid \lambda \in \mathcal{Z}(R^\times)\}$ when A is a free R-module, but may be larger when A is not free. This is a projective action of the Torelli group. To lift it to a linear action, we compose (3.8) with any section of the central extension $\widetilde{\mathfrak{T}}(\Sigma) \to \mathfrak{T}(\Sigma)$, which exists by Lemma 3.2. □

To prove Lemma 3.2, we will need a lemma describing the behaviour of the crossed homomorphism $\mathfrak{d}$ with respect to increasing genus. Consider the inclusion of surfaces $\Sigma_{g,1} \subseteq \Sigma_{h,1}$ given by boundary connected sum with $\Sigma_{h-g,1}$. This induces an inclusion of mapping class groups

$$\mathfrak{M}(\Sigma_{g,1}) \hookrightarrow \mathfrak{M}(\Sigma_{h,1}) \tag{3.11}$$

by extending diffeomorphisms by the identity on $\Sigma_{h-g,1}$.

LEMMA 3.4 ([**BPS21**, §5.2]). *The diagram*

$$\begin{array}{ccc} \mathfrak{M}(\Sigma_{g,1}) & \xhookrightarrow{(3.11)} & \mathfrak{M}(\Sigma_{h,1}) \\ \mathfrak{d}\big\downarrow & & \big\downarrow\mathfrak{d} \\ H^1(\Sigma_{g,1};\mathbb{Z}) & \longrightarrow & H^1(\Sigma_{h,1};\mathbb{Z}) \end{array} \tag{3.12}$$

commutes, where the bottom arrow is the map induced by the inclusion $\Sigma_{g,1} \hookrightarrow \Sigma_{h,1}$ *on* $H_1(-;\mathbb{Z})$*, conjugated by Poincaré duality.*

PROOF OF LEMMA 3.2. We begin by showing that it suffices to prove the statement for all sufficiently large g; we will then be able to assume $g \geq 3$ in the rest of the proof. For $g < h$, consider the inclusion of Torelli groups

$$\iota\colon \mathfrak{T}(\Sigma_{g,1}) \hookrightarrow \mathfrak{T}(\Sigma_{h,1}). \tag{3.13}$$

We claim that the pullback of the central extension $\widetilde{\mathfrak{T}}(\Sigma_{h,1})$ along (3.13) is $\widetilde{\mathfrak{T}}(\Sigma_{g,1})$. To see this, recall from Remark 3.1 that the central extension $\widetilde{\mathfrak{T}}(\Sigma_{g,1})$ of $\mathfrak{T}(\Sigma_{g,1})$ is represented by the 2-cocycle $(f,f') \mapsto \mathfrak{d}(f)^\sharp.\mathfrak{d}(f')^\sharp$. Similarly, the pullback of the central extension $\widetilde{\mathfrak{T}}(\Sigma_{h,1})$ of $\mathfrak{T}(\Sigma_{h,1})$ along the inclusion (3.13) is represented by the 2-cocycle $(f,f') \mapsto \mathfrak{d}(\iota(f))^\sharp.\mathfrak{d}(\iota(f'))^\sharp$. Lemma 3.4, together with the fact that the map $H_1(\Sigma_{g,1};\mathbb{Z}) \to H_1(\Sigma_{h,1};\mathbb{Z})$ preserves the intersection form, implies that these 2-cocycles are equal. Thus triviality of $\widetilde{\mathfrak{T}}(\Sigma_{h,1})$ will imply triviality of $\widetilde{\mathfrak{T}}(\Sigma_{g,1})$ for any $g < h$. For the remainder of this proof, we assume that $g \geq 3$ and abbreviate $\Sigma_{g,1}$ to Σ, as usual.

By [**BCRR20**, Lemma A.1(xiii)] and homological stability [**Wah13**, Theorem 1.2], the canonical surjection $\mathfrak{M}(\Sigma) \twoheadrightarrow Sp(H)$ induces an isomorphism on $H^2(-;\mathbb{Z})$ when $g \geq 3$. It follows that the inclusion $\mathfrak{T}(\Sigma) \hookrightarrow \mathfrak{M}(\Sigma)$ induces the trivial map on $H^2(-;\mathbb{Z})$. This means that every $\mathbb{Z}$-central extension of $\mathfrak{M}(\Sigma)$ becomes trivial when restricted to $\mathfrak{T}(\Sigma)$. To prove the lemma, it will therefore suffice to show that $\widetilde{\mathfrak{T}}(\Sigma)$ is the restriction of a $\mathbb{Z}$-central extension defined on the whole mapping class group $\mathfrak{M}(\Sigma)$. Recall from Remark 3.1 that the central extension $\widetilde{\mathfrak{T}}(\Sigma)$ of $\mathfrak{T}(\Sigma)$ is represented by the 2-cocycle c' given by $c'(f,f') = \mathfrak{d}(f)^\sharp.\mathfrak{d}(f')^\sharp$. We therefore just have to show that the 2-cocycle c' extends to $\mathfrak{M}(\Sigma)$.

Now, there is a 2-cocycle c on $\mathfrak{M}(\Sigma)$, defined by Morita [**Mor89b**], given by the formula $c(f,f') = \mathfrak{d}(f^{-1})^\sharp.\mathfrak{d}(f')^\sharp$. By general properties of crossed homomorphisms, we have $\mathfrak{d}(f^{-1})^\sharp = -f_*^{-1}(\mathfrak{d}(f)^\sharp)$, and so we may rewrite this as

$$c(f,f') = -f_*^{-1}(\mathfrak{d}(f)^\sharp).\mathfrak{d}(f')^\sharp = -\mathfrak{d}(f)^\sharp.f_*(\mathfrak{d}(f')^\sharp), \tag{3.14}$$

where for the second equality we have used the fact that the automorphism f_* of H preserves the intersection form. Restricted to the Torelli group, we have $f_* = \mathrm{id}$, so $c(f,f') = -\mathfrak{d}(f)^\sharp.\mathfrak{d}(f')^\sharp = -c'(f,f')$ for $f,f' \in \mathfrak{T}(\Sigma)$. Thus the 2-cocycle c' extends to the 2-cocycle $-c$ on $\mathfrak{M}(\Sigma)$. □

4. Representations of Earle-Morita subgroups

We now identify another large subgroup of the mapping class group $\mathfrak{M}(\Sigma)$ on which we may construct linear representations without passing to a central extension. We will do this in the setting where we take coefficients in the *Schrödinger representation* of $\mathcal{H}$ and the subgroup under consideration is the *Earle-Morita subgroup* of $\mathfrak{M}(\Sigma)$.

Recall from §2.1 that the *Earle-Morita subgroup* $\mathrm{Mor}(\Sigma) \subseteq \mathfrak{M}(\Sigma)$ is the kernel of the crossed homomorphism $\mathfrak{d}\colon \mathfrak{M}(\Sigma) \to H^1(\Sigma;\mathbb{Z})$ defined by Morita [**Mor89a**], and that this crossed homomorphism coincides with the one associated to the action $\mathfrak{M}(\Sigma) \to \mathrm{Aut}^+(\mathcal{H}) \cong Sp(H) \ltimes H^1(\Sigma;\mathbb{Z})$ from Proposition 1.3. We also recall, from Remark 2.2, that this crossed homomorphism, and its kernel $\mathrm{Mor}(\Sigma)$, depend on the parametrisation of the surface Σ.

An important representation of the Heisenberg group is the *Schrödinger representation*, which is parametrised by a non-zero real number $\hbar$ (called the Planck constant). It is given by the right action $\Pi_\hbar$ of $\mathcal{H}$ on the Hilbert space $W := L^2(\mathbb{R}^g)$ determined by the following formula:

$$\left[\Pi_\hbar\left(k, x = \sum_{i=1}^{g} p_i a_i + q_i b_i\right)\psi\right](s) = e^{i\hbar\frac{k-p\cdot q}{2}} e^{i\hbar p\cdot s}\psi(s-q). \tag{4.1}$$

In fact, this is an action of the *continuous* Heisenberg group $\mathcal{H}_\mathbb{R}$, which is the central extension of $H_\mathbb{R} := H_1(\Sigma;\mathbb{R})$ by $\mathbb{R}$ corresponding to the intersection form. There is a natural inclusion $\mathcal{H} \subset \mathcal{H}_\mathbb{R}$. The Schrödinger representation is a unitary action on $W = L^2(\mathbb{R}^g)$, so it may be written as

$$\Pi_\hbar\colon \mathcal{H}_\mathbb{R} \longrightarrow U(W). \tag{4.2}$$

We recall also that the group $\mathrm{Aut}^+(\mathcal{H}_\mathbb{R})$ of automorphisms acting trivially on the centre of $\mathcal{H}_\mathbb{R}$ decomposes as $\mathrm{Aut}^+(\mathcal{H}_\mathbb{R}) \cong Sp(H_\mathbb{R}) \ltimes H^1(\Sigma;\mathbb{R})$, similarly to the decomposition of $\mathrm{Aut}^+(\mathcal{H})$ described in §2.1. From these decompositions we see that there is a natural inclusion $\mathrm{Aut}^+(\mathcal{H}) \subset \mathrm{Aut}^+(\mathcal{H}_\mathbb{R})$.

4.1. Untwisting on a central extension of the mapping class group. We first recall from [**BPS21**, §5] how to untwist the twisted representation (1.11) on the *stably universal central extension* of $\mathfrak{M}(\Sigma)$ when the $\mathcal{H}$-representation V is the Schrödinger representation. In §4.2–§4.4 we then explain how to untwist on the Earle-Morita subgroup *without* passing to a central extension.

As recalled in [**BPS21**, §5.1], an immediate corollary of the *Stone-von Neumann theorem* (see for example [**LV80**, p. 19]) is the following.

COROLLARY 4.1 (of the Stone-von Neumann theorem). *Fix a positive real number $\hbar$ and let $\rho\colon \mathcal{H}_\mathbb{R} \to U(W)$ be an irreducible unitary representation whose restriction to the centre $\mathbb{R} \subset \mathcal{H}_\mathbb{R}$ is given by $\rho(t,0) = e^{\hbar i t/2}.\mathrm{id}_W$. Then there is an element $u \in U(W)$, unique up to rescaling by an element of S^1, such that $\rho = u.\Pi_\hbar.u^{-1}$.*

In particular, we may apply this result to the representation $\rho := \Pi_\hbar \circ \varphi$, for any $\varphi \in \mathrm{Aut}(\mathcal{H}_\mathbb{R})$. Sending φ to the element $u \in U(W)/S^1 = PU(W)$ provided by Corollary 4.1 defines a homomorphism

$$T\colon \mathrm{Aut}(\mathcal{H}_\mathbb{R}) \longrightarrow PU(W), \tag{4.3}$$

which is the *Segal-Shale-Weil projective representation.* Restricting to $\mathrm{Aut}^+(\mathcal{H}) \subset \mathrm{Aut}^+(\mathcal{H}_\mathbb{R}) \subset \mathrm{Aut}(\mathcal{H}_\mathbb{R})$, we may compose it with the action $\Psi\colon \mathfrak{M}(\Sigma) \to \mathrm{Aut}^+(\mathcal{H})$ from Proposition 1.3 to obtain a projective representation

$$\mathfrak{M}(\Sigma) \longrightarrow PU(W) \tag{4.4}$$

of the mapping class group. This is the key ingredient for untwisting the twisted representations of the mapping class group constructed in §1.3.

Definition 4.2. Let $\overline{\mathfrak{M}}(\Sigma)$ denote the central extension of $\mathfrak{M}(\Sigma)$ by S^1 pulled back from the central extension $U(W)$ of $PU(W)$ along (4.4). By construction, the projective representation (4.4) lifts to a linear representation

$$\overline{\mathfrak{M}}(\Sigma) \longrightarrow U(W) \tag{4.5}$$

on this central extension.

Definition 4.3. For $g \geq 4$, the mapping class group $\mathfrak{M}(\Sigma)$ is perfect and we have $H_2(\mathfrak{M}(\Sigma); \mathbb{Z}) \cong \mathbb{Z}$, so it has a universal central extension with kernel $\mathbb{Z}$. Let us denote this extension by $\widetilde{\mathfrak{M}}(\Sigma)$. For $h \geq g \geq 4$, the pullback of $\widetilde{\mathfrak{M}}(\Sigma_{h,1})$ along the inclusion (3.11) is $\widetilde{\mathfrak{M}}(\Sigma_{g,1})$. Thus we may define, for $g \geq 1$, the *stably universal central extension* $\widetilde{\mathfrak{M}}(\Sigma_{g,1})$ of $\mathfrak{M}(\Sigma_{g,1})$ to be the pullback of the universal central extension $\widetilde{\mathfrak{M}}(\Sigma_{h,1})$ of $\mathfrak{M}(\Sigma_{h,1})$ along the inclusion (3.11), for any $h \geq 4$.

We note that there is a canonical morphism of central extensions

$$\widetilde{\mathfrak{M}}(\Sigma) \longrightarrow \overline{\mathfrak{M}}(\Sigma). \tag{4.6}$$

When $g \geq 4$ this morphism exists and is unique by universality of $\widetilde{\mathfrak{M}}(\Sigma)$. For $g \leq 3$ it may be pulled back from the $g \geq 4$ case via the inclusion (3.11), as explained in [**BPS21**, §5.3].

Using these ingredients, we showed in [**BPS21**, §5.3] how to untwist the twisted representation (1.11) of $\mathfrak{M}(\Sigma)$ when taking coefficients in the Schrödinger representation W of $\mathcal{H}$, after passing to the stably universal central extension $\widetilde{\mathfrak{M}}(\Sigma)$. Let us recall briefly how this works, following the philosophy of Remark 1.6. In the notation of §1.3, the construction of the twisted representation (1.11) provides isomorphisms

$$\mathcal{V}_n\big({}_{f_{\mathcal{H}}}V\big) \longrightarrow \mathcal{V}_n(V) \tag{4.7}$$

for each $f \in \mathfrak{M}(\Sigma)$. (This is simply (1.10) with $\tau = \mathrm{id}_{\mathcal{H}}$.) For each $\bar{f} \in \overline{\mathfrak{M}}(\Sigma)$ lifting $f \in \mathfrak{M}(\Sigma)$, its image under (4.5) is a unitary automorphism of W that intertwines the left Schrödinger action of $\mathcal{H}$, as long as we twist the action on the codomain by $f_{\mathcal{H}}$. In other words, it is a (unitary) isomorphism of left $\mathcal{H}$-representations of the form $W \cong {}_{f_{\mathcal{H}}}W$. This is a direct consequence of the defining property of the Segal-Shale-Weil projective representation from Corollary 4.1. This isomorphism of coefficients induces an isomorphism $\mathcal{V}_n(W) \cong \mathcal{V}_n({}_{f_{\mathcal{H}}}W)$, which we may compose with (4.7) (for $V = W$) to obtain:

Theorem 4.4 ([**BPS21**, §5]). *For any $n \geq 2$, there is a representation*

$$\overline{\mathfrak{M}}(\Sigma) \longrightarrow GL(\mathcal{V}_n(W)) \tag{4.8}$$

induced by the natural action of the mapping class group on (1.9) *with coefficients in the Schrödinger representation $V = W$. Via the morphism* (4.6)*, we may view this as a representation of the stably universal central extension of $\mathfrak{M}(\Sigma)$.*

The purpose of this section is to show that, when we restrict to the Earle-Morita subgroup $\mathrm{Mor}(\Sigma) \subseteq \mathfrak{M}(\Sigma)$ and take coefficients in the Schrödinger representation $V = W$, we may obtain a representation of $\mathrm{Mor}(\Sigma)$ itself, *without* passing to any central extension.

We shall do this as follows. We first recall the *metaplectic extension* $\widehat{\mathfrak{M}}(\Sigma)$ of the mapping class group, which is an extension by $\mathbb{Z}/2$. Denoting by $\widehat{\mathrm{Mor}}(\Sigma)$ and

$\overline{\mathrm{Mor}}(\Sigma)$ the restrictions of $\widehat{\mathfrak{M}}(\Sigma)$ and $\overline{\mathfrak{M}}(\Sigma)$ to the Earle-Morita subgroup, we show that $\overline{\mathrm{Mor}}(\Sigma)$ contains $\widehat{\mathrm{Mor}}(\Sigma)$ and that $\widehat{\mathrm{Mor}}(\Sigma)$ is a trivial extension. It will then follow that we may restrict (4.8) to $\widehat{\mathrm{Mor}}(\Sigma) \subset \overline{\mathrm{Mor}}(\Sigma) \subset \overline{\mathfrak{M}}(\Sigma)$ and pre-compose with a section of the trivial extension $\widehat{\mathrm{Mor}}(\Sigma)$ to obtain a representation of the Earle-Morita subgroup $\mathrm{Mor}(\Sigma)$.

4.2. Metaplectic extensions. We first consider two extensions of the symplectic group $Sp(H_{\mathbb{R}})$.

DEFINITION 4.5. Recall that the fundamental group of $Sp(H_{\mathbb{R}}) \cong Sp_{2g}(\mathbb{R})$ is infinite cyclic. It therefore has a unique connected double covering group, which is called the *metaplectic group*, denoted by $Mp(H_{\mathbb{R}})$.

DEFINITION 4.6. Consider the restriction of the projective representation (4.3) to the subgroup $Sp(H_{\mathbb{R}}) \subset Sp(H_{\mathbb{R}}) \ltimes H^1(\Sigma;\mathbb{Z}) \cong \mathrm{Aut}^+(\mathcal{H}_{\mathbb{R}}) \subset \mathrm{Aut}(\mathcal{H}_{\mathbb{R}})$, which is a projective representation

$$Sp(H_{\mathbb{R}}) \longrightarrow PU(W) \tag{4.9}$$

of the symplectic group. It is in fact this restriction that is more usually referred to by the name *Segal-Shale-Weil projective representation.* Denote by $\overline{Sp}(H_{\mathbb{R}})$ the pullback of the central extension $U(W)$ of $PU(W)$. This is a central extension of $Sp(H_{\mathbb{R}})$ by S^1.

The main technical result of this subsection is the following:

PROPOSITION 4.7. *There is an inclusion $Mp(H_{\mathbb{R}}) \subset \overline{Sp}(H_{\mathbb{R}})$ of central extensions, restricting to the inclusion $\mathbb{Z}/2 \cong \{\pm 1\} \subset S^1$ on fibres.*

Before proving this, we record its implications under pulling back to the mapping class group. We first define the relevant extensions of the mapping class group and its subgroups.

DEFINITION 4.8. Denote by $\mathfrak{s}\colon \mathfrak{M}(\Sigma) \to Sp(H)$ the standard symplectic action of the mapping class group on $H = H_1(\Sigma;\mathbb{Z})$. The *metaplectic extension* $\widehat{\mathfrak{M}}(\Sigma)$ of $\mathfrak{M}(\Sigma)$ is defined to be its central extension by $\mathbb{Z}/2$ given by pulling back the $\mathbb{Z}/2$-central extension $Mp(H_{\mathbb{R}}) \to Sp(H_{\mathbb{R}})$ along $\mathfrak{s}$ and the inclusion $Sp(H) \subset Sp(H_{\mathbb{R}})$.

DEFINITION 4.9. For a subgroup $G \subseteq \mathfrak{M}(\Sigma)$, we denote the restrictions of the central extensions $\overline{\mathfrak{M}}(\Sigma)$ and $\widehat{\mathfrak{M}}(\Sigma)$ to G by $\overline{G}$ and $\widehat{G}$ respectively.

COROLLARY 4.10. *There is an inclusion $\widehat{\mathrm{Mor}}(\Sigma) \subset \overline{\mathrm{Mor}}(\Sigma)$ of central extensions, restricting to the inclusion $\mathbb{Z}/2 \cong \{\pm 1\} \subset S^1$ on fibres.*

PROOF. By definition, the central extension $\overline{\mathfrak{M}}(\Sigma)$ of $\mathfrak{M}(\Sigma)$ is pulled back from the central extension $U(W)$ of $PU(W)$ along the top row of the following diagram.

$$\begin{array}{ccccccc}
 & \overset{\Psi}{\nearrow} & \mathrm{Aut}^+(\mathcal{H}) & \hookrightarrow & \mathrm{Aut}^+(\mathcal{H}_{\mathbb{R}}) & \xrightarrow{T} & PU(W) \\
 & & \wr\Vert & & \wr\Vert & & \\
\mathfrak{M}(\Sigma) & \xrightarrow[(\mathfrak{s},\mathfrak{d})]{} & Sp(H) \ltimes H^1(\Sigma;\mathbb{Z}) & \hookrightarrow & Sp(H_{\mathbb{R}}) \ltimes H^1(\Sigma;\mathbb{R}) & & \\
\uparrow & & \uparrow & & \uparrow & & \\
\mathrm{Mor}(\Sigma) & \xrightarrow{\mathfrak{s}} & Sp(H) & \hookrightarrow & Sp(H_{\mathbb{R}}) & &
\end{array}$$

Since this diagram commutes (for the bottom-left square this is because $\mathfrak{d} \equiv 0$ on $\mathrm{Mor}(\Sigma)$), it follows that its restriction $\overline{\mathrm{Mor}}(\Sigma)$ to the Earle-Morita subgroup is the pullback of the central extension $\overline{Sp}(H_{\mathbb{R}})$ of $Sp(H_{\mathbb{R}})$ along the bottom row of the diagram. On the other hand, the metaplectic extension $\widehat{\mathrm{Mor}}(\Sigma)$ is by definition the pullback of $Mp(H_{\mathbb{R}})$ along the bottom row of the diagram. Thus the inclusion of central extensions $Mp(H_{\mathbb{R}}) \subset \overline{Sp}(H_{\mathbb{R}})$ of $Sp(H_{\mathbb{R}})$ from Proposition 4.7 pulls back to the desired inclusion $\widehat{\mathrm{Mor}}(\Sigma) \subset \overline{\mathrm{Mor}}(\Sigma)$ of central extensions of $\mathrm{Mor}(\Sigma)$. □

REMARK 4.11. The argument above does *not* show that the metaplectic extension includes into the S^1-extension pulled back via the Segal-Shale-Weil representation on the *full* mapping class group. This is because the inclusion of central extensions essentially arises at the level of the symplectic group (Proposition 4.7), so we have to restrict to the kernel of $\mathfrak{d}$ to ensure that the S^1-extension pulls back via the symplectic group.

PROOF OF PROPOSITION 4.7. Let us first slightly rewrite the statement in notation that makes the dependence on g explicit: our goal is to prove that, over the group $Sp_{2g}(\mathbb{R})$, there is an embedding of central extensions $Mp_{2g}(\mathbb{R}) \hookrightarrow \overline{Sp}_{2g}(\mathbb{R})$ (which must necessarily restrict to the inclusion $\mathbb{Z}/2 \cong \{\pm 1\} \subset S^1$ on fibres).

We first show that it suffices to prove this statement for all g sufficiently large; we will then be able to assume for the rest of the proof that $g \geq 4$, which is the stable range for (co)homology of degree at most 2 for $Sp_{2g}(\mathbb{R})$ and $\mathfrak{M}(\Sigma_{g,1})$. For any $g < h$ there is an inclusion map $Sp_{2g}(\mathbb{R}) \hookrightarrow Sp_{2h}(\mathbb{R})$ given by extending symplectic automorphisms of $\mathbb{R}^{2g}$ by the identity on $\mathbb{R}^{2h-2g}$. We claim that the pullbacks of $Mp_{2h}(\mathbb{R})$ and of $\overline{Sp}_{2h}(\mathbb{R})$ under this inclusion are $Mp_{2g}(\mathbb{R})$ and $\overline{Sp}_{2g}(\mathbb{R})$ respectively. For the metaplectic central extensions this follows from the fact that the induced map $\pi_1(Sp_{2g}(\mathbb{R})) \cong \mathbb{Z} \to \mathbb{Z} \cong \pi_1(Sp_{2h}(\mathbb{R}))$ is an isomorphism and the metaplectic double covering corresponds to the unique index-2 subgroup of π_1. For $\overline{Sp}$, note that the Segal-Shale-Weil projective representations in genus g and h fit into a commutative square as follows:

$$\begin{array}{ccccc} Sp_{2g}(\mathbb{R}) & \xrightarrow{(4.9)} & PU(L^2(\mathbb{R}^g)) & \twoheadleftarrow & U(L^2(\mathbb{R}^g)) \\ \big\downarrow & & \big\downarrow & & \big\downarrow \\ Sp_{2h}(\mathbb{R}) & \xrightarrow{(4.9)} & PU(L^2(\mathbb{R}^h)) & \twoheadleftarrow & U(L^2(\mathbb{R}^h)) \end{array} \tag{4.10}$$

The right-hand side square of this diagram arises as follows. We consider $L^2(\mathbb{R}^g)$ as the closed subspace of $L^2(\mathbb{R}^h)$ consisting of those L^2-functions that factor through $\mathbb{R}^h = \mathbb{R}^g \times \mathbb{R}^{h-g} \twoheadrightarrow \mathbb{R}^g$. Any closed subspace of a Hilbert space has an orthogonal complement, so we may extend unitary automorphisms of $L^2(\mathbb{R}^g)$ by the identity on this complement to obtain a homomorphism $U(L^2(\mathbb{R}^g)) \to U(L^2(\mathbb{R}^h))$, which descends to the projective unitary groups, forming a pullback square. By definition, $\overline{Sp}_{2g}(\mathbb{R})$ is the pullback along (4.9) of the extension $U(L^2(\mathbb{R}^g))$ of $PU(L^2(\mathbb{R}^g))$. Commutativity of (4.10) then implies that the pullback of $\overline{Sp}_{2h}(\mathbb{R})$ along the inclusion is $\overline{Sp}_{2g}(\mathbb{R})$. Thus the existence of an embedding $Mp_{2h}(\mathbb{R}) \hookrightarrow \overline{Sp}_{2h}(\mathbb{R})$ will imply the existence of an embedding $Mp_{2g}(\mathbb{R}) \hookrightarrow \overline{Sp}_{2g}(\mathbb{R})$ for $g < h$. We henceforth assume that $g \geq 4$ in this proof (this is only needed in the last paragraph).

First, we recall from [**LV80**, §1.7] that a particular choice of cocycle

$$\omega_{Sp} \colon Sp_{2g}(\mathbb{R}) \times Sp_{2g}(\mathbb{R}) \longrightarrow S^1,$$

representing the central extension $\overline{Sp_{2g}}(\mathbb{R})$, takes values in the finite cyclic subgroup $\mathbb{Z}/8 \subseteq S^1$, so there is an embedding of central extensions $Sp_{2g}(\mathbb{R})^{(8)} \hookrightarrow \overline{Sp_{2g}}(\mathbb{R})$, for a certain $\mathbb{Z}/8$-central extension $Sp_{2g}(\mathbb{R})^{(8)}$ of $Sp_{2g}(\mathbb{R})$. Moreover, this central extension is classified by the element $-[\tau].8\mathbb{Z} \in H^2(Sp_{2g}(\mathbb{R});\mathbb{Z}/8)$, the reduction modulo 8 of the element $-[\tau] \in H^2(Sp_{2g}(\mathbb{R});\mathbb{Z})$ represented by the negative of the *Maslov cocycle* τ (see formula 1.7.7 on page 70 of [**LV80**]).

Second, we also recall from [**LV80**, §1.7] that there is a function

$$s\colon Sp_{2g}(\mathbb{R}) \longrightarrow \mathbb{Z}/4 \subseteq S^1$$

such that $\omega_{Sp}(g,h)^2 = s(g)^{-1}s(h)^{-1}s(gh)$ (formula 1.7.8 on page 70 of [**LV80**]). It follows that the subset of $Sp_{2g}(\mathbb{R})^{(8)}$ of those pairs (t,g) for which $t^2 = s(g)$ is a subgroup. The projection onto $Sp_{2g}(\mathbb{R})$ restricted to this subgroup is a double covering, and so this subgroup must either be the trivial covering $Sp_{2g}(\mathbb{R}) \times \mathbb{Z}/2$ or the metaplectic covering $Mp_{2g}(\mathbb{R})$.

To finish the proof, we just have to show that it cannot be the trivial covering. Suppose for a contradiction that it is. Then $Sp_{2g}(\mathbb{R})^{(8)}$ admits a section, so it is a trivial extension and we must have $[\tau].8\mathbb{Z} = 0 \in H^2(Sp_{2g}(\mathbb{R});\mathbb{Z}/8)$. However, the pullback of $[\tau]$ along the projection $\mathfrak{M}(\Sigma) \to Sp_{2g}(\mathbb{R})$, also denoted by $[\tau]$, is precisely 4 times a generator of $H^2(\mathfrak{M}(\Sigma);\mathbb{Z}) \cong \mathbb{Z}$ (here we are using the assumption that $g \geq 4$). Thus $[\tau].8\mathbb{Z} \in H^2(\mathfrak{M}(\Sigma);\mathbb{Z}/8) \cong \mathbb{Z}/8$ is non-zero. Hence we must have $[\tau].8\mathbb{Z} \neq 0$ already in $H^2(Sp_{2g}(\mathbb{R});\mathbb{Z}/8)$. This completes the proof. □

4.3. Triviality of an extension. The last ingredient that we will need is the following.

PROPOSITION 4.12. *The $\mathbb{Z}/2$-central extension $\widehat{\mathrm{Mor}}(\Sigma)$ of $\mathrm{Mor}(\Sigma)$ is trivial.*

PROOF. We first note that it suffices to prove this statement for all sufficiently large g. This is because the inclusion of mapping class groups $\mathfrak{M}(\Sigma_{g,1}) \hookrightarrow \mathfrak{M}(\Sigma_{h,1})$ restricts to an inclusion of Earle-Morita subgroups $\mathrm{Mor}(\Sigma_{g,1}) \hookrightarrow \mathrm{Mor}(\Sigma_{h,1})$ (as an immediate consequence of Lemma 3.4), and the pullback of $\widehat{\mathrm{Mor}}(\Sigma_{h,1})$ along this inclusion is $\widehat{\mathrm{Mor}}(\Sigma_{g,1})$, for any $g < h$. This last statement follows from the fact that the pullback of $Mp_{2h}(\mathbb{R})$ along $Sp_{2g}(\mathbb{R}) \hookrightarrow Sp_{2h}(\mathbb{R})$ is $Mp_{2g}(\mathbb{R})$, which was explained during the proof of Proposition 4.7. We now assume that $g \geq 4$ for the rest of the proof.

Recall from the proof of Proposition 4.7 that there is an embedding of central extensions $Mp_{2g}(\mathbb{R}) \hookrightarrow Sp_{2g}(\mathbb{R})^{(8)}$, where $Sp_{2g}(\mathbb{R})^{(8)}$ is a certain central extension of $Sp_{2g}(\mathbb{R})$ by $\mathbb{Z}/8$. Pulling back along the symplectic action $\mathfrak{M}(\Sigma) \to Sp_{2g}(\mathbb{R})$, we obtain an embedding of central extensions $\widehat{\mathfrak{M}}(\Sigma) \hookrightarrow \mathfrak{M}(\Sigma)^{(8)}$, where $\mathfrak{M}(\Sigma)^{(8)}$ is classified by $-[\tau].8\mathbb{Z} \in H^2(\mathfrak{M}(\Sigma);\mathbb{Z}/8) \cong \mathbb{Z}/8$. Now, $H^2(\mathfrak{M}(\Sigma);\mathbb{Z})$ is infinite cyclic, generated by the first Chern class c_1, and we have $[\tau] = 4c_1$. There is also a cocycle $c\colon \mathfrak{M}(\Sigma) \times \mathfrak{M}(\Sigma) \to \mathbb{Z}$ defined by Morita [**Mor89b**] given by the formula $c(f,f') = \mathfrak{d}(f^{-1})^\sharp.\mathfrak{d}(f')^\sharp$ (see also the proof of Lemma 3.2) and we have $[c] = 12c_1$ in $H^2(\mathfrak{M}(\Sigma);\mathbb{Z})$. Thus, in particular, we have $3[\tau] = [c]$. Since $\mathrm{Mor}(\Sigma) = \ker(\mathfrak{d})$, Morita's cocycle c vanishes on $\mathrm{Mor}(\Sigma)$, and so after restricting to the Earle-Morita subgroup we have $3[\tau] = [c] = 0 \in H^2(\mathrm{Mor}(\Sigma);\mathbb{Z})$. Reducing modulo 8 we therefore have $3[\tau].8\mathbb{Z} = 0 \in H^2(\mathrm{Mor}(\Sigma);\mathbb{Z}/8)$. But this cohomology group is a $\mathbb{Z}/8$-module, and 3 is invertible modulo 8, so we may divide by 3 and deduce that $[\tau].8\mathbb{Z} = 0$ in $H^2(\mathrm{Mor}(\Sigma);\mathbb{Z}/8)$. Hence the restriction $\mathrm{Mor}(\Sigma)^{(8)}$ of $\mathfrak{M}(\Sigma)^{(8)}$ to the Earle-Morita subgroup $\mathrm{Mor}(\Sigma)$ is a trivial extension. It therefore follows from the embedding $\widehat{\mathrm{Mor}}(\Sigma) \hookrightarrow \mathrm{Mor}(\Sigma)^{(8)}$ that $\widehat{\mathrm{Mor}}(\Sigma)$ is also a trivial extension. □

REMARK 4.13. In summary, we have considered, in this subsection and the previous one, three nested central extensions $Mp(H_{\mathbb{R}}) \subset Sp(H_{\mathbb{R}})^{(8)} \subset \overline{Sp}(H_{\mathbb{R}})$ of the symplectic group $Sp(H_{\mathbb{R}})$ with fibres $\mathbb{Z}/2 \subset \mathbb{Z}/8 \subset S^1$. Clearly they are either all trivial or all non-trivial. We have seen that their pullbacks along the symplectic action $\mathfrak{M}(\Sigma) \to Sp(H_{\mathbb{R}})$ are non-trivial (and hence they must also be non-trivial to begin with), but their further pullbacks (restrictions) to the Earle-Morita subgroup $\mathrm{Mor}(\Sigma) \subset \mathfrak{M}(\Sigma)$ are trivial.

4.4. Untwisted representations of Earle-Morita subgroups. We may now conclude with the main result of this section:

THEOREM 4.14. *For any $n \geq 2$, there is a representation*

$$\mathrm{Mor}(\Sigma) \longrightarrow GL(\mathcal{V}_n(W)) \tag{4.11}$$

induced by the natural action of the mapping class group on the twisted Borel-Moore homology group (1.9) *with coefficients in the Schrödinger representation $V = W$.*

PROOF. By Theorem 4.4, such a representation is defined on the central extension $\overline{\mathfrak{M}}(\Sigma)$ of the mapping class group by S^1. By Corollary 4.10, the restriction of this extension to $\mathrm{Mor}(\Sigma) \subset \mathfrak{M}(\Sigma)$ contains the metaplectic extension $\widehat{\mathrm{Mor}}(\Sigma)$, so we may further restrict to this subgroup. By Proposition 4.12, the central extension $\widehat{\mathrm{Mor}}(\Sigma)$ of $\mathrm{Mor}(\Sigma)$ is trivial, i.e., it admits a section. Hence, composing with any such section, we obtain the desired representation (4.11). □

References

[BCRR20] Dave Benson, Caterina Campagnolo, Andrew Ranicki, and Carmen Rovi, *Signature cocycles on the mapping class group and symplectic groups*, J. Pure Appl. Algebra **224** (2020), no. 11, 106400, 49, DOI 10.1016/j.jpaa.2020.106400. MR4104487

[Bel04] Paolo Bellingeri, *On presentations of surface braid groups*, J. Algebra **274** (2004), no. 2, 543–563, DOI 10.1016/j.jalgebra.2003.12.009. MR2043362

[BPS21] Christian Blanchet, Martin Palmer, and Awais Shaukat, *Heisenberg homology on surface configurations*, arXiv:2109.00515, 2021.

[Bre02] Tara Elise Brendle, *The Torelli group and representations of mapping class groups*, ProQuest LLC, Ann Arbor, MI, 2002. Thesis (Ph.D.)–Columbia University. MR2703274

[Chi72] D. R. J. Chillingworth, *Winding numbers on surfaces. II*, Math. Ann. **199** (1972), 131–153, DOI 10.1007/BF01431419. MR321091

[DM22] Marco De Renzi and Jules Martel, *Homological construction of quantum representations of mapping class groups*, arXiv:2212.10940, 2022.

[Ear78] Clifford J. Earle, *Families of Riemann surfaces and Jacobi varieties*, Ann. of Math. (2) **107** (1978), no. 2, 255–286, DOI 10.2307/1971144. MR499328

[GM01] Juan González-Meneses, *New presentations of surface braid groups*, J. Knot Theory Ramifications **10** (2001), no. 3, 431–451, DOI 10.1142/S0218216501000949. MR1825967

[Joh79] Dennis L. Johnson, *Homeomorphisms of a surface which act trivially on homology*, Proc. Amer. Math. Soc. **75** (1979), no. 1, 119–125, DOI 10.2307/2042686. MR529227

[Joh80] Dennis Johnson, *An abelian quotient of the mapping class group $\mathcal{I}_g$*, Math. Ann. **249** (1980), no. 3, 225–242, DOI 10.1007/BF01363897. MR579103

[Joh85] Dennis Johnson, *The structure of the Torelli group. III. The abelianization of $\mathcal{T}$*, Topology **24** (1985), no. 2, 127–144, DOI 10.1016/0040-9383(85)90050-3. MR793179

[Kun09] Yusuke Kuno, *A combinatorial formula for Earle's twisted 1-cocycle on the mapping class group $\mathcal{M}_{g,*}$*, Math. Proc. Cambridge Philos. Soc. **146** (2009), no. 1, 109–118, DOI 10.1017/S0305004108001680. MR2461871

[LV80] Gérard Lion and Michèle Vergne, *The Weil representation, Maslov index and theta series*, Progress in Mathematics, vol. 6, Birkhäuser, Boston, MA, 1980, DOI 10.1016/0012-365x(79)90168-7. MR573448

[Mor89a] Shigeyuki Morita, *Families of Jacobian manifolds and characteristic classes of surface bundles. I* (English, with French summary), Ann. Inst. Fourier (Grenoble) **39** (1989), no. 3, 777–810, DOI 10.5802/aif.1188. MR1030850

[Mor89b] Shigeyuki Morita, *Families of Jacobian manifolds and characteristic classes of surface bundles. II*, Math. Proc. Cambridge Philos. Soc. **105** (1989), no. 1, 79–101, DOI 10.1017/S0305004100001389. MR966142

[Mor93] Shigeyuki Morita, *The extension of Johnson's homomorphism from the Torelli group to the mapping class group*, Invent. Math. **111** (1993), no. 1, 197–224, DOI 10.1007/BF01231286. MR1193604

[Sco70] G. P. Scott, *Braid groups and the group of homeomorphisms of a surface*, Proc. Cambridge Philos. Soc. **68** (1970), 605–617, DOI 10.1017/s0305004100076593. MR268889

[Tra92] Rolland Trapp, *A linear representation of the mapping class group $\mathcal{M}$ and the theory of winding numbers*, Topology Appl. **43** (1992), no. 1, 47–64, DOI 10.1016/0166-8641(92)90153-Q. MR1141372

[Wah13] Nathalie Wahl, *Homological stability for mapping class groups of surfaces*, Handbook of moduli. Vol. III, Adv. Lect. Math. (ALM), vol. 26, Int. Press, Somerville, MA, 2013, pp. 547–583. MR3135444

Université Paris Cité & Sorbonne Université, CNRS, IMJ-PRG, F-75006 Paris, France
Email address: christian.blanchet@imj-prg.fr

Simion Stoilow Mathematical Institute of the Romanian Academy, Bucharest, Romania
Email address: mpanghel@imar.ro

Abdus Salam School of Mathematical Sciences, Lahore, Pakistan
Email address: m.awais_shaukat@live.com

Contemporary Mathematics
Volume **813**, 2025
https://doi.org/10.1090/conm/813/16291

Flat principal 2-group bundles and flat string structures

Daniel Berwick-Evans, Emily Cliff, Laura Murray, Apurva Nakade, and Emma Phillips

ABSTRACT. For a weak 2-group $\mathcal{G}$, we construct a bicategory of flat $\mathcal{G}$-bundles over differentiable stacks as a localization of a functor bicategory. This description is amenable to explicit geometric constructions. For example, we show that flat $\mathcal{G}$-bundles can be described in terms of ordinary G-bundles together with a trivialization of a certain 2-gerbe. This specializes to a characterization of flat string structures on vector bundles over differentiable stacks.

1. Introduction

The *string group* is the smooth categorical central extension

$$1 \to \mathrm{pt} /\!\!/ U(1) \to \mathrm{String}_n \to \mathrm{Spin}_n \to 1, \qquad n \in \mathbb{N} \tag{1.1}$$

classified by the image of the fractional Pontryagin class under the isomorphism $\frac{p_1}{2} \in \mathrm{H}^4(B\,\mathrm{Spin}_n; \mathbb{Z}) \simeq \mathrm{H}^3_{\mathrm{SM}}(\mathrm{Spin}_n; U(1))$, where H_{SM} denotes Segal–Mitchison cohomology [**38**]. The smooth extension (1.1) is in the bicategory of smooth 2-groups, defined as group objects in smooth stacks. Every smooth 2-group has an underlying discrete 2-group [**38**, page 28]. In particular, the discrete version of (1.1) yields the discrete string group String^δ_n (see §A).

DEFINITION 1.1 (Definition 5.23). A *flat string structure* on a flat vector bundle V is a lift

$$\begin{array}{ccc} & & \mathrm{pt} /\!\!/ \mathrm{String}^\delta_n \\ & \nearrow & \downarrow \\ X & \xrightarrow[V]{} & \mathrm{pt} /\!\!/ \mathrm{O}^\delta_n \end{array} \tag{1.2}$$

where O^δ_n is the orthogonal group O_n with the discrete topology, and the arrow labeled V is the map classifying the frame bundle of the flat vector bundle V.

String structures were born out of physicists' work on the 2-dimensional supersymmetric sigma model [**21, 54**], and have been codified in various mathematical

2020 *Mathematics Subject Classification.* Primary 18F20, 55R15, 55R65, 18F15, 22A22; Secondary 20J99, 55N34.

The first author was supported in part by NSF Grant DMS 2205835.
The second author was supported in part by NSERC under Discovery Grant RGPIN-2022-04104 and FRQNT Research Support for New Academics under Grant 327706.
The third author was supported in part by NSF Grant DMS 2316646.

ways, e.g., [**10**, **46**]. Among these approaches, our Definition 1.1 is a flat version of Schommer-Pries's notion in [**38**] using the language of 2-group principal bundles. Another approach to string structures is Waldorf's definition involving trivializations of the Chern–Simons 2-gerbe [**50**]. The following connects these a priori distinct approaches in the flat case; we note that flat string structures are the objects of a bicategory.

THEOREM 1.2 (Theorem 5.28). *The bicategory of flat string structures on a vector bundle* $V \to X$ *is equivalent to the bicategory of trivializations of* $\mathrm{CS}_V \to X$ *where* CS_V *is the flat Chern–Simons 2-gerbe of* V.

Theorem 1.2 can be situated within a general framework of flat 2-group bundles. To explain this, note that the continuous homomorphism $\mathrm{Spin}_n^\delta \to \mathrm{Spin}_n$ induces a map

$$\mathrm{H}^3_{\mathrm{SM}}(\mathrm{Spin}_n; U(1)) \to \mathrm{H}^3(\mathrm{Spin}_n^\delta; U(1)) \tag{1.3}$$

from Segal–Mitchison cohomology to group cohomology of the discrete group Spin_n^δ (see Lemma A.1). The discrete extension underlying (1.1) is classified by an ordinary group cohomology class in $\mathrm{H}^3(\mathrm{Spin}_n^\delta; U(1))$, given by the image of $\frac{p_1}{2}$ under (1.3). More generally, a discrete group G, discrete abelian group A, and 3-cocycle $\alpha \in Z^3(G; A)$ for trivial G-action on A determine a discrete 2-group $\mathcal{G}$ sitting in the categorical central extension

$$1 \to \mathrm{pt} /\!\!/ A \to \mathcal{G} \to G \to 1 \tag{1.4}$$

classified by $[\alpha] \in \mathrm{H}^3(G; A)$. This gives a generalization of (1.2) for the moduli of lifts of a flat G-bundle P to flat $\mathcal{G}$-bundles,

$$\begin{array}{ccc} & & \mathrm{pt} /\!\!/ \mathcal{G} \\ & \nearrow & \downarrow \\ X & \xrightarrow[P]{} & \mathrm{pt} /\!\!/ G. \end{array} \tag{1.5}$$

Letting P vary, the collection of dashed arrows is the bicategory $\mathsf{Bun}_{\mathcal{G}}(X)$ of $\mathcal{G}$-bundles on a differentiable stack X, and there is a forgetful functor $\mathsf{Bun}_{\mathcal{G}}(X) \to \mathsf{Bun}_G(X)$ that extracts the ordinary G-bundle $P \to X$.

REMARK 1.3. Since the bundles considered in (1.5) use the discrete topologies on G and A, they are necessarily flat. However, to underscore the parallel with Definition 1.1, we often call a lift (1.5) a *flat* $\mathcal{G}$-bundle.

We prove the following generalization of Theorem 1.2.

THEOREM 1.4 (Theorem 4.22). *A flat* $\mathcal{G}$*-bundle on* X *is equivalent to a* G*-bundle* $P \to X$ *with a trivialization of the 2-gerbe* $\lambda_{P,\alpha}$ *determined by* P *and* α.

REMARK 1.5. For 2-groups with trivial associator (e.g., *strict* 2-groups), the 2-gerbe λ_{P_α} is canonically trivial; a change of trivialization is then a choice of (1-)gerbe. This agrees with previous descriptions of strict 2-group bundles, e.g., [**3**], [**34**, §7.1]. Theorem 1.4 therefore reveals a subtle distinction: the theory of strict 2-group bundles reduces to the geometry of gerbes, whereas (weak) 2-group bundles involve the geometry of 2-gerbes and their trivializations.

1.1. String structures from homotopy theory. For a topological space X, isomorphism classes of n-dimensional (metrized) vector bundles $V \to X$ are in bijection with homotopy classes of maps $X \to BO_n$ to the classifying space BO_n. This leads to homotopical notions of orientations, spin structures, and string structures corresponding to lifts,

$$
\begin{array}{ccccc}
 & & B\,\mathrm{String}_n & & \\
 & \nearrow & \downarrow & & \\
 & & B\,\mathrm{Spin}_n & \xrightarrow{\frac{p_1}{2}} & \mathrm{K}(\mathbb{Z},4) \\
 & \nearrow & \downarrow & & \\
 & & B\,\mathrm{SO}_n & \xrightarrow{w_2} & \mathrm{K}(\mathbb{Z}/2,2) \\
 & \nearrow & \downarrow & & \\
X & \xrightarrow[V]{} & BO_n & \xrightarrow{w_1} & \mathrm{K}(\mathbb{Z}/2,1).
\end{array}
\tag{1.6}
$$

In the above diagram,

$$\cdots \to B\,\mathrm{String}_n \to B\,\mathrm{Spin}_n \to B\,\mathrm{SO}_n \to BO_n \tag{1.7}$$

is the Whitehead tower of BO_n, i.e., each successive space in the tower (1.7) is obtained by killing the lowest-degree homotopy group of the previous space. In the cases of interest, these spaces admit equivalent descriptions as bundles classified by the maps w_1, w_2, and $p_1/2$ given by the Stiefel–Whitney classes and the fractional Pontryagin class.

In (1.6), the notation BG for various groups G is not entirely precise, as we now explain. By construction, the spaces in the tower (1.7) are only defined up to homotopy equivalence. The notation is justified by the fact that the spaces $B\,\mathrm{SO}_n$ and $B\,\mathrm{Spin}_n$ have the same homotopy type as the classifying spaces of the Lie groups SO_n and Spin_n, respectively. This leads to differential-geometric descriptions of lifts in (1.6) in terms of SO_n- and Spin_n-principal bundles on X. As an added bonus, this geometry leads to the "correct" definition of G-equivariant orientations and G-equivariant spin structures via G-actions on the corresponding SO_n- and Spin_n-principal bundles.

On the other hand, the homotopy type $B\,\mathrm{String}_n$ in (1.7) does not admit a description in terms of the classifying space of a Lie group [**38**, page 3]. It does admit various models as the classifying space of an infinite-dimensional group [**2, 5, 9, 18, 30, 45, 46, 49**], but these descriptions are not well-suited to differential geometric applications. Instead, the most promising model for $B\,\mathrm{String}_n$ is as the classifying space of the (weak) smooth 2-group String_n [**38**, §1]. This identifies $B\,\mathrm{String}_n$ with a classifying space for higher categorical bundles, namely 2-group principal bundles for the smooth 2-group String_n. One of the primary goals of this paper is to gain a better understanding of the geometry of such String_n-bundles and string structures via Theorem 1.2 and Conjecture 1.6 below.

1.2. String structures and topological modular forms. One reason for pursuing a deeper understanding of string geometry comes from the Stolz–Teichner program [**46, 47**]. It is expected that a certain (yet to be defined mathematically) class of 2-dimensional field theories furnishes a geometric model for the cohomology theory of topological modular forms (TMF). Physical reasoning predicts that string structures encode anomaly cancellation data for these field theories [**21**, **54**]. The most compelling evidence for Stolz and Teichner's proposal is that string structures are orientation data for TMF: vector bundles with string structure have Thom isomorphisms in TMF. However, a differential geometric description of TMF has evaded discovery for over 30 years. This is partly because there are so many seemingly distinct mathematical approaches to 2-dimensional quantum field theory (e.g., [**1**, **12**, **13**, **15**, **19**, **40**, **41**, **46**, **47**]), and partly because it has been difficult to construct a map from geometric objects to TMF.

An important development in this vein is Lurie's universal property that characterizes TMF [**24**, §5.5]. The key idea is that of a *2-equivariant cohomology theory*, which is a higher-categorical equivariant refinement involving 2-groups acting on topological stacks [**24**, §5.1-5.3]. Lurie sketches the construction of a 2-equivariant refinement of TMF, and further describes how TMF satisfies a universal property among 2-equivariant cohomology theories. This suggests a possible road to constructing a comparison map between TMF and field theories, utilizing Lurie's universal property. Furthermore, 2-equivariance is naturally accommodated in only a few current paradigms for mathematical quantum field theory. This cuts down the zoo of choices, and makes is harder to guess the "wrong" definition of quantum field theory that incorporates 2-equivariance.

The geometric origins of 2-equivariance are undoubtedly wrapped up in the string group and string structures, understood in terms of the geometry of 2-group principal bundles. Combined with the geometry of anomaly cancellation in physics, these ideas come together in a (as-yet undeveloped) theory of *equivariant* string structures, e.g., for vector bundles defined on quotient stacks $M /\!\!/ G$. Definition 1.1 accommodates such examples in the case of flat vector bundles, and Conjecture 1.6 below indicates the anticipated generalization to the non-flat case.

1.3. 2-group principal bundles. The majority of the previous literature on 2-group principal bundles concerns *strict* 2-groups, implying that the associator is trivial e.g., see [**3**, **4**, **34**, **39**, **55**]. Crucially, however, the string group (1.1) is not strict. Hence, a robust theory of string structures requires a better developed theory of 2-group principal bundles for weak 2-groups. In the previous approaches to principal $\mathcal{G}$-bundles for a weak 2-group $\mathcal{G}$ (e.g., [**6**, **23**, **31**, **32**, **38**]), a solid global theory was developed, but the bicategory of $\mathcal{G}$-bundles on a manifold M is somewhat difficult to access directly.

A main goal of this paper is to develop an explicit presentation of the bicategory $\mathsf{Bun}_{\mathcal{G}}(X)$ of flat principal $\mathcal{G}$-bundles over a differentiable stack X for $\mathcal{G}$ a weak 2-group. This presentation specializes to the standard Čech description of flat A-gerbes when $\mathcal{G} = \mathrm{pt} /\!\!/ A$, see §3.6. Similarly, when $\mathcal{G} = G$ is an ordinary group we obtain the description of flat G-bundles in terms of a cover and gluing data, see Example 4.13. Furthermore, flat $\mathcal{G}$-bundles in the sense of this paper give examples of principal $\mathcal{G}$-bundles in the sense of [**38**], see Lemma 4.16. The explicit description of this bicategory $\mathsf{Bun}_{\mathcal{G}}(X)$ is what allows for the proofs of Theorems 1.2 and 1.4 via concrete computations.

Ultimately, one would hope for a generalization of our main results for arbitrary categorical central extensions in the bicategory of smooth 2-groups, and in particular for the string group (1.1).

CONJECTURE 1.6. *For a smooth categorical central extension*

$$1 \to \mathrm{pt} /\!\!/ A \to \mathcal{G} \to G \to 1$$

of a Lie group G classified by $[\alpha] \in \mathrm{H}^3_{\mathrm{SM}}(G; A)$, a $\mathcal{G}$-bundle on a smooth stack X is equivalent to a G-bundle $P \to X$ with a trivialization of the 2-gerbe $\lambda_{P,\alpha} \to X$ determined by P and the Segal–Mitchison cocycle α.

In particular, string structures on a vector bundle $V \to X$ are equivalent to trivializing sections $\sigma\colon X \to \mathrm{CS}_V$ for CS_V the Chern-Simons 2-gerbe of V.

The current paper is the first in a program of the authors to prove Conjecture 1.6. We anticipate this will reveal the correct notion of *string connection* on a String_n-principal bundle, an important step in the Stolz–Teichner program to construct a geometric model for the string orientation of TMF [**46**].

The flatness assumptions of the current paper simplify some of the stacky geometry expected in Conjecture 1.6, while also retaining the anticipated categorical complexities of the complete picture. In particular, studying $\mathcal{G}$-bundles for a weak 2-group $\mathcal{G}$ reveals that the nontrivial associator α changes the flavor of the geometry rather dramatically when compared to bundles for strict 2-groups; see Remark 1.5. Similar behavior must persist for principal bundles for arbitrary smooth 2-groups.

1.4. Outline and conventions. We begin in §2 with a review of 2-groups. In particular, we define a 2-group $\mathcal{G}(G, A, \alpha)$ associated to a discrete group G, a discrete abelian group A with an action of G, and a 3-cocycle α. In the remainder of the paper, we focus on the case where the G-action on A is trivial. In §3, we introduce a bicategory of principal $\mathcal{G}$-bundles on a Lie groupoid $X = \{X_1 \rightrightarrows X_0\}$. Initially we formalize this notion via the bicategory $\mathsf{Fun}(X, \mathrm{pt} /\!\!/ \mathcal{G})$, but in the case of a 2-group $\mathcal{G}(G, A, \alpha)$, we unwind the data to give an equivalent bicategory, which we will denote $\mathsf{Bun}^{\mathrm{pre}}_{\mathcal{G}}(X)$. These bicategories have as objects principal $\mathcal{G}$-bundles that trivialize over the objects X_0 of X. Hence, the assignment $X \mapsto \mathsf{Bun}^{\mathrm{pre}}_{\mathcal{G}}(X)$ defines a prestack over the bicategory $\mathsf{LieGrpd}$ of Lie groupoids and smooth functors; the superscript "pre" emphasizes the fact that this prestack is (almost never) a stack.

In §4, we extend the assignment $X \mapsto \mathsf{Bun}^{\mathrm{pre}}_{\mathcal{G}}(X)$ to a stack. To this end, we first define a bicategory $\mathsf{Bun}^{\mathrm{pre}}_{\mathcal{G}}$ which is fibered in 2-groupoids over $\mathsf{LieGrpd}$ and localize with respect to the class of 1-morphisms covering essential equivalences in $\mathsf{LieGrpd}$. This results in a bicategory $\mathsf{Bun}_{\mathcal{G}}$ over the localized bicategory $\mathsf{LieGrpd}[\mathsf{W}^{-1}]$ of differential stacks, where W is the class of essential equivalences of Lie groupoids. By construction, the homotopy fiber of $\mathsf{Bun}_{\mathcal{G}}$ at X is a bicategory $\mathsf{Bun}_{\mathcal{G}}(X)$ of principal $\mathcal{G}$-bundles on X. As an application of this construction, we study the natural functor $\pi : \mathsf{Bun}_{\mathcal{G}}(X) \to \mathsf{Bun}_G(X)$; this enables us to characterize a principal $\mathcal{G}$-bundle $\mathcal{P}$ in terms of its underlying G-bundle $P = \pi(\mathcal{P})$ and the additional data of a trivialization of a 2-gerbe $\lambda_{P,\alpha}$ (Theorem 1.4/4.22).

In §5 we turn our attention to structures on flat vector bundles over a geometric stack. We show that orientations, flat spin structures, and flat string structures can equivalently be phrased in terms of lifts as in (1.2), or as trivializations of line bundles, gerbes, or 2-gerbes in the respective cases. This proves Theorem 1.2.

Throughout, a functor between bicategories will mean a pseudofunctor, i.e., a particular type of lax 2-functor.

A group G can be viewed as a groupoid with a single object pt and morphisms G, which we denote by $\mathrm{pt} /\!\!/ G$, where multiplication in G corresponds to composition in the category. Given a group G acting on a set S on the right, the *action groupoid* has objects the elements of S, morphisms $S \times G$ (where the source map is the projection and the target map is the action map), and composition is determined by multiplication in G. We denote this groupoid by $S \mathbin{/\!\!/} G$.

1.5. Acknowledgements. This paper began in conversations at the 2019 Mathematics Research Community, "Geometric Representation Theory and Equivariant Elliptic Cohomology." We thank the AMS and NSF for supporting this program, as well as Eric Berry and Joseph Rennie who contributed to this project in its early stages. We also thank Matt Ando, Nora Ganter, Connor Grady, Eugene Lerman, Christopher Schommer-Pries, Nat Stapleton, and Stephan Stolz for stimulating conversations on these ideas. We also thank the anonymous referees for their helpful comments. This material is based upon work supported by the National Science Foundation under Grant Number DMS 1641020. The first author was partially supported by the National Science Foundation under Grant Number DMS 2205835. The second author was partially supported by the Natural Sciences and Engineering Research Council of Canada under Discovery Grant RGPIN-2022-04104 and the Fonds de Recherche du Québec (Nature et Technologie) Research Support for New Academics under Grant 327706. The third author was partially supported by the National Science Foundation under Grant Number DMS 2316646.

2. Review of 2-groups

In this section, we introduce some foundational terminology and examples of 2-groups. The example of crucial interest in later sections of the paper is the 2-group $\mathcal{G}(G, A, \alpha)$ associated to a discrete group G, an abelian group A (with trivial G-action), and a 3-cocycle α as in Remark 2.13; however, in this section we consider the general case of a nontrivial G-action on A. None of the definitions or results in this section are original; standard references include [**2**, **38**]

2.1. Basic definitions.

Definition 2.1. A (discrete) *2-group* is a monoidal groupoid $(\mathcal{G}, \otimes, \mathbb{1})$ where every object is (weakly) $\otimes$-invertible, meaning for every object x there exists an object x^{-1} and isomorphisms $x \otimes x^{-1} \simeq \mathbb{1} \simeq x^{-1} \otimes x$. A *1-homomorphism* between 2-groups is a lax monoidal functor. A *2-homomorphism* between 1-homomorphisms is a lax monoidal transformation.

Proposition 2.2. *The collection of 2-groups, 1-homomorphisms, and 2-homomorphisms has the structure of a bicategory.*

We often use the abbreviated notation $\mathcal{G}$ for a 2-group, and adopt the following terminology that generalizes similar language for ordinary groups. Let pt denote the terminal category. A 2-group has a *unit* functor $\mathbb{1}\colon \text{pt} \to \mathcal{G}$ and a *multiplication* functor $\mu\colon \mathcal{G} \times \mathcal{G} \to \mathcal{G}$. It also has an *associator* natural transformation α and *unitor* natural transformations ϵ_l, ϵ_r fitting into the diagrams

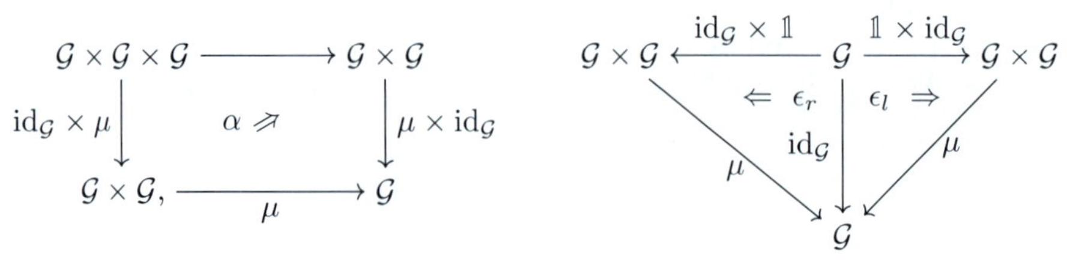

and satisfying standard properties (e.g., the pentagon axiom). A 1-homomorphism is the data of a functor $\varphi\colon \mathcal{G} \to \mathcal{G}'$ and natural transformation γ,

$$\begin{array}{ccc} \mathcal{G} \times \mathcal{G} & \xrightarrow{\varphi \times \varphi} & \mathcal{G}' \times \mathcal{G}' \\ {\scriptstyle \mu'}\big\downarrow & \gamma \nearrow & \big\downarrow{\scriptstyle \mu} \\ \mathcal{G}, & \xrightarrow[\varphi]{} & \mathcal{G}' \end{array}$$

making φ into a monoidal functor via the standard properties, including a commuting cube intertwining the associators for $\mathcal{G}$ and $\mathcal{G}'$ using γ. A 2-homomorphism is a natural transformation $(\varphi, \gamma) \Rightarrow (\varphi', \gamma')$ satisfying compatibility properties.

There are a few variations of Definition 2.1 in the literature, which we record below.

DEFINITION 2.3 ([**2**, Definition 29]). A 2-group with trivial associator and unitors is called a *strict 2-group.*

DEFINITION 2.4 ([**2**, Definition 7]). A *coherent 2-group* $\mathcal{G}$ is a 2-group equipped with the additional data of an object x^{-1} for each $x \in \mathcal{G}$ together with specified isomorphisms $x \otimes x^{-1} \simeq \mathbb{1} \simeq x^{-1} \otimes x$ satisfying the axioms of an adjoint equivalence.

2.2. Examples of 2-groups.

EXAMPLE 2.5. An ordinary group G determines a 2-group whose underlying groupoid has only identity morphisms with monoidal structure given by the group multiplication on G. A homomorphism of groups is equivalent data to a monoidal functor between the corresponding monoidal categories. Hence, the category of groups and homomorphisms admits a faithful embedding into the bicategory of 2-groups.

EXAMPLE 2.6. For an abelian group A, the groupoid $\mathrm{pt} /\!\!/ A$ has the structure of a 2-group, where the monoidal structure on morphisms is determined by multiplication in A.

EXAMPLE 2.7. For any monoidal category $(\mathsf{C}, \otimes, \mathbb{1})$, the *Picard 2-group* of C, denoted $\mathsf{Pic}(\mathsf{C})$, is the monoidal subgroupoid of C with objects the $\otimes$-invertible objects of C and morphisms the isomorphisms. In the case that $(\mathsf{C}, \otimes, \mathbb{1}) = (\mathsf{Vect}_{\mathbb{K}}, \otimes, \mathbb{K})$ is the (symmetric) monoidal category of vector spaces over $\mathbb{K}$ with tensor product, then $\mathsf{Pic}(\mathsf{Vect}_{\mathbb{K}})$ is the category of 1-dimensional $\mathbb{K}$-vector spaces.

EXAMPLE 2.8. Given a category C, the functor category $\mathrm{Fun}(\mathsf{C}, \mathsf{C})$ is a (strict) monoidal category with monoidal structure coming from composition of functors. The subcategory

$$\mathsf{Aut}(\mathsf{C}) := \mathsf{Pic}(\mathrm{Fun}(\mathsf{C}, \mathsf{C}))$$

is the 2-group of automorphisms of C. Composition of functors is strictly associative and unital; hence the associator and unitors for this 2-group are trivial.

An early example of 2-groups came from the concept of crossed modules, first formulated by Whitehead in [**52, 53**].

EXAMPLE 2.9. A *crossed module* is the data of homomorphisms $t\colon H \to G$ and $a\colon G \to \mathsf{Aut}(H)$ between (ordinary) groups such that the diagrams commute

$$
\begin{array}{ccc}
H \times H & \xrightarrow{t \times \mathrm{id}_H} & G \times H \\
 & {\scriptstyle \mathrm{Ad}} \searrow \quad \swarrow {\scriptstyle a} & \\
 & H &
\end{array}
\qquad\qquad
\begin{array}{ccc}
G \times H & \xrightarrow{\quad a \quad} & H \\
{\scriptstyle \mathrm{id}_G \times t}\big\downarrow & & \big\downarrow{\scriptstyle t} \\
G \times G & \xrightarrow[\quad \mathrm{Ad} \quad]{} & G,
\end{array}
\tag{2.1}
$$

where Ad is the adjoint action and (in a mild abuse of notation) the arrows labeled a are determined by the homomorphism a. This determines a 2-group with underlying groupoid the action groupoid $G /\!\!/ H$ for the (left) H-action on G via t. The monoidal structure on this groupoid comes from the multiplication in G and the semidirect product $H \rtimes G$ determined by a. The associator for this monoidal structure is trivial.

The bicategories of strict 2-groups and crossed modules are equivalent; see [**2**].

EXAMPLE 2.10. Given any 2-group $\mathcal{G}$ and (1-)category C, the category of functors from C to $\mathcal{G}$ is naturally a 2-group.

2.3. Classification of 2-groups as categorical extensions. For a 2-group $\mathcal{G}$, let $G := \pi_0(\mathcal{G})$ denote the isomorphism classes of objects in $\mathcal{G}$, $A := \mathsf{Aut}(\mathbb{1})$ denote the automorphism group of the unit object $\mathbb{1}\colon \mathrm{pt} \to \mathcal{G}$, and $\iota\colon \mathrm{pt} /\!\!/ A \to \mathcal{G}$ the inclusion of the automorphisms of the unit object. We refer to G as the ordinary group underlying $\mathcal{G}$. Note that we have the sequence of 1-homomorphisms

$$
1 \to \mathrm{pt} /\!\!/ A \xrightarrow{\iota} \mathcal{G} \xrightarrow{\pi_0} G \to 1, \tag{2.2}
$$

where G is an ordinary group regarded as a 2-group with group structure inherited from the monoidal structure on $\mathcal{G}$. The set A has a pair of multiplications given by composition or multiplication in $\mathcal{G}$; by the Eckmann–Hilton argument these agree and are commutative. Furthermore, there is a G-action on A from conjugating an automorphism $a \in \mathsf{Aut}(\mathbb{1})$ by id_g for an object g of $\mathcal{G}$,

$$
g \mapsto \mathrm{id}_{g^{-1}} \otimes (a \otimes \mathrm{id}_g).
$$

Finally, the associator of $\mathcal{G}$ determines a 3-cocycle $\alpha\colon G \times G \times G \to A$, where the pentagon axiom implies that α satisfies the cocycle condition. In analogy to the structures in ordinary extensions of a group G by an abelian group A, we call (2.2) a *categorical extension* of G by $\mathrm{pt} /\!\!/ A$. When the G-action on A is trivial, (2.2) is a *categorical central extension.*

PROPOSITION 2.11. *A 2-group $\mathcal{G}$ is determined up to 1-isomorphism by*

(1) *a group $G = \pi_0(\mathcal{G})$, the isomorphim classes of objects in $\mathcal{G}$;*
(2) *an abelian group $A = \mathsf{Aut}(1_G)$;*
(3) *a G-action on A, $G = \pi_0(\mathcal{G}) \to \mathsf{Aut}(A)$;*
(4) *a class $[\alpha] \in \mathrm{H}^3(G; A)$.*

PROOF. This follows from [**2**, Theorem 43] (based on [**43**]); for central extensions, it is an immediate consequence of [**38**, Theorem 99]. For the reader's convenience we review the construction on objects.

Above we already described how to extract data (1)–(4) from a 2-group. Next we describe how to construct a 2-group from the data (1)–(4). Start with the groupoid $\{G \ltimes A \rightrightarrows G\}$ where the source and target are the projections. Define a monoidal structure via multiplication in G and $G \ltimes A$. The 3-cocyle α gives the associator. The left and right unitors are determined by α and the requirement that the triangle identities hold. This yields a 2-group $\mathcal{G}(G, A, \alpha)$. □

REMARK 2.12. For trivial G-action on A, a 3-cocycle $\alpha \in Z^3(G; A)$ is *normalized* if

$$\alpha(1_G, g_1, g_2) = \alpha(g_1, 1_G, g_2) = \alpha(g_1, g_2, 1_G) = 1_A,$$

where $1_G \in G$ and $1_A \in A$ are the identity elements. Every 3-cocycle is cohomologous to a normalized one. Hereafter we shall assume that all 3-cocycles are normalized.

REMARK 2.13. Let $\mathcal{G} = \mathcal{G}(G, A, \alpha)$ be a 2-group determined by data (G, A, α) as in Proposition 2.11, where A is a trivial G-module and α is a normalized 3-cocycle. In this case the left and right unitors are trivial. Furthermore, this 2-group can be endowed with a coherent structure (see Definition 2.4), where the monoidal inverse of an object $g \in G$ is given by its multiplicative inverse g^{-1}, and the unit and counit structure morphisms are

$$i_g = 1_A : 1_G \to g \otimes g^{-1}, \quad e_g = \alpha(g^{-1}, g, g^{-1}) : g^{-1} \otimes g \to 1_G \tag{2.3}$$

for all objects $g \in G$, i.e., the unit i is the identity and counit e is determined by α. (Here we use that $g^{-1} \otimes g$ and $g \otimes g^{-1}$ are both equal to the object 1_G, so i_g and e_g are both elements of $\mathsf{Aut}(1_G) = A$.) Coherent 2-groups of this form will be the main examples of interest in this paper.

3. Flat principal 2-group bundles on a Lie groupoid

In this section, we define for each 2-group $\mathcal{G}$ and Lie groupoid X a bicategory $\mathsf{Bun}_{\mathcal{G}}^{\mathrm{pre}}(X)$ of flat principal $\mathcal{G}$-bundles on X admitting trivializations over the objects X_0 of the Lie groupoid X.

3.1. Background: Lie groupoids and essential equivalences. The material in this subsection is classical, e.g., see the references [**22**, **26**, **28**]. Readers already familiar with Lie groupoids may want to skip to Section 3.2.

DEFINITION 3.1 (e.g., [**22**, Defn 2.11]). A *Lie groupoid* is a groupoid internal to the category of manifolds whose source and target maps are submersions. We use the notation $X = \{X_1 \rightrightarrows X_0\}$, where X_0 is the manifold of objects and X_1 is the manifold of morphisms. We use the notation $s, t : X_1 \to X_0$ for the source and target maps, and $c : X_1 \times_{X_0} X_1 \to X_1$ for the composition. For $n \geq 2$, we denote by $X_n = X_1 \times_{X_0} \ldots \times_{X_0} X_1$, the manifold of sequences of n composable morphisms in X.

A *(smooth) functor* between Lie groupoids $X = \{X_1 \rightrightarrows X_0\}$ and $Y = \{Y_1 \rightrightarrows Y_0\}$ is a functor between their underlying categories that is a smooth map on objects $f_0 : X_0 \to Y_0$ and a smooth map on morphisms $f_1 : X_1 \to Y_1$. A *(smooth) natural transformation* $\eta : f \Rightarrow g$ between functors is a natural transformation between underlying functors with the property that the map $\eta : X_0 \to Y_1$ is smooth.

EXAMPLE 3.2. Given a compact Lie group G acting smoothly on a manifold M on the right, the action groupoid $M /\!\!/ G$ has objects M, morphisms $M \times G$, source map the projection, target map the action map, and composition determined by multiplication in G. Regarding M as a Lie groupoid with only identity morphisms, there is a smooth functor $M \to M /\!\!/ G$.

Given a smooth functor $f\colon X \to Y$ between Lie groupoids, consider the diagrams of smooth manifolds

$$\begin{array}{ccc} Y_1 \times_{Y_0} X_0 & \xrightarrow{p_2} & X_0 \\ {\scriptstyle p_1}\downarrow & & \downarrow{\scriptstyle f_0} \\ Y_1 & \xrightarrow[s]{} & Y_0 \end{array} \qquad \begin{array}{ccc} X_1 & \xrightarrow{f_1} & Y_1 \\ {\scriptstyle s\times t}\downarrow & & \downarrow{\scriptstyle s\times t} \\ X_0 \times X_0 & \xrightarrow[f_0\times f_0]{} & Y_0 \times Y_0. \end{array} \tag{3.1}$$

DEFINITION 3.3 (e.g., [**22**, Definition 3.5]). A smooth functor $f\colon X \to Y$ is an *essential equivalence* if $Y_1 \times_{Y_0} X_0 \to Y_0$ on the left in (3.1) is a surjective submersion, and the diagram on the right in (3.1) is a pullback.

REMARK 3.4. The two conditions in Definition 3.3 are smooth versions of the conditions for a functor to be essentially surjective and fully faithful.

EXAMPLE 3.5 ([**37**], Section 2.1). Suppose that f is a smooth functor $f\colon X \to Y$ for which there exists a smooth functor $g\colon Y \to X$ with $f \circ g \simeq \mathrm{id}$ and $g \circ f \simeq \mathrm{id}$. Then f is an essential equivalence.

EXAMPLE 3.6 ([**26**, Example 1.1.9], [**22**, Example 3.29]). Let $P \to X$ be a principal G-bundle. Then there is an essential equivalence $P /\!\!/ G \to X = P/G$. It is invertible in the sense of Example 3.5 if and only if $P \cong X \times G$ is the trivial principal G-bundle.

EXAMPLE 3.7 ([**26**, Example 1.1.8]). Given a smooth manifold M and a surjective submersion $Y \to M$, set $Y^{[k]}$ to be the k-fold fibered product of Y over M, e.g., $Y^{[2]} = Y \times_M Y$. The *Čech groupoid* associated with $Y \to M$ is

$$\check{C}(Y) := \{Y^{[2]} \rightrightarrows Y\} \tag{3.2}$$

with source and target maps given by projection to the factors, and composition given by the map $Y^{[3]} \to Y^{[2]}$ that projects out the middle factor in the fibered product. As a special case of this example, given an open cover $\{U_i\}$ of M, the map $\coprod_i U_i \to M$ is a surjective submersion, and we use the common notation for intersections, e.g., $\coprod U_{ij} = \coprod U_i \bigcap U_j = Y^{[2]}$ and $\coprod U_{ijk} = \coprod U_i \bigcap U_j \bigcap U_k = Y^{[3]}$. In this case, we use the notation $\check{C}(U_i)$ for the Čech groupoid associated to the cover. Returning to the general case of a surjective submersion $Y \to M$, there is an essential equivalence

$$\check{C}(Y) \to M \tag{3.3}$$

from the Čech groupoid to the manifold M viewed as a Lie groupoid M.

Conversely, suppose that $f\colon X \to M$ is an essential equivalence from a Lie groupoid X to a manifold M. Since $Y_0 = Y_1 = M$ in the notation of the diagrams (3.1), we deduce from the first diagram that the morphism $f_0\colon X_0 \to M$ is a surjective submersion, and from the second that X_1 is homeomorphic to $X_0 \times_M X_0$. In other words, the only essential equivalences of a Lie groupoid into a manifold are isomorphic to those coming from Čech groupoids.

EXAMPLE 3.8 ([**22**, Proposition 2.26]). Generalizing the previous example, given a Lie groupoid X and a surjective submersion $f\colon Y \to X_0$, define the *pullback groupoid* f^*X with objects $(f^*X)_0 = Y$ and morphisms the pullback

$$\begin{array}{ccc} (f^*X)_1 & \longrightarrow & X_1 \\ \downarrow & & \downarrow{\scriptstyle s\times t} \\ Y\times Y & \xrightarrow[f\times f]{} & X_0\times X_0 \end{array}$$

i.e., for any pair $x, y \in Y$ we take the morphisms between $f(x)$ and $f(y)$ in X_1. There is an evident smooth functor $f^*Y \to X$; it is an essential equivalence.

3.2. Motivation: Ordinary G-bundles in the language of Lie groupoids.

LEMMA 3.9. *Given a Lie group G and a surjective submersion $u\colon Y \to M$, a smooth functor*

$$\check{C}(Y) \to \mathrm{pt}/\!\!/G. \tag{3.4}$$

determines a principal G-bundle $P \to M$ with a trivialization of the pullback bundle $u^*P \to Y$. More precisely, a smooth functor (3.4) *determines a (strictly) commuting square of Lie groupoids*

$$\begin{array}{ccc} (Y^{[2]}\times G \rightrightarrows Y\times G) & \xrightarrow{\sim} & P \\ \downarrow & & \downarrow \\ \check{C}(Y) & \xrightarrow[\sim]{} & M \end{array} \tag{3.5}$$

where $P \to M$ is a principal G-bundle, the vertical arrows are the natural projections, and the horizontal arrows are essential equivalences. Furthermore, a natural transformation of functors (3.4) *determines an isomorphism of principal bundles $P \xrightarrow{\sim} P'$ over M.*

PROOF. Cocycle data for a G-bundle is a surjective submersion $u\colon Y \to M$ and a transition function $g\colon Y^{[2]} \to G$ satisfying the cocycle condition

$$g(y_1, y_2)\cdot g(y_2, y_3) = g(y_1, y_3) \text{ for all } (y_1, y_2, y_3) \in Y^{[3]}; \tag{3.6}$$

e.g., see [**33**]. This is precisely the data and property of a smooth functor (3.4) between Lie groupoids. For cocycle data g, g' defining a pair of G-bundles relative to the same open cover Y, an isomorphism between them is the data of a map $h\colon Y \to G$ satisfying

$$h(y_1)\cdot g(y_1, y_2)\cdot h(y_2)^{-1} = g'(y_1, y_2) \text{ for all } (y_1, y_2) \in Y^{[2]}. \tag{3.7}$$

This is the same data and property as a smooth natural transformation of functors. □

Let $\mathsf{Bun}_G(M)$ denote the category of G-bundles on M. An object of this category can be specified by a choice of surjective submersion $u\colon Y \to M$ and a smooth functor (3.4), which is equivalent to a zig-zag

$$M \xleftarrow{\sim} \check{C}(Y) \to \mathrm{pt}/\!\!/G \tag{3.8}$$

where the left arrow is an essential equivalence. We would like to understand the generalization of this category to the case where the base is allowed to be an arbitrary Lie groupoid $X = \{X_1 \rightrightarrows X_0\}$. It is easy to generalize the presentation of objects in terms of covers and smooth functors (3.8); see Definition 3.10 below. However, it becomes more awkward to describe the morphisms between such objects, as these can involve pairs of distinct essential equivalences, allowing for e.g., refinements of covers.

We describe a resolution to this issue that readily generalizes to 2-group bundles. First consider the category $\mathsf{Bun}_G^{\mathrm{pre}}(X) := \mathsf{Fun}(X, \mathrm{pt}/\!\!/G)$ of G-bundles on the Lie groupoid X that trivialize on objects. Since G-bundles pull back, letting X vary gives a fibered category $\mathsf{Bun}_G^{\mathrm{pre}}$ with a forgetful functor to Lie groupoids. Localizing this category with respect to essential equivalences of Lie groupoids (i.e., stackifying) gives a category Bun_G whose objects are zig-zags (3.8); this is the desired category of G-bundles over smooth stacks [**37**]. We explain this carefully in the setting of principal bundles for a 2-group $\mathcal{G}$ in §4, which reduces to this setting in the case $\mathcal{G} = G$.

3.3. Defining flat principal 2-group bundles over Lie groupoids.

The following definition generalizes ordinary G-bundles as described in Lemma 3.9. The smoothness condition below is the requirement that the structure maps (e.g., maps on objects and morphisms) are smooth relative to the Lie groupoid X and the discrete topology on the 2-group $\mathcal{G}$, see Remark 3.13. These structure maps are spelled out in detail in Lemmas 3.15 and 3.16.

Definition 3.10. For a Lie groupoid $X = \{X_1 \rightrightarrows X_0\}$, a *(flat) $\mathcal{G}$-bundle on X* is a zig-zag

$$X \overset{\sim}{\leftarrow} Y \overset{\mathcal{P}}{\to} \mathrm{pt}/\!\!/\mathcal{G} \tag{3.9}$$

for an essential equivalence $Y \overset{\sim}{\to} X$, and a smooth functor from Y to the 1-object bicategory $\mathrm{pt}/\!\!/\mathcal{G}$ associated to the monoidal category $\mathcal{G}$.

The goal of the next two sections is to construct a bicategory whose objects are $\mathcal{G}$-bundles (3.9). This bicategory is constructed by localizing (i.e., stackifying) the following intermediate definition of $\mathcal{G}$-bundle; we construct this localization in §4.

Definition 3.11. For a Lie groupoid $X = \{X_1 \rightrightarrows X_0\}$, a *(flat) $\mathcal{G}$-bundle on X that trivializes on X_0* is a smooth functor

$$\mathcal{P}\colon X \to \mathrm{pt}/\!\!/\mathcal{G} \tag{3.10}$$

from X to the 1-object bicategory $\mathrm{pt}/\!\!/\mathcal{G}$ associated to the monoidal category $\mathcal{G}$. These are the objects of the bicategory of smooth functors, natural transformations, and modifications [**20**, Theorem 4.4.11]

$$\mathsf{Bun}_{\mathcal{G}}^{\mathrm{pre}}(X) := \mathsf{Fun}(X, \mathrm{pt}/\!\!/\mathcal{G}). \tag{3.11}$$

The superscript "pre" refers to the fact that the assignment $X \mapsto \mathsf{Bun}_{\mathcal{G}}^{\mathrm{pre}}(X)$ is a prestack that is (almost never) a stack.

Remark 3.12. For readers familiar with smooth stacks, note the following repackaging of Definition 3.11. When considering the smooth stack $[X]$ underlying a Lie groupoid X, a $\mathcal{G}$-bundle in the sense of Definition 3.11 is a $\mathcal{G}$-bundle that trivializes when pulled back along the smooth functor $X_0 \to X$, i.e., trivializes on the chosen atlas for $[X]$; compare [**38**, Definition 70] or [**7**, Proposition 4.12].

REMARK 3.13. Since the objects and morphisms of $\mathcal{G}$ have the discrete topology, a smooth functor (3.10) factors through connected components on objects and morphisms of X, and hence is equivalent to an ordinary functor (with no smoothness condition),

$$\{\pi_0 X_1 \rightrightarrows \pi_0 X_0\} \to \mathrm{pt} /\!\!/ \mathcal{G}$$

where the source is the groupoid of connected components, regarded as a 2-groupoid, and the target is a 2-groupoid.

LEMMA 3.14. *Let $G := \pi_0\mathcal{G}$ be the ordinary group underlying $\mathcal{G}$. There is a natural functor*

$$\mathsf{Bun}^{\mathrm{pre}}_{\mathcal{G}}(X) \to \mathsf{Bun}^{\mathrm{pre}}_{G}(X) \tag{3.12}$$

from flat $\mathcal{G}$-bundles over X that trivialize over X_0 to (flat) $G = \pi_0(\mathcal{G})$-bundles on X that trivialize on X_0.

PROOF. The source and target of (3.12) are defined as functor categories, and the induced functor between them is

$$u_*\colon \mathsf{Fun}(X, \mathrm{pt} /\!\!/ \mathcal{G}) \to \mathsf{Fun}(X, \mathrm{pt} /\!\!/ G), \qquad u\colon \mathrm{pt} /\!\!/ \mathcal{G} \xrightarrow{\pi_0} \mathrm{pt} /\!\!/ G \tag{3.13}$$

for u the map of bicategories induced by the monoidal functor $\mathcal{G} \to \pi_0\mathcal{G} = G$. $\square$

3.4. $\mathcal{G}$-bundles as categorical 1-cocycles. In [**6**, §2.5.1], $\mathcal{G}$-bundles are defined as 1-cocycles valued in the monoidal category $\mathcal{G}$. Definition 3.11 recovers this, as demonstrated by the following lemma.

LEMMA 3.15. *Given a Lie groupoid $X = \{X_1 \rightrightarrows X_0\}$ and a 2-group $\mathcal{G}$, a flat $\mathcal{G}$-bundle over X that trivializes on X_0 is the data of*

(1) *a smooth (and hence locally constant) functor $\widehat{\rho}\colon X_1 \to \mathcal{G}$, where X_1 is viewed as a Lie groupoid with only identity morphisms, and $\mathcal{G}$ is viewed as a Lie groupoid with the discrete topology on objects and morphisms;*
(2) *smooth (and hence locally constant) natural transformations $\widehat{\gamma}$ and $\widehat{\epsilon}$*

$$\begin{array}{ccc} X_2 & \xrightarrow{\ c\ } & X_1 \\ {\scriptstyle p_1^*\widehat{\rho}\times p_2^*\widehat{\rho}}\downarrow & \widehat{\gamma}\ \swarrow & \downarrow{\scriptstyle\widehat{\rho}} \\ \mathcal{G}\times\mathcal{G}, & \xrightarrow[\ \mu\ ]{} & \mathcal{G} \end{array} \qquad \begin{array}{ccc} X_0 & \xrightarrow{\ u\ } & X_1 \\ \downarrow & \widehat{\epsilon}\ \swarrow & \downarrow{\scriptstyle\widehat{\rho}} \\ * & \xrightarrow[\ e\ ]{} & \mathcal{G}, \end{array} \tag{3.14}$$

These data must satisfy unity constraints: the composition of 2-morphisms in

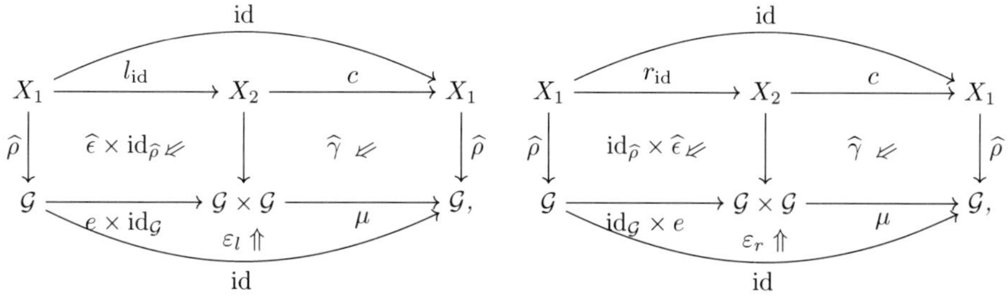

equals the identity 2-morphism on $\widehat{\rho}$. Above, $l_{\mathrm{id}}, r_{\mathrm{id}}\colon X_1 \to X_2$ are $l_{\mathrm{id}}(f) = \mathrm{id}\circ f$ and $r_{\mathrm{id}}(f) = f\circ\mathrm{id}$, respectively. The data (1) and (2) are also required to satisfy *an associativity constraint in the form of a 2-commuting cube,*

(3.15)

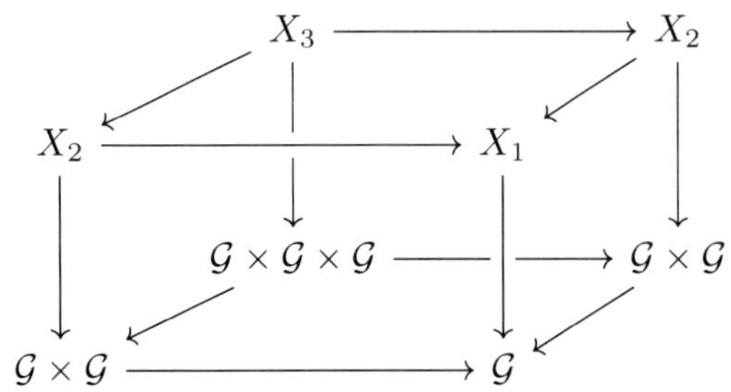

where the top face commutes on the nose (by associativity of composition in X), the bottom face has 2-commuting data given by the associator α for $\mathcal{G}$, and the sides carry pullbacks of 2-commuting data γ.

PROOF. This follows from unpacking the definition of a functor, see [**20**, §4.1]. Smoothness of $\widehat{\rho}$, $\widehat{\gamma}$ and $\widehat{\epsilon}$ is equivalent to local constancy, as the objects and morphisms of $\mathcal{G}$ have the discrete topology. □

The 1- and 2-morphisms between $\mathcal{G}$-bundles from [**6**, §2.5.1] are stated in terms of coboundaries for $\mathcal{G}$-valued 1-cocycles. This generalizes the usual cocycle description (3.7) of an isomorphism between ordinary G-bundles as follows.

LEMMA 3.16. *A 1-morphism between flat $\mathcal{G}$-bundles $(\widehat{\rho_1},\widehat{\gamma_1},\widehat{\epsilon_1})$ and $(\widehat{\rho_2},\widehat{\gamma_2},\widehat{\epsilon_2})$ over X that trivialize on X_0 is*

1. *a smooth (and hence locally constant) functor $\widehat{h}\colon X_0\to\mathcal{G}$*
2. *a smooth (and hence locally constant) natural transformation $\widehat{\eta}$,*

$$\begin{array}{ccc} X_1 & \xrightarrow{\;t^*\widehat{h}\times\widehat{\rho_1}\;} & \mathcal{G}\times\mathcal{G} \\ {\scriptstyle \widehat{\rho_2}\times s^*\widehat{h}}\big\downarrow & \widehat{\eta}\;\Nearrow & \big\downarrow{\scriptstyle\mu} \\ \mathcal{G}\times\mathcal{G} & \xrightarrow[\;\mu\;]{} & \mathcal{G}\end{array} \tag{3.16}$$

These data are required to satisfy a compatibility condition; see [**20**, §4.2].

A 2-morphism is a smooth natural isomorphism between functors $\widehat{\omega}\colon\widehat{h_1}\Rightarrow\widehat{h_2}$ satisfying the condition that the square commutes:

$$\begin{array}{ccc} \widehat{\rho_2}\cdot s^*\widehat{h_1} & \xrightarrow{\;\widehat{\eta_1}\;} & t^*\widehat{h_1}\cdot\widehat{\rho_1} \\ {\scriptstyle s^*\widehat{\omega}}\big\downarrow & & \big\downarrow{\scriptstyle t^*\widehat{\omega}} \\ \widehat{\rho_2}\cdot s^*\widehat{h_2} & \xrightarrow[\;\widehat{\eta_2}\;]{} & t^*\widehat{h_2}\cdot\widehat{\rho_1}\end{array} \tag{3.17}$$

PROOF. The description of 1-morphisms and 2-morphisms follows from unpacking the definitions of natural transformations and modifications, see [**20**, §4.2] and [**20**, §4.4] respectively. □

3.5. 2-group bundles as Čech cocycles. In the special case where the 2-group $\mathcal{G}$ is a central extension of the form $\mathcal{G} = \mathcal{G}(G, A, \alpha)$ as in Remark 2.13, then one can unpack the data of a $\mathcal{G}$-bundle in terms of group cohomology data.

PROPOSITION 3.17. *Let $\mathcal{G} = \mathcal{G}(G, A, \alpha)$ be as in Remark 2.13. A flat $\mathcal{G}$-bundle over a Lie groupoid $X = \{X_1 \rightrightarrows X_0\}$ that trivializes on X_0 is given by the same data as a pair (ρ, γ) where*

(1) *$\rho\colon X_1 \to G$ is a locally constant map satisfying the (ordinary) cocycle condition*

$$c^*\rho = p_1^*\rho \cdot p_2^*\rho\colon X_2 \to G \tag{3.18}$$

for $p_1, p_2, c\colon X_2 \to X_1$ the projections and the composition;

(2) *$\gamma\colon X_2 \to A$ is a locally constant map satisfying the conditions*

$$\rho^*\alpha = d\gamma\colon X_3 \to A, \tag{3.19}$$

$$\gamma(\phi, \mathrm{id}_{s(\phi)}) = \gamma(\mathrm{id}_{t(\psi)}, \psi), \quad \forall(\phi, \psi) \in X_2 \tag{3.20}$$

where d is the Čech differential on A-valued cochains, and

$$s(f) = x = t(g) \in X_0.$$

PROOF. By construction, the objects of $\mathcal{G}$ are the elements of the discrete group G. Hence a locally constant functor $\widehat{\rho}\colon X_1 \to \mathcal{G}$ is the same as a locally constant map $X_1 \to G$. Since $\mathcal{G}$ is skeletal (meaning objects in $\mathcal{G}$ are isomorphic if and only if they are equal), the existence of a natural transformation (3.14) requires ρ to satisfy the cocycle condition on X_2. Since the morphisms of $\mathcal{G}$ are $G \times A$, natural transformations (3.14) are maps $\gamma\colon X_2 \to A$ and $\epsilon\colon X_0 \to A$. The condition (3.15) is precisely (3.19). Since $\mathcal{G}$ is defined using a normalized 3-cocycle, the conditions on $\epsilon\colon X_0 \to A$ are

$$(s^*\epsilon) \cdot (l_{\mathrm{id}}^*\gamma) = 1\colon X_1 \to A, \quad (t^*\epsilon) \cdot (r_{\mathrm{id}}^*\gamma) = 1\colon X_1 \to A \tag{3.21}$$

where $1\colon X_1 \to A$ denotes the constant function to the unit of A. This implies that ϵ is determined by γ. Furthermore, (3.21) imposes the condition on γ,

$$\gamma(\phi, \mathrm{id}_x)^{-1} = \epsilon(x) = \gamma(\mathrm{id}_x, \psi)^{-1}, \qquad (\phi, \psi) \in X_2$$

where $s(\phi) = x = t(\psi)$. This verifies (3.20). □

PROPOSITION 3.18. *Let $\mathcal{G} = \mathcal{G}(G, A, \alpha)$ be as in Remark 2.13 and let $X = \{X_1 \rightrightarrows X_0\}$ be a Lie groupoid. Suppose that $(\widehat{h}, \widehat{\eta})\colon (\widehat{\rho}_1, \widehat{\gamma}_1) \to (\widehat{\rho}_2, \widehat{\gamma}_2)$ is a 1-morphism between flat principal $\mathcal{G}$-bundles that trivialize on X_0 which are specified by data $(\rho_i, \gamma_i), i = 1, 2$ as in Proposition 3.17. Then $(\widehat{h}, \widehat{\eta})$ is given by the same data as a pair (h, η), where*

(1) *$h\colon X_0 \to G$ is a locally constant map such that*

$$\rho_2 \cdot s^*h = t^*h \cdot \rho_1\colon X_1 \to G; \tag{3.22}$$

(2) $\eta\colon X_1 \to A$ *is a locally constant map such that the following diagram commutes, for any pair of composable morphisms* $(\phi, \psi) \in X_2$*:*

$$
\begin{array}{ccc}
\rho_2(\phi\circ\psi)s^*h(\psi) & \xrightarrow{\eta(\phi\circ\psi)} & t^*h(\phi)\rho_1(\phi\circ\psi) \\
\Big\downarrow{\scriptstyle \gamma_2(\phi,\psi)\times id} & & \Big\downarrow{\scriptstyle id\times\gamma_1(\phi,\psi)} \\
(\rho_2(\phi)\rho_2(\psi))s^*h(\psi) & & t^*h(\phi)(\rho_1(\phi)\rho_1(\psi)) \\
\Big\downarrow{\scriptstyle \alpha} & & \Big\uparrow{\scriptstyle \alpha} \\
\rho_2(\phi)(\rho_2(\psi)s^*h(\psi)) & & (t^*h(\phi)\rho_1(\phi))\rho_1(\psi) \\
\Big\downarrow{\scriptstyle id\times\eta(\psi)} & & \Big\uparrow{\scriptstyle \eta(\phi)\times id} \\
\rho_2(\phi)(t^*h(\psi)\rho_1(\psi)) & \xleftarrow[\alpha]{} & (\rho_2(\phi)s^*h(\phi))\rho_1(\psi).
\end{array}
\tag{3.23}
$$

PROOF. Because the objects of $\mathcal{G}$ are given by the discrete group G, a locally constant functor $\widehat{h}\colon X_0 \to \mathcal{G}$ is the same as a locally constant map $h\colon X_0 \to G$.

Now consider the data of the natural transformation $\widehat{\eta}\colon \mu\circ(\widehat{\rho_2}\times s^*h) \Rightarrow \mu\circ(t^*h\times\widehat{\rho_1})$ of functors $X_1 \to \mathcal{G}$. It consists of a smooth map from the objects of X_1 to the morphisms of $\mathcal{G}$, that is from $X_1 \to G\times A$, providing a family of isomorphisms

$$\widehat{\eta}_\phi\colon \widehat{\rho_2}(\phi)\cdot h(s(\phi)) \to h(t(\phi))\cdot\widehat{\rho_1}(\phi)) \in \mathcal{G},\ \phi\in X_1.$$

Because $\mathcal{G}$ is skeletal with automorphisms given by A, we must have $\rho_2(\phi)\cdot h(s(\phi)) = h(t(\phi))\rho_1(\phi) \in G$, and the data of $\widehat{\eta}_\phi$ is an element $\eta(\phi)\in A$.

The compatibility condition specifies that η behaves well with respect to composition of morphisms $(\phi,\psi)\in X_2$. □

REMARK 3.19. Considering $\eta(\phi)$ as the data of an isomorphism $\rho'(\phi)\cdot h(s(\phi)) \to h(t(\phi))\rho(\phi)$ in $\mathcal{G}$ for each $\phi\in X_1$ motivates the formulas for composition of two 1-morphisms. Indeed, given

$$(\rho_1,\gamma_1) \xrightarrow{(h_1,\eta_1)} (\rho_2,\gamma_2) \xrightarrow{(h_2,\eta_2)} (\rho_3,\gamma_3),$$

we have for every $\phi\colon x\to x'$ in X isomorphisms $\eta_i(\phi)\colon \rho_{i+1}(\phi)\cdot h_i(x) \to h_i(x')\cdot\rho_i(\phi)$ for $i=1,2$. This yields the following chain of isomorphisms:

$$
\begin{aligned}
&\rho_3(\phi)\cdot(h_2(x)\cdot h_1(x)) \xrightarrow{\alpha^{-1}(\rho_3(\phi),h_2(x),h_1(x))} (\rho_3(\phi)\cdot h_2(x))\cdot h_1(x) \xrightarrow{\eta_2(\phi)} \\
&(h_2(x')\cdot\rho_2(\phi))\cdot h_1(x) \xrightarrow{\alpha(h_2(x'),\rho_x(\phi),h_1(x))} h_2(x')\cdot(\rho_2(\phi)\cdot h_1(x)) \xrightarrow{\eta_1(\phi)} \\
&h_2(x')\cdot(h_1(x')\cdot\rho_1(\phi)) \xrightarrow{\alpha^{-1}(h_2(x'),h_1(x'),\rho_1(\phi))} (h_2(x')\cdot h_1(x'))\cdot\rho_1(\phi).
\end{aligned}
$$

In other words, we have a 1-morphism $(h_{21},\eta_{21})\colon(\rho_1,\gamma_1)\to(\rho_3,\gamma_3)$ given by

$$
\begin{aligned}
h_{21}\colon X_0 &\to G \\
x &\mapsto h_2(x)h_1(x); \\
\eta_{21}\colon X_1 &\to A \\
\phi &\mapsto \eta_2(\phi)\eta_1(\phi)\frac{\alpha(h_2(x'),\rho_2(\phi),h_1(x))}{\alpha(\rho_3(\phi),h_2(x),h_1(x))\alpha(h_2(x'),h_1(x'),\rho_1(\phi))}.
\end{aligned}
$$

This is the composition $(h_2,\eta_2)\circ(h_1,\eta_1)$.

These satisfy the compatibility conditions of Proposition 3.18 and hence do indeed give a 1-morphism as desired; this uses multiple iterations of the cocycle condition for α, the cocycle condition for ρ_i in equation (3.18), $i = 1, 2, 3$, the conditions for h_j in equation (3.22), and the conditions for η_j in equation (3.23), $j = 1, 2$.

PROPOSITION 3.20. *Let $\mathcal{G} = \mathcal{G}(G, A, \alpha)$ be as in Remark 2.13 and let $X = \{X_1 \rightrightarrows X_0\}$ be a Lie groupoid. Suppose that we have two flat principal $\mathcal{G}$-bundles $(\widehat{\rho}_i, \widehat{\gamma}_i), i = 1, 2$, specified by data (ρ_i, γ_i) as in Proposition 3.17, and two 1-morphisms $(\widehat{h}_i, \widehat{\eta}_i)\colon (\widehat{\rho}_1, \widehat{\gamma}_1) \to (\widehat{\rho}_2, \widehat{\gamma}_2)$ of principal $\mathcal{G}$-bundles, specified by data (h_i, η_i) as in Proposition 3.18. Then a 2-morphism $\widehat{\omega}\colon (\widehat{h}_1, \widehat{\eta}_1) \to (\widehat{h}_2, \widehat{\eta}_2)$ can exist only if $h_1 = h_2$. Furthermore, the data of $\widehat{\omega}$ is just a morphism*

$$\omega\colon X_0 \to A,$$

satisfying

$$\frac{\omega(x)}{\omega(x')} = \frac{\eta_1(\phi)}{\eta_2(\phi)}$$

for $\phi\colon x \to x' \in X_1$.

REMARK 3.21. Horizontal and vertical compositions of 2-morphisms are both given by pointwise multiplication in A. This is consistent with the interpretation of $\omega(x)$ as an isomorphism $h_1(x) \to h_2(x)$ in $\mathcal{G}$ for each $x \in X_0$.

DEFINITION 3.22. A pair (ρ, γ) is a *normalized* $\mathcal{G}$-bundle if

$$\gamma(\phi, \mathrm{id}_x) = \gamma(\mathrm{id}_x, \psi) = 1_A$$

for all $x \in X_0$ and $\phi, \psi \in X_1$ with $s(\phi) = x = t(\psi)$.

The following result is inspired by the "semi-strictification" results of [**6**, Propositions 25 and 26]; we are able to give a simpler proof here because our 2-group is a skeletal category.

LEMMA 3.23. *Let $\mathcal{G} = \mathcal{G}(G, A, \alpha)$ be as in Remark 2.13 and $X = \{X_1 \rightrightarrows X_0\}$ be a Lie groupoid. Let (ρ, γ) be as in Proposition 3.17. Then (ρ, γ) is isomorphic to a normalized $\mathcal{G}$-bundle.*

PROOF. We define a new pair (ρ, γ') which will be a normalized $\mathcal{G}$-bundle over X, isomorphic to the original pair (ρ, γ).

We keep ρ as before, but we define $\gamma'\colon X_2 \to A$ by $\gamma'(\phi, \psi) = \gamma(\phi, \psi)\epsilon(s(\phi))$. (As before, $\epsilon(s(\phi)) = \gamma(\phi, \mathrm{id}_{s(\phi)})^{-1} = \gamma(\mathrm{id}_{t(\psi)}, \psi)^{-1}$.) It is then clear that $\gamma'(\mathrm{id}_x, \psi) = \gamma(\mathrm{id}_x, \psi)\epsilon(x) = 1_A$ and $\gamma'(\phi, \mathrm{id}_x) = \gamma(\phi, \mathrm{id}_x)\epsilon(x) = 1_A$, so the normalization condition is satisfied.

We need to show that $d\gamma' = \rho^*\alpha$, so let us take $(\phi, \psi, \xi) \in X_3$. We have

$$\begin{aligned} d\gamma'(\phi, \psi, \xi) = \frac{\gamma'(\psi, \xi)\gamma'(\phi, \psi \circ \xi)}{\gamma'(\phi \circ \psi, \xi)\gamma'(\phi, \psi)} &= \frac{\gamma(\psi, \xi)\epsilon(s(\psi))\gamma(\phi, \psi \circ \xi)\epsilon(s(\phi))}{\gamma(\phi \circ \psi, \xi)\epsilon(s(\phi \circ \psi))\gamma(\phi, \psi)\epsilon(s(\phi))} \\ &= d\gamma(\phi, \psi, \xi)\frac{\epsilon(s(\psi))\epsilon(s(\phi))}{\epsilon(s(\psi))\epsilon(s(\phi))} = \alpha(\rho(\phi), \rho(\psi), \rho(\xi)), \end{aligned}$$

as desired. Therefore (ρ, γ') is a normalized $\mathcal{G}$-bundle. It remains to construct an isomorphism $(\rho, \gamma) \to (\rho, \gamma')$.

We take $h\colon X_0 \to G$ to be the constant function $h(x) = 1_G$, and we take $\eta\colon X_1 \to A$ to be $\eta(\phi) = \epsilon(s(\phi))^{-1}$. Because $h(x) = 1_G$ and α is normalized, the condition (3.23) for η simplifies to

$$\frac{\eta(\phi)\eta(\psi)}{\eta(\phi\circ\psi)} = \frac{\gamma(\phi,\psi)}{\gamma'(\phi,\psi)}, (\phi,\psi) \in X_2.$$

It is easy to see that with our definitions of η and γ', both sides of the above equation give $\epsilon(s(\phi))^{-1}$, and hence $(1_G, \eta)$ gives the desired isomorphism $(\rho,\gamma) \to (\rho,\gamma')$. □

To summarize, starting with the bicategory $\mathsf{Bun}^{\mathrm{pre}}_{\mathcal{G}}(X) = \mathsf{Fun}(X, \mathrm{pt}/\!\!/\mathcal{G})$ from Definition 3.11, for $\mathcal{G}$ of a preferred form (Remark 2.13) we have shown that objects are the same as data (ρ,γ), 1-morphisms are the same as data (h,η) and 2-morphisms are the same as data ω. Furthermore, there is an equivalent full and faithful subcategory of normalized pairs (ρ,γ). Hereafter, we will use the notation $\mathsf{Bun}^{\mathrm{pre}}_{\mathcal{G}}(X)$ to denote this description of the bicategory in terms of explicit normalized Čech data, whereas $\mathsf{Fun}(X, \mathrm{pt}/\!\!/\mathcal{G})$ will refer to Definition 3.11. With this convention the above discussion gives a canonical equivalence $\mathsf{Bun}^{\mathrm{pre}}_{\mathcal{G}}(X) \xhookrightarrow{\simeq} \mathsf{Fun}(X, \mathrm{pt}/\!\!/\mathcal{G})$.

LEMMA 3.24. *In terms of Čech cocycle data, the 2-functor* (3.12), $\mathsf{Bun}^{\mathrm{pre}}_{\mathcal{G}}(X) \to \mathsf{Bun}^{\mathrm{pre}}_{G}(X)$, *sends* (ρ,γ) *to* ρ, *and* (h,η) *to* h.

PROOF. As in Lemma 3.14, the morphism $\mathcal{G} \to G$ of 2-groups induces a 2-functor $\mathrm{pt}/\!\!/\mathcal{G} \to \mathrm{pt}/\!\!/G$ of bicategories, which in turn induces a 2-functor

$$\mathsf{Bun}^{\mathrm{pre}}_{\mathcal{G}}(X) \xhookrightarrow{\simeq} \mathsf{Fun}(X, \mathrm{pt}/\!\!/\mathcal{G}) \to \mathsf{Fun}(X, \mathrm{pt}/\!\!/G) = \mathsf{Bun}^{\mathrm{pre}}_{G}(X)$$

of bicategories with the following values

$$\begin{aligned} \textbf{On objects: } & (\rho,\gamma) \mapsto \rho; \\ \textbf{On 1-morphisms: } & (h,\eta) \mapsto \eta; \\ \textbf{On 2-morphisms: } & \omega \mapsto *. \end{aligned}$$

This proves the lemma. □

3.6. Flat gerbes as $\mathrm{pt}/\!\!/A$-bundles. Recall the 2-group $\mathrm{pt}/\!\!/A$ from Example 2.6.

DEFINITION 3.25 ([**38**, Example 73] and [**7**, §4-5]). For a Lie groupoid X, define the *bicategory of flat A-gerbes that trivialize on X_0* as

$$\mathsf{Gerbe}^{\mathrm{pre}}_{A}(X) := \mathsf{Bun}^{\mathrm{pre}}_{\mathrm{pt}/\!\!/A}(X).$$

In light of Propositions 3.17, 3.18, 3.20, and Lemma 3.23, the bicategory $\mathsf{Gerbe}^{\mathrm{pre}}_{A}(X)$ can be described as follows:

- **Objects:** given by $\gamma\colon X_2 \to A$ a locally constant function such that $d\gamma = 1_A$, and $\gamma(\phi,\psi) = 1_A$ for $(\phi,\psi) \in X_2$ with either ϕ or ψ equal to $\mathrm{id}_x, x \in X_0$.
- **1-morphisms $\gamma_1 \to \gamma_2$:** given by $\eta\colon X_1 \to A$ a locally constant function such that for any $(\phi,\psi) \in X_2$

$$\frac{\eta(\phi)\eta(\psi)}{\eta(\phi\circ\psi)} = \frac{\gamma_1(\phi,\psi)}{\gamma_2(\phi,\psi)}.$$

2-morphisms $\eta_1 \to \eta_2$: given by $\omega\colon X_0 \to A$ such that for any $\phi\colon x \to x'$ in X_1,

$$\frac{\omega(x)}{\omega(x')} = \frac{\eta_1(\phi)}{\eta_2(\phi)}.$$

In particular, if M is a manifold and $Y \to M$ is a surjective submersion, a flat A-gerbe on $\check{C}(Y)$ in the sense of Definition 3.25 is given by a locally constant function $\gamma\colon \check{C}(Y)_2 \cong Y^{[3]} \to A$. More geometrically, γ defines an isomorphism $p_{13}^*P \to p_{12}^*P \otimes_A p_{23}^*P$ between flat A-bundles over $Y^{[3]}$ where $P = Y^{[2]} \times A \to Y^{[2]}$ is the trivial A-bundle. The cocycle condition for γ implies a compatibility condition among the pullbacks of this isomorphism to $Y^{[4]}$. This recovers an A-gerbe in the classical definition of [**29**] with the additional condition that the A-bundles on $Y^{[2]}$ are flat.

4. The bicategory of flat principal 2-group bundles

For the remainder of the paper, $\mathcal{G} = \mathcal{G}(G, A, \alpha)$ is a 2-group as in Remark 2.13, i.e., a categorical central extension of G by $\mathrm{pt} /\!\!/ A$ classified by $[\alpha] \in \mathrm{H}^3(G; A)$.

In the previous section we constructed the bicategory $\mathsf{Bun}_{\mathcal{G}}^{\mathrm{pre}}(X)$ for a Lie groupoid X. In this section, we first construct the bicategory $\mathsf{Bun}_{\mathcal{G}}^{\mathrm{pre}}$ whose objects are flat $\mathcal{G}$-bundles on a (varying) Lie groupoid X that are trivial on X_0, and whose 1-morphisms are maps of $\mathcal{G}$-bundles covering functors between Lie groupoids. We then localize $\mathsf{Bun}_{\mathcal{G}}^{\mathrm{pre}}$ to obtain a bicategory $\mathsf{Bun}_{\mathcal{G}}$ whose objects are zig-zags (3.9). The (homotopy) fiber of $\mathsf{Bun}_{\mathcal{G}}$ at a Lie groupoid X therefore recovers the correct bicategory of flat $\mathcal{G}$-bundles over the smooth stack $[X]$. We show that when $\mathcal{G} = G$ is an ordinary group and X is a manifold, this construction recovers the usual category of principal G-bundles. Similarly when $\mathcal{G} = \mathrm{pt} /\!\!/ A$ is a one-object 2-group, we recover the bicategory of gerbes. We also compare our flat $\mathcal{G}$-bundles to principal $\mathcal{G}$-bundles in the sense of [**38**, Definition 70].

4.1. Pulling back $\mathcal{G}$-bundles. Let $X = \{X_1 \rightrightarrows X_0\}, Y = \{Y_1 \rightrightarrows Y_0\}$ be Lie groupoids, and let $f = (f_1\colon X_1 \to Y_1, f_0\colon X_0 \to Y_0)$ be a smooth functor $X \to Y$. For any $n \geq 2$, we let $f_n\colon X_n \to Y_n$ denote the induced map.

DEFINITION 4.1. We define a functor

$$f^*\colon \mathsf{Bun}_{\mathcal{G}}^{\mathrm{pre}}(Y) \to \mathsf{Bun}_{\mathcal{G}}^{\mathrm{pre}}(X).$$

as follows:

On objects: $f^*(\rho, \gamma) = (\rho \circ f_1, \gamma \circ f_2)$;

On 1-morphisms: $f^*(h, \eta) = (h \circ f_0, \eta \circ f_1)$;

On 2-morphisms: $f^*\omega = \omega \circ f_0$.

It is straightforward to check that this is a well-defined functor of bicategories. Moreover, for composable smooth functors f, f', we have a strict equality $(f \circ f')^* = (f')^* \circ f^*$.

REMARK 4.2. From now on we will leave out the subscripts on the components of f whenever they are clear from context, writing for example $f^*(\rho, \gamma) = (\rho \circ f, \gamma \circ f)$. Instead, the subscripts f_1 and f_2 will denote two different smooth functors, as in the following proposition.

PROPOSITION 4.3. *Let $\tau\colon f_1 \Rightarrow f_2$ be a natural transformation of smooth functors. It induces a natural transformation of functors of bicategories $\widetilde{\tau}\colon f_1^* \Rightarrow f_2^*$.*

PROOF. Let (ρ,γ) be an object in $\mathsf{Bun}_{\mathcal{G}}^{\mathrm{pre}}(Y)$. We need to define a 1-morphism $\widetilde{\tau}_{(\rho,\gamma)}\colon f_1^*(\rho,\gamma) \to f_2^*(\rho,\gamma)$ in $\mathsf{Bun}_{\mathcal{G}}^{\mathrm{pre}}(X)$. For $x \in X_0$, we have $\tau(x) \in Y_1$ giving an isomorphism in the groupoid Y:

$$\tau(x)\colon f_1(x) \to f_2(x).$$

We use this to define $\widetilde{\tau}_{(\rho,\gamma)}$ to be the pair (h_τ, η_τ), where

$$h_\tau\colon X_0 \to G, \quad x \overset{h_\tau}{\longmapsto} \rho(\tau(x)); \qquad \eta_\tau\colon X_1, \to A \quad \phi \overset{\eta_\tau}{\longmapsto} \frac{\gamma(\tau(y), f_1(\phi))}{\gamma(f_2(\phi), \tau(x))}.$$

Furthermore, given a 1-morphism $(h,\eta)\colon (\rho_1,\gamma_1) \to (\rho_2,\gamma_2)$ in $\mathsf{Bun}_{\mathcal{G}}^{\mathrm{pre}}(Y)$, we need a 2-morphism ω in $\mathsf{Bun}_{\mathcal{G}}^{\mathrm{pre}}(X)$ making the following diagram commute:

$$\begin{array}{ccc} f_1^*(\rho_1,\gamma_1) & \xrightarrow{\widetilde{\tau}_{(\rho_1,\gamma_1)}} & f_2^*(\rho_1,\gamma_1) \\ \big\downarrow{\scriptstyle f_1^*(h,\eta)} & & \big\downarrow{\scriptstyle f_2^*(h,\eta)} \\ f_1^*(\rho_2,\gamma_2) & \xrightarrow{\widetilde{\tau}_{(\rho_2,\gamma_2)}} & f_2^*(\rho_2,\gamma_2). \end{array}$$

We take $\omega\colon X_0 \to G, \omega(x) = \eta(\tau(x))$. It is straightforward to check that these data define a natural transformation as required. $\square$

With the structure of pullback, the bicategories $\mathsf{Bun}_{\mathcal{G}}^{\mathrm{pre}}(X)$ fit together into a bicategory $\mathsf{Bun}_{\mathcal{G}}^{\mathrm{pre}}$ over Lie groupoids fibered in 2-groupoids as follows.

DEFINITION 4.4. Define the bicategory $\mathsf{Bun}_{\mathcal{G}}^{\mathrm{pre}}$ as follows.

Objects: a Lie groupoid X, and a pair $(\rho,\gamma) \in \mathsf{Bun}_{\mathcal{G}}^{\mathrm{pre}}(X)$. We often denote this data by $(\rho,\gamma)_X$ to emphasize the base Lie groupoid.

1-morphisms: a smooth functor $f\colon X \to Y$ of Lie groupoids, together with a 1-morphism $(h,\eta)\colon (\rho_1,\gamma_1) \to f^*(\rho_2,\gamma_2)$ in $\mathsf{Bun}_{\mathcal{G}}^{\mathrm{pre}}(X)$.

2-morphisms: given 1-morphisms $(f_1,h_1,\eta_1), (f_2,h_2,\eta_2)\colon (\rho_1,\gamma_1)_X \to (\rho_2,\gamma_2)_Y$, a 2-morphism $(f_1,h_1,\eta_1) \to (f_2,h_2,\eta_2)$ is given by a natural transformation $\tau\colon f_1 \Rightarrow f_2$ of smooth functors, together with a 2-morphism ω in $\mathsf{Bun}_{\mathcal{G}}^{\mathrm{pre}}(X)$ between the 1-morphisms

$$\widetilde{\tau}_{(\rho_2,\gamma_2)} \circ (h_1,\eta_1) \text{ and } (h_2,\eta_2)\colon (\rho_1,\gamma_1) \to f_2^*(\rho_2,\gamma_2).$$

We spell out some of the compatibility conditions implied by the above definition. A 1-morphism $(\rho_1,\gamma_1)_X \to (\rho_2,\gamma_2)_Y$ is the data of a triple (f,h,η), where for $\phi\colon x \to x'$ a morphism in X we view $\eta(\phi) \in A$ as giving an isomorphism $\rho_2(\phi)h(x) \to h(x')\rho_1(\phi)$ in $\mathcal{G}$. This leads to the condition on h

$$\rho_2(f(\phi))h(x) = h(x')\rho_1(\phi) \in G.$$

Furthermore, the compatibility condition for η in (3.23) becomes

$$\frac{\eta(\phi)\eta(\psi)}{\eta(\phi\circ\psi)} = \frac{\gamma_1(\phi,\psi)}{\gamma_2(f(\phi),f(\psi))}\frac{\alpha(\rho_2(f(\phi)),h(x'),\rho_1(\psi))}{\alpha(\rho_2(f(\phi)),\rho_2(f(\psi)),h(x))\alpha(h(x''),\rho_1(\phi),\rho_1(\psi))},$$

for $x \xrightarrow{\psi} x' \xrightarrow{\phi} x''$ in X.

Given two composable 1-morphisms $(f_1, h_1, \eta_1)\colon (\rho_1, \gamma_1)_X \to (\rho_2, \gamma_2)_Y$ and $(f_2, h_2, \eta_2)\colon (\rho_2, \gamma_2)_Y \to (\rho_3, \gamma_3)_Z$, the composition is $(f_2 \circ f_1, f_1^*(h_2, \gamma_2) \circ (h_1, \gamma_1))$. Explicitly, its components are the following three maps:

$$f_2 \circ f_1 \colon X \to Z; \qquad \widetilde{h}\colon X_0 \to G, \qquad x \overset{\tilde{h}}{\mapsto} h_2(f_1(x))h_1(x);$$

$$\widetilde{\eta}\colon X_1 \to A,$$

$$(\phi\colon x \to x')$$
$$\overset{\tilde{\eta}}{\mapsto} \eta_2(f_1(\phi))\eta_1(\phi)\frac{\alpha(h_2(f_1(x')), \rho_2(f_1(\phi)), h_1(x))}{\alpha(\rho_2(f_1(\phi)), h_2(f_1(x)), h_1(x))\alpha(h_2(f_1(x')), h_1(x'), \rho_1(\phi))}.$$

The following observation is simple but useful:

LEMMA 4.5. *Every 1-morphism $(f, h, \eta)\colon (\rho_1, \gamma_1)_X \to (\rho_2, \gamma_2)_Y$ in $\mathsf{Bun}_{\mathcal{G}}^{\mathrm{pre}}$ factors as a composition of a morphism in $\mathsf{Bun}_{\mathcal{G}}^{\mathrm{pre}}(X)$ and one of the form $(f, 1_G, 1_A) = (f, \mathrm{id}_{f^*(\rho_2,\gamma_2)})$:*

$$(\rho, \gamma)_X \xrightarrow{(\mathrm{id}_X, h, \eta)} (\rho_2 \circ f, \gamma_2 \circ f)_X \xrightarrow{(f, 1_G, 1_A)} (\rho_2, \gamma_2)_Y.$$

We now unwind the data and conditions satisfied by 2-morphisms. Given two 1-morphisms $(f_1, h_1, \eta_1), (f_2, h_2, \eta_2)$ between $(\rho_1, \gamma_1)_X \to (\rho_2, \gamma_2)_Y$, a 2-morphism $(f_1, h_1, \eta_1) \to (f_2, h_2, \eta_2)$ is a pair (τ, ω) where $\tau\colon f_1 \Rightarrow f_2$ is a natural transformation such that $\rho_2(\tau(x))h_1(x) = h_2(x)$ for all $x \in X_0$; furthermore $\omega\colon X_0 \to A$ provides a family of isomorphisms witnessing this identity, satisfying

$$\frac{\omega(x)}{\omega(x')} = \frac{\eta_1(\phi)}{\eta_2(\phi)} \frac{\gamma_2(\tau(x'), f_1(\phi))}{\gamma_2(f_2(\phi), \tau(x))} \frac{\alpha(\rho_2(\tau(x')), \rho_2(f_1(\phi)), h_1(x))}{\alpha(\rho_2(\tau(x')), h_1(x'), \rho_1(\phi))\alpha(\rho_2(f_2(\phi)), \rho_2(\tau(x)), h_1(x))}, \tag{4.1}$$

for $\phi\colon x \to x'$ in X.

Suppose we also have two 1-morphisms $(f_3, h_3, \eta_3), (f_4, h_4, \eta_4)\colon (\rho_2, \gamma_2)_Y \to (\rho_3, \gamma_3)_Z$ and a 2-morphism $(\tau_2, \omega_2)\colon (f_3, h_3, \eta_3) \to (f_4, h_4, \eta_4)$. The horizontal composition $(\tau_2, \omega_2) \star (\tau, \omega)$ is a 2-morphism $(f_3, h_3, \eta_3) \circ (f_1, h_1, \eta_1) \to (f_4, h_4, \eta_4) \circ (f_2, h_2, \eta_2)$. It is the pair consisting of the natural transformation $\tau_2 \star \tau\colon f_3 \circ f_1 \Rightarrow f_4 \circ f_2$ given by the horizontal composition of τ_2 and τ, and the locally constant map $\omega_2 \star \omega\colon X_0 \to A$ given by

$$\omega_2 \star \omega(x) = \omega_2(f_1(x))\omega(x)\eta_4(\tau(x))\gamma_3(f_4(\tau(x)), \tau_2(f_1(x))) \tag{4.2}$$
$$\frac{\alpha(\rho_3(f_4(\tau(x))), \rho_3(\tau_2(f_1(x))), h_3(f_1(x))h_1(x))\alpha(h_4(f_2(x)), \rho_2(\tau(x)), h_1(x))}{\alpha(\rho_3(\tau_2(f_1(x))), h_3(f_1(x)), h_1(x))\alpha(\rho_3(f_4(\tau(x))), h_4(f_1(x)), h_1(x))}.$$

Similarly, given a third 1-morphism $(f_3, h_3, \eta_3)\colon (\rho_1, \gamma_1)_X \to (\rho_2, \gamma_2)_Y$ and a 2-morphism $(\tau_2, \omega_2)\colon (f_2, h_2, \eta_2) \to (f_3, h_3, \eta_3)$, we obtain the vertical composition $(\tau_2, \omega_2) \circ (\tau, \omega)\colon (f_1, h_1, \eta_1) \to (f_3, h_3, \eta_3)$. It is the pair consisting of the natural transformation $\tau_2 \circ \tau\colon f_1 \to f_3$ given by the vertical composition of τ_2 and τ, and the locally constant map $\omega_2 \circ \omega\colon X_0 \to A$ given by

$$(\omega_2 \circ \omega)(x) = \omega_2(x)\omega(x)\alpha(\rho_2(\tau_2(x)), \rho_2(\tau(x)), h_1(x))\gamma_2(\tau_2(x), \tau(x)).$$

REMARK 4.6. The above defines a prestack of $\mathcal{G}$-bundles on Lie groupoids. There are various machines that implement 2-stackification; we will use the calculus of right fractions from [**37**], though see also [**33**].

THEOREM 4.7. *The bicategory $\mathsf{Bun}_{\mathcal{G}}^{\mathrm{pre}}$ admits a calculus of right fractions with respect to the class of 1-morphisms $\mathcal{W}$ of the form (f, h, η) where f is an essential equivalence. Let $\mathsf{Bun}_{\mathcal{G}} := \mathsf{Bun}_{\mathcal{G}}^{\mathrm{pre}}[\mathcal{W}^{-1}]$ denote the localized bicategory.*

We defer the proof of Theorem 4.7 to Appendix C.

REMARK 4.8. By construction, the bicategory $\mathsf{Bun}_{\mathcal{G}}$ has the property that for any object $(\rho,\gamma)_X$ and for any essential equivalence $f\colon Y \to X$, the 1-morphism $(f, 1_G, 1_A)\colon f^*(\rho,\gamma) \to (\rho,\gamma)$ is an equivalence.

REMARK 4.9. The 2-functor $\mathsf{Bun}_{\mathcal{G}}^{\mathrm{pre}} \to \mathsf{LieGrpd}$ sends 1-morphisms in $\mathcal{W}$ to essential equivalences. By the universal property of localization [**37**, Theorem 21], we obtain a 2-functor $\mathsf{Bun}_{\mathcal{G}} \to \mathsf{LieGrpd}[W^{-1}]$, where W is the class of essential equivalences. The bicategory $\mathsf{LieGrpd}[W^{-1}]$ is equivalent to the bicategory of differentiable stacks. Furthermore, morphisms in $\mathsf{LieGrpd}[W^{-1}]$ can be described using *bibundles* [**22, 38**]; we do not use these structures here.

4.2. The bicategory $\mathsf{Bun}_{\mathcal{G}}(X)$.

DEFINITION 4.10. For a Lie groupoid X, define the bicategory $\mathsf{Bun}_{\mathcal{G}}(X)$ of *principal $\mathcal{G}$-bundles on X* as the homotopy fiber of $\mathsf{Bun}_{\mathcal{G}} \to \mathsf{LieGrpd}[\mathsf{W}^{-1}]$ at the object $X \in \mathsf{LieGrpd}[\mathsf{W}^{-1}]$.

COROLLARY 4.11. *The bicategory of principal $\mathcal{G}$-bundles forms a 2-stack over the bicategory* $\mathsf{LieGrpd}[W^{-1}]$.

PROOF. After localization, the homotopy fibers of $\mathsf{Bun}_{\mathcal{G}} \to \mathsf{LieGrpd}[\mathsf{W}^{-1}]$ necessarily satisfy descent. □

By definition, an object of the homotopy fiber is a pair $(f\colon Y \to X, (\rho,\gamma)_Y)$ consisting of an invertible 1-morphism f in $\mathsf{LieGrpd}[\mathsf{W}^{-1}]$ and an object $(\rho,\gamma)_Y$ of $\mathsf{Bun}_{\mathcal{G}}$. The invertible morphism f in $\mathsf{LieGrpd}[\mathsf{W}^{-1}]$ is a zig-zag of essential equivalences, $Y \overset{g}{\leftarrow} Z \to X$, and the composition $f \circ g\colon Z \to X$ in $\mathsf{LieGrpd}[W^{-1}]$ is naturally isomorphic to the right leg of this zig-zag. Then following Remark 4.8, the pair $(f\colon Y \to X, (\rho,\gamma)_Y)$ is equivalent to the pair $(f \circ g, g^*(\rho,\gamma))$. Therefore, $\mathsf{Bun}_{\mathcal{G}}(X)$ is equivalent to a bicategory whose objects are zig-zags

$$X \xleftarrow{f} Y \xrightarrow{(\rho,\gamma)} \mathrm{pt}/\!\!/\mathcal{G}. \tag{4.3}$$

with f an essential equivalence (c.f. Definition 3.10). We will denote the zig-zag by $(f, (\rho,\gamma)_Y)$ below.

Similarly, a 1-morphism $(f_1, (\rho_1,\gamma_1)_{Y_1}) \to (f_2, (\rho_2,\gamma_2)_{Y_2})$ in the homotopy fiber is given up to 2-isomorphism by the following data: $(Z, g_1, g_2, \tau, (h,\eta))$, where $g_i\colon Z \to Y_i, i = 1,2$ are essential equivalences, τ is a smooth natural transformation $f_1 \circ g_1 \Rightarrow f_2 \circ g_2$, and (h,η) is a 1-morphism in $\mathsf{Bun}_{\mathcal{G}}^{\mathrm{pre}}(Z)$ between $g_1^*(\rho_1,\gamma_1)$ and $g_2^*(\rho_2,\gamma_2)$.

Finally, let $(Z_j, g_{1j}, g_{2j}, \tau_j, (h_j,\eta_j)), j = 3,4$ be two 1-morphisms of this form between objects $(f_1, (\rho_1,\gamma_1)_{Y_1})$ and $(f_2, (\rho_2,\gamma_2)_{Y_2})$. Unwinding the definitions, we see that a 2-morphism between them is given by data $(Z, g_3, g_4, \sigma_1, \sigma_2, \omega)$, where $g_j\colon Z \to Z_j$ are essential equivalences, $\sigma_i\colon g_{i3} \circ g_3 \Rightarrow g_{i4} \circ g_4$, $i = 1,2$ are smooth natural transformations such that $\sigma_2 \circ \tau_3 = \tau_4 \circ \sigma_1$, and ω is a 2-morphism in $\mathsf{Bun}_{\mathcal{G}}^{\mathrm{pre}}(Z)$ between $\tilde{\sigma}_2 \circ g_3^*(h_3,\eta_3)$ and $g_4^*(h_4,\eta_4) \circ \tilde{\sigma}_1$. Two sets of data of the form $(Z, g_3, g_4, \sigma_1, \sigma_2, \omega)$ represent the same 2-morphism if they agree "upon refinement", that is, after being pulled back along compatible essential equivalences.

In particular, all 1-morphisms and 2-morphisms are invertible.

NOTATION 4.12. Hereafter, $\mathsf{Bun}_{\mathcal{G}}(X)$ will denote the bicategory equivalent to the one in Definition 4.10 whose objects are zig-zags (4.3), and 1-morphisms and 2-morphisms are described above.

We complete this subsection with a comparison of the bicategory $\mathsf{Bun}_{\mathcal{G}}(X)$ with previously studied special cases.

EXAMPLE 4.13. Let G be an ordinary group, viewed as a 2-group, and let M be a smooth manifold regarded as a Lie groupoid. We check that the bicategory $\mathsf{Bun}_G(M)$ just defined is in fact an ordinary category, and agrees with the usual category of principal G-bundles in terms of Čech cocycle data, as discussed in §3.2. Recall from Example 3.7 that the only essential equivalences to M are of the form $\widetilde{f}\colon \check{C}(Y)\to M$ for $f\colon Y\to M$ a surjective submersion. With this in mind, we see that an object of $\mathsf{Bun}_G(M)$ is given by a pair $(f\colon Y\to M, \rho\colon Y\times_M Y\to G)$ consisting of a surjective submersion and a locally constant 1-cocycle; i.e., a principal G-bundle on the manifold M.

To unpack the notion of 1-morphisms, we first note that if $\widetilde{g}\colon \widetilde{Z}\to \check{C}(Y)$ is an essential equivalence, then $\widetilde{f}\circ\widetilde{g}$ is an essential equivalence from the Lie groupoid $\widetilde{Z}$ to the manifold M. Then as in Example 3.7, if we let $Z=\widetilde{Z}_0$ and $g=\widetilde{g}_0$ we have that $f\circ g\colon Z\to M$ is a surjective submersion, $\widetilde{Z}$ is the associated Čech groupoid $\check{C}(Z)$, and $\widetilde{g}$ is the essential equivalence $\check{C}(Z)\to\check{C}(Y)$ induced by the smooth map g. We also note that the category of morphisms from a Lie groupoid into a manifold is in fact a set; there are no non-identity 2-morphisms. These observations allow us to see that a morphism in $\mathsf{Bun}_G(M)$ from (f_1,ρ_1) to (f_2,ρ_2) consists of a smooth manifold Z with smooth morphisms $g_i\colon Z\to Y_i$ for $i=1,2$, such that $f_1\circ g_1=f_2\circ g_2$ is a surjective submersion $Z\to Y$, together with a locally constant 0-cochain $h\colon Z\to A$ such that $\rho_2(g_2(z),g_2(z'))h(z')=h(z)\rho_1(g_1(z),g_1(z'))$. (Here $(z,z')\in Z\times_M Z$ is viewed as a morphism $z'\to z$ in $\check{C}(Z)$.) In other words, h is a 1-morphism of principal G-bundles between the pullbacks of ρ_1 and ρ_2 to Z.

Finally, there are no non-identity 2-morphisms.

EXAMPLE 4.14. Let G and H be ordinary groups and $X=M /\!\!/ H$ an action groupoid. We claim that $\mathsf{Bun}_G(M /\!\!/ H)$ is the category whose objects are flat principal G-bundles $p\colon P\to M$ where P is equipped with an H-action compatible with the H-action on M and commuting with the G-action i.e., H-equivariant G-bundles on M. Indeed, the smooth functor $M\to M /\!\!/ H$ of Lie groupoids induces a pullback functor

$$\mathsf{Bun}_G(M /\!\!/ H)\to \mathsf{Bun}_G(M), \tag{4.4}$$

and so Example 4.13 extracts an ordinary G-bundle on M from an object in $\mathsf{Bun}_G(M /\!\!/ H)$. Conversely, lifting an object $P\in\mathsf{Bun}_G(M)$ along the functor (4.4) requires descent data for P along the map $M\to M /\!\!/ H$. When P is defined relative to a surjective submersion $f\colon Y\to M$, this descent data is equivalent to a G-bundle over the pullback groupoid $f^*(M /\!\!/ H)$ (see Example 3.8) that trivializes on $Y\to M\to M /\!\!/ H$. A straightforward computation verifies that this descent data does indeed determine an H-action on P.

EXAMPLE 4.15. Applying the localization construction to Example 3.6 for a fixed abelian Lie group A, we see that gerbes on a Lie groupoid form a bicategory $\mathsf{Bun}_{\mathrm{pt} /\!\!/ A}(X)=\mathsf{Gerbe}_A(X)$. Concretely, an object is given by the data of a pair

$(f\colon Y \xrightarrow{\sim} X, \gamma\colon Y_2 \to A)$ where f is a smooth essential equivalence and γ defines a gerbe on Y trivial over Y_0.

The constructions of §4 recover the fact that the assignment $X \mapsto \mathsf{Gerbe}_A(X)$ is a 2-stack, as was already known (e.g., see [**34**, §5]).

Finally, we indicate how our definition of flat $\mathcal{G}$-bundles compares with [**38**, Definition 70]. Let $(f\colon Y \to X, (\rho,\gamma)_Y)$ be a flat principal $\mathcal{G}$-bundle. We define a Lie groupoid $\mathcal{P}$ with objects $P_0 = Y_0 \times G$ and morphisms $P_1 = Y_1 \times G \times A$, where $(\phi\colon y \to y', g, a)$ has source (y, g) and target $(y', \rho(\phi)g)$. Composition is given by

$$(\phi, \rho(g), a) \circ (\psi, g, b) = (\phi \circ \psi, g, ab\gamma(\phi,\psi)\alpha(\rho(\phi), \rho(\psi), g)),$$

and the identity morphism of an object (y, g) is $(\mathrm{id}_y, g, 1_A)$.

LEMMA 4.16. *This defines a Lie groupoid* $\mathcal{P} = \{P_1 \rightrightarrows P_0\}$ *over* X*, equipped with a smooth functor* $\mathsf{act}\colon \mathcal{P} \times \mathcal{G} \to \mathcal{P}$*.*

PROOF. The properties of a Lie groupoid follow from the cocycle and normalization conditions on α, ρ and γ. The smooth functor $\pi\colon \mathcal{P} \to X$ is given by $(y, g) \mapsto f(y)$ on objects and $(\phi, g, a) \mapsto f(\phi)$ on morphisms.

The smooth functor act is given on objects by $((y, g), h) \mapsto (y, gh)$ and on morphisms by $((\phi, g, a), (h, b)) \mapsto (\phi, gh, \frac{ab}{\alpha(\rho(\phi), g, h)})$. Functoriality also follows from the properties of α, ρ, and γ. □

To complete the comparison with [**38**], one must extend the functor act to an action of the 2-group $\mathcal{G}$ on $\mathcal{P}$ with a equivariant isomorphism over the (trivializing) cover $Y \to X$. This is not difficult, but requires introducing higher categorical actions that are beyond the scope of the present paper.

4.3. From $\mathcal{G}$-bundles to trivializations of 2-gerbes. There is a natural forgetful functor from principal $\mathcal{G}$-bundles to principal bundles for the underlying ordinary group G. The goal of this section is to describe the fibers of this forgetful functor; this will use the language of 2-gerbes reviewed in §B.

LEMMA 4.17. *For a 2-group* $\mathcal{G} = \mathcal{G}(G, A, \alpha)$ *and a Lie groupoid* X*, the functor* (3.12), $\mathsf{Bun}^{\mathrm{pre}}_{\mathcal{G}}(X) \to \mathsf{Bun}^{\mathrm{pre}}_{G}(X)$*, induces a forgetful functor*

$$\pi\colon \mathsf{Bun}_{\mathcal{G}}(X) \to \mathsf{Bun}_G(X) \tag{4.5}$$

from the bicategory of $\mathcal{G}$*-bundles on* X *to the category of* G*-bundles on* X*.*

PROOF. The functor (3.12) sends the class $\mathcal{W}$ of 1-morphisms (f, h, η) with f an essential equivalence to the class W of 1-morphisms (f, h) with f an essential equivalence. By the universal property of localization [**37**, Theorem 21], we obtain a functor $\mathsf{Bun}_{\mathcal{G}} \to \mathsf{Bun}_G$, as follows:

On objects: $(\rho, \gamma)_X \mapsto \rho_X$;

On 1-morphisms: $(f, h, \eta) \mapsto (f, h)$;

On 2-morphisms: $(\tau, \omega) \mapsto \tau$.

For any Lie groupoid X, we obtain a functor between homotopy fibers yielding the desired forgetful functor (4.5); its values on objects are as above, while its values on zig-zags giving 1- and 2-morphisms are given by applying the above formulas to the components of the zig-zag. □

The following lemma shows that (4.5) satisfies a path-lifting property: morphisms in $\mathsf{Bun}_G(X)$ can be lifted to 1-morphisms in $\mathsf{Bun}_{\mathcal{G}}(X)$.

LEMMA 4.18. *Let $(f\colon Y \to X, (\rho_1, \gamma_1)_Y)$ be an object of $\mathsf{Bun}_{\mathcal{G}}(X)$, and suppose that (f, ρ_2) is an object of $\mathsf{Bun}_G(X)$ defined relative to the same essential equivalence $f\colon Y \to X$. If $h\colon \rho_1 \to \rho_2$ is a 1-morphism in $\mathsf{Bun}_G^{\mathrm{pre}}(Y)$, then there exists locally constant cochains $\gamma_2\colon Y_2 \to A$ and $\eta\colon Y_1 \to A$ such that $(\rho_2, \gamma_2)_Y$ is a principal $\mathcal{G}$-bundle, and (h, η) is a 1-isomorphism $(\rho_1, \gamma_1) \to (\rho_2, \gamma_2)$ in $\mathsf{Bun}_{\mathcal{G}}^{\mathrm{pre}}(Y)$.*

PROOF. The existence of $h\colon \rho_1 \to \rho_2$ tells us that for $\phi\colon y \to y'$ in the Lie groupoid Y, we have an equality $\rho_2(\phi) = (h(y')\rho_1(\phi))\, h(y)^{-1}$ in G.

To lift ρ_2 to a pair $(\rho_2, \gamma_2) \in \mathsf{Bun}_{\mathcal{G}}^{\mathrm{pre}}(Y)$, we need to provide for every composable pair of morphisms $y \xrightarrow{\psi} y' \xrightarrow{\phi} y''$ in Y an isomorphism

$$\gamma_2(\phi, \psi)\colon \rho_2(\phi \circ \psi) \to \rho_2(\phi)\rho_2(\psi)$$

in $\mathcal{G}$. That is, we need an isomorphism

$$(h(y'')\rho_1(\phi \circ \psi))\, h(y)^{-1} \to \left[(h(y'')\rho_1(\phi))\, h(y')^{-1}\right] \left[(h(y')\rho_1(\psi))\, h(y)^{-1}\right].$$

The associator, the isomorphism $\gamma_1\colon \rho_1(\phi \circ \psi) \to \rho_1(\phi) \circ \rho_1(\psi)$, and the morphism $e_{h(y')}\colon h(y')^{-1}h(y') \to 1_G$ from (2.3) provide exactly the ingredients needed to do this, and we take

$$\gamma_2(\phi, \psi) = \gamma_1(\phi, \psi)\frac{\alpha(h(y'), h(y')^{-1}, h(y')\rho_1(\psi))\alpha(\rho_2(\phi), h(y')\rho_1(\psi), h(y)^{-1})}{\alpha(h(y''), \rho_1(\phi), \rho_1(\psi))\alpha(h(y'')\rho_1(\phi), h(y')^{-1}, h(y')\rho_1(\psi))}.$$

The cocycle condition on α and the compatibility condition $d\gamma_1 = \rho_1^*\alpha$ yield the analogous condition $d\gamma_2 = \rho_2^*\alpha$, so that (ρ_2, γ_2) is indeed an object of $\mathsf{Bun}_{\mathcal{G}}^{\mathrm{pre}}(Y)$.

It remains to extend $h\colon \rho_1 \to \rho_2$ to an isomorphism $(h, \eta)\colon (\rho_1, \gamma_1) \to (\rho_2, \gamma_2)$. For $\phi\colon y \to y'$ in Y we need to specify an isomorphism $\eta(\phi)\colon \rho_2(\phi)h(y) \to h(y')\rho_1(\phi)$. Just as before, we have all the ingredients we need to do this:

$$\begin{aligned}\rho_2(\phi)h(y) = \left[(h(y')\rho_1(\phi))h(y)^{-1}\right] h(y) &\xrightarrow{\alpha} (h(y')\rho_1(\phi))(h(y)^{-1}h(y)) \\ &\xrightarrow{e_{h(y)}} (h(y')\rho_1(\phi))1_G = h(y')\rho_1(\phi).\end{aligned}$$

That is, we take $\eta(\phi) = \frac{\alpha(h(y')\rho_1(\phi), h(y)^{-1}, h(y))}{\alpha(h(y), h(y)^{-1}, h(y))}$. Applying the cocycle condition to the tuple $(\rho_2(\phi), h(y), h(y)^{-1}, h(y))$, we obtain $\eta(\phi) = \alpha(\rho_2(\phi), h(y), h(y)^{-1})$. Again, we use the cocycle condition on α, the compatibility condition of γ_1, and the definition of γ_2 to show that (h, η) is indeed an isomorphism $(\rho_1, \gamma_1) \to (\rho_2, \gamma_2)$ as claimed. □

We are now ready to describe the fibers of (4.5) in terms of 2-gerbes; we refer to §B for background on 2-gerbes. Our main examples are the following.

EXAMPLE 4.19. Let G be a group and A an abelian group. Then a 3-cocycle $\alpha \in Z^3(G; A)$ determines a 2-gerbe on $\mathrm{pt}/\!\!/G$. We will again denote this 2-gerbe by α.

EXAMPLE 4.20. Let X be a Lie groupoid. Given a flat G-bundle on X

$$X \xleftarrow{\sim} Y \xrightarrow{\rho} \mathrm{pt}/\!\!/G, \tag{4.6}$$

define a 2-gerbe by

$$Y_3 \xrightarrow{\rho^{\times 3}} G \times G \times G \xrightarrow{\alpha} A. \tag{4.7}$$

We use the notation $\rho^*\alpha$ for this 2-gerbe that trivializes over the (fixed) cover $Y \to X$, compare Definition B.1.

Next we describe how $\rho^*\alpha$ changes under isomorphisms of G-bundles, and in particular, under changing the cover $Y \to X$: given an isomorphism between principal bundles, we construct an explicit 1-isomorphism between the corresponding 2-gerbes. Let $Z = Y_1 \times_X Y_2$ and $h\colon Z_0 \to G$ be the data of the isomorphism $(f_1\colon Y_1 \to X, \rho_1) \to (f_2\colon Y_2 \to X, \rho_2)$. Let $p_i\colon Z \to Y_i$, $i = 1, 2$ denote the projections out of the fibered product. For $z \xrightarrow{\psi} z' \xrightarrow{\phi} z''$ in Z, define

$$\gamma(\phi, \psi) = \frac{\alpha(\rho_2(p_2(\phi)), h(z'), \rho_1(p_1(\psi)))}{\alpha(\rho_2(p_2(\phi)), \rho_2(p_2(\psi)), h(z))\alpha(h(z''), \rho_1(p_1(\phi)), \rho_1(p_1(\psi)))}.$$

It follows from the cocycle condition and the compatibility condition between h, ρ_1 and ρ_2 that γ gives the claimed 1-morphism $\rho_1^*\alpha \to \rho_2^*\alpha$. Therefore, for any principal G-bundle P, we obtain a flat 2-gerbe which, up to isomorphism, does not depend on the choice of cocycle data (f, ρ). We denote this flat 2-gerbe by $\lambda_{P,\alpha}$.

Proposition 4.21. *Let $\mathcal{G} = \mathcal{G}(G, A, \alpha)$ be a 2-group and let X be a Lie groupoid. For $\rho\colon X_1 \to G$ an object of $\mathsf{Bun}_G^{\mathrm{pre}}(X)$, the homotopy fiber $\mathsf{fib}(\rho)$ of $\mathsf{Bun}_{\mathcal{G}}^{\mathrm{pre}}(X) \to \mathsf{Bun}_G^{\mathrm{pre}}(X)$ over ρ is canonically equivalent to the bicategory of trivializations of the 2-gerbe $\rho^*\alpha$, constructed in Example* 4.20. *We emphasize that this is a 2-gerbe that trivializes on a fixed cover $Y \to X$, see Definition* B.1.

Proof. An object of $\mathsf{fib}(\rho)$ is given by an object (ρ_1, γ_1) of $\mathsf{Bun}_{\mathcal{G}}^{\mathrm{pre}}(X)$ together with an isomorphism $h_1\colon \rho_1 \to \rho$ in $\mathsf{Bun}_G^{\mathrm{pre}}(X)$. A 1-morphism $((\rho_1, \gamma_1), h_1) \to ((\rho_2, \gamma_2), h_2)$ is given by a 1-morphism $(h, \eta)\colon (\rho_1, \gamma_1) \to (\rho_2, \gamma_2)$ such that the morphism $h\colon \rho_1 \to \rho_2$ is compatible with the morphisms $h_i\colon \rho_i \to \rho$ in $\mathsf{Bun}_G^{\mathrm{pre}}(X)$. This compatibility condition amounts to the requirement that $h = h_2^{-1}h_1$ as cochains $X_0 \to G$. Finally, given two 1-morphisms $(h, \eta_1), (h, \eta_2)\colon (\rho_1, \gamma_1) \to (\rho_2, \gamma_2)$, a 2-morphism in $\mathsf{fib}(\rho)$ is just a 2-morphism $\omega\colon (h, \eta_1) \to (h, \eta_2)$.

Given an object $((\rho_1, \gamma_1), h)$, Lemma 4.18 provides locally constant cochains $\gamma\colon X_2 \to G$ and $\eta\colon X_1 \to A$ such that (h, η) gives an isomorphism $(\rho_1, \gamma_1) \to (\rho, \gamma)$. Then $((\rho, \gamma), 1_G)$ is an object of $\mathsf{fib}(\rho)$, and $(h, \eta)\colon ((\rho_1, \gamma_1), h) \to ((\rho, \gamma), 1_G)$. So every object in $\mathsf{fib}(\rho)$ is canonically isomorphic to one of the form $((\rho, \gamma), 1_G)$, and γ satisfies $d\gamma = \rho^*\alpha$. In other words, γ provides a trivialization of the 2-gerbe $\rho^*\alpha$.

Now let $((\rho, \gamma_1), 1_G), ((\rho, \gamma_2), 1_G)$ be two objects of $\mathsf{fib}(\rho)$. A 1-morphism between them is a pair (h, η) with $h = 1_G^{-1}1_G = 1_G$; that is, it is just a locally constant cochain $\eta\colon X_1 \to A$ satisfying the condition (3.23), which simplifies to $d\eta = \frac{\gamma_1}{\gamma_2}$, since $h = 1$. Thus, η is a 2-morphism between the two trivializations γ_1 and γ_2.

Finally, let $(1_G, \eta_1), (1_G, \eta_2)$ give two 1-morphisms in $\mathsf{fib}(\rho)$; a 2-morphism between them is a locally constant cochain $\omega\colon X_0 \to A$ such that $d\omega = \frac{\eta_2}{\eta_1}$. This is a 3-morphism between the 2-morphisms η_1, η_2 in the 3-category $\mathsf{2Gerbe}_A^{\mathrm{pre}}(X)$. □

We finally consider the homotopy fiber of the functor $\pi\colon \mathsf{Bun}_{\mathcal{G}}(X) \to \mathsf{Bun}_G(X)$ over a principal G-bundle P on X. We refer to it as the *bicategory of $\mathcal{G}$-bundles with underlying G-bundle P*, and denote it by $\mathsf{Bun}_{\mathcal{G}}(X)_P$.

Theorem 4.22. *Given the data as above, there is a canonical equivalence between the bicategory $\mathsf{Bun}_{\mathcal{G}}(X)_P$ of $\mathcal{G}$-bundles with underlying flat G-bundle P and the bicategory $\mathsf{Triv}(\lambda_{P,\alpha})$ of trivializations of the flat 2-gerbe $\lambda_{P,\alpha}$. Furthermore, in* *the event that these two bicategories are non-empty, there is a non-canonical equivalence between the bicategory* $\mathsf{Bun}_{\mathcal{G}}(X)_P$ *and the bicategory* $\mathsf{Gerbe}_A(X)$ *of gerbes on* X*, where an equivalence is determined by a choice of principal* $\mathcal{G}$*-bundle lifting* P*.*

PROOF. Let us choose cocycle data $(f\colon Y\to X,\rho)$ in $\mathsf{Bun}_G(X)$ representing the G-bundle P. Stackification is a left adjoint, so commutes with homotopy fiber products. Hence $\mathsf{Bun}_{\mathcal{G}}(X)_P$ is canonically equivalent to the stackification of the homotopy fiber $\mathsf{fib}(\rho)$ studied in Proposition 4.21. Combined with Lemma B.5, we obtain a sequence of equivalences of bicategories

$$\mathsf{Bun}_{\mathcal{G}}(X)_P\simeq \mathsf{Triv}(f,\rho^*\alpha)\simeq \mathsf{Gerbe}_A(X), \tag{4.8}$$

where the first equivalence is canonical and the second depends upon the choice of a trivialization $(g\colon Z\to Y,\gamma)$ of $(f,\rho^*\alpha)$, or equivalently, a lift $(g\circ f,g^*\rho,\gamma)$ of (f,ρ). □

REMARK 4.23. We can describe the functor from $\mathsf{Bun}_{\mathcal{G}}(X)_P$ to $\mathsf{Triv}(f,\rho^*\alpha)$ more concretely. An object of $\mathsf{Bun}_{\mathcal{G}}(X)_P$ is a pair, consisting of an object (f_1,ρ_1,γ_1) in $\mathsf{Bun}_{\mathcal{G}}(X)$ and an isomorphism between (f_1,ρ_1) and (f,ρ) in $\mathsf{Bun}_G(X)$. Similarly, a 1-morphism is a pair consisting of a 1-morphism between the objects of $\mathsf{Bun}_{\mathcal{G}}(X)$, and a 2-morphism providing compatibility data on the level of 1-morphisms in $\mathsf{Bun}_G(X)$. Finally, a 2-morphism is a 2-morphism in $\mathsf{Bun}_{\mathcal{G}}(X)$ which satisfies a compatibility condition in $\mathsf{Bun}_G(X)$.

We claim that every object is isomorphic to one consisting of a pair where the object of $\mathsf{Bun}_{\mathcal{G}}(X)$ is of the form $(f\circ g,g^*\rho,\gamma_1)$ for some essential equivalence $g\colon Z\to Y$, and the 1-morphism in $\mathsf{Bun}_G(X)$ corresponds to $\mathrm{id}\colon g^*\rho\to g^*\rho$ in $\mathsf{Bun}_G(Z)$. To prove this claim, we first remark that every object of $\mathsf{Bun}_{\mathcal{G}}(X)_P$ is isomorphic to a pair where the $\mathcal{G}$-bundle is of the form $(f_1=f\circ g,\rho_1,\gamma_1)$ for some essential equivalence $g\colon Z\to Y$, and the 1-morphism in $\mathsf{Bun}_G(X)$ is given by a 1-morphism $h\colon \rho_1\to g^*\rho$ in $\mathsf{Bun}_G(Z)$. Furthermore, it follows from Lemma 4.18 that up to canonical isomorphism we can also replace the $\mathcal{G}$-bundle by one of the form $(f_1=f\circ g,g^*\rho,\gamma)$, and the 1-morphism by the identity as claimed. This object of $\mathsf{Bun}_{\mathcal{G}}(X)_P$ corresponds to a trivialization (g,γ) of $\rho^*\alpha$.

Given a principal $\mathcal{G}$-bundle represented by data $(f\colon Y\to X,(\rho,\gamma_1)_Y)$ and an A-gerbe $(f\colon Y\to X,\gamma_2)$, we can "twist" the principal bundle by the gerbe to obtain a new principal bundle,

$$(f\colon Y\to X,(\rho,\gamma_1\cdot\gamma_2)_Y); \tag{4.9}$$

indeed, we have

$$d(\gamma_1\gamma_2)=(d\gamma_1)(d\gamma_2)=(\rho^*\alpha)\cdot 1_G=\rho^*\alpha.$$

We remark that this twisting does not change the underlying G-bundle, which is still determined by ρ. Furthermore, the twisting can be defined even when the original $\mathcal{G}$-bundle and the gerbe are defined relative to two different surjective submersions $f_1\colon Y_1\to X$ and $f_2\colon Y_2\to X$: it suffices to pull the data back to $Y_1\times_X Y_2$ before multiplying the 2-cochains as in (4.9).

PROPOSITION 4.24. *The twisting operation* (4.9) *extends to a functor*

$$F\colon \mathsf{Bun}_{\mathcal{G}}(X)\times \mathsf{Gerbe}_A(X)\to \mathsf{Bun}_{\mathcal{G}}(X). \tag{4.10}$$

The functor $\pi\colon \mathsf{Bun}_{\mathcal{G}}\to \mathsf{Bun}_G$ *is invariant under* F *in the sense that* $\pi\circ F\cong\pi$*.*

PROOF. We will describe a functor

$$F^{\mathrm{pre}}\colon \mathsf{Bun}_{\mathcal{G}}^{\mathrm{pre}}(X) \times \mathsf{Gerbe}_A^{\mathrm{pre}}(X) \to \mathsf{Bun}_{\mathcal{G}}^{\mathrm{pre}}(X) \tag{4.11}$$

which will induce the desired functor F upon stackification.

We know that on objects F^{pre} must behave as in the formula (4.9):

$$F^{\mathrm{pre}}((\rho, \gamma_1), \gamma_2) = (\rho, \gamma_1 \cdot \gamma_2). \tag{4.12}$$

Similarly, given 1-morphisms $(h, \eta_1)\colon (\rho_1, \gamma_1) \to (\rho_2, \gamma_2)$ in $\mathsf{Bun}_{\mathcal{G}}^{\mathrm{pre}}(X)$ and $\eta_2\colon \gamma_3 \to \gamma_4$ in $\mathsf{Gerbe}_A^{\mathrm{pre}}(X)$ it is easy to see that setting

$$F^{\mathrm{pre}}((h, \eta_1), \eta_2) = (h, \eta_1 \cdot \eta_2)$$

gives a 1-morphism from $(\rho_1, \gamma_1 {\cdot} \gamma_3)$ to $(\rho_2, \gamma_2 {\cdot} \gamma_4)$. The value of F^{pre} on 2-morphisms is similarly determined by pointwise multiplication in A. It is straightforward to check that this defines a functor, and that for $\pi^{\mathrm{pre}}\colon \mathsf{Bun}_{\mathcal{G}}^{\mathrm{pre}}(X) \to \mathsf{Bun}_{G}^{\mathrm{pre}}(X)$ we have $\pi^{\mathrm{pre}} \circ F^{\mathrm{pre}} = \pi^{\mathrm{pre}}$.

The desired functor F and invariance $\pi \circ F \cong \pi$ follow upon stackification. □

In fact, this functor F can be extended to a (weak) action of the symmetric monoidal bicategory $\mathsf{Gerbe}_A(X)$ on the bicategory $\mathsf{Bun}_{\mathcal{G}}(X)$ making it into a torsor over $\mathsf{Bun}_G(X)$. Spelling this out in detail requires introducing natural transformations and further higher compatibility data, and is beyond the scope of this paper. However, what is relevant for our purposes is that this higher categorical action is free and transitive on fibers, inducing an equivalence of $\mathsf{Gerbe}_A(X)$ with each non-empty fiber. More precisely, we have the following:

THEOREM 4.25. *For any object $\mathcal{P}$ of $\mathsf{Bun}_{\mathcal{G}}(X)$ with underlying G-bundle P, the restriction of F to $\mathsf{Gerbe}_A(X) \cong \{\mathcal{P}\} \times \mathsf{Gerbe}_A(X)$ provides the equivalence of $\mathsf{Gerbe}_A(X)$ with the bicategory $\mathsf{Bun}_{\mathcal{G}}(X)_P$ of Theorem 4.22.*

PROOF. Choose cocycle data $(Y \to X, (\rho, \gamma)_Y)$ classifying $\mathcal{P}$. Restricting the functor F^{pre} from the proof of Proposition 4.24, we obtain a functor

$$\{(\rho, \gamma)\} \times \mathsf{Gerbe}_A^{\mathrm{pre}}(Y) \to \mathsf{Bun}_{\mathcal{G}}^{\mathrm{pre}}(Y), \tag{4.13}$$

that induces the functor of interest upon stackification. Comparing the formulas given in the proofs of Lemma B.5 and Proposition 4.24, one recovers the functor of Theorem 4.22. □

5. Flat string structures

In this section we apply Theorem 4.22 to analyze flat string structures in the sense of Definition 1.1. The material in Sections 5.1–5.3 is well-known, though the presentation below is tailored to fit our purposes. We begin with a recollection of flat bundles in Section 5.1, before translating flat orientations and spin structures into this language in Sections 5.2–5.3. Section 5.4 then uses the language of 2-groups to give a similar formulation of flat string structures.

5.1. Background: Flat vector bundles. Let M be a manifold. A metrized vector bundle $V \to M$ with a compatible connection ∇ is flat if its curvature $F = \nabla \circ \nabla \in \Omega^2(M; \mathrm{End}(V))$ vanishes. This can be rephrased in terms of principal bundles: a bundle is flat if and only if its O_n-frame bundle admits a reduction of structure group to an O_n^δ-bundle, where O_n^δ is the orthogonal group O_n endowed with the discrete topology. By standard principal bundle theory, such a reduction of structure group is equivalent to admitting a Čech cocycle whose transition data uses only *constant* O_n-valued functions. This generalizes to flat bundles on Lie groupoids as follows; compare Example 5.2 below.

DEFINITION 5.1. The groupoid of *vector bundles* on a Lie groupoid X is the groupoid of maps $X \to \mathrm{pt} /\!\!/ \mathrm{O}_n$ in $\mathsf{LieGrpd}[W^{-1}]$. A *flat structure* on a vector bundle is a lift in $\mathsf{LieGrpd}[W^{-1}]$,

$$\begin{array}{ccc} & & \mathrm{pt} /\!\!/ \mathrm{O}_n^\delta \\ & \nearrow & \downarrow \\ X & \xrightarrow[V]{} & \mathrm{pt} /\!\!/ \mathrm{O}_n. \end{array} \tag{5.1}$$

The groupoid of *flat vector bundles on* X is the groupoid of maps $X \to \mathrm{pt} /\!\!/ \mathrm{O}_n^\delta$ in $\mathsf{LieGrpd}[W^{-1}]$, which we identify with the (1-)groupoid $\mathsf{Bun}_{\mathrm{O}_n^\delta}(X)$ from the previous section.

EXAMPLE 5.2. When $X = M$ is a manifold, the dashed arrow in (5.1) is a zig-zag

$$M \xleftarrow{\sim} \check{C}(Y) \xrightarrow{V} \mathrm{pt} /\!\!/ \mathrm{O}_n^\delta$$

for an essential equivalence determined by a surjective submersion $Y \to M$. Hence, (5.1) recovers a standard definition of flat bundles in terms of locally constant transition functions; see Definition 3.10.

EXAMPLE 5.3. A *flat manifold* is a Riemannian manifold M whose curvature vanishes, i.e., such that the tangent bundle TM is equipped with a flat structure. More generally, for a Lie group G acting on a manifold M, a *G-equivariant flat structure on M* is a lift

$$\begin{array}{ccc} & & \mathrm{pt} /\!\!/ \mathrm{O}_n^\delta \\ & \nearrow & \downarrow \\ M /\!\!/ G & \xrightarrow[TM]{} & \mathrm{pt} /\!\!/ \mathrm{O}_n \end{array}$$

where TM classifies the G-equivariant frame bundle of M, compare Example 4.14.

5.2. Orientations of flat bundles. The inclusion $\mathrm{SO}_n^\delta \hookrightarrow \mathrm{O}_n^\delta$ determines a functor

$$\mathsf{Bun}_{\mathrm{SO}_n^\delta}(X) \to \mathsf{Bun}_{\mathrm{O}_n^\delta}(X) \tag{5.2}$$

for any Lie groupoid X.

DEFINITION 5.4. Given a flat vector bundle $V \in \mathsf{Bun}_{\mathrm{O}_n^\delta}(X)$ the category of *orientations* on V is the homotopy fiber of (5.2).

Unwinding the definition, an orientation on V is 2-commutative triangle in $\mathsf{LieGrpd}[W^{-1}]$

$$\begin{array}{ccc} & & \mathrm{pt} /\!\!/ \mathrm{SO}_n^\delta \\ & \nearrow & \downarrow \\ X & \xrightarrow[V]{} & \mathrm{pt} /\!\!/ \mathrm{O}_n^\delta \end{array} \tag{5.3}$$

where we emphasize that 2-commutativity is additional data beyond the dashed arrow. Similarly, an isomorphism between oriented vector bundles is a 2-morphism between dashed arrows where the 2-commutative data is required to satisfy the evident compatibility condition.

EXAMPLE 5.5. When $X = M$ is a manifold, the dashed arrow in (5.3) specifies a Čech groupoid $\check{C}(Y)$ and a SO_n^δ-valued cocycle for a principal bundle defined relative to the surjective submersion $Y \to M$. The 2-commutativity data in (5.3) gives an isomorphism (possibly after refining the submersion) with the O_n^δ-valued cocycle specifying the given flat vector bundle.

DEFINITION 5.6. The *orientation double cover* associated to a flat vector bundle is the $\{\pm 1\}$-principal bundle on X, denoted $\mathsf{or}_V \to X$, classified by the composition in $\mathsf{LieGrpd}[W^{-1}]$,

$$X \xrightarrow{V} \mathrm{pt} /\!\!/ \mathrm{O}_n^\delta \xrightarrow{\det} \mathrm{pt} /\!\!/ \{\pm 1\} \tag{5.4}$$

where the rightmost arrow is induced by the determinant homomorphism.

REMARK 5.7. Explicitly, the composition (5.4) gives an essential equivalence $Y \xrightarrow{\sim} X$ and a functor $Y \to \mathrm{pt} /\!\!/ \{\pm 1\}$ determined by a map $Y_1 \to \{\pm 1\}$ that reads off whether the cocycle for V preserves or reverses orientation.

DEFINITION 5.8. A *trivialization* of the orientation double cover is a section $X \to \mathsf{or}_V$. Equivalently, a trivialization is a 2-commuting diagram in $\mathsf{LieGrpd}[W^{-1}]$,

$$\begin{array}{ccccc} & & & & \mathrm{pt} \\ & \nearrow^{\text{(dashed)}} & & & \downarrow \\ X & \xrightarrow[V]{} & \mathrm{pt} /\!\!/ \mathrm{O}_n^\delta & \xrightarrow[\det]{} & \mathrm{pt} /\!\!/ \{\pm 1\}. \end{array} \tag{5.5}$$

The *category of vector bundles on X with trivialized orientation double cover* has objects diagrams (5.5) and morphisms isomorphisms between maps $X \to \mathrm{pt} /\!\!/ \mathrm{O}_n^\delta$ that are compatible with the trivialization of or_V.

LEMMA 5.9. *The category $\mathsf{Bun}_{\mathrm{SO}_n^\delta}(X)$ of oriented flat vector bundles is equivalent to the category of flat vector bundles with trivialized orientation double cover.*

PROOF. Since $\mathrm{pt} /\!\!/ \mathrm{SO}_n^\delta \to \mathrm{pt} /\!\!/ \mathrm{O}_n^\delta$ is the pullback of $\mathrm{pt} \to \mathrm{pt} /\!\!/ \{\pm 1\}$ along the morphism $\det \colon \mathrm{pt} /\!\!/ \mathrm{O}_n^\delta \to \mathrm{pt} /\!\!/ \{\pm 1\}$, the orientation double cover fits into a pullback diagram as follows:

$$\begin{array}{ccccc} \mathsf{or}_V & \longrightarrow & \mathrm{pt} /\!\!/ \mathrm{SO}_n^\delta & \longrightarrow & \mathrm{pt} \\ \downarrow & \lrcorner & \downarrow & \lrcorner & \downarrow \\ X & \longrightarrow & \mathrm{pt} /\!\!/ \mathrm{O}_n^\delta & \longrightarrow & \mathrm{pt} /\!\!/ \{\pm 1\}. \end{array} \tag{5.6}$$

The statement of the lemma then follows from the universal property of the pullback

$$\begin{array}{ccccc} X & & \text{- - - - - - - - - - - - - - - - -} & \searrow & \\ & \ddots & & & \\ \mathrm{id} \Big\downarrow & & \mathsf{or}_V & \longrightarrow & \mathrm{pt} /\!\!/ \mathrm{SO}_n^\delta \\ & & \downarrow & \nearrow^{\text{(dashed)}} & \downarrow \\ & & X & \xrightarrow[V]{} & \mathrm{pt} /\!\!/ \mathrm{O}_n \end{array} \tag{5.7}$$

where the dashed arrows are the data of the oriented vector bundle $X \to \mathrm{pt} /\!/ \mathrm{O}_n^\delta$, and the dotted arrow is the data of a trivializing section of or_V. Hence, a lift determines a trivializing section $X \to \mathsf{or}_V$ and conversely. Similarly, an isomorphism between lifts determines an isomorphism between trivializations of or_V. □

REMARK 5.10. We observe that the smooth functor $\mathrm{pt} /\!/ \mathrm{SO}_n^\delta \to \mathrm{pt} /\!/ \mathrm{O}_n^\delta$ is not a fibration of groupoids. However, it factors as a composition of smooth functors

$$\mathrm{pt} /\!/ \mathrm{SO}_n^\delta \xrightarrow{\sim} (\mathrm{pt} \textstyle\coprod \mathrm{pt}) /\!/ \mathrm{O}_n^\delta \to \mathrm{pt} /\!/ \mathrm{O}_n^\delta \tag{5.8}$$

where $(\mathrm{pt} \coprod \mathrm{pt}) /\!/ \mathrm{O}_n^\delta$ is the action groupoid for the O_n^δ-action on $\mathrm{pt} \coprod \mathrm{pt} \simeq \{\pm 1\}$ through the determinant homomorphism. In (5.8), the first arrow is an essential equivalence and the second arrow is a fibration of groupoids. Hence, replacing $\mathrm{pt} /\!/ \mathrm{SO}_n^\delta$ in (5.6) with $(\mathrm{pt} \coprod \mathrm{pt}) /\!/ \mathrm{O}_n^\delta$, we find that the pullback exists in Lie groupoids and is the claimed orientation double cover.

REMARK 5.11. In light of Lemma 5.9, an orientation of a vector bundle V is a trivialization of a $\{\pm 1\}$-bundle on X. The category of trivializations of a $\{\pm 1\}$-bundle is equivalent to a set. Hence orientations of a flat vector bundle V naturally form a set.

EXAMPLE 5.12. For a flat G-manifold M, a *G-equivariant orientation* is a lift

$$\begin{array}{ccc} & & \mathrm{pt} /\!/ \mathrm{SO}_n^\delta \\ & \nearrow & \downarrow \\ M /\!/ G & \xrightarrow[TM]{} & \mathrm{pt} /\!/ \mathrm{O}_n^\delta \end{array}$$

where TM classifies the G-equivariant frame bundle of M. By Lemma 5.9, an equivariant orientation is equivalent to a G-invariant trivialization of the orientation double cover of TM, i.e., a unit norm G-invariant section of $\Lambda^{\mathrm{top}} TM$ over M.

5.3. Flat spin structures. The double cover $\mathrm{Spin}_n^\delta \to \mathrm{SO}_n^\delta$ postcomposed with the inclusion $\mathrm{SO}_n^\delta \hookrightarrow \mathrm{O}_n^\delta$ determines a functor

$$\mathsf{Bun}_{\mathrm{Spin}_n^\delta}(X) \to \mathsf{Bun}_{\mathrm{O}_n^\delta}(X) \tag{5.9}$$

for any Lie groupoid X.

DEFINITION 5.13. Given a flat vector bundle $V \in \mathsf{Bun}_{\mathrm{O}_n^\delta}(X)$ the category of *flat spin structures* on V is the homotopy fiber of (5.9).

Hence, a spin structure on V is 2-commutative triangle in $\mathsf{LieGrpd}[W^{-1}]$

$$\begin{array}{ccc} & & \mathrm{pt} /\!/ \mathrm{Spin}_n^\delta \\ & \nearrow & \downarrow \\ X & \xrightarrow[V]{} & \mathrm{pt} /\!/ \mathrm{O}_n^\delta. \end{array} \tag{5.10}$$

The functor (5.9) factors as

$$\mathsf{Bun}_{\mathrm{Spin}_n^\delta}(X) \to \mathsf{Bun}_{\mathrm{SO}_n^\delta}(X) \to \mathsf{Bun}_{\mathrm{O}_n^\delta}(X).$$

Hence, there is a natural functor from the category of vector bundles with spin structure to the category of oriented vector bundles.

EXAMPLE 5.14. In analogy to Example 5.5, when $X = M$ is a manifold the dashed arrow in (5.10) specifies a Čech cocycle for a Spin_n^δ-principal bundle that (up to refinement) lifts the given cocycle for the flat vector bundle V.

REMARK 5.15. Flat spin structures have been studied in the context of index theory, e.g., see [**14**, **25**, **36**].

DEFINITION 5.16. The *Stiefel–Whitney gerbe of a flat vector bundle* V, denoted SW_V, is the pullback in $\mathsf{LieGrpd}[W^{-1}]$

$$\begin{array}{ccc} \mathrm{SW}_V & \longrightarrow & \mathrm{pt} /\!/ \mathrm{Spin}_n^\delta \\ \downarrow & \lrcorner & \downarrow \\ \mathsf{or}_V & \longrightarrow & \mathrm{pt} /\!/ \mathrm{SO}_n^\delta \end{array} \tag{5.11}$$

of the fibration of groupoids $\mathrm{pt} /\!/ \mathrm{Spin}_n^\delta \to \mathrm{pt} /\!/ \mathrm{SO}_n^\delta$. This defines a Lie groupoid over X via the composition

$$\mathrm{SW}_V \to \mathsf{or}_V \to X. \tag{5.12}$$

A *trivialization* of $\mathrm{SW}_V \to X$ is a section $X \to \mathrm{SW}_V$ (considered in $\mathsf{LieGrpd}[W^{-1}]$). Trivializations are naturally the objects in a category whose morphisms are isomorphisms between sections.

REMARK 5.17. A trivializing section $X \to \mathrm{SW}_V$ induces a trivialization of or_V. We note that one can ask for an intermediate trivialization of $\mathrm{SW}_V \to \mathsf{or}_V$ over the double cover or_V. Below, we will only consider trivializations of the composite (5.12).

REMARK 5.18. The equivalence between trivializations of gerbes and sections in $\mathsf{LieGrpd}[W^{-1}]$ is standard; e.g., see [**7**, 4.2].

LEMMA 5.19. *The Stiefel–Whitney gerbe is a $\mathbb{Z}/2$-gerbe on the Lie groupoid or_V, i.e., a $(\mathrm{pt} /\!/ \mathbb{Z}/2)$-bundle determined by a 2-functor $\mathrm{SW}_V\colon \mathsf{or}_V \to \mathrm{pt} /\!/ (\mathrm{pt} /\!/ (\mathbb{Z}/2))$.*

PROOF. We describe the pullback (5.11) explicitly, using that the lax pullback along a fibration of (Lie) groupoids $\mathrm{pt} /\!/ \mathrm{Spin}_n^\delta \to \mathrm{pt} /\!/ \mathrm{SO}_n^\delta$ is equivalent to the strict pullback. Suppose that the map $\mathsf{or}_V \to \mathrm{pt} /\!/ \mathrm{SO}_n^\delta$ in $\mathsf{LieGrpd}[W^{-1}]$ is determined by the zig-zag of smooth functors $\mathsf{or}_V \xleftarrow{\sim} Y \to \mathrm{pt} /\!/ \mathrm{SO}_n^\delta$. Then SW_V has Y_0 as objects, and morphisms are the pullback (in smooth manifolds)

$$\begin{array}{ccc} \widehat{Y}_1 & \longrightarrow & \mathrm{Spin}_n^\delta \\ \downarrow & \lrcorner & \downarrow \\ Y_1 & \longrightarrow & \mathrm{SO}_n^\delta \end{array} \tag{5.13}$$

where $\widehat{Y}_1 \to Y_1$ is a principal $\mathbb{Z}/2$-bundle.

But since the topology on SO_n^δ is discrete, the map $Y_1 \to \mathrm{SO}_n^\delta$ is locally constant, and hence $\widehat{Y}_1 \to Y_1$ is the trivial $\mathbb{Z}/2$-principal bundle. This implies that data of SW_V is the essential equivalence $Y \xrightarrow{\sim} \mathsf{or}_V$ and a $\mathbb{Z}/2$-valued 2-cocycle $Y_1 \times_{Y_0} Y_1 \to \mathbb{Z}/2$. This proves the lemma. □

REMARK 5.20. When $X = M$ is a manifold, the 2-cocycle determining SW_V is a standard cocycle construction of the second Stiefel–Whitney class $w_2(V)$. More generally, the Stiefel–Whitney gerbe sits in a fiber sequence

$$\mathrm{pt} /\!\!/ \mathbb{Z}/2 \to \mathrm{SW}_V \to \mathsf{or}_V$$

that is a $\mathbb{Z}/2$-*central extension* in the sense of [**7**, §4-5]. Such central extensions are examples of twists for Thom classes in real K-theory, see [**16**, 1.92 and 1.161].

PROPOSITION 5.21. *The category of flat spin structures on V is equivalent to the category of trivializations of $SW_V \to X$.*

PROOF. Consider the diagram in $\mathsf{LieGrpd}[W^{-1}]$ of iterated pullbacks,

$$\begin{array}{ccc} X & \dashrightarrow & \\ & \searrow^{\cdots} & \\ & \mathrm{SW}_V & \longrightarrow & \mathrm{pt} /\!\!/ \mathrm{Spin}_n^\delta \\ \mathrm{id} & \downarrow & & \downarrow \\ & \mathsf{or}_V & \longrightarrow & \mathrm{pt} /\!\!/ \mathrm{SO}_n \\ & \downarrow & & \downarrow \\ & X & \xrightarrow[V]{} & \mathrm{pt} /\!\!/ \mathrm{O}_n \end{array} \tag{5.14}$$

where the dashed arrows are the data of a spin structure on V, and the dotted arrow is the data of a trivializing section. By the universal property of the pullback, these are equivalent data. □

EXAMPLE 5.22. For a flat G-manifold M, a *G-equivariant flat spin structure* is a lift

$$\begin{array}{ccc} & & \mathrm{pt} /\!\!/ \mathrm{Spin}_n^\delta \\ & \nearrow & \downarrow \\ M /\!\!/ G & \xrightarrow[TM]{} & \mathrm{pt} /\!\!/ \mathrm{O}_n^\delta \end{array} \tag{5.15}$$

where TM classifies the G-equivariant frame bundle of M. From Example 5.12, this lift determines a G-invariant orientation on M. By inspection of (5.15), an equivariant spin structure is the data of a Spin_n^δ-principal bundle $P \to M$ covering the frame bundle of TM where P is furthermore equipped with a G-action and the map $P \to M$ is G-equivariant. An equivariant spin structure is the data of an orientation and a trivialization of the Stiefel–Whitney gerbe of TM (e.g., see [**51**, Example 4.1.4]), together with equivariant descent data for these trivializations along the map $M \to M /\!\!/ G$ in $\mathsf{LieGrpd}[W^{-1}]$.

5.4. Flat string structures. The 1-homomorphism of 2-groups $\mathrm{String}_n^\delta \to \mathrm{Spin}_n^\delta \to \mathrm{O}_n^\delta$ determines a 2-functor

$$\mathsf{Bun}_{\mathrm{String}_n^\delta}(X) \to \mathsf{Bun}_{\mathrm{O}_n^\delta}(X) \tag{5.16}$$

for any Lie groupoid X.

DEFINITION 5.23. Given a flat vector bundle $V \in \mathsf{Bun}_{\mathrm{O}_n^\delta}(X)$ the bicategory of *flat string structures* on V is the homotopy fiber of (5.16).

Hence, a string structure on V is lifting 2-functor

$$\begin{array}{ccc} & & \mathrm{pt} /\!\!/ \mathrm{String}_n^\delta \\ & \nearrow & \downarrow \\ X & \xrightarrow[V]{} & \mathrm{pt} /\!\!/ \mathrm{O}_n^\delta. \end{array}$$

Similarly to above, a string structure on V determines a spin structure and an orientation.

We recall that there is a 3-cocycle $\mathrm{Spin}_n^\delta \times \mathrm{Spin}_n^\delta \times \mathrm{Spin}_n^\delta \to U(1)$ that classifies the discrete string 2-group (A.1); we denote this 3-cocycle by CS^δ. By Proposition 2.11, CS^δ is unique up to coboundary.

DEFINITION 5.24. The (flat) *Chern–Simons 2-gerbe* on $\mathrm{pt} /\!\!/ \mathrm{Spin}_n^\delta$ is specified by the 3-cocycle CS^δ on Spin_n^δ.

REMARK 5.25. We observe that (by construction) the fractional Pontryagin class maps to the class underlying the 3-cocycle CS^δ,

$$\begin{aligned} \left[\frac{p_1}{2}\right] \in \mathrm{H}^4(B\,\mathrm{Spin}_n; \mathbb{Z}) \simeq \mathrm{H}^3_{\mathrm{SM}}(\mathrm{Spin}_n; U(1)) &\xrightarrow{\delta^*} \mathrm{H}^3_{\mathrm{SM}}(\mathrm{Spin}_n^\delta; U(1)) \\ &\simeq \mathrm{H}^3(B\,\mathrm{Spin}_n^\delta; U(1)) \ni [\mathrm{CS}^\delta]. \end{aligned}$$

Hence, the Chern–Simons 2-gerbe in sense of Waldorff [**50**, §2.1] pulls back to the flat Chern–Simons 2-gerbe of Definition 5.24.

DEFINITION 5.26. The *Chern–Simons 2-gerbe of a flat vector bundle* V, denoted CS_V, is the 2-gerbe over the Lie groupoid SW_V gotten by pulling back the flat Chern–Simons 2-gerbe along the top horizontal arrow $\mathrm{SW}_V \to \pi /\!\!/ \mathrm{Spin}_n^\delta$ in (5.11) (see Example 4.20 for a formula for this pullback). A *trivialization* of $\mathrm{CS}_V \to X$ is a trivializing section $\sigma\colon X \to \mathrm{SW}_V$ and a trivialization of the 2-gerbe $\sigma^* \mathrm{CS}_V$ over X. We observe that trivializations of the Chern–Simons 2-gerbe form a (possibly empty) bicategory.

REMARK 5.27. One can rephrase the above in the geometry of 2-stacks. The full 3-category of 2-stacks is understood, but technical, e.g., see [**35**]. Here we will only indicate the relevance for (geometric) string structures. Consider the iterated pullback diagram in 2-stacks

$$\begin{array}{ccc} \mathrm{CS}_V & \longrightarrow & \mathrm{pt} /\!\!/ \mathrm{String}_n^\delta \\ \downarrow & \lrcorner & \downarrow \\ \mathrm{SW}_V & \longrightarrow & \mathrm{pt} /\!\!/ \mathrm{Spin}_n^\delta \\ \downarrow & \lrcorner & \downarrow \\ \mathsf{or}_V & \longrightarrow & \mathrm{pt} /\!\!/ \mathrm{SO}_n^\delta \\ \downarrow & \lrcorner & \downarrow \\ X & \xrightarrow[V]{} & \mathrm{pt} /\!\!/ \mathrm{O}_n^\delta \end{array} \tag{5.17}$$

where dashed arrow is a trivializing section of $\mathrm{CS}_V \to X$. By the universal property of the pullback, such a trivialization is equivalent to a map of 1-stacks $X \to \mathrm{SW}_V$ together with a trivialization of a 2-gerbe, as in Definition 5.26. The advantage of

Definition 5.26 is that we need only work with the bicategory of trivializations of a 2-gerbe, avoiding 3-categories.

THEOREM 5.28. *Given a flat vector bundle $V \to X$, the bicategory of flat string structures on V is equivalent to the bicategory of trivializations of* $\mathrm{CS}_V \to X$.

PROOF. This follows from Proposition 5.21 and Theorem 4.22. □

REMARK 5.29. We note that there are two possible versions of string structure on a flat spin vector bundle. One involves (potentially non-flat) trivializations of the Chern–Simons 2-gerbe, whereas the second only considers flat trivializations. In terms of the extensions (A.1) and the associated 2-group principal bundles, this distinction corresponds to whether one takes the usual topology on $U(1)$ as a Lie group, or its discrete topology.

EXAMPLE 5.30. For a flat G-manifold M, a *G-equivariant flat string structure* is a lift

$$\begin{array}{ccc} & & \mathrm{pt} /\!\!/ \mathrm{String}_n^\delta \\ & \nearrow & \downarrow \\ M /\!\!/ G & \xrightarrow[TM]{} & \mathrm{pt} /\!\!/ \mathrm{O}_n^\delta \end{array} \tag{5.18}$$

where TM classifies the G-equivariant frame bundle of M. By inspection of (5.18), an equivariant string structure is a G-equivariant String_n^δ-bundle on M whose underlying Spin_n^δ-bundle is the spin bundle of M for the underlying flat spin structure (see Example 5.22). An equivariant string structure is therefore the data of an orientation, spin structure, and trivialization of the (nonequivariant) Chern–Simons 2-gerbe of TM (see [**51**, §5.1] or [**50**, §2.1]), together with equivariant descent data. An explicit (though tedious) exercise using Theorem 4.22 gives an explicit description of this descent data; one recovers a flat version of Sharpe's heterotic B-field from [**42**, Equations 4 and 5].

Appendix A. The discrete string 2-group

The forgetful functor from geometric stacks to (discete) groupoids provides a discrete 2-group underlying any smooth 2-group [**38**, page 28]. Below we give an explicit description of the discrete 2-group underlying the string 2-group, String_n, i.e., the categorical central extension (1.1) [**38**, Theorem 2].

For a Lie group G and abelian Lie group A, smooth categorical central extensions

$$1 \to \mathrm{pt} /\!\!/ A \to \mathcal{G} \to G \to 1$$

are classified (up to 1-isomorphism) by $\mathrm{H}^3_{\mathrm{SM}}(G; A)$, the 3rd Segal–Mitchison cohomology group of G with values in A [**38**, Theorem 1]. This cohomology is computed by a double complex that depends on a simplicial cover of the nerve $BG_\bullet$.

LEMMA A.1. *When G has the discrete topology, the Segal–Mitchison cohomology $\mathrm{H}^\bullet_{\mathrm{SM}}(G; A)$ is isomorphic to the usual group cohomology of G valued in A.*

PROOF. In fact, the Segal–Mitchison bicomplex of G with coefficients in A is quasi-isomorphic to the complex computing the usual group cohomology of G. To construct this quasi-isomorphism, take the open cover of G given by its points; this determines a simplicial cover of $BG_\bullet$. The resulting double complex is exact for the

Čech differential, where the remaining differential is the one from group cohomology (e.g., see [**38**, page 35]). Hence we obtain a canonical quasi-isomorphism between the Segal–Mitchison complex and and the usual complex computing the group cohomology of G. □

The categorical central extensions (and the Segal–Mitchison complex) are functorial in the groups G and A [**38**, page 4]. This provides a map from a discrete extension to the string extension,

$$\begin{array}{ccccccccc} 1 & \longrightarrow & \mathrm{pt}/\!\!/U(1)^\delta & \longrightarrow & \mathrm{String}_n^\delta & \longrightarrow & \mathrm{Spin}_n^\delta & \longrightarrow & 1 \\ & & \downarrow & & \downarrow & & \downarrow & & \\ 1 & \longrightarrow & \mathrm{pt}/\!\!/U(1) & \longrightarrow & \mathrm{String}_n & \longrightarrow & \mathrm{Spin}_n & \longrightarrow & 1 \end{array} \tag{A.1}$$

where Spin_n^δ and $U(1)^\delta$ are the respective groups with the discrete topology. To spell out the map of extensions (A.1), by Lemma A.1 the Segal–Mitchison cocycle defining the lower extension determines (by precomposing with $\mathrm{Spin}_n^\delta \to \mathrm{Spin}_n$) a 3-cocycle $\mathrm{CS}^\delta\colon (\mathrm{Spin}_n^\delta)^{\times 3} \to U(1)$ classifying the upper extension. Here we use that any map $\mathrm{CS}^\delta\colon (\mathrm{Spin}_n^\delta)^{\times 3} \to U(1)$ is smooth; hence we may also take the discrete topology on $U(1)$.

Definition A.2. The *discrete string group* is the 2-group String_n^δ in (A.1).

Appendix B. Overview of 2-gerbes

Bundle 2-gerbes are the natural categorification of the 2-stack of gerbes, which in turn are a categorification of the 1-stack of line bundles. Early work on 2-gerbes includes [**8, 11, 17, 44**]. Modern approaches using ∞-stacks have streamlined the theory considerably; a non-exhaustive list of references are [**23**, §7.2.2], [**31, 32**], and [**35**]. The full apparatus of ∞-stacks is both unnecessary and beyond the scope of this paper. For our purposes it suffices to define 2-gerbes in terms of Čech data analogous to that appearing in the definition of A-gerbes in Examples 3.6 and 4.15. The abelian group A will be omitted from the notation when it is clear from context.

Definition B.1. Let X be a Lie groupoid. A flat *A-2-gerbe on X that trivializes on X_0* is given by a locally constant Čech 3-cocycle $\lambda\colon X_3 \to A$. In particular, the *trivial* 2-gerbe is given by the constant map $1_A\colon X_3 \to A$.

A *1-isomorphism* $\lambda_1 \to \lambda_2$ between a pair of such 2-gerbes is a locally constant 2-cochain $\gamma\colon X_2 \to A$ such that $d\gamma = \frac{\lambda_2}{\lambda_1}$: that is, for $(\phi, \psi, \theta) \in X_3$,

$$\frac{\gamma(\psi, \theta), \gamma(\phi, \psi \circ \theta)}{\gamma(\phi \circ \psi, \theta)\gamma(\phi, \psi)} = \frac{\lambda_2(\phi, \psi, \theta)}{\lambda_1(\phi, \psi, \theta)}.$$

In particular, a *trivialization* of a 2-gerbe λ that trivializes on X_0 is a 1-isomorphism $1_A \to \lambda$, i.e. a locally constant 2-cochain $\gamma\colon X_2 \to A$ such that $d\gamma = \lambda$.

A *2-isomorphism* $\gamma_1 \to \gamma_2$ is a locally constant 1-cochain $\eta\colon X_1 \to A$ with $d\eta = \frac{\gamma_1}{\gamma_2}$:

$$\frac{\eta(\psi)\eta(\phi)}{\eta(\phi \circ \psi)} = \frac{\gamma_1(\phi, \psi)}{\gamma_2(\phi, \psi)} \text{ for } (\phi, \psi) \in X_2.$$

A *3-isomorphism* $\eta_1 \to \eta_2$ is a locally constant 0-cochain $\omega\colon X_0 \to A$ with $d\omega = \frac{\eta_2}{\eta_1}$:

$$\frac{\omega(x')}{\omega(x)} = \frac{\eta_2(\phi)}{\eta_1(\phi)} \text{ for } \phi\colon x \to x' \text{ in } X.$$

This forms a strict 3-category, with composition induced from multiplication in A. We denote this 3-category by $\mathsf{2Gerbe}_A^{\mathrm{pre}}(X)$.

PROPOSITION B.2. *The assignment $X \mapsto \mathsf{2Gerbe}_A^{\mathrm{pre}}(X)$ forms a prestack.*

PROOF. It is clear that a smooth functor $f\colon X \to Y$ induces a pullback $f^*\colon \mathsf{2Gerbe}_A^{\mathrm{pre}}(Y) \to \mathsf{2Gerbe}_A^{\mathrm{pre}}(X)$ by precomposition of cochains, analogously to Definition 4.1.

It remains to show that a natural transformation $\tau\colon f_1 \Rightarrow f_2$ of smooth functors induces a natural transformation of pullbacks $\tau^*\colon f_1^* \Rightarrow f_2^*$. Let λ be an object in $\mathsf{2Gerbe}_A^{\mathrm{pre}}(Y)$. We define a 1-morphism $\tau^*_\lambda\colon f_1^*(\lambda) \to f_2^*(\lambda)$ by

$$\tau^*_\lambda(\phi, \psi) = \frac{\lambda(f_2(\phi), \tau(x'), f_1(\psi)}{\lambda(f_2(\phi), f_2(\psi), \tau(x))\lambda(\tau(x''), f_1(\phi), f_1(\psi))} \text{ for } x \xrightarrow{\psi} x' \xrightarrow{\phi} x'' \text{ in } X.$$

The 3-cocycle condition on λ implies that $d\lambda(\phi, \psi, \theta) = \frac{\lambda(f_2(\phi), f_2(\psi), f_2(\theta))}{\lambda(f_1(\phi), f_1(\psi), f_1(\theta))}$ for all (ϕ, ψ, θ) in X_3, as required. It is natural in λ and hence defines the required natural transformation. □

With this established, we adopt the following definition of 2-gerbe, e.g., see [**31**, Definition 4.48].

DEFINITION B.3. Let X be a Lie groupoid. A *flat 2-gerbe on X* is an essential equivalence $f\colon Y \xrightarrow{\sim} X$ and a locally constant 3-cocycle $\lambda\colon Y_3 \to A$. A *1-isomorphism* $(f_1\colon Y_1 \to X, \lambda_1) \to (f_2\colon Y_2 \to X, \lambda_2)$ of 2-gerbes is given by a tuple $(Z, g_1, g_2, \tau, \gamma)$, where Z is a Lie groupoid, $g_1\colon Z \to Y_1$ and $g_2\colon Z \to Y_2$ are smooth functors, τ is a smooth natural transformation $f_1 \circ g_1 \to f_2 \circ g_2$, and $\gamma\colon Z_2 \to A$ gives a 1-isomorphism $g_1^*(\lambda_1) \to g_2^*(\lambda_2)$. That is, for $(\phi, \psi, \theta) \in Z_3$, we have

$$\frac{\gamma(\psi, \theta)\gamma(\phi, \psi \circ \theta)}{\gamma(\phi \circ \psi, \theta)\gamma(\phi, \psi)} = \frac{\lambda_2(g_2(\phi), g_2(\psi), g_2(\theta))}{\lambda_1(g_1(\phi), g_1(\psi), g_1(\theta))}.$$

Similarly, we have 2- and 3-morphisms between 2-gerbes defined analogously using Definition B.1 together with refinements of essential equivalences $Y \xrightarrow{\sim} X$. The collection of 2-gerbes on X forms a $(3,1)$-category $\mathsf{2Gerbe}_A(X)$.

REMARK B.4. More formally, the 3-category $\mathsf{2Gerbe}_A(X)$ is the stackification of the 3-category $\mathsf{2Gerbe}_A^{\mathrm{pre}}(X)$. This stackification arises from a higher-categorical localization of $\mathsf{2Gerbe}_A^{\mathrm{pre}}(X)$, e.g., see [**35**] for an exposition that generalizes [**37**].

In light of the previous remark, morphisms between a pair of 2-gerbes form a (possibly empty) bicategory. In particular, trivializations of a given flat 2-gerbe $(f\colon Y \to X, \lambda)$ are the objects of the (possibly empty) bicategory $\mathrm{Triv}(f, \lambda)$. This bicategory itself can be identified with the 2-stackification of the bicategory of trivializations of λ in $\mathsf{2Gerbe}_A^{\mathrm{pre}}(Y)$. Up to isomorphism, the objects of $\mathrm{Triv}(f, \lambda)$ are of the form $(g\colon Z \to Y, \gamma\colon 1_A \to g^*\lambda)$.

LEMMA B.5. *Let $(f\colon Y \to X, \lambda)$ be a flat 2-gerbe on X. If it is non-empty, the bicategory* $\mathrm{Triv}(f,\lambda)$ *of trivializations of* (f,λ) *is non-canonically equivalent to the bicategory* $\mathsf{Gerbe}_A(X)$*, where a choice of equivalence* $\mathrm{Triv}(f,\lambda) \simeq \mathsf{Gerbe}_A(X)$ *is determined by a choice of trivialization* (g,γ) *of* (f,λ).

PROOF. It suffices to show that the choice of trivialization γ of $g^*\lambda$ determines an equivalence (in fact an isomorphism) of the bicategory of trivializations of $g^*\lambda$ in $2\mathsf{Gerbe}_A^{\mathrm{pre}}(Z)$ with the bicategory $\mathsf{Gerbe}_A^{\mathrm{pre}}(Z)$; upon 2-stackification this yields the desired equivalence.

Let γ_1 be any other trivialization of $g^*\lambda$; then as in definition B.1 we have $d\gamma = d\gamma_1 = g^*\lambda$, so $d\frac{\gamma_1}{\gamma} = 1$, and $\frac{\gamma_1}{\gamma}$ is an object of $\mathsf{Gerbe}_A^{\mathrm{pre}}(Z)$. This gives the bijection on the level of objects.

If γ_1, γ_2 are two trivializations of $g^*\lambda$ and η is a 2-morphism between them, then we have $\frac{\eta(\psi)\eta(\phi)}{\eta(\phi\circ\psi)} = \frac{\gamma_1}{\gamma_2} = \frac{\gamma_1/\gamma}{\gamma_2/\gamma}$. So η is also a 1-morphism between the gerbes γ_1/γ and γ_2/γ.

Finally, if η_1, η_2 are two 2-morphisms between trivializations of $g^*\lambda$, and ω is a 3-morphism between them, we have $d\omega = \frac{\eta_2}{\eta_1}$, which implies that ω is also a 2-morphism between the η_i's viewed as 1-morphisms in $\mathsf{Gerbe}_A^{\mathrm{pre}}(Z)$.

This completes the proof. □

Appendix C. Proof of Theorem 4.7

Recall that Theorem 4.7 states that the bicategory $\mathsf{Bun}_{\mathcal{G}}^{\mathrm{pre}}$ admits a calculus of right fractions with respect to the class of 1-morphisms $\mathcal{W}$ of the form (f,h,η) where f is an essential equivalence. Proving this requires verification of conditions BF1-BF5 from [**37**, §2.1]; we take the equivalent (diagrammatic) version of BF4 from [**48**]. Performing this localization yields the bicategory $\mathsf{Bun}_{\mathcal{G}} := \mathsf{Bun}_{\mathcal{G}}^{\mathrm{pre}}[\mathcal{W}^{-1}]$, the stack of $\mathcal{G}$-bundles on Lie groupoids.

LEMMA C.1. *Let* $(f,h,\eta) : (\rho_1,\gamma_1)_X \to (\rho_2,\gamma_2)_Y$ *be a 1-morphism in* $\mathsf{Bun}_{\mathcal{G}}^{\mathrm{pre}}$. *If* (f,h,η) *is an equivalence, then* f *is an essential equivalence. This is* [**37**, BF1].

PROOF. If (f,h,η) is an equivalence, there exists (f_2,h_2,η_2) such that $(f,h,\eta)\circ(f_2,h_2,\eta_2) \cong id_{(\rho_2,\gamma_2)_Y}$ and $(f_2,h_2,\eta_2)_Y\circ(f,h,\eta) \cong id_{(\rho_1,\gamma_1)_X}$; in particular $f\circ f_2 \cong \mathrm{id}_Y$ and $f_2\circ f \cong \mathrm{id}_X$. Hence f is an equivalence of Lie groupoids and (in particular) an essential equivalence. □

LEMMA C.2. *Given* $(f_4,h_4,\eta_4) : (\rho_1,\gamma_1)_X \to (\rho_3,\gamma_3)_Z, (f_3,h_3,\eta_3) : (\rho_2,\gamma_2)_Y \to (\rho_3,\gamma_3)_Z$ *in* $\mathsf{Bun}_{\mathcal{G}}^{\mathrm{pre}}$, *where* (f_4,h_4,η_4) *is in* $\mathcal{W}$, *there exists the data filling the diagram below:*

$$\begin{array}{ccc} (\rho_0,\gamma_0)_V & \xrightarrow{(f_1,h_1,\eta_1)} & (\rho_2,\gamma_2)_Y \\ {\scriptstyle (f_2,h_2,\eta_2)}\downarrow & \overset{(\tau,\omega)}{\Rightarrow}\,\cong & \downarrow{\scriptstyle (f_3,h_3,\eta_3)} \\ (\rho_1,\gamma_1)_X & \xrightarrow[(f_4,h_4,\eta_4)]{} & (\rho_3,\gamma_3)_Z \end{array} \tag{C.1}$$

where (f_1,h_1,η_1) *is in* $\mathcal{W}$ *and* (τ,ω) *is a 2-isomorphism. This is BF3 of* [**37**].

PROOF. The desired data in terms of the Lie groupoid structure is given by the lax pullback V of X and Y over Z, see [**27**]. This pullback has objects $V_0 := X_0 \times_{Z_0} Z_1 \times_{Z_0} Y_0$ and morphisms $V_1 := eq\Big(X_1 \times (V_0\times V_0) \times Y_1 \rightrightarrows Z_1\Big)$. Unpacking this, objects are of the form (x, ϕ, y) where $\phi : f_4(x) \to f_3(y)$ in Z_1. A morphism Φ from (x, ϕ, y) to (x', ϕ', y') consists of $\Phi_X \in X_1, \Phi_Y \in Y_1$ such that $f_3(\Phi_Y) \circ \phi = \phi' \circ f_4(\Phi_X)$ in Z_1.

We denote the natural projections from V onto Y and X by f_1, f_2 respectively. The natural transformation $\tau \colon f_4 \circ f_2 \to f_3 \circ f_1$ in the diagram (C.1) is given by $V_0 \to Z_1$, $(x, \phi, y) \mapsto \phi$. The equality $f_3(\phi_Y)\circ\phi = \phi'\circ f_4(\phi_X)$ verifies the naturality condition for τ.

By Lemma 4.5, the lower horizontal arrow in (C.1) factors as

$$\begin{array}{ccccc} & & & & (\rho_2, \gamma_2)_Y \\ & & & & \big\downarrow{\scriptstyle (f_3, h_3, \eta_3)} \\ (\rho_1, \gamma_1)_X & \xrightarrow[(1, h_4, \eta_4)]{} & (f_4^*(\rho_3, \gamma_3))_X & \xrightarrow[(f_4, 1, 1)]{} & (\rho_3, \gamma_3)_Z . \end{array} \tag{C.2}$$

We will construct the square (C.1) in two steps, via squares over each of the horizontal arrows in (C.2). First take the square

$$\begin{array}{ccc} (f_1^*(\rho_2, \gamma_2))_V & \xrightarrow{(f_1, 1, 1)} & (\rho_2, \gamma_2)_Y \\ {\scriptstyle (f_2, h, \eta)}\big\downarrow & & \big\downarrow{\scriptstyle (f_3, h_3, \eta_3)} \\ (f_4^*(\rho_3, \gamma_3))_X & \xrightarrow[(f_4, 1, 1)]{} & (\rho_3, \gamma_3)_Z \end{array} \tag{C.3}$$

where (h, η) is the composition of the 1-morphisms

$$f_1^*(\rho_2, \gamma_2) \xrightarrow{f_1^*(h_3, \eta_3)} f_1^* f_3^*(\rho_3, \gamma_3) \xrightarrow{(h_\tau, \eta_\tau)} f_2^* f_4^*(\rho_3, \gamma_3)$$

for (h_τ, η_τ) the value of the natural transformation $\tilde{\tau} \colon (f_3 \circ f_1)^* \Rightarrow (f_4 \circ f_2)^*$ on (ρ_3, γ_3), see Proposition 4.3. The 2-isomorphism filling (C.3) square is $(\tau, \omega := 1_A)$. To construct the remaining square whose lower horizontal arrow is the leftmost arrow in (C.2), consider

$$\begin{array}{ccc} & & (f_1^*(\rho_2, \gamma_2))_V \\ & & \big\downarrow{\scriptstyle (f_2, h, \eta)} \\ (\rho_1, \gamma_1)_X & \xrightarrow[(1, h_4, \eta_4)]{} & (f_4^*(\rho_3, \gamma_3))_X . \end{array}$$

As $(1, h_4, \eta_4)$ admits an inverse, consider the composition

$$f_1^*(\rho_2, \gamma_2) \xrightarrow{(f_2, h, \eta)} f_4^*(\rho_3, \gamma_3) \xrightarrow{(1, h_4, \eta_4)^{-1}} (\rho_1, \gamma_1),$$

which admits a factoring as in Lemma 4.5

$$\begin{array}{ccc} f_1^*(\rho_2, \gamma_2) & \longrightarrow & (\rho_1, \gamma_1). \\ & {\scriptstyle (1, h_1, \eta_1)^{-1}} \searrow \quad \nearrow {\scriptstyle (f_2, 1, 1)} & \\ & f_2^*(\rho_1, \gamma_1) & \end{array}$$

Taking $(\rho_0, \gamma_0)_V := (f_2^*(\rho_1, \gamma_1))_V$, we obtain

$$\begin{array}{ccc} (\rho_0, \gamma_0)_V & \xrightarrow{(1,h_1,\eta_1)} & (f_1^*(\rho_2, \gamma_2))_V \\ {\scriptstyle (f_2,1,1)}\downarrow & & \downarrow{\scriptstyle (f_2,h,\eta)} \\ (\rho_1, \gamma_1)_X & \xrightarrow[(1,h_4,\eta_4)]{} & (f_4^*(\rho_3, \gamma_3))_X \end{array}$$

where the 2-commutativity data comes from the uniquely specified 2-morphism $(1, h_1, \eta_1) \simeq ((1, h_1, \eta_1)^{-1})^{-1}$ determined by (2.3). Composing these two squares gives the desired data for the square in BF3 of [**37**]. $\square$

LEMMA C.3. *Suppose the data in the diagram below (in $\mathsf{Bun}_G^{\mathrm{pre}}$)*

$$(\rho_1, \gamma_1)_X \underset{(f_2,h_2,\eta_2)}{\overset{(f_1,h_1,\eta_1)}{\rightrightarrows}} (\rho_2, \gamma_2)_Y \xrightarrow{(f_3,h_3,\eta_3)} (\rho_3, \gamma_3)_Z,$$

where (f_3, h_3, η_3) is in $\mathcal{W}$, and where there exists a 2-morphism $(\tau, \omega) : (f_3, h_3, \eta_3) \circ (f_1, h_1, \eta_1) \Rightarrow (f_3, h_3, \eta_3) \circ (f_2, h_2, \eta_2)$. Then there exists a 2-morphism $(\tilde{\tau}, \tilde{\omega}) : (f_1, h_1, \eta_1) \Rightarrow (f_2, h_2, \eta_2)$. This is the first part of BF4 of [**37**]*; alternatively it is condition 1-Frc of* [**48**]

PROOF. Note that τ gives a map $\tau : X_0 \to Z_1$. Because f_3 is an essential equivalence, see (3.1), we have the following pullback diagram

$$\begin{array}{ccccc} X_0 & & \xrightarrow{\quad\tau\quad} & & \\ & \overset{\tilde{\tau}}{\searrow} & & & \\ & & Y_1 & \longrightarrow & Z_1 \\ & {\scriptstyle f_1\times f_2}\searrow & \downarrow & & \downarrow \\ & & Y_0 \times Y_0 & \longrightarrow & Z_0 \times Z_0. \end{array}$$

For the naturality conditions $\tilde{\tau}$ satisfies, see [**37**] Section 4.

Define $\tilde{\omega} := \omega : X_0 \to A$. To check that this satisfies the compatibility conditions needed to define the desired 2-morphism $(\tilde{\tau}, \tilde{\omega})$ (from (4.1)) note that the conditions to have the original 2-morphism (τ, ω) give that

$$\begin{aligned} h_3(f_2(x))h_2(x) &= \rho_3(\tau(x))h_3(f_1(x))h_1(x) \\ &= h_3(f_2(x))\rho_2(\tau(x))h_1(x) \end{aligned}$$

So $\rho_2(\tau(x))h_1(x) = h_2(x)$. Additionally,

$$\frac{\tilde{\omega}(x)}{\tilde{\omega}(x')} = \frac{\omega(x)}{\omega(x')} = \frac{\eta_{31}(\phi)}{\eta_{32}(\phi)} \frac{\gamma_3(\tau(x'), f_{31}(\phi))}{\gamma_3(f_{32}(\phi), \tau(x)}$$
$$\frac{\alpha(\rho_3(\tau(x')), \rho_3(f_{31}(\phi)), h_3(f_1(x))h_1(x))}{\alpha(\rho_3(\tau(x')), h_3(f_1(x'))h_1(x'), \rho_1(\phi))\alpha(\rho_3(f_{32}(\phi)), \rho_3(\tau(x)), h_3(f_1(x))h_1(x)}$$
$$= \frac{\eta_1(\phi)}{\eta_2(\phi)} \frac{\gamma_2(\tilde{\tau}(x'), f_1(\phi))}{\gamma_2(f_2(\phi), \tilde{\tau}(x))}$$
$$\frac{\alpha(\rho_2(\tilde{\tau}(x')), \rho_2(f_1(\phi)), h_1(x))}{\alpha(\rho_2(\tilde{\tau}(x')), h_1(x'), \rho_1(\phi))\alpha(\rho_2(f_2(\phi)), \rho_2(\tilde{\tau}(x)), h_1(x))}$$

for $\phi\colon x \to x'$ in X. This calculation involves repeated application of the cocycle condition for α and the conditions for $\rho_i, h_i, i = 1, 2, 3$. This shows that $\tilde{\omega} := \omega : X_0 \to A$ satisfies the necessarily compatibility conditions to be a 2-morphism between the desired 1-morphisms. $\square$

LEMMA C.4. *Given the data in the diagram below (in* $\mathsf{Bun}_{\mathcal{G}}^{\mathrm{pre}}$*),*

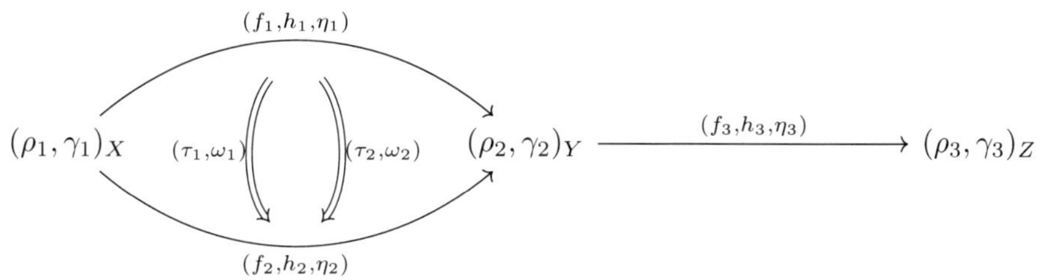

where f_3 *is an essential equivalence and* $(f_3, h_3, \eta_3)\star(\tau_1, \omega_1) = (f_3, h_3, \eta_3)\star(\tau_2, \omega_2)$*, then* $(\tau_1, \omega_1) = (\tau_2, \omega_2)$*. This is the second part of BF4 of* [**37**]*; alternatively it is condition 2-Frc of* [**48**].

PROOF. Because f_3 is an essential equivalence, we obtain the pullback square

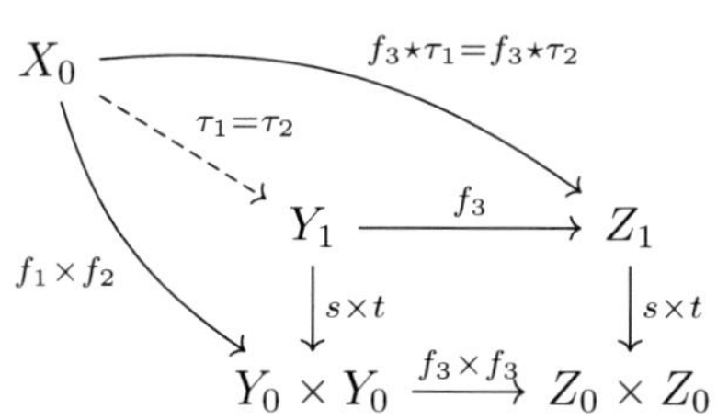

By the universal property, the dashed arrow from $X_0 \to Y_1$ is unique; both τ_1, τ_2 fill that arrow, so $\tau_1 = \tau_2 : X_0 \to Y_1$.

Let us denote $\mathrm{id}_{(f_3,h_3,\eta_3)}$ by (τ_3, ω_3), so $\tau_3 : Y_0 \to Z_1$ sends $y \mapsto \mathrm{id} : f_3(y) \to f_3(y)$, and $\omega_3 : Y_0 \to A$ is the constant map to 1_A. From the equation for horizontal composition given in (4.2) and using the fact that α is normalized,

$$\begin{aligned}\omega_3 \star \omega_1(x) =& \omega_3(f_1(x))\omega_1(x)\eta_3(\tau_1(x))\gamma_3(f_3(\tau_1(x)),\tau_2(f_1(x)))\\ &\frac{\alpha(\rho_3(f_3(\tau_1(x))),\rho_3(\tau_3(f_1(x))),h_2(f_1(x))h_1(x))}{\alpha(\rho_3(\tau_3(f_1(x))),h_2(f_1(x)),h_1(x))}\\ &\frac{\alpha(h_3(f_2(x)),\rho_2(\tau_1(x)),h_1(x))}{\alpha(\rho_3(f_3(\tau_1(x))),h_3(f_1(x)),h_1(x))}\\ =& \,\omega_1(x)\eta_3(\tau_1(x))\gamma_3(f_3(\tau_1(x)),\tau_1(f_1(x)))\\ &\frac{\alpha(\rho_3(f_3(\tau_1(x))),1_G,h_2(f_1(x))h_1(x))\alpha(h_3(f_2(x)),\rho'(\tau_1(x)),h_1(x))}{\alpha(1_G,h_2(f_1(x)),h_1(x))\alpha(\rho_3(f_3(\tau_1(x))),h_3(f_1(x)),h_1(x))}\\ =& \,\omega_1(x)\eta_3(\tau_1(x))\gamma_3(f_3(\tau_1(x)),\tau_3(f_1(x)))\\ &\frac{\alpha(h_3(f_2(x)),\rho_2(\tau_1(x)),h_1(x))}{\alpha(\rho_3(f_3(\tau_1(x))),h_3(f_1(x)),h_1(x))}\end{aligned}$$

$$\begin{aligned}\omega_3 \star \omega_2(x) =& \omega_3(f_1(x))\omega_2(x)\eta_3(\tau_2(x))\gamma_3(f_3(\tau_2(x)),\tau_3(f_1(x)))\\ &\frac{\alpha(\rho_3(f_3(\tau_2(x))),\rho_3(\tau_3(f_1(x))),h_2(f_1(x))h_1(x))}{\alpha(\rho_3(\tau_3(f_1(x))),h_2(f_1(x)),h_1(x))}\\ &\frac{\alpha(h_3(f_2(x)),\rho_2(\tau_2(x)),h_1(x))}{\alpha(\rho_3(f_3(\tau_2(x))),h_3(f_1(x)),h_1(x))}\\ =& \,\omega_2(x)\eta_3(\tau_2(x))\gamma_3(f_3(\tau_2(x)),\tau_3(f_1(x)))\\ &\frac{\alpha(h_3(f_2(x)),\rho_2(\tau_2(x)),h_1(x))}{\alpha(\rho_3(f_3(\tau_2(x))),h_3(f_1(x)),h_1(x))}\end{aligned}$$

Since $\omega_3 \star \omega_1 = \omega_3 \star \omega_2$ and $\tau_1 = \tau_2$, the remaining terms in the above equations cancel out, giving $\omega_1 = \omega_2 : X_0 \to A$. □

PROOF OF THEOREM 4.7. Note that the composition of 1-morphisms in $\mathcal{W}$ is again in $\mathcal{W}$; this is condition BF2 of [**37**]. If there is a 2-isomorphism between two 1-morphisms where one of the 1-morphisms in is $\mathcal{W}$, then the other 1-morphism is also in $\mathcal{W}$; this is condition BF5 of [**37**]. Both these conditions are satisfied trivially. Together with lemmas C.1, C.2, C.3, and C.4, this completes the proof. □

References

[1] N. A. Baas, B. I. Dundas, and J. Rognes, *Two-vector bundles and forms of elliptic cohomology*, Topology, geometry and quantum field theory, London Math. Soc. Lecture Note Ser., vol. 308, Cambridge Univ. Press, Cambridge, 2004, pp. 18–45, DOI 10.1017/CBO9780511526398.005. MR2079370

[2] J. C. Baez and A. D. Lauda, *Higher-dimensional algebra. V. 2-groups*, Theory Appl. Categ. **12** (2004), 423–491. MR2068521

[3] J. C. Baez and U. Schreiber, *Higher gauge theory*, Categories in algebra, geometry and mathematical physics, Contemp. Math., vol. 431, Amer. Math. Soc., Providence, RI, 2007, pp. 7–30, DOI 10.1090/conm/431/08264. MR2342821

[4] J. C. Baez and D. Stevenson, *The classifying space of a topological 2-group*, Algebraic topology, Abel Symp., vol. 4, Springer, Berlin, 2009, pp. 1–31, DOI 10.1007/978-3-642-01200-6_1. MR2597732

[5] J. C. Baez, D. Stevenson, A. S. Crans, and U. Schreiber, *From loop groups to 2-groups*, Homology Homotopy Appl. **9** (2007), no. 2, 101–135. MR2366945

[6] T. K. Bartels, *Higher gauge theory: 2-bundles*, ProQuest LLC, Ann Arbor, MI, 2006. Thesis (Ph.D.)–University of California, Riverside. MR2709030

[7] K. Behrend and P. Xu, *Differentiable stacks and gerbes*, J. Symplectic Geom. **9** (2011), no. 3, 285–341. MR2817778

[8] L. Breen, *On the classification of* 2*-gerbes and* 2*-stacks* (English, with English and French summaries), Astérisque **225** (1994), 160. MR1301844

[9] S. Bunk, *Principal* ∞*-bundles and smooth string group models*, Hamburger Beitraege Nr. **858** (2020).

[10] U. Bunke, *String structures and trivialisations of a Pfaffian line bundle*, Comm. Math. Phys. **307** (2011), no. 3, 675–712, DOI 10.1007/s00220-011-1348-0. MR2842963

[11] A. L. Carey, M. K. Murray, and B. L. Wang, *Higher bundle gerbes and cohomology classes in gauge theories*, J. Geom. Phys. **21** (1997), no. 2, 183–197, DOI 10.1016/S0393-0440(96)00014-9. MR1427865

[12] K. Costello, *A geometric construction of the Witten genus, I*, Proceedings of the International Congress of Mathematicians. Volume II, Hindustan Book Agency, New Delhi, 2010, pp. 942–959. MR2827826

[13] K. Costello, *A geometric construction of the Witten genus, I*, Proceedings of the International Congress of Mathematicians. Volume II, Hindustan Book Agency, New Delhi, 2010, pp. 942–959. MR2827826

[14] K. Dekimpe, M. Sadowski, and A. Szczepański, *Spin structures on flat manifolds*, Monatsh. Math. **148** (2006), no. 4, 283–296, DOI 10.1007/s00605-005-0367-3. MR2234081

[15] C. L. Douglas and A. G. Henriques, *Topological modular forms and conformal nets*, Mathematical foundations of quantum field theory and perturbative string theory, Proc. Sympos. Pure Math., vol. 83, Amer. Math. Soc., Providence, RI, 2011, pp. 341–354, DOI 10.1090/pspum/083/2742433. MR2742433

[16] D. Freed, *Lectures on twisted K-theory and orientifolds*, lectures at ESI, Vienna, 2012.

[17] P. Gajer, *Geometry of Deligne cohomology*, Invent. Math. **127** (1997), no. 1, 155–207, DOI 10.1007/s002220050118. MR1423029

[18] A. Henriques, *Integrating* L_∞*-algebras*, Compos. Math. **144** (2008), no. 4, 1017–1045, DOI 10.1112/S0010437X07003405. MR2441255

[19] P. Hu and I. Kriz, *Conformal field theory and elliptic cohomology*, Adv. Math. **189** (2004), no. 2, 325–412, DOI 10.1016/j.aim.2003.11.012. MR2101224

[20] N. Johnson and D. Yau, *2-dimensional categories*, Oxford University Press, Oxford, 2021, DOI 10.1093/oso/9780198871378.001.0001. MR4261588

[21] T. P. Killingback, *World-sheet anomalies and loop geometry*, Nuclear Phys. B **288** (1987), no. 3-4, 578–588, DOI 10.1016/0550-3213(87)90229-X. MR892061

[22] E. Lerman, *Orbifolds as stacks?*, Enseign. Math. (2) **56** (2010), no. 3-4, 315–363, DOI 10.4171/LEM/56-3-4. MR2778793

[23] J. Lurie, *Higher topos theory*, Annals of Mathematics Studies, vol. 170, Princeton University Press, Princeton, NJ, 2009, DOI 10.1515/9781400830558. MR2522659

[24] J. Lurie, *A survey of elliptic cohomology*, Algebraic topology, Abel Symp., vol. 4, Springer, Berlin, 2009, pp. 219–277, DOI 10.1007/978-3-642-01200-6_9. MR2597740

[25] R. Lutowski and B. Putrycz, *Spin structures on flat manifolds*, J. Algebra **436** (2015), 277–291, DOI 10.1016/j.jalgebra.2015.03.037. MR3348474

[26] K. C. H. Mackenzie, *General theory of Lie groupoids and Lie algebroids*, London Mathematical Society Lecture Note Series, vol. 213, Cambridge University Press, Cambridge, 2005, DOI 10.1017/CBO9781107325883. MR2157566

[27] I. Moerdijk, *The classifying topos of a continuous groupoid. I*, Trans. Amer. Math. Soc. **310** (1988), no. 2, 629–668, DOI 10.2307/2000984. MR973173

[28] I. Moerdijk, *Orbifolds as groupoids: an introduction*, `arXiv:math/0203100`, 2002.

[29] M. K. Murray, *Bundle gerbes*, J. London Math. Soc. (2) **54** (1996), no. 2, 403–416, DOI 10.1093/acprof:oso/9780199534920.003.0012. MR1405064

[30] T. Nikolaus, C. Sachse, and C. Wockel, *A smooth model for the string group*, Int. Math. Res. Not. IMRN **16** (2013), 3678–3721, DOI 10.1093/imrn/rns154. MR3090706

[31] T. Nikolaus, U. Schreiber, and D. Stevenson, *Principal ∞-bundles: General theory*, J. Homotopy Relat. Struct. **10** (2015), no. 4, 749–801, DOI 10.1007/s40062-014-0083-6. MR3423073

[32] T. Nikolaus, U. Schreiber, and D. Stevenson, *Principal ∞-bundles: Presentations*, J. Homotopy Relat. Struct. **10** (2015), no. 3, 565–622, DOI 10.1007/s40062-014-0077-4. MR3385700

[33] T. Nikolaus and C. Schweigert, *Equivariance in higher geometry*, Adv. Math. **226** (2011), no. 4, 3367–3408, DOI 10.1016/j.aim.2010.10.016. MR2764891

[34] T. Nikolaus and K. Waldorf, *Four equivalent versions of nonabelian gerbes*, Pacific J. Math. **264** (2013), no. 2, 355–419, DOI 10.2140/pjm.2013.264.355. MR3089401

[35] J. Nuiten, *Higher stacks as a category of fractions*, preprint, 2016.

[36] F. Pfäffle, *The Dirac spectrum of Bieberbach manifolds*, J. Geom. Phys. **35** (2000), no. 4, 367–385, DOI 10.1016/S0393-0440(00)00005-X. MR1780760

[37] D. A. Pronk, *Etendues and stacks as bicategories of fractions*, Compositio Math. **102** (1996), no. 3, 243–303. MR1401424

[38] C. J. Schommer-Pries, *Central extensions of smooth 2-groups and a finite-dimensional string 2-group*, Geom. Topol. **15** (2011), no. 2, 609–676, DOI 10.2140/gt.2011.15.609. MR2800361

[39] U. Schreiber and K. Waldorf, *Smooth functors vs. differential forms*, Homology Homotopy Appl. **13** (2011), no. 1, 143–203, DOI 10.4310/HHA.2011.v13.n1.a7. MR2803871

[40] G. Segal, *Elliptic cohomology*, Séminaire N. Bourbaki **695** (1988).

[41] G. Segal, *The definition of conformal field theory*, Topology, geometry and quantum field theory, London Math. Soc. Lecture Note Ser., vol. 308, Cambridge Univ. Press, Cambridge, 2004, pp. 421–577. MR2079383

[42] E. Sharpe, *Discrete torsion in perturbative heterotic string theory*, Phys. Rev. D (3) **68** (2003), no. 12, 126005, 7, DOI 10.1103/PhysRevD.68.126005. MR2071753

[43] H. Sinh, *Gr-categories*, Université Paris VII, doctoral thesis, 1975.

[44] D. Stevenson, *Bundle 2-gerbes*, Proc. London Math. Soc. (3) **88** (2004), no. 2, 405–435, DOI 10.1112/S0024611503014357. MR2032513

[45] S. Stolz, *A conjecture concerning positive Ricci curvature and the Witten genus*, Math. Ann. **304** (1996), no. 4, 785–800, DOI 10.1007/BF01446319. MR1380455

[46] S. Stolz and P. Teichner, *What is an elliptic object?*, Topology, geometry and quantum field theory, London Math. Soc. Lecture Note Ser., vol. 308, Cambridge Univ. Press, Cambridge, 2004, pp. 247–343, DOI 10.1017/CBO9780511526398.013. MR2079378

[47] S. Stolz and P. Teichner, *Supersymmetric field theories and generalized cohomology*, Mathematical foundations of quantum field theory and perturbative string theory, Proc. Sympos. Pure Math., vol. 83, Amer. Math. Soc., Providence, RI, 2011, pp. 279–340, DOI 10.1090/pspum/083/2742432. MR2742432

[48] P. Bustillo Vazquez, D. Pronk, and M. Szyld, *The three F's for bicategories I: Localization by fractions is exact*, 2021.

[49] K. Waldorf, *A construction of string 2-group models using a transgression-regression technique*, Analysis, geometry and quantum field theory, Contemp. Math., vol. 584, Amer. Math. Soc., Providence, RI, 2012, pp. 99–115, DOI 10.1090/conm/584/11588. MR3013040

[50] K. Waldorf, *String connections and Chern–Simons theory*, Trans. Amer. Math. Soc. **365** (2013).

[51] K. Waldorf, *Spin structures on loop spaces that characterize string manifolds*, Algebr. Geom. Topol. **16** (2016).

[52] J. H. C. Whitehead, *On adding relations to homotopy groups*, Ann. of Math. (2) **42** (1941), 409–428, DOI 10.2307/1968907. MR4123

[53] J.H.C. Whitehead, *Combinatorial homotopy. II*, Bull. Amer. Math. Soc. (1949).

[54] E. Witten, *The index of the Dirac operator in loop space*, Elliptic curves and modular forms in algebraic topology (Princeton, NJ, 1986), Lecture Notes in Math., vol. 1326, Springer, Berlin, 1988, pp. 161–181, DOI 10.1007/BFb0078045. MR970288

[55] C. Wockel, *Principal 2-bundles and their gauge 2-groups*, Forum Math. **23** (2011), no. 3, 565–610, DOI 10.1515/FORM.2011.020. MR2805195

Department of Mathematics, University of Illinois Urbana-Champaign, Champaign, Illinois 61820

Department of Mathematics, Université de Sherbrooke, Sherbrooke, Quebec J1K 2R1, Canada

Department of Mathematics & Computer Science, Providence College, Providence, Rhode Island 02918

Department of Mathematics, Northwestern University, Evanston, Illinois 60208

Department of Mathematics & Statistics, University of New Hampshire, Durham, New Hampshire 03824

Contemporary Mathematics
Volume **813**, 2025
https://doi.org/10.1090/conm/813/16292

Categorical center of higher genera

Jin-Cheng Guu

Abstract. The Witten-Reshetikhin-Turaev (WRT) 3D TQFT is a comprehensive theory that encompasses various quantum invariants of knots and 3-manifolds, such as the Jones polynomials. Despite its wide-ranging scope, the WRT theory is a subset of the Crane-Yetter (CY) 4D TQFT, which is a more extensive theory that can be extended to (co)dimension 2 (as the Karoubi completion of the skein category) and is conjectured to be extensible to dimension 0.

This study is focused on computing the Crane-Yetter theory in (co)dimension 2 (or the Karoubi completion of the skein category). Specifically, we aim to calculate the values of this theory for all smooth and oriented surfaces with at least one puncture. These computed values share similarities with and broaden the scope of the Drinfeld categorical center, which is present in the context of quantum group construction, higher categories, and factorization homologies. These values are thus referred to as the categorical center of higher genera.

Contents

1. Introduction
2. Preliminaries
3. Topological theory
4. Algebraic theory
5. Proof of the main theorem

1. Introduction

1.1. TQFTs as higher invariants. In essence, a TQFT assigns a vector space to a manifold, much like homology. However, it also assigns a linear transformation to the manifold of one higher dimension that 'bridges' the lower manifolds in a compatible manner. Interestingly, this results in a number-valued invariant for closed manifolds, since each closed manifold 'bridges' the empty manifolds, which must be assigned the trivial vector space.

This generalization from numbers to vector spaces does not stop here. By viewing vector spaces as objects in a category, one can bring this process to assigning objects in higher categories. An instance of such is called an extended topological quantum field theory (extended TQFT) [**Lur09**].

2020 *Mathematics Subject Classification.* Primary.

EXAMPLE 1.1 (Witten-Reshetikhin-Turaev model). One-dimensional and two-dimensional manifolds have been completely classified. While a presentation for three-dimensional manifolds exists through the use of knots and surgery theory (see [**Lic12**, chap.12] and [**Kir78**]), this classification is dependent on the classification of knots, which is a complicated and ongoing research field. Other presentations suffer from the same issue. As a result,invariants of other types for 3-manifolds are of great interest.

In the 1980s, a new kind of invariant emerged from the intersection of mathematics and physics, starting with Vaughan Jones' groundbreaking Jones polynomials. These invariants can be constructed using a variety of mathematical structures such as von Neumann algebras, quantum groups, and rational conformal theories, as described in [**CY93**, sec.1]. It is natural to wonder if these different types of invariants can be unified, and indeed they can. They all turn out to be special cases of the Witten-Reshetikhin-Turaev (WRT) model, which takes as its input algebraic data a modular tensor category, as discussed in [**BDSPV15**].

As an extended TQFT, WRT model assigns a number to each closed 3-manifold, a vector space to each closed 2-manifold, and a linear category to each closed 1-manifold. However, it does not go on and assign a linear 2-category in the 0-th dimension. See 1.5 for more discussion. ◇

EXAMPLE 1.2 (Turaev-Viro model). The Turaev-Viro (TV) model was originally introduced as an invariant for 3-manifolds using state sum construction [**TV92**]. It was then realized that the TV model provides a 3-dimensional TQFT.

Later, it was established in [**KB10**] that the TV model is actually a fully extended TQFT. This means that it assigns a number to each closed 3-manifold, a vector space to each closed 2-manifold, a linear category to each closed 1-manifold, and a linear 2-category to each 0-manifold (i.e., the point) in a compatible manner. It was also demonstrated that

$$TV_M(C) \simeq WRT_M(Z(C))$$

where M is a closed manifold of dimension 2 or 3, C is a spherical category and $Z(C)$ is its Drinfeld categorical center ([**Bal10**] and [**Bal11**]). ◇

EXAMPLE 1.3 (Crane-Yetter model). As a 4-dimensional analogue of the TV model, the Crane-Yetter (CY) model has premodular categories as its input algebraic data. It is the main topic of this paper, and will be treated in its own section 1.2. ◇

THEOREM 1.4 ($\partial CY = WRT$). *The WRT model is the boundary theory of CY, while the input algebraic data is a modular category. More precisely, for a modular category C and a 4-dimensional manifold W possibly with boundary,*

$$CY_W(C) = \kappa^{\sigma(W)} WRT_{\partial W}(C),$$

where σ denotes the signature and κ denotes a constant based on the input algebraic data (the modular category given in 2.13*)* [**BGIM07**, section 5.2.2]. *This extends to manifolds with colored graphs* [**BGIM07**, Theorem 2].

For a 3-dimensional manifold M possibly with boundary, the vector spaces $CY_M(C)$ and $WRT_{\partial M}(C)$ are widely believed to be equivalent. However, to our best knowledge a rigorous proof has not been provided. For 2-dimensional manifolds Σ with boundary, see 1.14. ◇

REMARK 1.5. In 1.1 we mentioned that the WRT model does not always extend to the 0-th dimension. There are two explanations for this fact.

(1) From the Turaev-Viro model perspective [**Bal10**], for WRT to extend to a point and to coincide with the Turaev-Viro theory, one needs the input modular category C to be the Drinfeld center of a spherical fusion category D, which is not always the case.
(2) From the Crane-Yetter (CY) model perspective, the WRT model is a boundary theory of CY (1.4). However, a single point is not the boundary of any 1-manifold: one needs at least two points.

From the second point of view of 1.1, one sees that the WRT model, though successful and fruitful, is expected to be a part of larger theory - the Crane-Yetter model, which is the main TQFT we will focus on in this work. ◇

1.2. The Crane-Yetter TQFT. Originated in [**CY93**], the Crane-Yetter model was first defined as a state-sum. In particular, for a triangulated 4-manifold W and a modular tensor category C, one define

$$CY_W(C) = \Sigma_c D^{(n_0-n_1)}\Pi_\sigma \dim c(\sigma)\Pi_t \dim c(t)\Pi_\xi \underline{15j}(c,\xi),$$

where c runs through all "colorings", n_i is the number of simplices of dimension i in the triangulation, σ runs through all triangles, t runs through all tetrahedra, ξ runs through all 4-simplices, and $\underline{15j}$ denotes the so called $15j$-symbols. The upshot is that the sum is independent of the choice of triangularization, therefore defines an invariant of 4-manifolds. This holds even when the C is a premodular category.

Later in [**CKY97**], it was known that this seemingly complicated sum can be expressed in terms of the Euler characteristic and the signature, both being old and well-known topological invariants. While this provides a combinatorial formula for the signature of 4-folds, it also means that the CY model with the input data being a modular tensor category C somehow trivializes. A reason why it trivializes is that the WRT model is the boundary theory of the CY model (1.4). Indeed, the input algebraic data for 3-manifolds (modular tensor categories) are too 'ideal' for 4-manifolds. Other types of algebraic data should be considered. Premodular categories are such examples, on which we focus in this paper.

On the other hand, the CY model is expected to be a fully extended TQFT. In particular, it gives a number to each closed 4-manifold, a vector space to each closed 3-manifold, a linear category to each closed 2-manifold (or a 2-manifold with boundary but with empty boundary condition) (see [**KT22**] and [**Tha21**] where it is defined as the Karoubi completion of the skein category), and conjecturally a linear 2-category to each closed 1-manifold (the circle), and conjecturally a linear 3-category to each closed 0-manifold (the point). To the best knowledge of the author, however, please note that the Crane-Yetter theory has not been written down as a k-extended TQFT in the sense of Lurie (i.e. as an $(\infty, k+1)$ functor), even for $k = 1$. Therefore, that the Crane-Yetter theory in (co)dimension 2 should be the Karoubi completion of the skein theory is also half-conjectural. In this work, however, we work with this construction anyways.

1.3. Previous work of CY in (co)dimension two. Recent developments of the CY model include its Hamiltonian formulation and its higher-codimensional aspect. The former is called the Walker-Wang model (cf [**WW12**] and [**Wal15**]).

However, while the relation between CY and the Walker-Wang model are commonly believed, to the best knowledge of the author, it has not been explicitly proved.

The later, on the other hand, is currently studied by A. Kirillov's school starting around 2018. In particular, Tham and Kirillov defined CY model in (co)dimension 2 using skein categories (first defined in [**Wal06**] and [**JF15**]), computed some examples, and proved the excision property. This in turn showed that the CY model coincides with the factorization homology in the special case that the input category is premodular.

REMARK 1.6. In a recent paper [**KT22**], it is stated in Corollary 7.6 that the Crane-Yetter theory in (co)dimension 2 coincides with the factorization homology given in [**BZBJ18**] for certain algebraic input data. Therefore, we expect that both approaches actually study the same theory. Please note that [**BZBJ18**] wrote their work earlier than [**KT22**]. However, we will still quote the results from [**KT22**] as the author is more familiar with it. ◇

In the rest of this section, we recall the results from Kirillov and Tham. The main statement of this paper will follow in the next section.

Σ	Disk	Cylinder	Sphere	1-punctured torus	General
$CY_\Sigma(C)$	C	$Z(C)$	$Mu(C)$	$Z^{el}(C)$	$Z_\sigma(C)$
		Drinfeld center	Muger center	Elliptic center	Categorical center of higher genera

THEOREM 1.7 ([**KT22**, Section 5]). *Let C be a premodular category, and $\Sigma = D^2$ be the open disk. Then as abelian categories*

$$CY_\Sigma(C) \simeq C.$$

◇

THEOREM 1.8 (Drinfeld center). [**KT22**, Example 8.2] *Let C be a premodular category, $Z(C)$ its Drinfeld categorical center, and $\Sigma = S^1 \times I = S^1 \times (0,1)$ be the cylinder. Then as abelian categories*

$$CY_\Sigma(C) \simeq Z(C).$$

*Moreover, as multifusion categories, the topological nature of Σ induces a so called reduced tensor product $\overline{\otimes}$ ([**Tha22**], [**Was19**]) for $Z(C)$. Indeed, stacking two cylinders together produces another cylinder $S^1 \times (0,2) \simeq S^1 \times I$.* ◇

Notice that the reduced tensor product is in general different from the usual tensor product of the Drinfeld center.

THEOREM 1.9 (Excision principle). [**Coo23**][**KT22**, Theorem 2.3]

Denote the Deligne tensor product by $\boxtimes$. Let Σ be an open surface with n punctures. Then the category $CY_\Sigma(C)$ has a structure of module category over $CY_\Sigma(S^1 \times I)^{\boxtimes n}$, which is $(Z(C), \overline{\otimes})^{\boxtimes n}$ by 1.8.

Let Σ_1 and Σ_2 be smooth oriented surfaces possibly with punctures. And let CY be Crane-Yetter theory in dimension two. Then we have an equivalence of abelian categories.

$$CY_{(\Sigma_1 \cup \Sigma_2)} \simeq CY_{\Sigma_1} \boxtimes_{CY_{(\Sigma_1 \cap \Sigma_2)}} CY_{\Sigma_2},$$

where $\boxtimes_D$ denotes the balanced (Deligne) tensor product over D [**DSPS19**]. ◇

REMARK 1.10. For a related result on the excision principle, see [**Coo23**] and [**BZBJ18**]. In particular, we mentioned above that the CY model is a special case of the factorization homology. In [**Coo23**], it is further shown that a factorization homology coincides with the skein category for ribbon categories, which are more general than premodular categories; it is also shown that the ribbon categories satisfy the excision principle, providing a more general result than the theorem above. Notice that in [**BJS21**], a fully-extended 4-dimensional TQFT (based on a braided fusion category) is constructed, and is expected to be an extension of the CY model. Please see discussions in section 1.5.2 and 1.5.3 therein. ◇

THEOREM 1.11 (Muger centralizer). [**KT22**, Corollary 8.5] *Let C be a premodular category, $Z'(C)$ its Muger centralizer, and $\Sigma = S^2$ be the 2-sphere. Then as abelian categories*

$$CY_\Sigma(C) \simeq Z'(C).$$

In particular, if C is modular, then the result trivializes as in

$$CY_\Sigma(C) \simeq (Vect).$$

◇

Tham defined for a premodular category C an associated category $Z^{el}(C)$, coined the elliptic Drinfeld center. Its objects are the triples (X, γ_1, γ_2), where X is an object of C and the γ_i's are half-braidings of X that satisfy certain relations. See [**Tha19**] for a full definition. The name is justified by the following theorem.

THEOREM 1.12 (elliptic Drinfeld center). [**KT22**, Corollary 9.5+6] *Let C be a premodular category, $Z^{el}(C)$ its elliptic Drinfeld center, and $\Sigma = \Sigma_{1,0}$ be a once-punctured torus. Then as abelian categories*

$$CY_\Sigma(C) \simeq Z^{el}(C).$$

In particular, if C is modular, then there is an equivalence of left $CY_{S^1\times I}(C)$-modules

$$C \simeq CY_{D^2}(C) \simeq CY_\Sigma(C).$$

◇

THEOREM 1.13 ([**Tha22**, Corollary 4.5]). *Let C be a premodular category, Z the Drinfeld center construction, $\overline{\otimes}$ the stacking tensor product, and $\Sigma = S^1 \times S^1$ be the standard torus. Then*

$$CY_\Sigma(C) \simeq Z((Z(C), \overline{\otimes})).$$

◇

One of the main theorem of [**KT22**] is that Crane-Yetter theorem in dimension two also trivializes when the input data is modular.

THEOREM 1.14 ([**KT22**, Remark 9.8]). *Let C be a modular category, and Σ an n-punctured surface of genus g. Then up to equivalence $CY_\Sigma(C)$ is independent of g. In fact, we have an equivalence of module categories over $(Z(C), \overline{\otimes})^{\boxtimes n}$*

$$CY_\Sigma(C) \simeq C^{\boxtimes n}.$$

Notice that when $n = 0$ the power is the category of finite dimensional vector spaces.
◇

An easy proof of this fact due to the author of this paper uses the excision principle and a basic equivalence of braided categories $C \boxtimes C^{bop} \simeq Z(C)$ [**EGNO15**], where C^{bop} denotes the same braided category as C but with the inverse braiding. Note that this supports the conjecture that the WRT model is the boundary theory of the CY model.

1.4. Main result: $CY \simeq Z$**.** The main result of this paper is an explicit calculation of the Crane-Yetter theory for all smooth oriented surfaces with at least one puncture. In particular, we obtain a description of the value on the 3-punctured 2-sphere which was missing in the literature. With the excision principle, this provides us several combinatorial constructions of the same category. Also with the excision principle, this provides us a fairly compact description for the values on **all** smooth oriented surfaces. In this section, we overview the statement and its consequences, leaving a detailed proof to section 5.

Crane-Yetter theory in dimension two is defined as a linear category whose spaces of morphisms are vector spaces presented by many complicated generators and relations. By a calculation we mean to describe it as a category whose description is much smaller. An analogy of this is that the first homology of the circle "calculated" to be the group of integers

$$H_1(S^1) \simeq \mathbb{Z}.$$

Given any open surface Σ, we present it by an oriented 2-disk and some segments of its boundary glued. Which segments are glued together are described by some combinatorial data, called the admissible gluings. With a choice of an admissible gluing σ, the categorical center of higher genera $Z = Z_\sigma$ is a specific category explicitly constructed in 4.11, with some basic properties given in 4.3. Roughly, given a premodular category C and an admissible gluing σ of rank n (i.e. the combinatorial datum that encodes a surface from a disk with $2n$ legs) , the categorical center of higher genera $Z = Z_\sigma(C)$ is a category with objects of the form $(X, \gamma_1, \dots, \gamma_n)$ where X is a C-object and the γ_i's are half-braidings satisfying specific relations. The main statement is that Z is equivalent to CY_Σ as a finite semisimple abelian category, and the equivalence is explicitly given.

THEOREM 1.15 (Main statement). *Let n be a nonnegative integer, σ an admissible gluing of rank n, Σ_σ the surface constructed from σ, C a premodular category, $Z_\sigma(C)$ the categorical center of higher genera of C with respect to σ, and $CY_{\Sigma_\sigma}(C)$ the Crane-Yetter theory (over C) of the surface Σ_σ.*

Then we have an (explicit) equivalence of finite semisimple abelian categories

$$CY_{\Sigma_\sigma}(C) \simeq Z_\sigma(C).$$

◇

Thus up to equivalence $Z_\sigma(C)$ only depends on C and the oriented topological type of the surface Σ_σ. A detailed proof of the main statement is given in section 5.

EXAMPLE 1.16 ($n = 0$). For $n = 0$, the surface Σ is the open disk, the categorical center of higher genera reduces to the underlying premodular category C, so the theorem recovers that $CY_\Sigma(C) \simeq C$ as shown in 1.7. ◇

EXAMPLE 1.17 ($n = 1$). For $n = 1$, the only possible surface is the cylinder, the categorical center of higher genera reduces to the Drinfeld center $Z(C)$. Hence, the theorem recovers that $CY_\Sigma(C) \simeq Z(C)$ as shown in 1.8. ◇

EXAMPLE 1.18 ($n = 2$). For $n = 2$, there are two possible surfaces: the 1-punctured torus and the 3-punctured disk. In the former case, the categorical center of higher genera reduces to the elliptic center $Z^{el}(C)$, so the theorem recovers that $CY_\Sigma(C) \simeq Z^{el}(C)$ as shown in 1.12. In the later case, the theorem provides a new result. ◇

REMARK 1.19. In $H_1(S^1) \simeq \mathbb{Z}$, one sees the algebra of the shape S^1 and the shape of the algebra $\mathbb{Z}$. Our main result should be viewed as a higher analogue. That is, one sees the (higher) algebra of the shape Σ_σ and the shape of the (higher) algebra Z_σ. ◇

REMARK 1.20. These categories have their tensor structures and module categorical structures [**EGNO15**] coming from their topological nature. This will be treated in the author's following work. ◇

REMARK 1.21. The smoothness condition is not necessary for our theory, but is included for the sake of simplicity. Indeed, later we will see that the Crane-Yetter theory in dimension 2 can be defined based on stringnets. With the smooth structure, it is easier to regulate how they meet each other. On the other hand, Crane-Yetter theory works also in the PL-setting, parallel to its 3-dimensional analogue, the Turaev-Viro theory. Curious readers are refer to a setup given in [**KB10**]. ◇

2. Preliminaries

2.1. Premodular categories. Experts familiar with modular categories can think of the premodular categories as those without the nondegenerate condition on its S-matrix. However, note that there are various conventions. Our conventions directly assumes a premodular category to be semisimple finite abelian. A complete and recommended source is [**EGNO15**]. Other useful sources are [**BK02**], [**Kas12**], [**Tur10**]. Examples of premodular categories can be found in section 2.2.

DEFINITION 2.1 (Braided Fusion Category & Premodular Category). A *braided fusion category* is a braided category whose underlying monoidal category is a fusion category. A *premodular category* is a braided fusion category equipped with a spherical structure. ◇

DEFINITION 2.2 (Muger center). Given a braided fusion category C with braided structure $c_{-,\star}$, we say an object X in C is *transparent* (and otherwise *opaque*) if

$$c_{-,X} \circ c_{X,-} = id_{X\otimes -}.$$

We define the Muger center $Mu(C)$ of C to be the full tensor subcategory of C consisting of transparent objects. Note that in some other literature, the Muger center is also called a Muger centralizer or an E_2-center. ◇

Recall that if $c_{-,\star}$ is a braided structure of a braided fusion category C, then $c^{-1}_{\star,-}$ is also a braided structure for the underlying fusion category. This produces an opposite braided fusion category, which we denote by C^{bop}. Directly by the definition of a (Drinfeld) categorical center, there is a tautological functor from $C \boxtimes C^{bop}$ to $Z(C)$.

DEFINITION 2.3 (Tautological functor F). Given a braided fusion category C, there is a natural functor $C \boxtimes C^{bop} \xrightarrow{F} Z(C)$ that maps each object $X \boxtimes Y$ to $(X \otimes Y, (id_X \otimes c^{-1}_{Y,-}) \circ (c_{-,X} \otimes id_Y))$ and each morphism $(f \boxtimes g)$ to $(f \otimes g)$. ◇

DEFINITION 2.4 (Factorizable category). Given a braided fusion category C, if its tautological functor F is an equivalence of categories, we say that C is *factorizable*, and call any of its inverse functor a factorization of the Drinfeld center $Z(C)$. ◇

Notice that the structure of $Z(C)$ is in general opaque. For example, even the fusion ring of $Z(C)$ is hard to identify. Factorizability reduces the complexity of $Z(C)$ to that of C.

DEFINITION 2.5 (Complete set of simple objects). Let C be a premodular category. By a complete set of simple objects $O(C)$ we mean a set $O(C) = \{i, j, \ldots\}$ of simple objects in C that exhausts all simple types and that satisfies $(i \neq j) \Rightarrow (i \not\simeq j)$. Define its dual set to be

$$O(C)^\star = \{i^\star \mid i \in O(C)\},$$

where $i^\star$ denotes the (left) dual object of i. ◇

Notice that by the axiom of premodular category, any $O(C)$ is a finite set. From now on, we assume that any premodular category C comes with a fixed complete set of simple objects $O(C)$.

DEFINITION 2.6 (S-matrix). Let C be a premodular category with the braided structure c. The S-matrix of C is defined by

$$S := (s_{XY})_{X,Y \in O(C)}$$

where $s_{XY} = Tr(c_{Y,X} c_{X,Y}) \in \Bbbk$, where Tr denotes the (left) quantum trace that depends on the spherical structure of C. ◇

DEFINITION 2.7 (Modular Category). [**EGNO15**, 8.13.14] A modular category is a premodular category C whose S-matrix is non-degenerate. ◇

FACT 2.8 (Characterization of Modularity). [**EGNO15**, 8.20.12 and 8.19.3] The following conditions are equivalent for a premodular category C:

(1) C is modular.
(2) $Mu(C) \simeq (Vect)$
(3) C is factorizable.

◇

In particular, a premodular category C is modular if and only if its only transparent simple object is the trivial object. Moreover, the fact indicates that modularity and factorizability are equivalent for premodular categories. Thus modularity reduces the complexity of the structure of $Z(C)$. This is desirable for algebraists but not for topologists: The power of the topological quantum field theory in 4D is largely reduced by modularity exactly due to this fact.

2.2. Examples. In this section we provide examples for premodular categories.

EXAMPLE 2.9 (Finite group). Let G be a finite group. Then the category $Rep(G)$ of finite-dimensional linear representations of G over $\Bbbk$ has a natural structure of a premodular category. $\diamond$

EXAMPLE 2.10 (Drinfeld double). Let G be a finite group and $D(G)$ its Drinfeld double over $\Bbbk$. Then the category $Rep(D(G))$ of finite-dimensional linear representations of $D(G)$ over $\Bbbk$ has a natural structure of a premodular category. $\diamond$

EXAMPLE 2.11 (Crossed module). Let $\mathfrak{X}$ be a finite 2-group (or called a finite crossed-module) [**Ban10**]. Then the category $Rep(\mathfrak{X})$ of finite-dimensional linear representations of $\mathfrak{X}$ over $\Bbbk$ has a natural structure of a premodular category. $\diamond$

REMARK 2.12. Let G be a finite group. Both G and $D(G)$ can be viewed as special cases of finite crossed modules. Hence, 2.11 generalizes 2.9 and 2.10. $\diamond$

EXAMPLE 2.13 (Quantum group). In the case $\Bbbk = \mathbb{C}$, let $\mathfrak{g}$ be a semisimple Lie algebra and q a root of unity. The semisimplified category $Rep(U_q(\mathfrak{g}))$ of the category of finite-dimensional representations of the quantum group $U_q(\mathfrak{g})$ has a natural structure of a premodular category. For explicit constructions, see [**Kas12**] and [**BK02**]. $\diamond$

EXAMPLE 2.14 (Even part of the quantum $\mathfrak{sl}_2$). In the case $\Bbbk = \mathbb{C}$, let q be a root of unity. The semisimplified category $C = Rep(U_q(\mathfrak{sl}_2))$ of the category of finite-dimensional representations of the quantum group $U_q(\mathfrak{sl}_2)$ has a structure of a modular category. The even part C_0 of C has a structure of a premodular category [**KO01**]. $\diamond$

2.3. Graphical calculus. We will use the technique of graphical calculus ([**BK02**] and [**Kas12**]) while dealing with premodular categories.

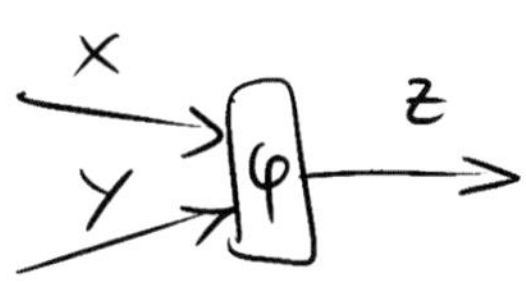

$X \otimes Y \xrightarrow{\phi} Z$

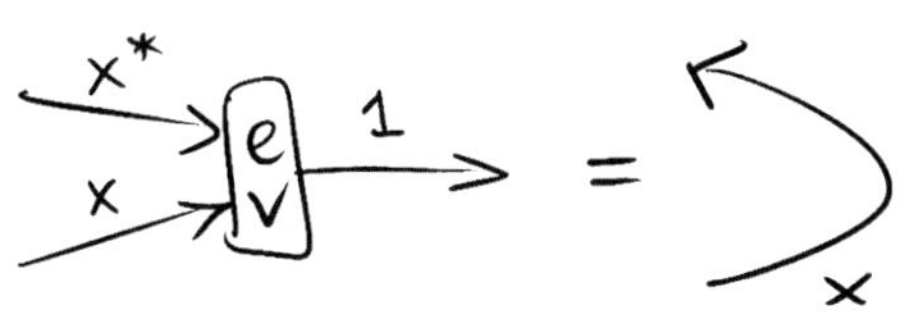

$X^\star \otimes X \xrightarrow{ev_X} 1$

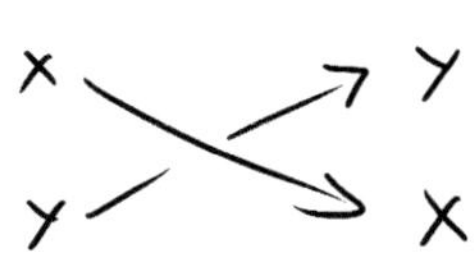

$X \otimes Y \xrightarrow{c_{X,Y}} Y \otimes X$

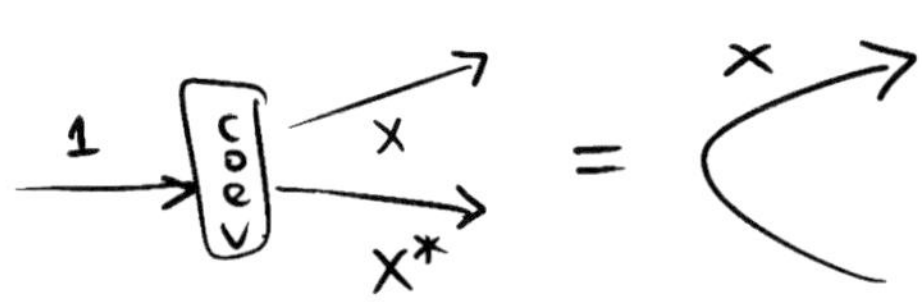

$1 \xrightarrow{coev_X} X \otimes X^\star$

An advantage of this is that many equalities among morphisms can be proved graphically, thanks to the work of Reshetikhin and Turaev [**Res**] [**BK02**, Theorem 2.3.10]. For example, to prove

$$eval_Y \circ c_{X,Y} \circ c_{X,Y^\star} \circ c_{X,Y} \circ coev_Y = c_{X,Y},$$

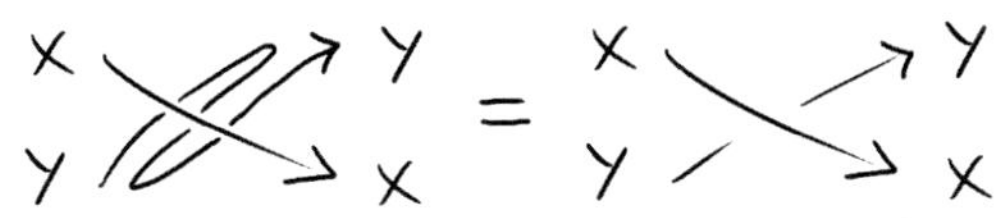

it suffices to establish an isotopy of ribbon tangles (a trivial task), and translate the procedure back into the algebraic equations. Such feature of graphical calculus provides sophisticated quantum link invariants (e.g. Jones polynomials of all Lie types and beyond).

In the rest of the section, we provide some useful lemmas and notations for graphical calculus.

LEMMA 2.15. *Let C be a premodular category with spherical structure a. Let X, Y be C-objects. Define a pairing of $\Bbbk$-linear spaces*

$$Hom_C(X,Y) \otimes Hom_C(Y,X) \xrightarrow{(,)} \Bbbk$$

that sends $\phi \otimes \psi$ to

$$Tr(\psi \circ \phi) = eval_X \circ ((a_X \circ \psi \circ \phi) \otimes 1_{X^\star}) \circ coev_X \in End_C(\mathbb{1}) \simeq \Bbbk.$$

Then the pairing is nondegenerate by the semisimplicity of C, identifying the linear space with its linear dual $Hom_C(Y,X) \simeq Hom_C(X,Y)^\star$ ◇

Define the Casimir element

$$\omega_{X,Y} := \Sigma_i \phi_i \otimes \phi^i \in Hom_C(X,Y) \otimes Hom_C(Y,X)$$

where the ϕ_i's is any basis of the former multiplicand and the ϕ^i's is its dual basis under the identification given in 2.15. Graphically, we use dummy variables ϕ and $\phi^\star$ as a short-hand notation:

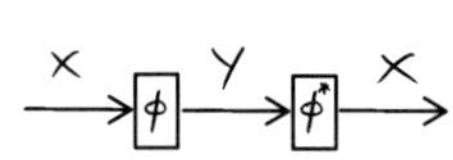

LEMMA 2.16. *Let C be a premodular category and W be a C-object. Then*

$$1_W = \Sigma_{i \in \mathcal{O}(C)} \Sigma_l dim(i) \phi^l \circ \phi_l,$$

where the ϕ_l's and the ϕ^l's form a pair of dual bases for the vector spaces $Hom_C(W,i)$ and $Hom_C(i,W)$ respectively, and $dim(i)$ denotes the (left) quantum trace of id_i.◇

NOTATION 2.17 (regular color). *Let C be a premodular category and $O(C)$ a complete (up to isomorphism) set of simple objects of C. The notation Ω means two things. While it is used as a C-object, it means the regular color $\oplus_{i \in \mathcal{O}(C)} i$; while it is used as a C-morphism $\oplus_{i \in \mathcal{O}(C)} dim(i) id_i : \oplus_{i \in \mathcal{O}(C)} i \to \oplus_{i \in \mathcal{O}(C)} i$. We also denote $\Sigma_{i \in \mathcal{O}(C)} dim(i)^2$ by $dim(\Omega)$, which is nonzero* [**ENO**]. *See the side warning in 4.13 as well.* ◇

With this shorthand notation Ω, we can present lemma 2.16 graphically by

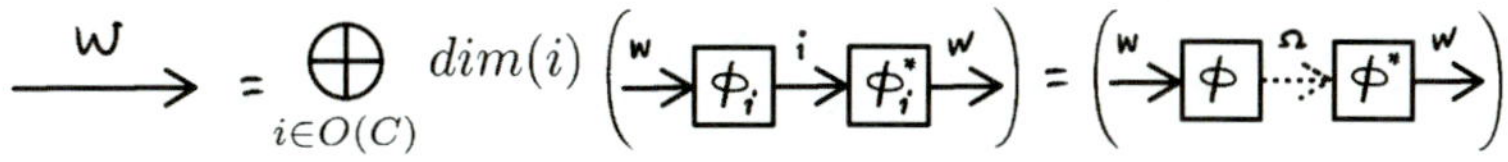

LEMMA 2.18 (Sliding lemma). *Let C be a premodular category. Then the following morphisms are all equal, where Ω is the shorthand notation given in* 2.17. ◇

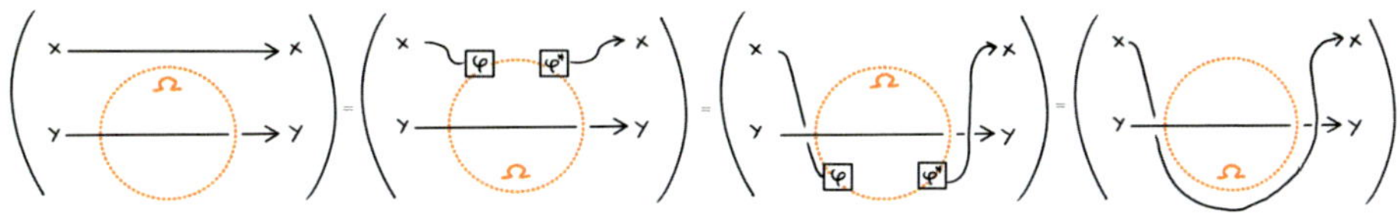

PROOF. Here we only provide the essence of the proof. Apply 2.16 locally with $W = X \otimes \Omega$. Use isotopy and the naturality of the braidings. And then apply 2.16 locally again with $W = \Omega \otimes X$. This proof is incomplete because the ϕ's in the second and the third quantity do not have the same signature. To complete this proof, one has to take the passing through the cup (cap, resp.) of Ω into account and make suitable change to the third ϕ ($\phi^\star$, resp.) while maintaining them as a dual basis. ■

Heuristically, the moral of this lemma is that Ω protects everything "inside" it by making it transparent. Notice that in fact Y can be more general than an object - the lemma works even when Y is a puncture in the context of string nets in a 3-fold.

LEMMA 2.19 (Censorship of Opacity). [**Mue03**, Lemma 2.13] *Let C be a premodular category, $Mu(C)$ its Muger center, i a simple C-object, and $\lambda = dim(\Omega)\delta_{i\in Mu(C)}$, we have the following equality (the right-hand-side denotes a scalar multiplication).*

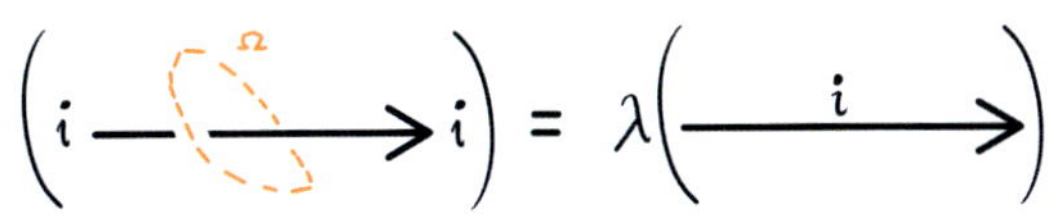

◇

Heuristically, Ω allows transparent objects to pass and nullifies all opaque objects. Note that by the characterization of modularity (2.8), only the identity object is allowed to pass when C is modular. Heuristically, this is the graphical reason why modular categories trivialize 4D TQFTs.

3. Topological theory

In this section, we describe the topological side of our main statement (1.15). We describe a string net (also called a tensor net or a tensor network) in section 3.1, and use it to define Crane-Yetter theory in dimension 2 and 3 in section 3.2. Then we provide a combinatorial description of an orientable smooth surface in section 3.3.

3.1. String nets. String nets originated from Penrose combinatorial description of a space-time [**Pen71**]. In dimension two, they were first explicitly written by the physicists Levin and Wen in [**LW05**]. For Crane-Yetter theory, however, we need string nets in dimension three. Following [**KT22**], we provide a formal definition for them in 3.10. Despite its lengthy and formal definition, a string net is roughly depicted the following picture: a 3-fold with embedded strands, each colored by an object in a fixed premodular category.

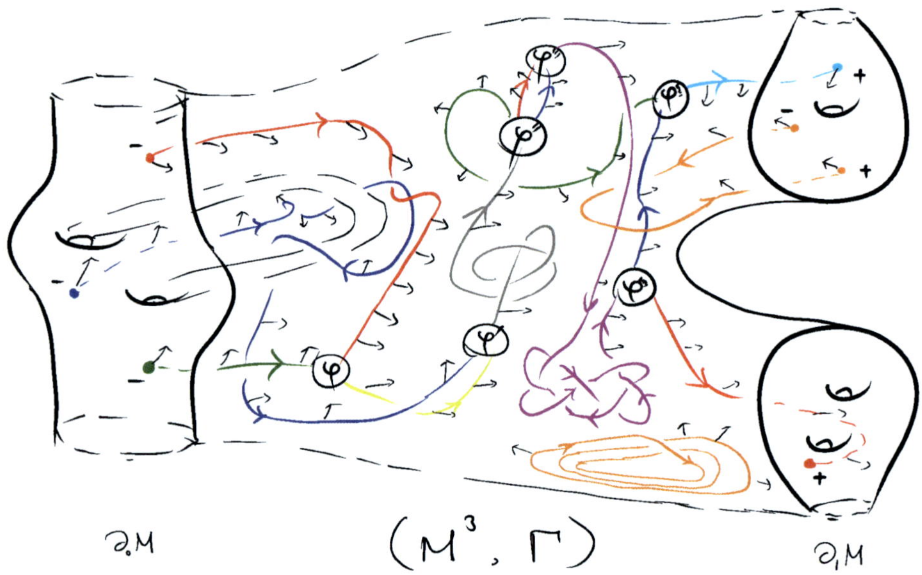

DEFINITION 3.1 (2-folds). A 2-fold is either a compact oriented smooth manifold without boundary of real dimension 2, or such a manifold with finitely many points removed (punctures). A 2-fold is also called a surface. ◇

DEFINITION 3.2 (Extended 2-folds). An extended 2-fold is a 2-fold M with the extra data

$$\{(p_1, v_1, or_1)\dots(p_n, v_n, or_n)\},$$

where $n < \infty$, the points $p_i \in M$ are disjoint to each other, the tangent vectors $v_i \in T_{p_i}M$ are nonzero, and the orientations or_i are in the set $\{+, -\}$. ◇

DEFINITION 3.3 (3-folds). A 3-fold is an oriented smooth manifold with boundary of real dimension 3. ◇

DEFINITION 3.4 (Framed arcs in a 3-fold). Let M be a 3-fold (3.3). An arc α in M is a smooth embedding of the standard interval $I = [0, 1]$ (with orientation from 0 to 1) into M. We require that if an end-point is sent by α to the boundary ∂M, then α has to intersect the boundary transversally.

A framing of an arc α in M is a non-vanishing smooth section s of the normal bundle of $\alpha(I) \subseteq M$. A framed arc is an arc equipped with a framing. ◇

REMARK 3.5. For our theory, the smoothness condition is not necessary but included for the sake of simplicity. Crane-Yetter theory works also in the PL-setting, parallel to its 3-dimensional analogue, the Turaev-Viro theory. Curious readers are referred to the setup given in [**KB10**].

The framing, on the other hand, is necessary. Such structure is expressed in slightly different way in related works. For example, in the context of skein modules, people use the notion of ribbons instead of that of arcs. The "width" of a ribbon corresponds to the normal vector from the section. ◇

DEFINITION 3.6 (Framed graphs in a 3-fold). Let M be a 3-fold (3.3). A framed graph Γ in M is a finite collection of framed arcs α_i 3.4 satisfying the following conditions.

- Denote the set of arcs of Γ by $E(\Gamma)$.
- The images of the embeddings α_i's do not intersect, except at their end-points, which must lie within the interior of M.
- Let p be a point in the interior of M. Denote $Out(p)$ ($In(p)$, resp.) be the set of the α_i's with $p = \alpha_i(0)$ ($\alpha_i(1)$, resp.). Denote the set of all arcs (edges) $E(p) := In(p) \cup Out(p)$. The directions of the tangent vectors of the α's that end at p must be different, i.e. for each $i \neq j$, there is no positive real number r such that $v_i = rv_j$. In the case where $E(p)$ is nonempty, we call p a vertex of Γ. Denote the set of vertices by $V(\Gamma)$.
- Let $\alpha \in E(\Gamma)$. If an end of α ends on the boundary of M, we call the end-point a boundary point of Γ. Denote the set of boundary points by $B(\Gamma)$.

◇

DEFINITION 3.7 (Extended 3-folds). An extended 3-fold is a pair of a 3-fold M and a framed graph Γ in M. ◇

Notice that an extended 3-fold (M, Γ) naturally induces an extended 2-fold $\partial(M, \Gamma) := (\partial M, \partial\Gamma)$, where $\partial\Gamma$ denotes the set

$$\{(p_i, v_i, or_i) \mid p_i \in B(\Gamma)\},$$

where v_i is naturally identified via

$$N_{p_i} = T_{p_i}M/T_{p_i}\alpha \simeq T_{p_i}(\partial M)$$

with the framing vector of Γ at the point p_i, and $or_i \in \{+, -\}$ is $+$ ($-$, resp.) if the framed arc α that passes through p_i is oriented such that $\alpha(1) = p_i$ ($\alpha(0) = p_i$, resp.).

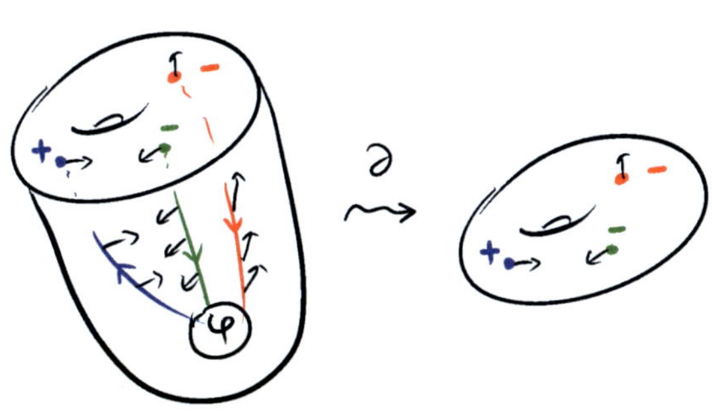

DEFINITION 3.8 (C-extended 2-folds). Given a premodular category C, a C-extended 2-fold, or a C-colored extended 2-fold, is an extended 2-fold with an extra

data: a C-object X_i is assigned to each oriented framed point (p_i, v_i, or_i). We call X_i the "color" assigned to the point p_i. ◇

To define a C-extended 3-fold, we need the Reshetikhin-Turaev theory ([**Tur10**], [**BK02**], [**Kas12**]) for genus-0 surfaces which we recall here. Let C be a premodular category. To each C-extended 2-fold

$$M = (M, \{(X_1, p_1, v_1, or_1) \dots (X_n, p_n, v_n, or_n)\})$$

diffeomorphic to a sphere, the first part of Reshetikhin-Turaev theory functorially assigns a vector space $RT(M)$ that is (non-canonically) isomorphic to

$$\langle X_1^\epsilon, \dots, X_n^\epsilon \rangle := Hom_C(\mathbb{1}, X_1^\epsilon \otimes \dots \otimes X_n^\epsilon),$$

where X_i^ϵ denotes X_i if $(or_i = +)$ or $X_i^\star$ if $(or_i = -)$.

Let C be a premodular category, and V be a real vector space of dimension 3. Let S be a finite collection of distinct framed oriented rays from the origin of V, with an assignment $S \xrightarrow{\phi} Obj(C)$. In this case, we say V has a finite collection of distinct C-colored rays. Then the Reshetikhin-Turaev theory for genus-0 surfaces naturally assigns a vector space $RT_C(V, S, \phi)$ to the sphere $(V - 0)/\mathbb{R}_+$.

Definition 3.9 (C-extended 3-folds). Let C be a premodular category, and M be an extended 3-fold (M, Γ). A C-coloring of (M, Γ) is an assignment as follows:

- To each arc α of Γ, assign a C-object $X(\alpha)$.
- After such assignment, to each vertex p of Γ, the tangent space at p naturally has a finite collection of C-colored rays (S, ϕ).
- The Reshetikhin-Turaev theory for genus-0 surfaces assigns a vector space $RT(p) := RT_C(T_p(M), S, \phi)$ as above. Note that $RT(p)$ is (non-canonically) isomorphic to
$$Hom_C(\otimes X_i, \otimes X_o)$$
where the X_i runs through the objects assigned to all incoming arcs, and the X_o runs through the objects assigned to all outgoing arcs.
- After such assignment, to each vertex p of γ assign a vector $v \in RT(p)$.

A C-extended 3-fold M is a 3-fold with a C-colored graph inside. This gives the boundary ∂M a C-extended surface structure. Conversely, we call such a C-colored graph a framed graph that satisfies the boundary condition posed by the C-extended surface ∂M. ◇

Let C be a premodular category. While the first part of Reshetikhin-Turaev theory for genus-0 surfaces assigns a vector space to a C-extended genus 0 surface, the second part of it assigns a C-extended 3-fold (M, Γ) diffeomorphic to a ball to a vector $RT(M, \Gamma) \in RT(\partial M)$.

Definition 3.10 (String nets in 3D). Let C be a premodular category and M a 3-fold whose boundary ∂M is a C-extended surface. Let F be the free vector space over $\mathbb{k}$ generated by all C-colored graphs that satisfy the boundary condition posed by ∂M. Let N be the subspace generated by either of the following

- The difference $\Gamma - \Gamma'$ of two C-colored graphs that are smoothly ambient isotopic to each other.
- A linear combination $v = \Sigma c_i \Gamma_i$ such that there exists a closed region $B \subseteq M$ diffeomorphic to a ball such that the vector assigned by the Reshetikhin-Turaev theory for genus-0 surfaces of $v|_B$ is the zero vector.

We call $S(M) := F/N$ the space of string nets of M with the given boundary condition, and we call an element of $S(M)$ a string net. ◇

3.2. Crane-Yetter theory in (co)dimension two (CY). We define the Crane-Yetter theory in dimension two in this subsubsection, following [**KT22**, section 5]. Let Σ be a smooth oriented 2-manifold and C a premodular category. To define $CY_\Sigma(C)$, we first recall the definition of a skein category $cy_\Sigma(C)$, which was first defined in [**Wal06**] and [**JF15**].

DEFINITION 3.11 ($cy_\Sigma(C)$, skein category). Given a premodular category C and a 2-fold Σ, we define the $\Bbbk$-linear category $cy_\Sigma(C)$ as follows. An object is a collection c of C-colored points and tangent vectors, such that (Σ, c) is a C-extended 2-fold. Given two objects c and c', the morphism space $Hom_{cy_\Sigma(C)}(c, c')$ between c and c' is defined to be the space of string nets for the 3-fold $\Sigma \times [0,1]$ satisfying the boundary condition $(\overline{c} \times \{0\}) \cup (c' \times \{1\})$, where $\overline{c}$ denotes the same collection of C-colored points as c does but with all orientations flipped. ◇

Two examples of morphisms in $cy_{\Sigma_{1,0}}(C)$ is depicted as follows.

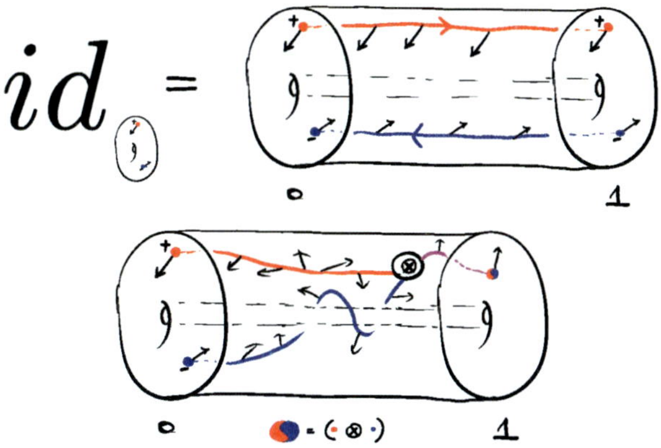

DEFINITION 3.12 (Karoubi envelope). Given an additive category C, its Karoubi envelope (Karoubi completion) $\mathrm{Kar}(C)$ is defined to be the category as follows. The objects are pairs (X, p), where $X \in \mathrm{Obj}(C)$ and $p \in Hom_C(X, X)$, such that $p^2 = p$. Given objects $\bar{X} = (X, p)$ and $\bar{Y} = (Y, q)$, the space of morphisms $Hom_{\mathrm{Kar}(C)}(\bar{X}, \bar{Y})$ is defined to be the subspace of $Hom_C(X, Y)$ consisting of those f such that $qfp = f$. ◇

The Karoubi envelope is the pre-abelian completion in our context.

DEFINITION 3.13 ($CY_\Sigma(C)$, Crane-Yetter theory in dimension 2). With the notations above, we define

$$CY_\Sigma(C) := \mathrm{Kar}(cy_\Sigma(C)). \tag{3.14}$$

◇

REMARK 3.15. The definition given in 3.13 was first given in [**KT22**, section 5]. That it extends the original Crane-Yetter theory is proved in [**Tha21**]. ◇

It is immediate from the definition that $cy_\Sigma(C)$ is additive. On the other hand, $CY_\Sigma(C)$ is in fact finite semisimple abelian for all surfaces with at least one puncture (cf 3.19, 4.19, 5.1). We conjecture that it holds for all closed surfaces too.

3.3. A presentation of surfaces (σ-construction). In this paper, we construct a surface Σ from the standard disk and an additional data $\sigma \in \mathrm{Adm}_{2n}$., give some examples, and prove that such construction produces all oriented surfaces with at least one puncture.

DEFINITION 3.16 (*Adm_{2n}*, admissible gluings). Let n be a nonnegative integer. An element σ in the permutation group S_{2n} on $2n$ elements is called an admissible gluing (of rank n), if σ satisfies the following conditions.

- σ has no fixed points.
- σ is an involution; i.e. $\sigma^2 = 1$.

We denote the subset of admissible gluings by $\mathrm{Adm}_{2n} \subseteq S_{2n}$. ◇

DEFINITION 3.17 (Σ_σ, σ-construction). For each admissible gluing $\sigma \in \mathrm{Adm}_{2n}$, we construct a smooth surface Σ_σ. Start from the standard oriented disk. We choose $2n$ closed segments with the same length from the boundary. To make the presentation easier, we emphasize them by drawing them like legs (without changing the diffeomorphism type), and we call them legs from now on.

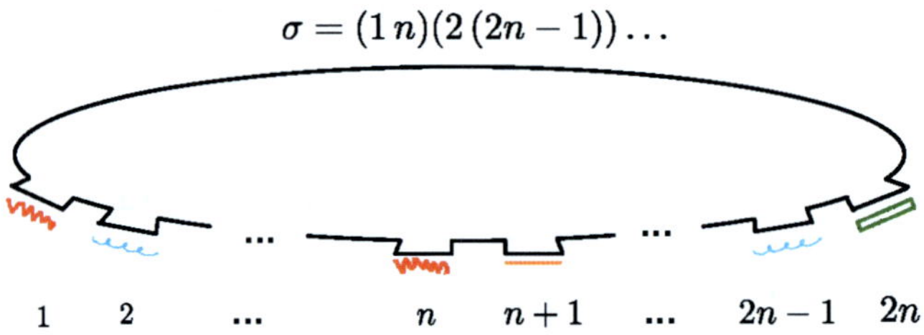

Glue the ends of the legs in pairs according to σ with the orientation preserved. Finally, remove the boundary to get an open surface. The result is denoted by Σ_σ. ◇

EXAMPLE 3.18. The only element (12) in Adm_2 constructs the cylinder $\Sigma_{(12)} \simeq$ Cylinder.

The elements $\sigma_{0,3} = (12)(34)$ and $\sigma_{1,1} = (13)(24)$ in Adm_4 construct a 3-punctured sphere Σ and a 1-punctured torus respectively.

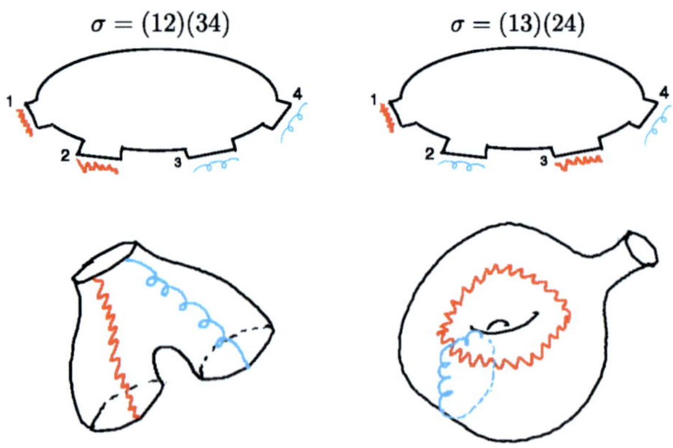

◇

The surfaces so constructed must have at least one puncture, and vice versa:

THEOREM 3.19. *The σ-constructions produce all oriented surfaces with at least one puncture (i.e. all compactifiable and orientable open surfaces).* $\diamond$

PROOF. Indeed, the admissible gluing $\sigma_{1,1} = (13)(24) \in \mathrm{Adm}_4$ gives an once-punctured torus $\Sigma_{\sigma_{1,1}}$. Similarly, the admissible gluing

$$\sigma_{2,1} = \sigma_{1,1} \circ (57)(68) = (13)(24)(57)(68) \in \mathrm{Adm}_8$$

gives an once-punctured surface of genus two. Following this fashion, for any $g \in \mathbb{N}$ one can construct an once-punctured surface of genus g by using the admissible gluing

$$\sigma_{g,1} = (13)(24)(57)(68)\ldots((4g-3)(4g-1))((4g-2)(4g)) \in \mathrm{Adm}_{4g}\,.$$

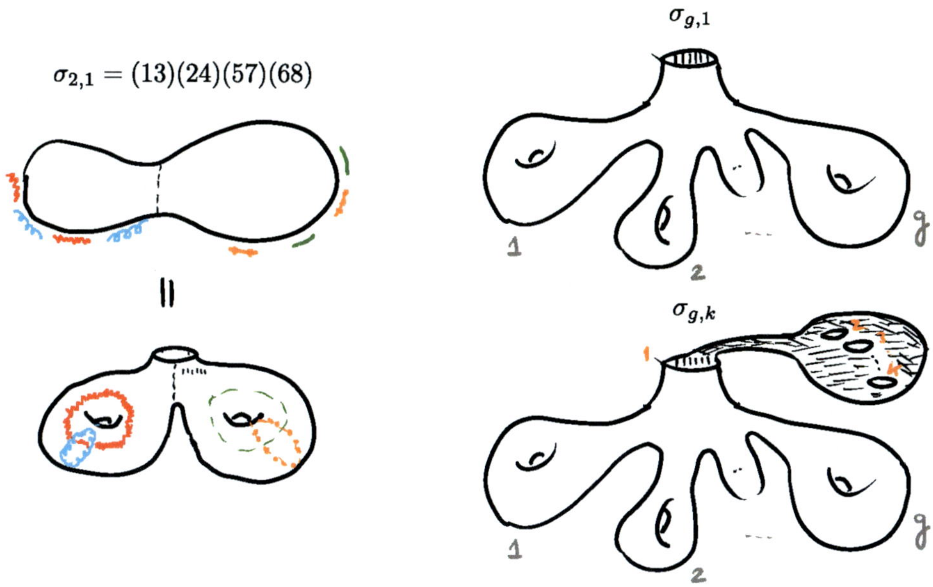

To add k punctures to the surface, use the admissible gluing

$$\sigma_{g,(k+1)} = \sigma_{g,1} \circ ((4g+1)(4g+2))((4g+3)(4g+4))\ldots((4g+2k-1)(4g+2k)) \in \mathrm{Adm}_{4g+2k}\,.$$

Then the statement follows from the well-known classification of oriented smooth surfaces. ∎

NOTATION 3.20 (σ-orbit). *We introduce a notation which is useful later. Let $\sigma \in \mathrm{Adm}_{2n}$. Denote $[i]$ to be the orbit $\{i, \sigma(i)\}$ of i under the action of σ, $[i]'$ the smaller number in the set $[i]$, and $[i]''$ the larger number in the set $[i]$. In particular, each $[i]$ has two elements, and the set $\{[1],[2],\ldots,[2n]\}$ has exactly n elements.*

Let $\sigma = (13)(24)$, *for example, we have*

$$\begin{aligned} [1] &= \{1,3\}, & [1]' &= 1, & [1]'' &= 3; \\ [2] &= \{2,4\}, & [2]' &= 2, & [2]'' &= 4; \\ [3] &= \{1,3\}, & [3]' &= 1, & [3]'' &= 3; \\ [4] &= \{2,4\}, & [4]' &= 2, & [4]'' &= 4. \end{aligned}$$

◇

4. Algebraic theory

In this section, we describe the algebraic side of our main statement (cf 1.15), namely the categorical center of higher genera Z, which generalizes the Drinfeld center construction. Its formal definition of Z is provided in section 4.2. Some properties are provided and proven in section 4.3: we show that Z is finite abelian semisimple, and that there is a strictly ambidextrous adjunction between $Z = Z(C)$ and the underlying premodular category C.

4.1. Motivation: Drinfeld categorical center. Abstract algebraic theories (groups, rings, modules.. etc) are ubiquitous in modern mathematics. Among the algebraic objects, the abelian ones are simpler, and are often first treated. One then builds the theory toward the generic cases. In group theory, for example, one can study a group G by starting with its center $Z(G) \subseteq G$ and then apply induction.

Drinfeld's categorical center is an analogue in the categorical setting. There, algebras are replaced by categorical algebras (more precisely, by monoidal categories [**EGNO15**]), and centers are replaced by categorical centers. As in the classical theory, the theory of the one side helps that of the other.

In contrast to the classical case, categorical centers need not be smaller nor easier. This is due to the fact that equalities are replaced by equivalences in the categorical settings. Therefore, the condition $ab = ba$ is replaced by $ab \simeq ba$. That is to say, a categorical commutativity not only remembers both sides being identified, but also **how** they are identified. Therefore, a typical object in the Drinfeld center $Z(C)$ is a pair $(X \in \mathrm{Obj}(C), \gamma)$, where γ is a half-braiding that encodes how X commutes with all the others. To be more precise, a **half-braiding** γ of X is a natural equivalence

$$(-) \otimes X \xrightarrow{\gamma} X \otimes (-)$$

satisfying some compatibility conditions [**EGNO15**]. It is worthwhile to mention that such construction has been successful in many contexts, e.g. representation theory, statistical physics, knot theory, .. etc.

Categorical center of higher genera $Z = Z_\sigma(C) = Z_{\Sigma_\sigma}(C)$, on the other hand, generalizes the Drinfeld center. Instead of remembering how X commutes with others, an object (X, γ) remembers how X commutes in **multiple** different ways. The amount of ways depends on the underlying surface $\Sigma = \Sigma_\sigma$. Therefore, an object of $Z_\sigma(C)$ is a pair (X, γ), where γ is a collection of half-braidings

$$\gamma = \{\gamma_1, \gamma_2, \ldots, \gamma_n\}.$$

However, extra conditions must be carefully imposed in order to keep track of the underlying topological data. In contrast to the case of Drinfeld center, multiple half-braidings give essentially infinite ways to fuse via tensors, e.g. $\gamma_1\gamma_2\gamma_2\gamma_1\gamma_1\gamma_1\gamma_2 \ldots$.

Therefore, suitable commutative relations among the half-braidings are needed. This is given in the formal definition of $Z_\sigma(C)$ as (:comm 1), (:comm 2), and (:comm 3) (cf 4.2).

Note that while the Drinfeld center construction requires the underlying category C to have monoidal structure, categorical center construction of higher genera requires C to at least have a braiding.

4.2. Categorical center of higher genera (Z). Throughout this section, let C be a premodular category and $\sigma \in \mathrm{Adm}_{2n}$ be an admissible gluing. We will define the categorical center of higher genera $Z_\sigma(C)$.

DEFINITION 4.1 (σ-pair). Define a σ-pair of C to be a pair (X, γ), where X is a C-object and γ is a set of half-braidings for X (cf [**EGNO15**])

$$\gamma = \{\gamma_{[1]}, \gamma_{[2]}, \dots, \gamma_{[2n]}\}$$

satisfying pairwise commutative relations given in 4.2. ◇

Recall that the notation $[i]$ from 3.20, and note that γ contains exactly n elements instead of $2n$.

DEFINITION 4.2 ((:comm), technical commutative relations). Let Z_1 and Z_2 be objects in C, and c be the braided structure of C (so $a \otimes b \xrightarrow{c_{a,b}} b \otimes a$). Given $[i]$ and $[j]$, there are three possible cases without loss of generality

- $[i]' < [i]'' < [j]' < [j]''$
- $[i]' < [j]' < [i]'' < [j]''$
- $[i]' < [j]' < [j]'' < [i]''$

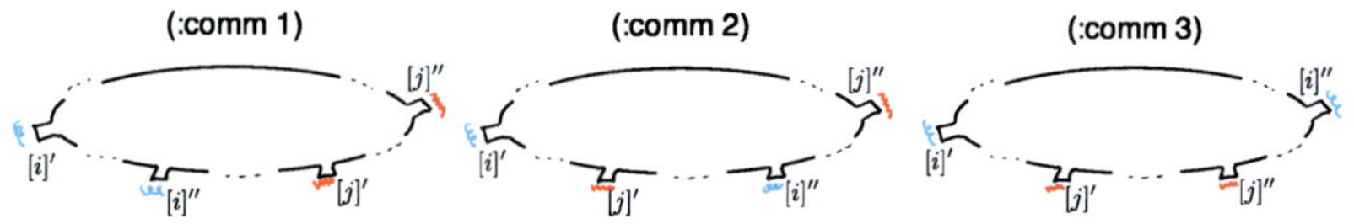

We will give the technical conditions that the γ's should obey, following by their graphical versions.

(1) In the first case, $\gamma_{[i]}$ and $\gamma_{[j]}$ are required to satisfy the following commutative relation (:comm 1), functorial in Z_1 and Z_2.

$$(\gamma_{[j],Z_2} \otimes 1)(1 \otimes (c_{Z_1,X} c_{X,Z_1} \gamma_{[i],Z_1})) \tag{4.3}$$

$$= (1 \otimes c_{Z_1,Z_2})((c_{Z_1,X} c_{X,Z_1} \gamma_{[i],Z_1}) \otimes 1)(1 \otimes \gamma_{[j],Z_2})(c^{(-1)}_{Z_1,Z_2} \otimes 1) \tag{4.4}$$

(2) In the second case, $\gamma_{[i]}$ and $\gamma_{[j]}$ are required to satisfy the following commutative relation (:comm 2), functorial in Z_1 and Z_2.

$$(\gamma_{[j],Z_2} \otimes 1)(1 \otimes \gamma_{[i],Z_1}) \tag{4.5}$$

$$= (1 \otimes c^{(-1)}_{Z_2,Z_1})(\gamma_{[i],Z_1} \otimes 1)(1 \otimes \gamma_{[j],Z_2})(c^{(-1)}_{Z_1,Z_2} \otimes 1) \tag{4.6}$$

(3) In the third case, $\gamma_{[i]}$ and $\gamma_{[j]}$ are required to satisfy the following commutative relation (:comm 3), functorial in Z_1 and Z_2.

$$(\gamma_{[j],Z_2} \otimes 1)(1 \otimes \gamma_{[i],Z_1}) \tag{4.7}$$

$$= (1 \otimes c_{Z_1,Z_2})(\gamma_{[i],Z_1} \otimes 1)(1 \otimes \gamma_{[j],Z_2})(c^{(-1)}_{Z_1,Z_2} \otimes 1) \tag{4.8}$$

Notice that the first and the third case are almost the same, which is not surprising given their topological meaning. To make them look alike, define

$$\tilde{\gamma}_{[i],-} = c_{-,X} c_{X,-} \gamma_{[i],-}.$$

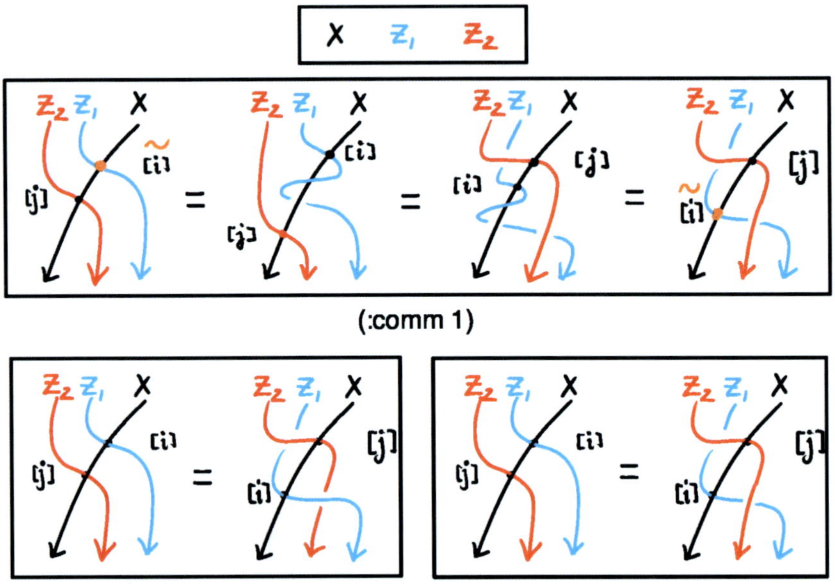

◇

DEFINITION 4.9 (σ-morphism). Given two σ-pairs $\bar{X} = (X, \gamma)$ and $\bar{Y} = (Y, \beta)$ of C, define $[\bar{X}, \bar{Y}]_\sigma$ to be the linear subspace of $Hom_C(X, Y)$ consisting of the morphisms $(X \xrightarrow{f} Y)$ compatible with all the half-braidings in the following sense. For any $Z \in C$, we have for each $1 \le i \le 2n$,

$$\beta_{[i],Z}(1 \otimes f) = (f \otimes 1)\gamma_{[i],Z}; \tag{4.10}$$

pictorially, it means

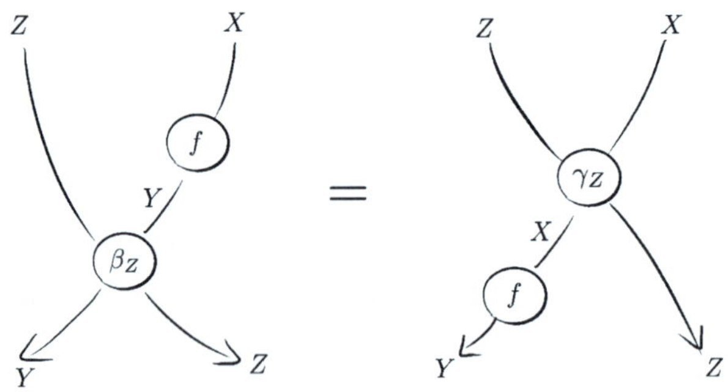

Finally, let the identity maps and the compositions be inherited from that of C. ◇

DEFINITION 4.11 (categorical center of higher genera). Let C be a premodular category, and $\sigma \in \mathrm{Adm}_{2n}$ an admissible gluing. The categorical center of higher genera $Z_\sigma(C)$ of C is defined to be the category with objects the σ-pairs of C and with morphisms the σ-morphisms. ◇

4.3. Properties of Z. In this section, we establish some basic properties of categorical centers of higher genera. In particular, we show that they are finite semisimple abelian categories, and that there is a strictly ambidextrous adjunction between it and the underlying premodular category C.

4.3.1. *Connecting functors.* In this section, we establish the relation between C and its categorical center of higher genera $Z_\sigma(C)$, where C is a premodular category and σ is an admissible gluing. More precisely, there exist two additive functors I_σ and F_σ.

$$I_\sigma : C \rightleftarrows Z_\sigma(C) : F_\sigma.$$

We will see that I_σ is both a right and a left adjoints of F_σ (thus vice versa) in section 4.3.2. Such a pair of adjunction is called a (strictly) ambidextrous adjunction in the literature.

DEFINITION 4.12 (forgetful functor). The forgetful functor

$$C \xleftarrow{F_\sigma} Z_\sigma C$$

is defined to send objects (X, γ) to X, and to send morphisms by inclusion (recall that the morphism space of $Z_\sigma(C)$ is defined as a subspace of that of C). Clearly, it is an additive functor. ◇

DEFINITION 4.13 (induction functor). The induction functor

$$C \xrightarrow{I_\sigma} Z_\sigma(C)$$

is more complicated, so will be defined step-by-step. Define $I_\sigma(X)$ to be (X_σ, γ), which is given below.

Denote by $\mathcal{O}(C)$ the set of isomorphism classes of simple objects of C. Let $o(C)$ be a set of representatives. To each C-object X, define another C-object

$$X_\sigma := \bigoplus_{o(C)^n} (X_1 \otimes X_2 \otimes \ldots \otimes X_n \otimes X \otimes X_{n+1} \otimes \ldots \otimes X_{2n-1} \otimes X_{2n}), \tag{4.14}$$

where X_k runs through $o(C)$ for those $k = [k]'$, and $X_k = X^\star_{([k]')}$ for those $k = [k]''$. Here, recall that the first case means that k is the smaller member in the set $[k]$, while the second case means that k is the larger member (cf 3.20). For example,

$$1_{(12)} = \oplus_{X \in o(C)} X \otimes X^\star.$$

Notice that it does not depend on the choice of $o(C)$ up to canonical isomorphisms. Similarly, neither does X_σ for general admissible gluings σ.

Next, to each k, we define the $[k]$-th half-braiding $\gamma_{[k]}$ for X_σ

$$(-) \otimes X_\sigma \xrightarrow{\gamma_{[k]}} X_\sigma \otimes (-) \tag{4.15}$$

as in the following pictures (note that each of the vertical dotted lines represents the regular color (notation 2.17)).

Side Warning. The Ω notations in the following graphics are sloppy in a harmless way. When observed without the half-braiding, they should not be interpreted as endomorphisms of $\Omega^{\otimes n} \otimes X \otimes \Omega^{\otimes n}$ but rather as those of X_σ. This usage of short-hand is reiterated later as well.

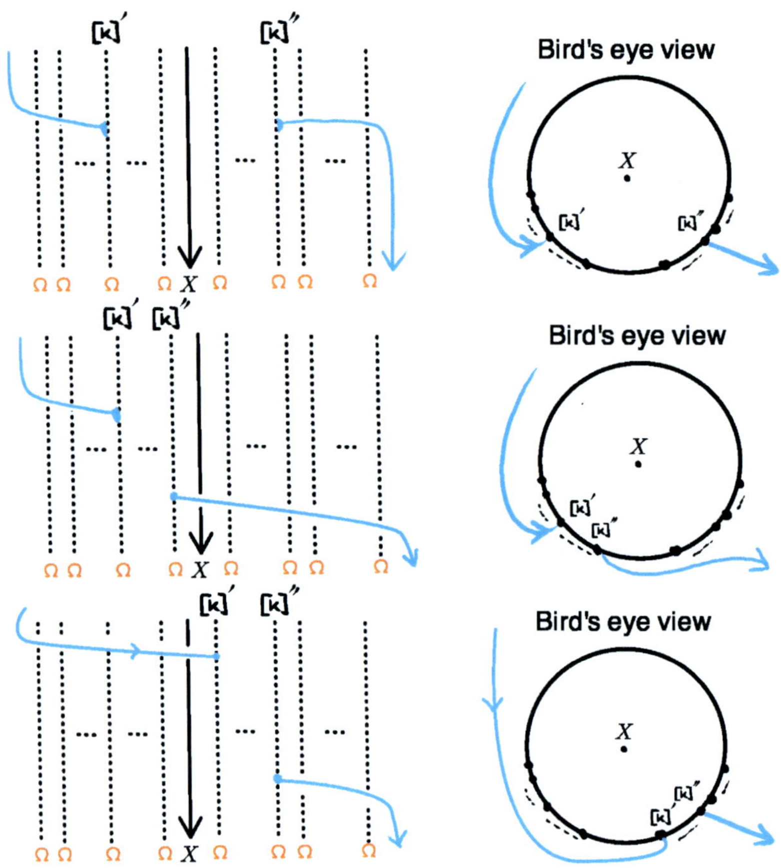

To be more precise, denote the braiding of C by β. The morphism in the first picture above is

$$\beta_{-,\Omega_{2n}} \circ \beta_{-,\Omega_{2n\text{-}1}} \circ \ldots \circ \beta_{-,\Omega_{[k]''+1}} \circ \phi^a \circ \phi_a \circ \beta_{-,\Omega_{[k]'\text{-}1}} \circ \ldots \circ \beta_{-,\Omega_2} \circ \beta_{-,\Omega_1}$$

where Ω_i denotes the ith component of Ω in the picture counted from the left, and where

$$\phi^a \otimes \phi_a \in Hom_C(\Omega_{[k]''}, \Omega_{[k]''} \otimes (-)) \otimes Hom_C((-) \otimes \Omega_{[k]'}, \Omega_{[k]'})$$

is the canonical element (a sum of any dual basis) given similarly as in 2.16.

We contend that so given (X_σ, γ) is indeed an object of $Z_\sigma(C)$. One needs to show that each element in γ is a half-braiding, and that γ satisfies the commutative relations (4.2). A proof of this can be found in the arXiv version. It remains to define the map on the morphism spaces $Hom_C(X, Y)$. Given a morphism $X \xrightarrow{f} Y$, one defines

$$I_\sigma(f) := \overline{f} := \mathrm{id}^{\otimes n} \otimes f \otimes \mathrm{id}^{\otimes n}. \tag{4.16}$$

To conclude, it remains to show that

- The morphism $\overline{f}$ is compatible with the half-braidings γ and β.
- The construction $(\overline{-})$ preserves the identities and the compositions.

The first point is shown in the arXiv version. The second point is clear. ◇

4.3.2. *Ambidextrous adjunction.* Both functors F_σ and I_σ are additive immediately by definition. We are ready to state and prove the main statement of this section.

THEOREM 4.17. *The functors*

$$I_\sigma : C \rightleftarrows Z_\sigma(C) : F_\sigma$$

so defined in 4.12 *and* 4.13 *are (strictly) ambidextrous adjoint to each other. In other words,* I_σ *is both a left adjoint and a right adjoint of* F_σ*, thus vice versa.* ◇

PROOF. We will prove that F_σ is right adjoint to I_σ. Namely, we need to show that for each C-object X and for each $Z_\sigma(C)$-object (Y, β), there is a vector space isomorphism

$$F : Hom_C(X, F_\sigma(Y, \beta)) \rightleftarrows Hom_{Z_\sigma(C)}(I_\sigma(X), (Y, \beta)) : G.$$

From this it is obvious that the left-adjunction can be proved verbatim by taking duals (or by flipping the graphs upside down). We will explicitly construct the maps F and G, and argue that each composition equals the identity map.

Given $\phi \in Hom_C(X, F_\sigma(Y, \beta))$, define its image on the other side to be

$$F(\phi) := \frac{1}{dim(\Omega)^n}(\widehat{\Pi}_{k\in\{[1],\ldots,[2n]\}}\beta^\star_{k,\Omega}) \circ (1 \otimes \ldots \otimes 1 \otimes \phi \otimes 1 \otimes \ldots \otimes 1),$$

where Ω is the shorthand notation given in 2.17, the index set $\{[1], \ldots, [2n]\}$ consists of exactly n elements instead of $2n$ (cf. 3.20), and the term $\widehat{\Pi}$ is explained below: The term $\widehat{\Pi}$ is a C-morphism $I_\sigma(Y) \to Y$. Each $\beta^\star_{[i],\Omega}$ is a C-morphism $\Omega \otimes Y \otimes \Omega \to Y$, induced from $\Omega \otimes Y \xrightarrow{\beta_{[i],\Omega}} Y \otimes \Omega$ by composing the evaluation map (note that $\Omega^\star \simeq \Omega$). So the $\beta^\star_{[i],\Omega}$'s are maps that kills the $[i]'$-th and the $[i]''$-th component of Ω by using $\beta_{[i]}$. However, depending on the combinatorial nature of $\sigma \in \mathrm{Adm}_{2n}$, one should insert suitable braidings for it to make sense. For example, if $n = 3$, $[1] = [4]$, $[2] = [6]$, and $[3] = [5]$, we define the $\widehat{\Pi}$ term as in the following diagram – the order of the $[i]$'s does not really matter, thanks to 4.2.

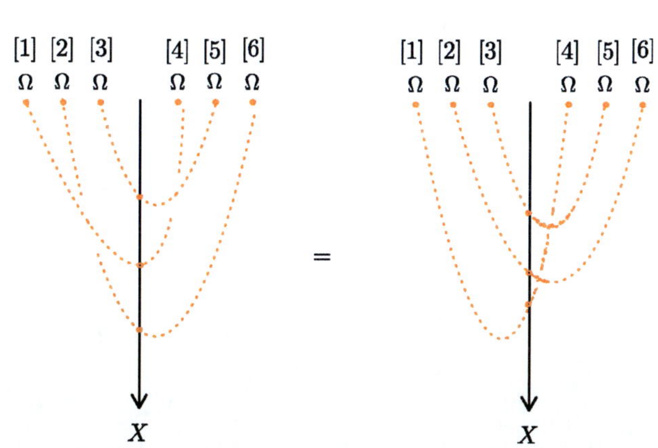

That $F(\phi)$ is indeed a morphism in $Z_\sigma(C)$ follows directly from the commutative relation 4.2, that half-braidings are by definition monoidal, and the sliding lemma 2.18.

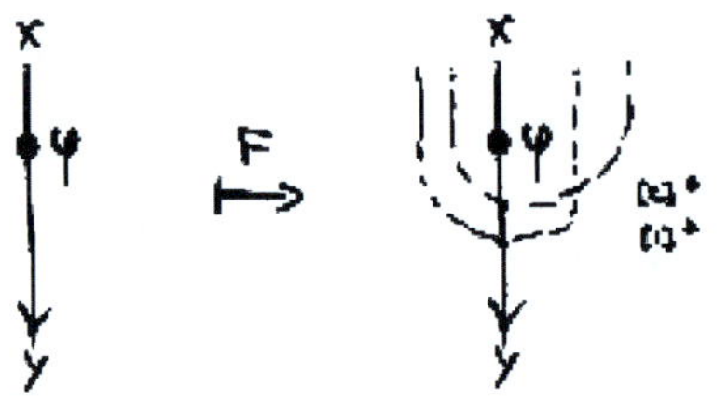

On the other hand, given $\psi \in Hom_{Z_\sigma(C)}(I_\sigma(X), (Y, \beta))$, define its image $G(\psi)$ on the other side to be as indicated in the graph below.

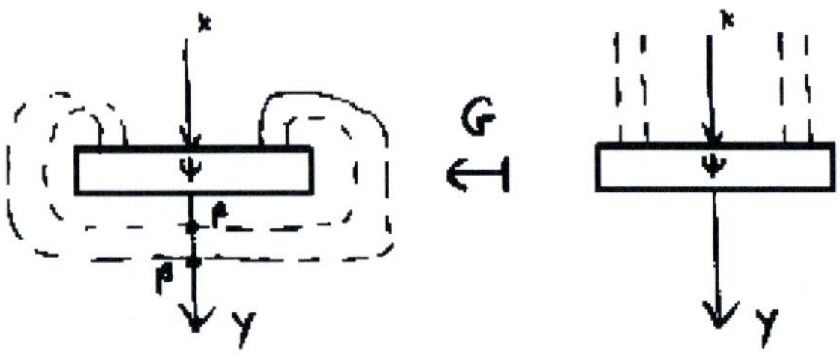

To prove that GF is the identity map, use the fact that the half-braidings are by definition functorial. Hence one can slide the Ω's out the axis. Finally, the product of the dimensions of the Ω's cancel with the denominator.

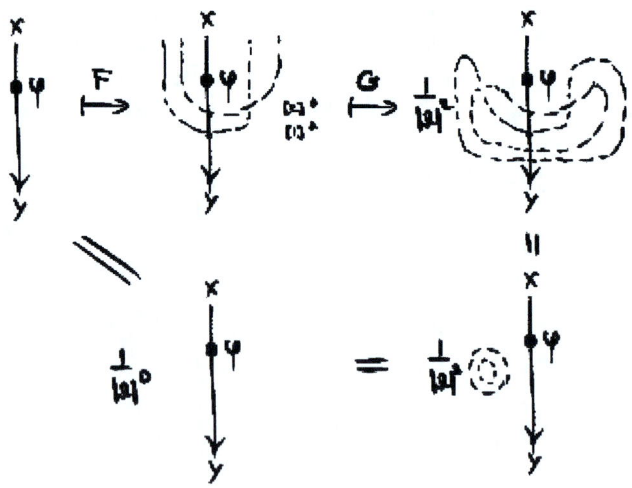

To prove that FG is the identity map, use the sliding lemma again. Then use the assumption that ψ is a $Z_\sigma(C)$-morphism to drag Ω down. Finally, slide the Ω's away from the axis as in the case for $GF = 1$.

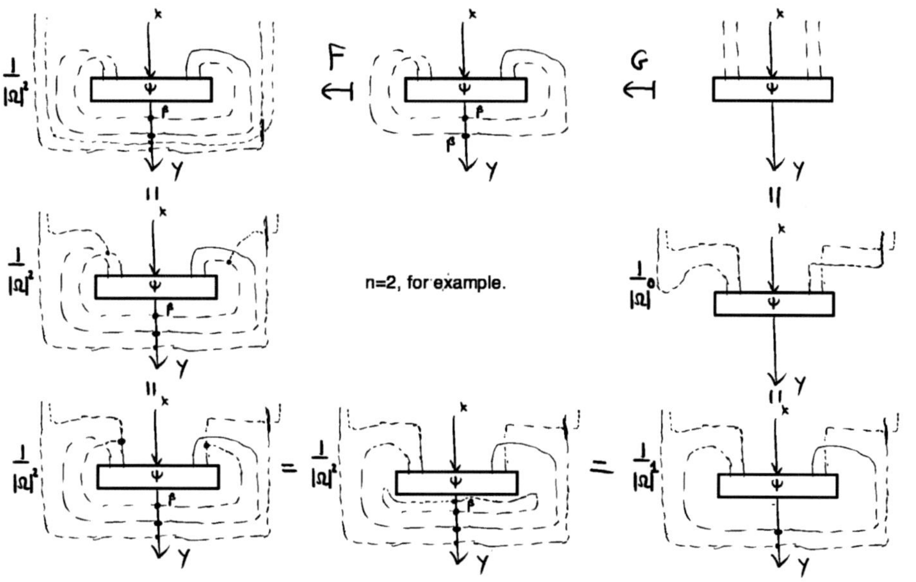

Note that this proof focuses on this particular setting, and it essentially relies on the fact that the components do not link together. One may wonder if this works for every other case. The full proof is tedious and non-essential, but if one really wants, here's a more complete sketch: Define a $[k] < [l]$ if $[k]' < [l]'$ (i.e. the smaller number in the set $[k]$ is smaller than the smaller number in the set $[l]$). And for the linear map F I can start drawing the Ω lines from the smallest $[k]$ amount $\{[1], \ldots, [2n]\}$ to the largest. Divide the height of the rectangle drawing region into n pieces. For the smallest $[k]$, the Ω line should stay in the top-most piece; for the second smallest $[k]$, the Ω line should stay in the top two pieces, where the cup should stay in the second piece from the top; and so on. This way, the Ω lines can be drawn almost separately to the most extent. In case where there should be crossings, let the larger $[k]$ to always go over the smaller $[k]$. Do the same thing for the linear map G, and it is obvious that there is no linking issue. ■

4.3.3. *Z is finite semisimple abelian.* In this section, we show that the categorical centers of higher general over premodular categories are finite semisimple abelian categories.

LEMMA 4.18 ((monadic) projection / idempotent). *Let $\sigma \in Adm_{2n}$ be an admissible gluing, C be a premodular category, $Z = Z_\sigma(C)$ be the categorical center of higher genera of C with respect to σ, and $\overline{X} = (X, \gamma)$ and $\overline{Y} = (Y, \beta)$ be Z-objects.*

Recall that the morphism space $Z(\overline{X}, \overline{Y})$ is by definition a subspace of $C(X, Y)$. Then there is a natural projection $\pi_{\gamma,\beta} \in End(Mor_C(X, Y))$ to the subspace $Z(\overline{X}, \overline{Y})$ that respects the composition: for any $(Z, \alpha) \in Z_\sigma(C)$, any $\phi \in Mor_C(X, Y)$

and any $\psi \in Mor_C(Y, Z)$

$$\pi_{\gamma,\alpha}(\psi \circ \phi) = \pi_{\beta,\alpha}(\psi) \circ \pi_{\gamma,\beta}(\phi).$$

◇

PROOF. The complete and formal proof is tedious and postponed to the arXiv version. Roughly, the statement follows from the monadic nature of the strictly ambidextrous adjunctions and a condition called the "unity trace condition". ■

THEOREM 4.19. *Let* C *be a premodular category,* $\sigma \in \mathrm{Adm}_{2n}$ *an admissible gluing. Then the categorical center of higher genera* $Z_\sigma(C)$ *is a finite semisimple abelian category.* ◇

PROOF. The complete and formal proof is tedious and thus postponed to the arXiv version. The main idea is to make heavy use of the projection. ■

5. Proof of the main theorem

In this section, we prove the main statement of this paper.

THEOREM 5.1 (Main Statement). *Let* C *be a premodular category,* $\sigma \in \mathrm{Adm}_{2n}$ *an admissible gluing,* $\Sigma = \Sigma_\sigma$ *the surface constructed from* σ*. Then the Crane-Yetter theory of* Σ_σ *over* C *and the categorical center of higher genera* $Z_\sigma(C)$ *are equivalent as* $\Bbbk$*-linear categories*

$$CY_{\Sigma_\sigma}(C) \simeq Z_\sigma(C).$$

As the $Z_\sigma(C)$*'s are proven to be finite semisimple abelian* 4.19*, the Crane-Yetter theory for each open surface is also a finite semisimple abelian category.* ◇

To stress the informal aspect again, we recall 1.19.

REMARK 5.2. In $H_1(S^1) \simeq \mathbb{Z}$, one sees the algebra of the shape S^1 and the shape of the algebra $\mathbb{Z}$. Our main result should be viewed as a higher analogue. That is, one sees the (higher) algebra of the shape Σ_σ and the shape of the (higher) algebra Z_σ. ◇

EXAMPLE 5.3 ($n = 0$). For $n = 0$, the surface Σ is the open disk, the categorical center of higher genera reduces to the underlying premodular category C, so the theorem recovers that $CY_\Sigma(C) \simeq C$ as shown in 1.7. ◇

EXAMPLE 5.4 ($n = 1$). For $n = 1$, the only possible surface is the cylinder, the categorical center of higher genera reduces to the Drinfeld center $Z(C)$. Hence, the theorem recovers that $CY_\Sigma(C) \simeq Z(C)$ as shown in 1.8. ◇

EXAMPLE 5.5 ($n = 2$). For $n = 2$, there are two possible surfaces: the 1-punctured torus and the 3-punctured disk. In the former case, the categorical center of higher genera reduces to the elliptic center $Z^{el}(C)$, so the theorem recovers that $CY_\Sigma(C) \simeq Z^{el}(C)$ as shown in 1.12. ◇

EXAMPLE 5.6 (General case). In the later case, the theorem provides a new result by providing an explicit value $Z_{0,3}(C)$ for the 3-punctured sphere. Together with the excision principle and the value for the disk and the cylinder, this allows us to actually compute the values for **all** oriented surfaces, in several different ways. For example, a 4-punctured sphere has values $Z_{0,4}(C)$ and $Z_{0,3}(C) \boxtimes_{(Z(C),\overline{\otimes})} Z_{0,3}(C)$. By the invariance of tqfts, we know that they are equivalent. We can then use this to compute the value for the surface of genus 2, say, and so on. ◇

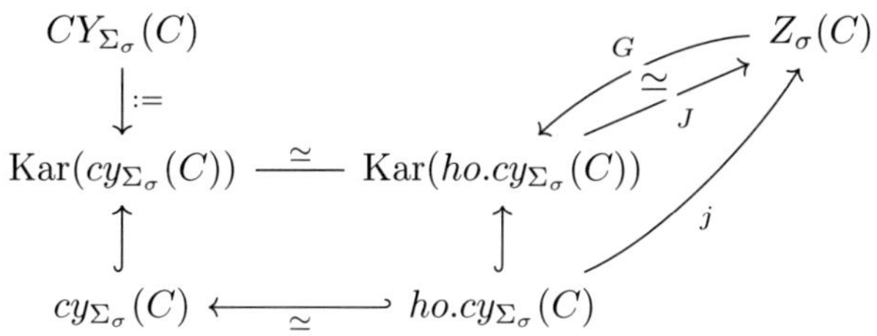

5.1. Strategy.

(1) (Condensation of string nets) By definition, $CY_{\Sigma_\sigma}(C)$ is the Karoubi envelope of $cy_{\Sigma_\sigma}(C)$. Find an equivalent subcategory $ho.cy_{\Sigma_\sigma}(C)$ of cy_{Σ_σ} by reducing topological data. Then of course

$$\mathrm{Kar}(cy_{\Sigma_\sigma}) \simeq \mathrm{Kar}(ho.cy_{\Sigma_\sigma}(C)).$$

(2) (top $\to$ alg) Construct a functor

$$ho.cy_{\Sigma_\sigma}(C) \xrightarrow{j} Z_\sigma(C),$$

and extend it to

$$\mathrm{Kar}(ho.cy_{\Sigma_\sigma}(C)) \xrightarrow{J} Z_\sigma(C).$$

(3) (top $\leftarrow$ alg) Construct a functor

$$\mathrm{Kar}(ho.cy_{\Sigma_\sigma}(C)) \xleftarrow{G} Z_\sigma(C).$$

(4) Argue that the compositions $J \circ G$ and $G \circ J$ are equivalent to the identity functors.
(5) Show that the equivalence is of finite semisimple abelian categories.

Note that in the previous sections the notation G was used as a linear map between two vector spaces. It should not be confused with our G here, which is a functor.

5.2. proof. In this section, we give the full proof of the main statement. Each subsection corresponds to each step in the outlined strategy.

5.2.1. *Reducing topological data.* In this section, we reduce the topological data by constructing a smaller yet equivalent subcategory

$$ho.cy_{\Sigma_\sigma}(C) \underset{\sim}{\overset{\subset}{\to}} cy_{\Sigma_\sigma}(C). \tag{5.7}$$

DEFINITION 5.8 ($ho.cy_{\Sigma_\sigma}(C)$). Let C be a premodular category and σ an admissible gluing. The subcategory $ho.cy_{\Sigma_\sigma}(C)$ of $cy_{\Sigma_\sigma}(C)$ is defined as follows.

Let p be the central point of the standard disk. An object of $ho.cy_{\Sigma_\sigma}(C)$ is defined to be the single C-colored point (p, X) for some $X \in \mathrm{Obj}(C)$. A morphism from (p, X) to (p, Y) is the equivalence class in which the following string net lives.

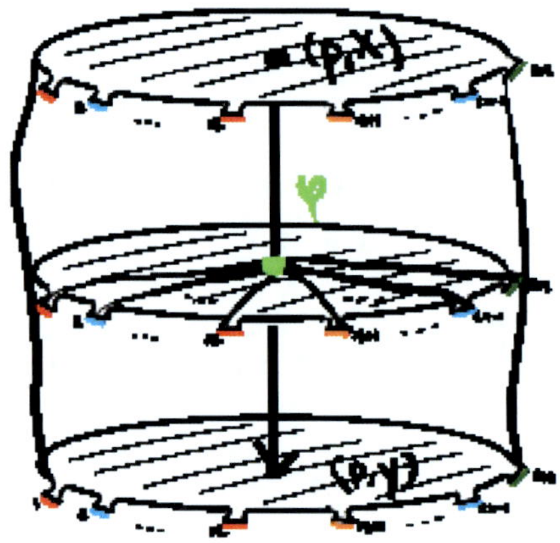

Clearly, $ho.cy_\sigma(C)$ is a subcategory of $cy_\sigma(C)$. ◇

THEOREM 5.9 (equivalence of reduction). *The inclusion functor*

$$\iota : ho.cy_{\Sigma_\sigma}(C) \xrightarrow{\subseteq} cy_{\Sigma_\sigma}(C)$$

is an equivalence of categories. Clearly, it is additive. ◇

PROOF. By a basic lemma in category theory, it is enough to show that ι is fully faithful and essentially surjective.

(Essentially surjective) Recall that a typical object of $cy_{\Sigma_\sigma}(C)$ is a finite collection of C-colored points on Σ_σ. It suffices to find an equivalent object of the form (p, X), for some $X \in \mathrm{Obj}(C)$. This can be done by the following reductions.

(1) Slightly push the points on the boundary into the smaller side.
(2) Compress the points from the legs into the disk.
(3) Then compress further for the points to stay in a small unit disk in the middle.
(4) Project the objects to a fixed line.
(5) Take their tensor products.

Each step above can be realized as an isomorphism in $CY_{\Sigma_\sigma}(C)$, so every object is isomorphic to an object in $ho.cy_{\Sigma_\sigma}(C)$.

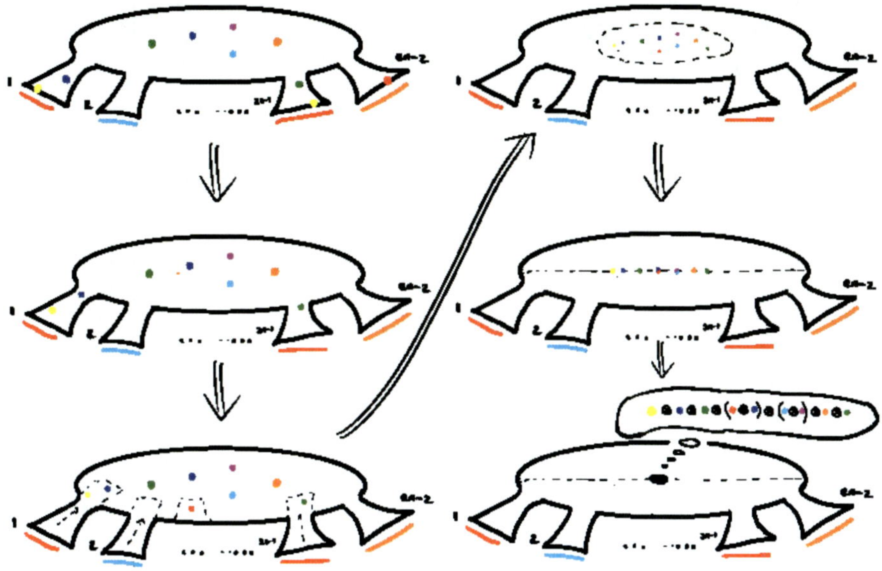

(Fully faithful) We ought to show that

$$Hom_{ho.cy_{\Sigma_\sigma}(C)}((p,X),(p,Y)) \xrightarrow{\iota} Hom_{cy_{\Sigma_\sigma}(C)}((p,X),(p,Y))$$

is an equivalence of vector spaces. Clearly, it is linear and injective, as the quotient relations on both sides are the same. To prove surjectivity, we have to show that any arbitrary string net with given boundary condition is equivalent to a stringnet given in the definition of $ho.cy_{\Sigma_\sigma}(C)$. This can be done by a similar compression process as in the proof of essential surjectivity.

(1) Push the stringnets away from the end of the legs.
(2) Push the stringnets away from the legs.
(3) Compress everything into a fixed central bar.
(4) Replace the stringnets through boundaries with one strand for each leg by taking tensor products by using the Reshetikhin-Turaev evaluation.
(5) Then compress vertically.
(6) Then finally replace the tangled mess in the middle by a morphism.

■

5.2.2. *topology → algebra.* In this section, we aim to construct a functor

$$\mathrm{Kar}(ho.cy_{\Sigma_\sigma}(C)) \xrightarrow{J} Z_\sigma(C).$$

As $Z_\sigma(C)$ is abelian (4.19), we only have to construct an additive functor

$$ho.cy_{\Sigma_\sigma}(C) \xrightarrow{j} Z_\sigma(C).$$

To define j, recall that a typical object of $ho.cy_{\Sigma_\sigma}(C)$ is a colored point (p, X), where p denotes the central point of the standard disk. Define its image under j to be $I_\sigma(X) = (X_\sigma, \gamma_\sigma)$ as in (4.14). A typical morphism from (p, X) to (p, Y) is a

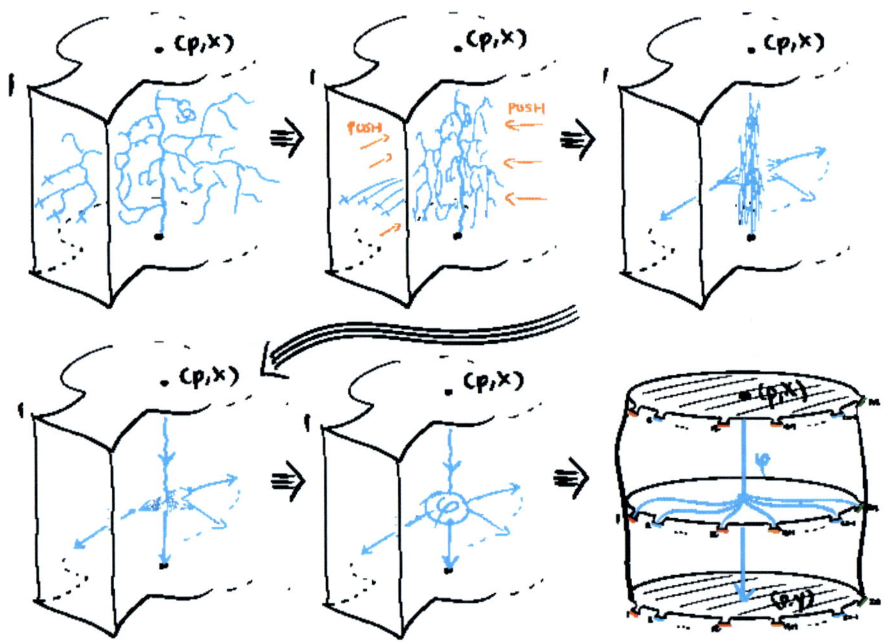

linear combination of the equivalence classes of the stringnets like Γ (given in the graphic below). Define the image of $[\Gamma]$ under j to be

$$\Gamma = \Gamma_{\varphi} \mapsto [\Gamma] \text{ in parallel as in:}$$

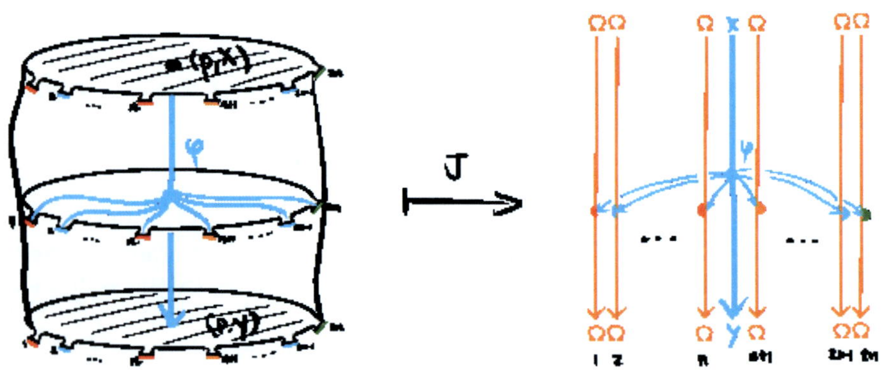

where Ω is the shorthand notation given in 2.17 crossings mean the braidings of C, and the nontrivial pairs of intertwiners are given in [**KB10**, (1.8)]. Extend the definition additively, and then we have our desired additive functor

$$ho.cy_{\Sigma_\sigma}(C) \xrightarrow{j} Z_\sigma(C).$$

We argue that j is well-defined on the morphism spaces. With the objects fixed, the domain is a linear space $S(M^3)$ (where ∂M^3 is colored suitably) given in definition 3.10. There, $S(M^3)$ is a quotient space F/N, where F is the linear space freely generated by C-colored graphs with suitable boundary condition. It would be enough to establish a map first from F, and argue that the map factors through

F/N. However, this is not easily done with rigor, because things in F may have coupons occurring in the boundary of the leg (a presentation of the surface has been fixed ahead). Instead, we have to pass to the subspace $F' \subset F$ that is freely generated by C-colored graphs having no coupons on the boundary of any leg. It is easy to see that F/N is isomorphic to $F'/(F' \cap N)$, by pushing the bad coupon away from the boundary of legs (thanks to N, which includes isotopic equivalent graphs). Now the plan is to define j from F', and argue that it factors through $F'/(F' \cap N)$. For any graph in F', just define the images as in the graphic above; indeed, there are no troubles on the boundary. This indeed factors through $F'/(F' \cap N)$ because

(1) If a linear combination locally evaluates to 0, then the corresponding Γ_ϕ evaluates to zero. Then the image is not altered by adding this.
(2) If any graph is altered by an isotopy, then there are two cases. In the first case, no coupon passes through the boundary of any leg. Then, the image does not change by definition. In the second case, some coupons pass through some boundary. We can deal with a coupon at a time. One can prove that $[\Gamma_\phi] = [\Gamma'_\phi]$ by pairing the dual basis and sliding the coupon to the corresponding strand.

Since the codomain is abelian (4.19), j extends to J naturally. In particular, for any object $((X,p),f)$ in $Kar(ho.cy_{\Sigma_\sigma}(C))$, $J((X,p),f)$ is defined to be the image of the morphism $[\Gamma_f]$ in $Z_\sigma(C)$.

5.2.3. *topology* $\leftarrow$ *algebra.* In this section, we construct a functor

$$\mathrm{Kar}(ho.cy_{\Sigma_\sigma}(C)) \overset{G}{\leftarrow} Z_\sigma(C).$$

Recall that a typical object in $Z_\sigma(C)$ is (X,γ), where $X \in \mathrm{Obj}(C)$ and γ is a set of half-braidings $\gamma = \{\gamma_{[1]}, \gamma_{[2]}, \dots \gamma_{[2n]}\}$ satisfying some relations 4.2. Define the image of (X,γ) under G to be $((p,X),\pi_\gamma)$, where p denotes the central point of the standard disk, and π_γ to be the equivalence class of the stringnets below multiplied by the scalar $\frac{1}{|\Omega|^n}$.

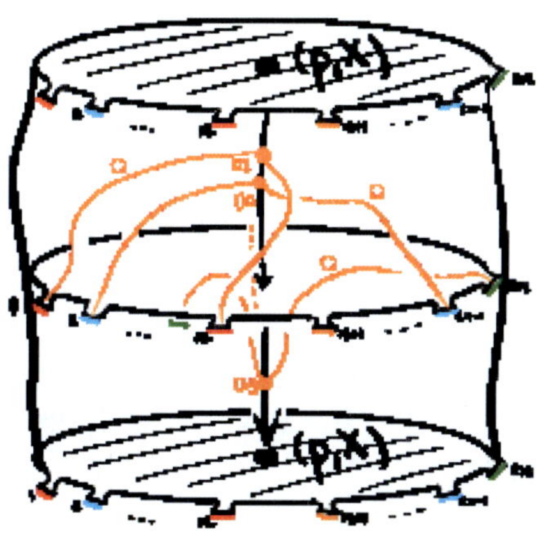

That π_γ is a projection follows from the commutative relations (4.2) and a graphical lemma [**Kir11**, (3.7)]. For morphisms, define the image of $(X,\gamma) \xrightarrow{f} (Y,\beta)$ under G to be the figure above

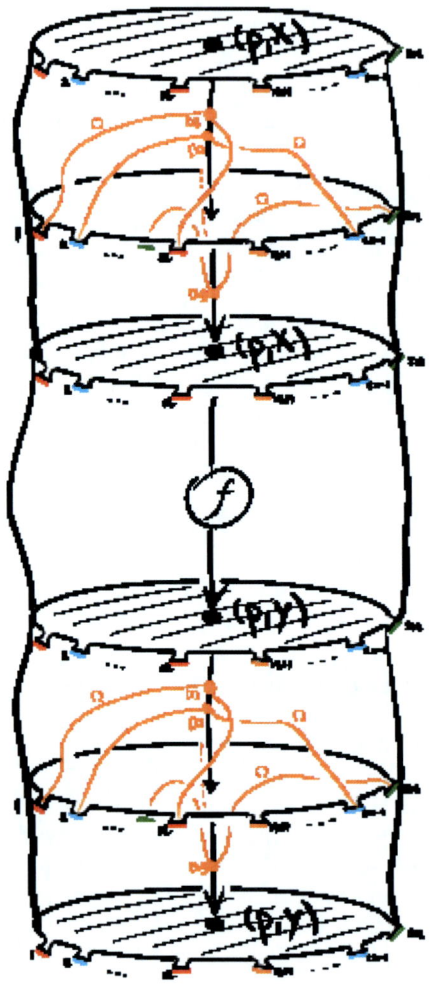

which is indeed a morphism in the Karoubi envelope because π_γ and π_β are idempotents.

5.2.4. *topology* $\leftrightarrow$ *algebra.* In this section, we will show that $J \circ G$ and $G \circ J$ are equivalent to identity functors. The complete proof requires a movie, which is hard to present in a paper. We only sketch the high-level idea, and provide details for a simple case in remark 5.10.

That $G \circ J \simeq 1$ follows directly from the same argument of [**Kir11**, Figure 15]. We just have to do it n times; this is possible because of the communicative relations are suitably chosen. On the other hand, we have $J \circ G \simeq 1$. Indeed, denote $(J \circ G)((X, \gamma)) = (X', \gamma')$. That $X' \simeq X$ follows directly from the sliding lemma 2.18, and that $\gamma' = \gamma$ follows from the sliding lemma, and the fact that half-braidings are by definition monoidal.

Finally, since G and J are additive, this proves the equivalence of both sides as abelian categories. Therefore, $CY_{\Sigma_\sigma}(C)$ and $Z_\sigma(C)$ are equivalent as finite semisimple abelian categories.

REMARK 5.10 (Equivalence for the cylinder case). Let C be a premodular category, $Z(C)$ its Drinfeld center, and Σ the cylinder. In this remark, we show in details that the two functors are inverse to each other:

$$J : Kar(ho.cy_{\Sigma}(C)) \longleftrightarrow Z(C) : G.$$

For convenience, we denote the 3D stringnet as a 2D graph.

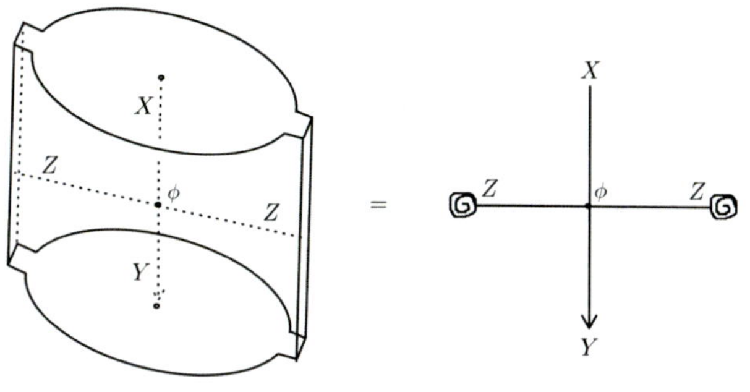

First, we prove that $J \circ G \simeq 1$. By definition, $J \circ G((X, \gamma))$ is $(Im([\Psi]), \gamma') \in Z(C)$, where Ψ is the morphism

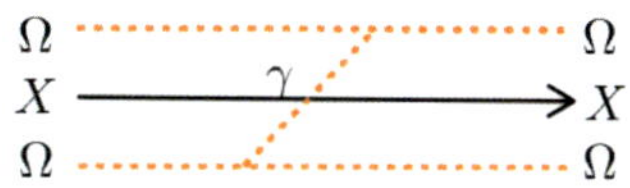

and γ' is the obvious half-braiding using a dual basis. We need to show that $Im([\Psi])$ is functorially isomorphic to X and that γ' is γ up to the isomorphism.

(Proof of $Im([\Psi]) \simeq X$). First denote the image as I. The morphism Ψ is equal to

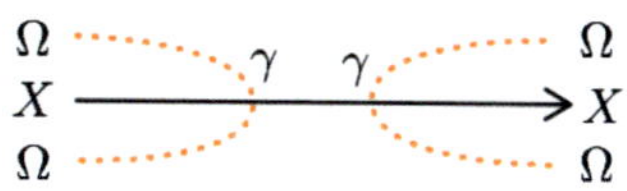

by the monoidal property of γ and the sliding lemma. Denote ϕ^i_j to be $\frac{\sqrt{d_i}\sqrt{d_j}}{D^2}$ times

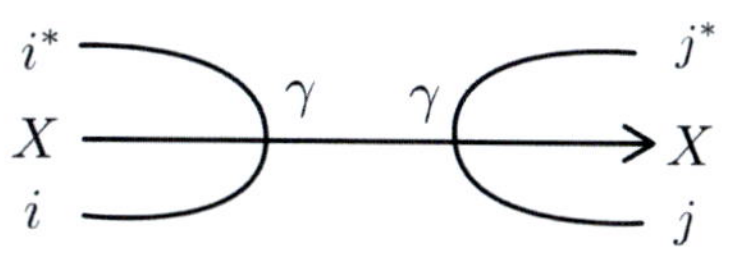

,

$\phi_j := \sum_i \phi^i_j$, $\phi^i := \sum_j \phi^i_j$. Then $\Psi = \sum_{i,j} \phi^i_j = \sum_j \phi_j$. We contend that $Im(\phi^0_j) \subseteq Im(\phi_j) \subseteq Im(\phi^0_j)$. The first containment follows from that ϕ^0_j is a direct summand of ϕ_j; the second containment follows from that $\phi_j = \phi^0_j \circ \phi_0$. Therefore, $Im(\phi^0_j) = Im(\phi_j)$. We then contend that $ker(\phi^0_j) = 0$. Indeed, ϕ^0_j is a composition of an isomorphism and $(coev \otimes 1)$, and the kernel of the later is clearly 0. Therefore $X = X/ker(\phi^0_j) \simeq Im(\phi^0_j) = Im(\phi_j)$. Also, since $\phi_j = \phi^0_j \circ \phi_0$ and $ker(\phi^0_j) = 0$, we have $ker(\phi_j) = ker(\phi_0)$ for each j. Thus $ker(\Psi) = ker(\sum_j \phi_j) = \cap_j ker(\phi_j) = ker(\phi_0)$, and thus $Im(\Psi) = Im(\phi_0) \simeq X$.

(Proof of $\gamma \simeq \gamma'$). It is given in the following graphic:

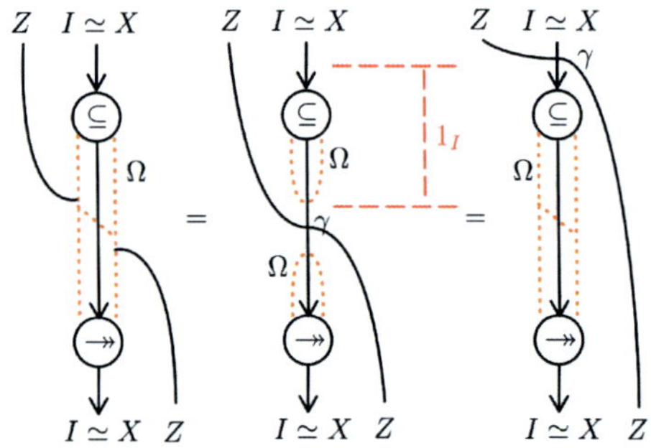

Therefore, $J \circ G \simeq 1_{Z(C)}$.

Second, we show that $G \circ J \simeq 1$. A typical object in $Kar(ho.cy_{\Sigma}(C))$ is an idempotent stringnet $\Gamma_{(X,\phi)}$:

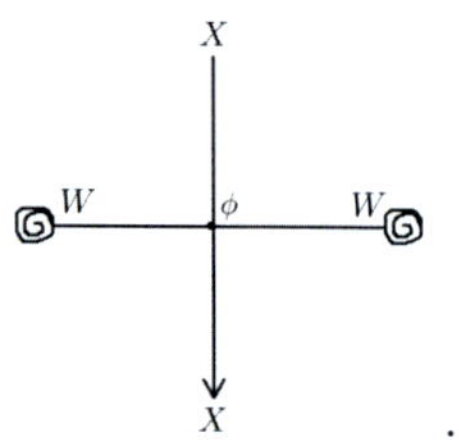

Denote I to be the image of

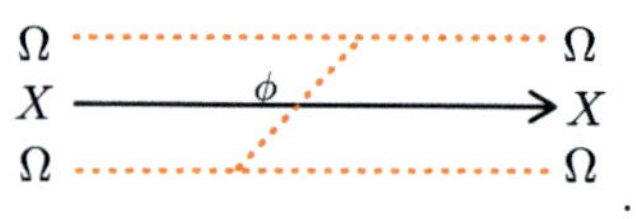

Now, $G \circ J(\Gamma_{(X,\phi)})$ (shorthanded as $\Gamma_{(I,\tilde{\phi})}$) is the following.

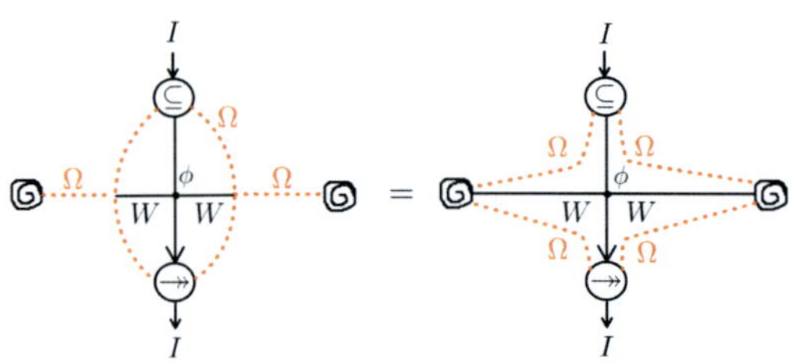

We must show that $\Gamma_{(X,\phi)} \simeq \Gamma_{(I,\tilde{\phi})}$ in $Kar(ho.cy_{\Sigma}(C))$. Define two stringnets $pr_{(X\to I)}$ and $pr_{(I\to X)}$:

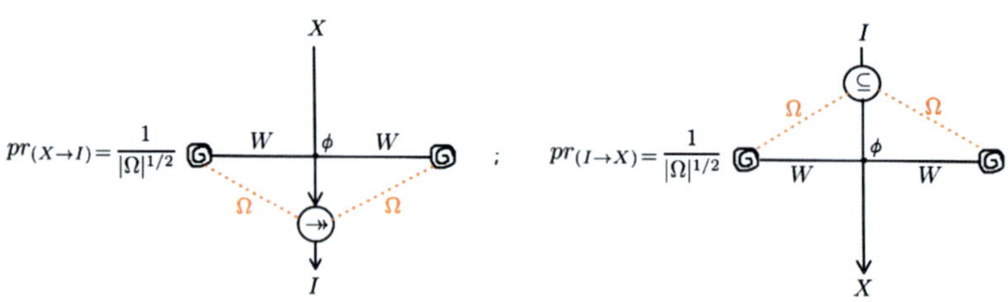

We contend that they are both morphisms in $Kar(ho.cy_\Sigma(C))$. Indeed, by the definition of Karoubi completion, we just have to show the equalities in $ho.cy_\Sigma(C)$:

$$\Gamma_{(I,\tilde{\phi})} \circ pr_{(X\to I)} \circ \Gamma_{(X,\phi)} = pr_{(X\to I)},$$
$$\Gamma_{(X,\phi)} \circ pr_{(I\to X)} \circ \Gamma_{(I,\tilde{\phi})} = pr_{(I\to X)}.$$

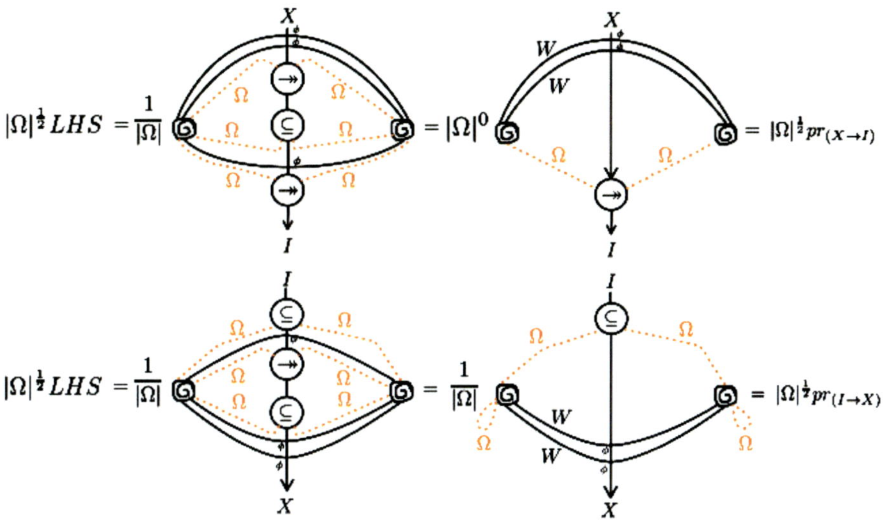

Finally, we need to show that $pr_{(I\to X)}$ and $pr_{(X\to I)}$ are inverse to each other in $Kar(ho.cy_\Sigma(C))$, i.e. we need to show the equalities in $ho.cy_\Sigma(C)$: $pr_{(I\to X)} \circ pr_{(X\to I)} = \Gamma_{(X,\phi)}$ and $pr_{(X\to I)} \circ pr_{(I\to X)} = \Gamma_{(I,\tilde{\phi})}$.

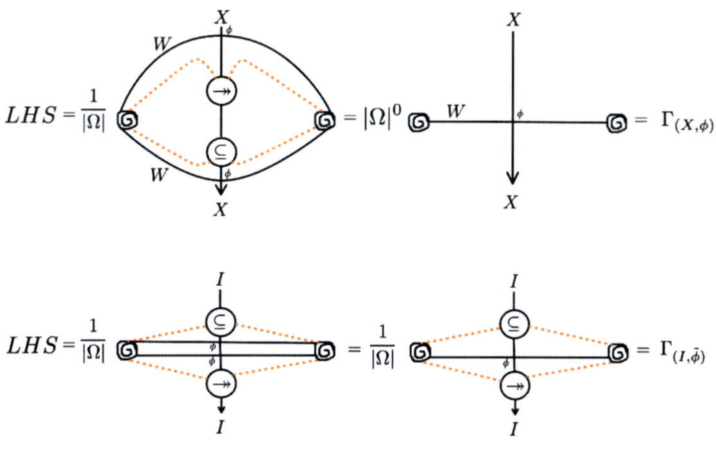

So we are done for the cylinder case. ◇

Acknowledgments

The author is deeply honored to have had the opportunity to contribute to such a beautiful theory. The author would like to express sincere gratitude to their advisor, Alexander Kirillov, for his invaluable guidance and immense patience, without which this work would not have been possible.

The author would also like to extend thanks to David Ben-Zvi, Thibault Décoppet, Stavros Garoufalidis, David Jordan, Ying-Hsuan Lin, Julia Plavnik, and Ying Hong Tham, as well as the committee members of Quantum Topology, for their valuable suggestions and insightful discussions.

Bibliography

[Bal10] Benjamin Balsam, *Turaev-Viro invariants as an extended TQFT. II.*

[Bal11] Benjamin Balsam, *Turaev-Viro invariants as an extended TQFT. III.*

[Ban10] P. Bantay, *Characters of crossed modules and premodular categories*, Moonshine: the first quarter century and beyond, London Math. Soc. Lecture Note Ser., vol. 372, Cambridge Univ. Press, Cambridge, 2010, pp. 1–11. MR2681773

[BDSPV15] Bruce Bartlett, Christopher L. Douglas, Christopher J. Schommer-Pries, and Jamie Vicary, *Modular categories as representations of the 3-dimensional bordism 2-category.*

[BGIM07] John W. Barrett, J. Manuel García-Islas, and João Faria Martins, *Observables in the Turaev-Viro and Crane-Yetter models*, J. Math. Phys. **48** (2007), no. 9, 093508, 18, DOI 10.1063/1.2759440. MR2355097

[BJS21] Adrien Brochier, David Jordan, and Noah Snyder, *On dualizability of braided tensor categories*, Compos. Math. **157** (2021), no. 3, 435–483, DOI 10.1112/s0010437x20007630. MR4228258

[BK02] Bojko Bakalov and Alexander Kirillov, *Lectures on tensor categories and modular functors*, Bull. London Math. Soc. **34** (2002), no. 3, 374–384.

[BZBJ18] David Ben-Zvi, Adrien Brochier, and David Jordan, *Integrating quantum groups over surfaces*, J. Topol. **11** (2018), no. 4, 874–917, DOI 10.1112/topo.12072. MR3847209

[CKY97] Louis Crane, Louis H. Kauffman, and David N. Yetter, *State-sum invariants of 4-manifolds*, J. Knot Theory Ramifications **6** (1997), no. 2, 177–234, DOI 10.1142/S0218216597000145. MR1452438

[Coo23] Juliet Cooke, *Excision of skein categories and factorisation homology*, Adv. Math. **414** (2023), Paper No. 108848, 51, DOI 10.1016/j.aim.2022.108848. MR4536120

[CY93] Louis Crane and David Yetter, *A categorical construction of 4D topological quantum field theories*, Quantum topology, Ser. Knots Everything, vol. 3, World Sci. Publ., River Edge, NJ, 1993, pp. 120–130, DOI 10.1142/9789812796387_0005. MR1273569

[DSPS19] Christopher L. Douglas, Christopher Schommer-Pries, and Noah Snyder, *The balanced tensor product of module categories*, Kyoto J. Math. **59** (2019), no. 1, 167–179, DOI 10.1215/21562261-2018-0006. MR3934626

[EGNO15] Pavel Etingof, Shlomo Gelaki, Dmitri Nikshych, and Victor Ostrik, *Tensor categories*, Mathematical Surveys and Monographs, vol. 205, American Mathematical Society, Providence, RI, 2015, DOI 10.1090/surv/205. MR3242743

[ENO] Pavel Etingof, Dmitri Nikshych, and Viktor Ostrik, *On fusion categories*, Ann. of Math. (2) **162** (2005), no. 2, 581–642, DOI 10.4007/annals.2005.162.581. MR2183279

[KO01] Yang-Hui He and Jun S. Song, *Of McKay correspondence, non-linear sigma-model and conformal field theory*, Adv. Theor. Math. Phys. **4** (2000), no. 4, 747–789, DOI 10.4310/ATMP.2000.v4.n4.a1. MR1867506

[JF15] Theo Johnson-Freyd, *Heisenberg-picture quantum field theory*, Representation theory, mathematical physics, and integrable systems, Progr. Math., vol. 340, Birkhäuser/Springer, Cham, 2021, pp. 371–409, DOI 10.1007/978-3-030-78148-4_13. MR4391354

[Kas12] Christian Kassel, *Quantum groups*, Graduate Texts in Mathematics, vol. 155, Springer-Verlag, New York, 1995, DOI 10.1007/978-1-4612-0783-2. MR1321145

[KB10] Alexander Kirillov and Benjamin Balsam, *Turaev-Viro invariants as an extended TQFT.*

[Kir78] Robion Kirby, *A calculus for framed links in S^3*, Invent. Math. **45** (1978), no. 1, 35–56, DOI 10.1007/BF01406222. MR467753

[Kir11] Alexander Kirillov, *String-net model of Turaev-Viro invariants.*

[KO02] Alexander Kirillov Jr. and Viktor Ostrik, *On a q-analogue of the McKay correspondence and the ADE classification of $\mathfrak{sl}_2$ conformal field theories*, Adv. Math. **171** (2002), no. 2, 183–227, DOI 10.1006/aima.2002.2072. MR1936496

[KT22] Alexander Kirillov Jr. and Ying Hong Tham, *Factorization homology and 4D TQFT*, Quantum Topol. **13** (2022), no. 1, 1–54, DOI 10.4171/qt/159. MR4404797

[Lic12] W. B. Raymond Lickorish, *An introduction to knot theory*, Graduate Texts in Mathematics, vol. 175, Springer-Verlag, New York, 1997, DOI 10.1007/978-1-4612-0691-0. MR1472978

[Lur09] Jacob Lurie, *On the classification of topological field theories*, Current developments in mathematics, 2008, Int. Press, Somerville, MA, 2009, pp. 129–280. MR2555928

[LW05] Michael A. Levin and Xiao-Gang Wen, *String-net condensation: A physical mechanism for topological phases*, Phys. Rev. B **71** (2005), 045110.

[Mue03] Michael Müger, *On the structure of modular categories*, Proc. London Math. Soc. (3) **87** (2003), no. 2, 291–308, DOI 10.1112/S0024611503014187. MR1990929

[Pen71] Roger Penrose, *Angular momentum: an approach to combinatorial space-time*, Living Reviews in Relativity. `http://www.livingreviews.org/lrr-2008-5` Carlo Rovelli, University Press, 1971, pp. 151–180.

[Res] N. Yu. Reshetikhin and V. G. Turaev, *Ribbon graphs and their invariants derived from quantum groups*, Comm. Math. Phys. **127** (1990), no. 1, 1–26. MR1036112

[Tha19] Ying Hong Tham, *The elliptic Drinfeld center of a premodular category*, `arXiv:2019.09511`, 2019.

[Tha21] Ying Hong Tham, *On the Category of Boundary Values in the Extended Crane-Yetter TQFT*, ProQuest LLC, Ann Arbor, MI, 2021. Thesis (Ph.D.)–State University of New York at Stony Brook. MR4336887

[Tha22] Ying Hong Tham, *Reduced tensor product on the Drinfeld center*, J. Algebra **608** (2022), 573–616, DOI 10.1016/j.jalgebra.2022.05.031. MR4447742

[Tur10] Vladimir G. Turaev, *Quantum invariants of knots and 3-manifolds*, Second revised edition, De Gruyter Studies in Mathematics, vol. 18, Walter de Gruyter & Co., Berlin, 2010, DOI 10.1515/9783110221848. MR2654259

[TV92] V. G. Turaev and O. Ya. Viro, *State sum invariants of 3-manifolds and quantum 6j-symbols*, Topology **31** (1992), no. 4, 865–902, DOI 10.1016/0040-9383(92)90015-A. MR1191386

[Wal06] Kevin Walker, *TQFTs*, early incomplete draft.

[Wal15] Kevin Walker, *How are the Walker-Wang TQFT and the Crane-Yetter TQFT related?*, 1 2015.

[WW12] Kevin Walker and Zhenghan Wang, *(3+1)-TQFTs and topological insulators*, Frontiers of Physics (2012).

[Was19] Thomas A. Wasserman, *The reduced tensor product of braided tensor categories containing a symmetric fusion category.*

Department of Mathematics and Statistics, University of Massachusetts Amherst, Amherst, Massachusetts 01002

Email address: `jguu@umass.edu`

Contemporary Mathematics
Volume **813**, 2025
https://doi.org/10.1090/conm/813/16293

Twisted homogeneous racks over the alternating groups

Joseph Vulakh

Abstract. An important step towards the classification of finite-dimensional pointed Hopf algebras is the classification of finite-dimensional Nichols algebras arising from braided vector spaces of group type. This question is fundamentally linked with the structure of algebraic objects called racks. Of particular interest to this classification is the type D condition on racks, a sufficient condition for a rack to not be the source of a finite-dimensional Nichols algebra. In this paper, we study the type D condition in simple racks arising from the alternating groups. Expanding upon previous work in this direction, we make progress towards a general classification of twisted homogeneous racks of type D by proving that several families of twisted homogeneous racks arising from alternating groups are of type D.

1. Introduction

The classification of finite-dimensional pointed Hopf algebras, that is, finite-dimensional Hopf algebras with all simple left or right comodules one-dimensional, has been the subject of a rich line of research. In [**MS**], several non-trivial examples of finite-dimensional pointed Hopf algebras were constructed from small symmetric and dihedral groups, and related quadratic algebras $\mathcal{E}_n$ were proved to be Nichols algebras. The connection between pointed Hopf algebras and Nichols algebras was made clearer by the introduction of a classification program for finite-dimensional pointed Hopf algebras in [**AS**]. This program proposes first classifying braided vector spaces V of group type such that the Nichols algebra $\mathfrak{B}(V)$ is finite dimensional, and then extending this result from Nichols algebras to pointed Hopf algebras constructed from $\mathfrak{B}(V)$ using the lifting method described therein.

In this paper, we consider the classification of pointed Hopf algebras using algebraic objects called racks, which generalize the properties of a group's conjugation action. Racks are important for this method because every Yetter-Drinfeld module over a finite group can be obtained from a finite rack through a braiding defined by 2-cocycles associated with the rack; this connection was uncovered in [**AG**], and a brief overview is given in Section 2. As a result, the first step of the program proposed in [**AS**] can be studied using racks.

Of particular importance to the classification of finite-dimensional pointed Hopf algebras are racks of type D, which, as explained in Section 2, provide a useful way of determining that certain Nichols algebras are infinite dimensional. It is therefore natural to attempt to classify simple racks of type D; this effort was

2020 *Mathematics Subject Classification.* Primary 16T05.

initiated in [**AFGV3**], where most conjugacy classes of the alternating groups and symmetric groups were determined to be of type D. The type D condition was later investigated for twisted homogeneous racks over the alternating groups in [**AFGV1**], for conjugacy classes of groups of Lie type in [**AFGV2, ACG**], for twisted conjugacy calsses of groups of Lie type in [**CGI**], for conjugacy classes of the sporadic simple groups in [**AFGV4**], and for twisted conjugacy classes over the sporadic simple groups in [**FV**]. See Section 3 for a brief summary of prior work as it pertains to the results of this paper.

We study the type D condition specifically for twisted homogeneous racks over the alternating groups, which are an important type of simple rack. Our results build on [**AFGV1**, Theorem 1.2], which showed that all twisted homogeneous racks not listed in Table 3.1 or Table 3.2 are of type D. In this paper, we resolve most cases left unsolved in [**AFGV1**]. Using the notation of [**AFGV1**], also explained in Section 3, our main result is as follows.

THEOREM 1.1. *For $n \geq 5$, a permutation $\ell \in \mathbb{A}_n$, an integer t greater than 1, and an automorphism θ of $\mathbb{A}_n$ given by conjugation by an element of $\mathbb{S}_n$, if the twisted homogeneous rack $\mathcal{C}_\ell$ of type $(\mathbb{A}_n, t, \theta)$ is not of type D, then $\mathcal{C}_\ell$ is described by one of the following:*

(1) $\theta = \mathrm{id}$ *and* $\ell = e$, *with* $\gcd(t, n!) = 1$.
(2) $\theta = \mathrm{id}$ *and* $\ell = e$, *with* $n = 5$ *and* $t = 2$.
(3) $\theta = \mathrm{id}$ *and* $\ell = e$, *with* $n = 6$ *and* $t = 2$.
(4) $\theta = \iota_{(1\ 2)}$ *and ℓ is of cycle type* $(1^{s_1}, 2^{s_2}, \ldots, n^{s_n})$ *with* $s_1 \leq 1$ *and* $s_2 = 0$.
(5) $\theta = \iota_{(1\ 2)}$ *and ℓ is of cycle type* $(1^{s_1}, 2^{s_2}, 4^{s_4})$ *with* $s_1 \leq 2$ *or* $s_2 \geq 1$, *and* $t = 2$.

Theorem 1.1 proves that all racks in Table 3.1 are of type D, leaving only the racks in Table 3.2 not known to be of type D. When $\theta = \mathrm{id} = \iota_e$, the only remaining cases are twisted homogeneous racks $\mathcal{C}_e$. It was found in [**AFGV1**] that the racks $\mathcal{C}_e$ of type $(\mathbb{A}_5, 2, \mathrm{id})$ and $(\mathbb{A}_6, 2, \mathrm{id})$ are not of type D, and it is not known whether these racks collapse. For the other twisted homogeneous racks with $\theta = \mathrm{id}$ listed in Table 3.2, it is still unknown whether they are of type D. The cases remaining when $\theta = \iota_{(1\ 2)}$ are more broad, and include racks $\mathcal{C}_\ell$ with $\ell(1\ 2)$ of arbitrarily large order. In addition, the case when θ is an exceptional automorphism of $\mathbb{A}_6$ has not yet been explored.

We discuss the technical background of our work in Section 2, where we provide definitions of the algebraic structures used in this paper. We review prior work on simple racks, and twisted homogeneous racks over the alternating groups in particular, in Section 3. Finally, in Section 4, we prove our main result, Theorem 1.1.

2. Preliminaries

We denote the group identity by e throughout this paper. In this section, we summarize the technical preliminaries of our work.

We first describe basic notions related to racks. See [**AG**] and [**AFGV3**] for more details. A *rack* is a pair $(X, \triangleright)$, with X a non-empty set and $\triangleright\colon X \times X \to X$

a function, such that:

(1) For all x in X, the map $\varphi_x\colon X \to X$ given by

$$\varphi_x(y) = x \rhd y$$

is a bijection.

(2) For all x, y, z in X,

$$x \rhd (y \rhd z) = (x \rhd y) \rhd (x \rhd z).$$

Additional definitions concerning racks are listed below.

- A *morphism* of racks is a function $f\colon X \to Y$ such that

$$f(x \rhd y) = f(x) \rhd f(y)$$

for all x, y in X.
- A *subrack* of a rack X is a non-empty subset Y such that $Y \rhd Y = Y$.
- A *decomposition* of a rack X is a family $(X_i)_{i \in I}$ of pairwise disjoint subracks of X such that $X = \coprod_{i \in I} X_i$ and $X \rhd X_i = X_i$ for all i in I.
- A rack X is *decomposable* if there exists a decomposition of X.
- A rack is *trivial* if it contains exactly one element.

An important construction of racks comes from group conjugation. Let G be a group, and let $x \rhd y = xyx^{-1}$ for x, y in G. Then $(G, \rhd)$ is a rack, and conjugacy classes of G are subracks.

More generally, we consider twisted conjugacy classes of the group G. Let $f \in \mathrm{Aut}(G)$, and consider the action of G on itself given by $y \rightharpoonup_f x = yxf(y^{-1})$. For x in G, the orbit $\mathcal{O}_x^{G,f}$ of x under this action is called the *twisted conjugacy class* of x and is a rack with operation $y \rhd_f z = yf(zy^{-1})$. We will refer to these racks as *twisted conjugacy classes of type* (G, f).

Another important rack construction is that of an *affine rack*. Let A be a finite abelian group, and let $g\colon A \to A$ be an automorphism. Define the affine rack (A, g) as the rack with underlying set A and rack operation $x \rhd y = x + g(y - x)$. Letting $f = \mathrm{id} - g$, this can be written as $x \rhd y = f(x) + g(y)$ or $x \rhd y = f(x - y) + y$.

We now consider cocycles and Nichols algebras. Let $(X_i)_{i \in I}$ be a decomposition of a rack X, and let $\mathbf{n} = (n_i)_{i \in I}$ be a family of positive integers. A 2-*cocycle of degree* $\mathbf{n}$ is a family $\mathbf{q} = (q_i)_{i \in I}$ of functions $q_i\colon X \times X_i \to \mathrm{GL}(n_i, \mathbb{C})$ satisfying

$$q_i(x, y \rhd z) q_i(y, z) = q_i(x \rhd y, x \rhd z) q_i(x, z)$$

for all $i \in I$, $x, y \in X$, $z \in X_i$.

Let $V = \bigoplus_{i \in I} \mathbb{C}X_i \otimes \mathbb{C}^{n_i}$, and denote the element of $\mathbb{C}X_i$ corresponding to an element x of X_i by e_x. For a family $\mathbf{q} = (q_i)_{i \in I}$ of functions $q_i\colon X \times X_i \to \mathrm{GL}(n_i, \mathbb{C})$, consider the linear map $c^{\mathbf{q}}\colon V \otimes V \to V \otimes V$ acting on the basis elements of $V \otimes V$ by

$$c^{\mathbf{q}}(e_x v \otimes e_y w) = e_{x \rhd y} q_i(x, y)(w) \otimes e_x v$$

for $x \in X_j$, $y \in X_i$, $v \in \mathbb{C}^{n_j}$, $w \in \mathbb{C}^{n_i}$. By [**AG**, Theorem 4.14], the map $c^{\mathbf{q}}$ satisfies the braid equation

$$(\mathrm{id} \otimes c^{\mathbf{q}})(c^{\mathbf{q}} \otimes \mathrm{id})(\mathrm{id} \otimes c^{\mathbf{q}}) = (c^{\mathbf{q}} \otimes \mathrm{id})(\mathrm{id} \otimes c^{\mathbf{q}})(c^{\mathbf{q}} \otimes \mathrm{id})$$

if and only if $\mathbf{q}$ is a 2-cocycle. Moreover, if $\mathbf{q}$ is a 2-cocycle, then there exists a group G such that V is a Yetter-Drinfeld module over G, and any Yetter-Drinfeld module over a finite group can be realized in this way. The Nichols algebra of the Yetter-Drinfeld module obtained in this way is denoted $\mathfrak{B}(X, \mathbf{q})$.

Let X be a rack, $\mathbf{q}$ a 2-cocycle, and V a vector space as above. The 2-cocycle $\mathbf{q}$ is *finite* if the image of q_i generates a finite subgroup of $\mathrm{GL}(n_i, \mathbb{C})$ for all $i \in I$. Let $g\colon X \to \mathrm{GL}(V)$ be the map given by

$$g_x(e_y w) = e_{x \triangleright y} q_i(x, y)(w)$$

for $i \in I$, $x \in X$, $y \in X_i$, $w \in \mathbb{C}^{n_i}$. Using the rack structure of $\mathrm{GL}(V)$ given by conjugation, the map g is a rack morphism. The cocycle $\mathbf{q}$ is *faithful* if g is injective. These notions are important for the following definition.

DEFINITION 2.1 ([**AFGV3**, Definition 2.2]). A finite rack X *collapses* if for every finite faithful cocycle $\mathbf{q}$ associated with any decomposition of X and any degree $\mathbf{n}$, the Nichols algebra $\mathfrak{B}(X, \mathbf{q})$ is infinite-dimensional.

We close this section by introducing the type D condition, which is the main focus of this paper.

DEFINITION 2.2. [**AFGV3**, Definition 3.5] A rack X is of *type D* if there exists a decomposable subrack $Y = R \coprod S$ of X such that

$$r \triangleright (s \triangleright (r \triangleright s)) \neq s \tag{2.1}$$

for some $r \in R$ and $s \in S$.

REMARK 2.3. When X is the rack arising from a conjugacy class $\mathcal{O}_x^G$ of a group G, Equation (2.1) is equivalent to the condition that $(rs)^2 \neq (sr)^2$. The condition that R and S are disjoint translates into the condition that r and s are not conjugates in the group generated by r and s.

The significance of this definition stems from the following result, which, loosely speaking, states that any Nichols algebra arising from a rack of type D is infinite-dimensional, allowing for a rack to be discarded from consideration as being the source of a finite-dimensional Nichols algebra.

THEOREM 2.4 ([**AFGV3**, Theorem 3.6]). *If X is a finite rack of type D, then X collapses.*

3. Prior Work

Of particular research interest are those racks which have no nontrivial quotients. We consider such racks in this section.

A rack X is *simple* if it is not trivial and for any rack morphism $f\colon X \to Y$, either $|Y| = 1$ or $|Y| = |X|$. Simple racks have been classified in [**AG**, Theorem 3.9–Theorem 3.12].

THEOREM 3.1 ([**AG**, Theorem 3.9–Theorem 3.12]). *For a simple rack X, exactly one of the following holds:*

(1) *$|X| = p$, where p is prime, and $X \cong \mathbb{F}_p$, with $x \triangleright y = y + 1$. Such racks are called* permutation racks.

(2) $|X| = p^t$, *with* p *a prime and* t *a positive integer, and* $X \cong (\mathbb{F}_p^t, T)$ *an affine rack, where* T *is the companion matrix of an irreducible monic polynomial in* $\mathbb{F}_p[X]$ *of degree* t *different from* X *and* $X - 1$.

(3) $|X|$ *is divisible by at least two distinct primes, and there exists a non-abelian simple group* L, *a positive integer* t, *and an automorphism* θ *in* $\operatorname{Aut}(L)$ *such that* X *is a twisted conjugacy class of* L^t *with respect to the automorphism*

$$x\colon L^t \to L^t, \quad x(\ell_1, \dots, \ell_t) = (\theta(\ell_t), \ell_1, \dots, \ell_{t-1}).$$

Racks produced by such a construction are called twisted homogeneous racks. *In the case when* $t = 1$, *the rack* X *is a twisted conjugacy class of a simple group.*

We refer to simple racks as in (3) as *twisted homogeneous racks of type* (L, t, θ). Following [**AFGV1**], we denote the twisted homogeneous rack of $(e, \dots, e, \ell)$ in L^t by $\mathcal{C}_\ell$. The power of this notation stems from the following theorem.

THEOREM 3.2 ([**AFGV1**, Proposition 3.3]). *An element* $(x_1, \dots, x_t)$ *of* L^t *is in* $\mathcal{C}_\ell$ *if and only if* $x_t x_{t-1} \cdots x_2 x_1 \in \mathcal{O}_\ell^{L,\theta}$.

Theorem 3.2 establishes a bijection between twisted homogeneous racks of type (L, t, θ) and twisted conjugacy classes of L and allows twisted homogeneous racks to be identified by a single element of L.

In this paper, we are interested in simple racks over the alternating groups. By Theorem 3.1, the twisted homogeneous racks of type $(\mathbb{A}_n, t, \theta)$ with $n \geq 5$ are simple. It is well-known that, for $n \neq 6$, the automorphisms of $\mathbb{A}_n$ are given by conjugation by an element of $\mathbb{S}_n$ and $\operatorname{Out}(\mathbb{A}_n) \cong C_2$ (see, for example, [**S**, Chapter 3.2]). We therefore consider only the automorphisms $\theta = \mathrm{id} = \iota_e$ and $\theta = \iota_{(1\ 2)}$, where ι_u acts on $\mathbb{A}_n$ by conjugation by u.

The case of $t = 1$ has been considered in [**AFGV3**] and [**AFGV2**]. Set u to equal e if $\theta = \mathrm{id}$ and $(1\ 2)$ if $\theta = \iota_{(1\ 2)}$. It is known that if the cycle type of xu is not in the following list, then $\mathcal{O}_x^{G,\theta}$ is of type D [**AFGV3**, Theorem 4.1] (see also [**AFGV2**, Theorem 6.1]):

(a) $(2, 3)$; (2^3); $(1^k, 2)$.
(b) (3^2); $(2^2, 3)$; $(1^k, 3)$; (2^4); $(1^2, 2^2)$; $(1, 2^2)$; $(1, p)$, (p), where p is prime.

Even if xu has cycle type in list (b), $\mathcal{O}_x^{G,\theta}$ is known to collapse [**AFGV3**, Theorem 1.1].

For the rest of this paper, we consider only the case of $t > 1$, which was first investigated in [**AFGV1**]. Many twisted homogeneous racks over the alternating groups were shown to be of type D in [**AFGV1**, Theorem 1.2], but many racks, listed in Table 3.1 and Table 3.2, remained not known to be of type D. Interestingly, two twisted homogeneous racks, listed as part of Table 3.2, were found *not* to be of type D in [**AFGV1**], and it is not known whether these racks collapse.

Theorem 1.1 resolves all cases listed in Table 3.1.

TABLE 3.1. Twisted homogeneous racks $\mathcal{C}_\ell$ of type $(\mathbb{A}_n, t, \iota_u)$, where $t > 1$ and $n \geq 5$, shown to be of type D by Theorem 1.1.

u	n	Type of ℓu	t	Resolved by
e	5	(1^5)	4	Proposition 4.1
	5	$(1, 2^2)$	4, odd	Proposition 4.2
	6	$(1^2, 2^2)$	odd	
	8	(2^4)	odd	
	any	$(1^{r_1}, 2^{r_2}, 4^{r_4})$ with $r_4 > 0$, $r_2 + r_4$ even	2	Proposition 4.3, Proposition 4.4
(1 2)	5	$(1^3, 2)$	2, 4	Proposition 4.5
	6	$(1^4, 2)$	2	Proposition 4.5
	6	(2^3)	2	Proposition 4.6
	7	$(1, 2^3)$	2, odd	Proposition 4.7
	8	$(1^2, 2^3)$	odd	
	10	(2^5)	odd	

TABLE 3.2. Twisted homogeneous racks $\mathcal{C}_\ell$ of type $(\mathbb{A}_n, t, \iota_u)$, where $t > 1$ and $n \geq 5$, not known to be of type D. Two of these racks were found not to be of type D in [**AFGV1**].

u	n	Type of ℓu	t	Type D
e	any	(1^n)	$\gcd(t, n!) = 1$	unknown
	5	(1^5)	2	not type D
	6	(1^6)	2	not type D
(1 2)	any	$(1^{s_1}, 2^{s_2}, \dots, n^{s_n})$ with $s_1 \leq 1$, $s_2 = 0$, $s_h \geq 1$ for some h with $3 \leq h \leq n$	any	unknown
		$(1^{s_1}, 2^{s_2}, 4^{s_4})$ with $s_1 \leq 2$ or $s_2 \geq 1$, $s_2 + s_4$ odd, $s_4 \geq 1$	2	unknown

4. Twisted Homogeneous Racks

In this section, we prove Theorem 1.1, considering first the twisted homogeneous racks arising from the identity automorphism in Section 4.1, and then those arising from $\iota_{(1\ 2)}$, the automorphism given by conjugation by (1 2), in Section 4.2.

4.1. Identity Automorphism. In this section, we prove that the racks listed in the first section of Table 3.1 are of type D. We begin with the case when $\ell = e$.

PROPOSITION 4.1. *For even t greater than or equal to 4 and $n \geq 5$, the twisted homogeneous rack $\mathcal{C}_e$ of type $(\mathbb{A}_n, t, \mathrm{id})$ is of type D.*

PROOF. Let $G = \{e, (1\ 2\ 3), (1\ 3\ 2), (1\ 2)(4\ 5), (2\ 3)(4\ 5), (1\ 3)(4\ 5)\}$. The group G is isomorphic to $\mathbb{S}_3$, with the isomorphism φ given by considering the action of G only on $\{1, 2, 3\}$.

Let R be the set of t-tuples $(g_1, \dots, g_t)$ of elements of G with $g_t g_{t-1} \cdots g_2 g_1 = e$ and each $\varphi(g_i)$ an even permutation, and let S be the set of t-tuples $(g_1, \dots, g_t)$ of elements of G with $g_t g_{t-1} \cdots g_2 g_1 = e$ and each $\varphi(g_i)$ an odd permutation. Let

$Y = R \cup S$. The set Y is a subset of $\mathcal{C}_e$ by Theorem 3.2, and because t is even, R and S are nonempty. Moreover, for two elements $a = (a_1, \dots, a_t)$ and $b = (b_1, \dots, b_t)$ of Y,

$$\begin{aligned} a \triangleright b &= (a_1, \dots, a_t) \triangleright (b_1, \dots, b_t) \\ &= \left(a_1 b_t a_t^{-1}, a_2 b_1 a_1^{-1}, \dots, a_t b_{t-1} a_{t-1}^{-1}\right), \end{aligned}$$

each element of which has the same parity upon mapping through φ as the corresponding b_i. Also, the product of the tuple elements of $a \triangleright b$ is

$$\begin{aligned} (a_t b_{t-1} a_{t-1}^{-1})(a_{t-1} b_{t-2} a_{t-2}^{-1}) \cdots (a_2 b_1 a_1^{-1})(a_1 b_t a_t^{-1}) &= a_t (b_{t-1} b_{t-2} \cdots b_1) b_t a_t^{-1} \\ &= a_t b_t^{-1} b_t a_t^{-1} \\ &= e, \end{aligned}$$

so the sets Y, R, and S are closed under the rack operation. Therefore, Y, R, and S are racks, and $Y = R \coprod S$.

Let $x = (1\ 2)(4\ 5)$ and $y = (2\ 3)(4\ 5)$. Now let $r = e \in R$ and $s = (x, x, y, y, \dots) \in S$, with each element of s except for the first two equal to y. Then

$$\begin{aligned} r \triangleright (s \triangleright (r \triangleright s)) &= r \triangleright (s \triangleright (y, x, x, y, \dots)) \\ &= r \triangleright (x, xyx^{-1}, y, yxy^{-1}, \dots) \\ &= \begin{cases} (yxy^{-1}, x, xyx^{-1}, y), & \text{if } t = 4 \\ (y, x, xyx^{-1}, y, yxy^{-1}, \dots) & \text{if } t > 4 \end{cases} \\ &\neq s, \end{aligned}$$

so $\mathcal{C}_\ell$ is of type D. □

We now consider the case when ℓ is an involution.

PROPOSITION 4.2. *For ℓ in $\mathbb{A}_n$ and of cycle type $(1^{r_1}, 2^{r_2})$, with r_2 even and positive, the twisted homogeneous rack $\mathcal{C}_\ell$ of type $(\mathbb{A}_n, t, \mathrm{id})$, with $t \geq 3$, is of type D.*

PROOF. Let

$$\begin{aligned} x &= \prod_{i=1}^{\frac{r_2}{2}} (4i-3\ 4i-2)(4i-1\ 4i) \\ y &= \prod_{i=1}^{\frac{r_2}{2}} (4i-3\ 4i-1)(4i-2\ 4i) \\ z &= \prod_{i=1}^{\frac{r_2}{2}} (4i-3\ 4i)(4i-2\ 4i-1). \end{aligned}$$

Then $\{e, x, y, z\}$ is isomorphic to the Klein four-group, which is abelian. Let Y be the set consisting of t-tuples $(c_1, \dots, c_t)$ with each c_i in $\{e, x, y, z\}$ and with $\prod_{i=1}^t c_i$ equal to x or y. By Theorem 3.2, Y is a subset of $\mathcal{C}_\ell$. Because the Klein four-group is abelian, the rack action in Y does not change the product of the elements of a tuple in Y. Therefore, letting R be the subset of Y consisting of tuples $(c_1, \dots, c_t)$ with $\prod_{i=1}^t c_i = x$ and S be the subset of Y consisting of tuples $(c_1, \dots, c_t)$ with $\prod_{i=1}^t c_i = y$, we find that R, S, and $Y = R \coprod S$ are racks.

Now let $r = (x, e, \dots) \in R$ and $s = (y, e, \dots) \in S$. These elements satisfy Equation (2.1) when $t \geq 3$. Indeed,

$$\begin{aligned} s \triangleright (r \triangleright s) &= s \triangleright (x, z, e, \dots) \\ &= (y, z, z, \dots), \end{aligned}$$

so $r \triangleright (s \triangleright (r \triangleright s))$ is equal to (y, z, z) if $t = 3$, or to $(x, z, z, z, \dots)$ if $t \geq 4$. In both cases, Equation (2.1) is satisfied. □

Proposition 4.2 resolves the first three rows of Table 3.1. In particular, it proves that racks $\mathcal{C}_\ell$ of type $(\mathbb{A}_n, t, \mathrm{id})$ with n equal to 5 or 6, the integer t equal to 4 or odd and greater than or equal to 3, and ℓ an involution of type $(1, 2^2)$ or $(1^2, 2^2)$ and racks $\mathcal{C}_\ell$ of type $(\mathbb{A}_8, t, \mathrm{id})$ with t odd and greater than or equal to 3 and ℓ an involution of type (2^4) are all of type D. We proceed to resolve the last row of the first section of Table 3.1.

PROPOSITION 4.3. *For ℓ in $\mathbb{A}_n$ and of cycle type $(1^{r_1}, 2^{r_2}, 4^{r_4})$, with $r_2, r_4 > 0$, the twisted homogeneous rack $\mathcal{C}_\ell$ of type $(\mathbb{A}_n, 2, \mathrm{id})$ is of type D.*

PROOF. Let $x = (1\ 2)(3\ 4\ 5\ 6)$ and $y = (1\ 2\ 3\ 6)(4\ 5)$, and let G be the subgroup of $\mathbb{A}_n$ generated by x and y. The action of G on the set $S = \{(1\,3\,5), (1\,5\,3), (2\,4\,6), (2\,6\,4)\}$ by conjugation induces a homomorphism φ into the cyclic group C_4, with x and y mapping to distinct generators. Indeed, the action of x is given by the cycle $((1\ 3\ 5)(2\ 4\ 6)(1\ 5\ 3)(2\ 6\ 4))$, and the action of y is given by the cycle $((1\ 3\ 5)(2\ 6\ 4)(1\ 5\ 3)(2\ 4\ 6))$.

Let π be an element of $\mathbb{A}_n$ acting only on numbers greater than 6 and with cycle type $(1^{r_1}, 2^{r_2-1}, 4^{r_4-1})$. Now let R be the set consisting of pairs (r_1, r_2) such that $r_1\pi^{-1}$ and r_2 are in G and $r_1\pi^{-1}$ and r_2 have product mapped to $\varphi(x)$ by φ. All elements of $\mathbb{A}_6$ which act on S in the same way x does have cycle type $(2, 4)$, so R is a subset of $\mathcal{C}_\ell$ by Theorem 3.2. Also, the rack operation in R is of the form $(g_1\pi, h_1) \triangleright (g_2\pi, h_2) = (g_1h_2h_1^{-1}\pi, h_1g_2g_1^{-1})$ and $\varphi(g_1h_2g_2g_1^{-1}) = \varphi(g_2h_2)$, so R is closed under the rack operation and hence is a rack. Similarly, let S be the subrack of $\mathcal{C}_\ell$ consisting of pairs (s_1, s_2) such that $s_1\pi^{-1}$ and s_2 are in G and $s_1\pi^{-1}$ and s_2 have product mapping to $\varphi(y)$.

The set $Y = R \coprod S$ is a subrack of $\mathcal{C}_\ell$. Finally,

$$\begin{aligned} (x\pi, e) \triangleright (y\pi, e) &= (x\pi, yx^{-1}) \\ (y\pi, e) \triangleright (x\pi, yx^{-1}) &= (y^2x^{-1}\pi, xy^{-1}) \\ (x\pi, e) \triangleright (y^2x^{-1}\pi, xy^{-1}) &= (x^2y^{-1}\pi, y^2x^{-2}) \\ &\neq (y\pi, e), \end{aligned}$$

with $(x\pi, e) \in R$ and $(y\pi, e) \in S$, so $\mathcal{C}_\ell$ is of type D. □

Proposition 4.3 covers all ℓ included within the last row of the first section of Table 3.1 satisfying $r_2 > 0$. Because r_4 must be even in order for ℓ to be in $\mathbb{A}_n$, the only case remaining for this row is when $r_2 = 0$ and $r_4 > 1$, addressed in the following proposition.

PROPOSITION 4.4. *For ℓ in $\mathbb{A}_n$ and of cycle type $(1^{r_1}, 4^{r_4})$, with $r_4 > 1$, the twisted homogeneous rack $\mathcal{C}_\ell$ of type $(\mathbb{A}_n, 2, \mathrm{id})$ is of type D.*

PROOF. This follows from the same reasoning as Proposition 4.3, but with $x = (1\ 2\ 3\ 4)(5\ 6\ 7\ 8)$ and $y = (1\ 6\ 7\ 8)(2\ 3\ 4\ 5)$ acting on the set $\{(1\ 3\ 5\ 7), (1\ 7\ 5\ 3), (2\ 4\ 6\ 8), (2\ 8\ 6\ 4)\}$. The element x acts by the cycle $((1\ 3\ 5\ 7)(2\ 4\ 6\ 8)(1\ 7\ 5\ 3)(2\ 8\ 6\ 4))$, and the element y acts by the cycle $((1\ 3\ 5\ 7)(2\ 8\ 6\ 4)(1\ 7\ 5\ 3)(2\ 4\ 6\ 8))$, giving the needed homomorphism $\varphi\colon G \to C_4$. □

4.2. Automorphism by Conjugation by a Transposition. Recall that the automorphism of $\mathbb{A}_n$ acting by conjugation by $(1\ 2)$ is denoted $\iota_{(1\ 2)}$. In this section, we prove that the twisted homogeneous racks in the second section of Table 3.1 are of type D. We begin with the case when $\ell(1\ 2)$ is a transposition.

PROPOSITION 4.5. *For even t and $n \geq 5$, the twisted homogeneous rack $\mathcal{C}_e$ of type $(\mathbb{A}_n, t, \iota_{(1\ 2)})$ is of type D.*

PROOF. Let $G = \{e, (1\ 2\ 3), (1\ 3\ 2), (1\ 2)(4\ 5), (2\ 3)(4\ 5), (1\ 3)(4\ 5)\}$. As observed in the proof of Proposition 4.1, the group G is isomorphic to $\mathbb{S}_3$, with the isomorphism φ given by considering the action of G only on $\{1, 2, 3\}$.

Let R be the set of t-tuples of elements g of G with $\varphi(g)$ an even permutation, let S be the set of t-tuples of elements g of G with $\varphi(g)$ an odd permutation, and let $Y = R \cup S$. Because each element of G with even image under φ is in $\mathcal{O}_e^{\mathbb{A}_n, \iota_{(1\ 2)}}$ and t is even, Y is a subset of $\mathcal{C}_e$ by Theorem 3.2. Moreover, for any two elements $a = (a_1, \ldots, a_t)$ and $b = (b_1, \ldots, b_t)$ of Y,

$$\begin{aligned} a \triangleright b &= (a_1, \ldots, a_t) \triangleright (b_1, \ldots, b_t) \\ &= \left(a_1 (1\ 2) b_t a_t^{-1} (1\ 2), a_2 b_1 a_1^{-1}, \ldots, a_t b_{t-1} a_{t-1}^{-1}\right), \end{aligned}$$

each element of which has the same parity upon mapping through φ as the corresponding b_i. Therefore, Y, R, and S are racks, and $Y = R \coprod S$.

Finally, let $r = e$ and $s = ((1\ 3)(4\ 5), \ldots, (1\ 3)(4\ 5))$ in $\mathcal{C}_e$. Then

$$\begin{aligned} r \triangleright (s \triangleright (r \triangleright s)) &= r \triangleright (s \triangleright ((2\ 3)(4\ 5), (1\ 3)(4\ 5), \ldots)) \\ &= r \triangleright ((1\ 3)(4\ 5), (1\ 2)(4\ 5), \ldots) \\ &= \begin{cases} ((1\ 2)(4\ 5), (1\ 3)(4\ 5)), & \text{if } t = 2 \\ ((2\ 3)(4\ 5), (1\ 3)(4\ 5), (1\ 2)(4\ 5), (1\ 3)(4\ 5), \ldots) & \text{if } t > 2 \end{cases} \\ &\neq s, \end{aligned}$$

so $\mathcal{C}_\ell$ is of type D, as claimed. □

We now address the case when $\ell(1\ 2)$ is a product of a larger number of transpositions, beginning with the case $t = 2$.

PROPOSITION 4.6. *For $\ell(1\ 2)$ in $\mathbb{S}_n$ and of type (2^3), the twisted homogeneous rack $\mathcal{C}_\ell$ of type $(\mathbb{A}_n, 2, \iota_{(1\ 2)})$ is of type D.*

PROOF. Let R and S be the subracks of $\mathcal{C}_\ell$ defined by Table 4.1, with each row corresponding to one element of each rack. Each row determines an element of each rack; the first column contains the first permutation of the pair, and the column corresponding to the rack (either R or S) contains the second permutation of the pair.

The racks R and S can also be described as follows. Let $x = (3\ 4)(5\ 6)$ and $y = (3\ 6)(4\ 5)$. For each element π of the alternating group acting on $\{3, 4, 5, 6\}$,

TABLE 4.1. Racks R and S used in the proof of Proposition 4.6. Each row describes an element of R and an element of S: if the permutations, reading across, are π, σ, and τ, then $(\pi, \sigma) \in R$ and $(\pi, \tau) \in S$.

First permutation	Second permutation	
	R	S
e	(3 4)(5 6)	(3 6)(4 5)
(4 5 6)	(3 5 6)	(3 4 5)
(4 6 5)	(3 6 5)	(3 5 4)
(3 4)(5 6)	e	(3 5)(4 6)
(3 4 5)	(3 4 6)	(4 5 6)
(3 4 6)	(3 4 5)	(3 5 6)
(3 5 4)	(3 6 4)	(4 6 5)
(3 5 6)	(4 5 6)	(3 4 6)
(3 5)(4 6)	(3 6)(4 5)	(3 4)(5 6)
(3 6 4)	(3 5 4)	(3 6 5)
(3 6 5)	(4 6 5)	(3 6 4)
(3 6)(4 5)	(3 5)(4 6)	e

let the element $(\pi, x\pi^{-1}x)$ be in R if x is not an involution, and let the element $(\pi, x\pi)$ be in R if x is an involution. The elements of S are characterized similarly using y.

The sets R and S so defined can be checked to be racks with $Y = R \coprod S$ also a rack. Moreover, setting $r = ((3\ 4)(5\ 6), e)$ and $s = ((3\ 4\ 5), (4\ 5\ 6))$, we obtain

$$\begin{aligned} r \triangleright (s \triangleright (r \triangleright s)) &= r \triangleright (s \triangleright ((3\ 4\ 6), (3\ 5\ 6))) \\ &= r \triangleright (e, (3\ 6)(4\ 5)) \\ &= ((3\ 5)(4\ 6), (3\ 4)(5\ 6)) \\ &\neq s, \end{aligned}$$

so $\mathcal{C}_\ell$ is of type D. □

We now address the case $t \geq 3$.

PROPOSITION 4.7. *For $\ell(1\ 2)$ in $\mathbb{S}_n$ and of type $(1^{r_1}, 2^{r_2})$, with r_2 odd and greater than* 1, *the twisted homogeneous rack $\mathcal{C}_\ell$ of type $(\mathbb{A}_n, t, \iota_{(1\ 2)})$, with $t \geq 3$, is of type D.*

PROOF. Because $\iota_{(1\ 2)}$ leaves permutations which act trivially on $\{1, 2\}$ fixed, the twisted homogeneous rack $\mathcal{C}'_{\ell'}$ of type $(\mathbb{A}_{\{3,\dots,n\}}, t, \mathrm{id})$ is a subrack of $\mathcal{C}_\ell$ for ℓ' of cycle type $(1^{r_1+2}, 2^{r_2-1})$ and leaving 1 and 2 fixed. Thus, the claim follows from Proposition 4.2. □

4.3. Proof of Main Result. In this section, we use the results of Section 4.1 and Section 4.2 to prove Theorem 1.1.

PROOF OF THEOREM 1.1. All cases not listed in Table 3.1 or Table 3.2 are resolved by [**AFGV1**, Theorem 1.2]. The case listed in the first section of Table 3.1 with $\ell = e$ is resolved by Proposition 4.1, the cases in this section with ℓ an involution are resolved by Proposition 4.2, and the last case listed in this section is resolved by Proposition 4.3 and Proposition 4.4. The first two cases listed in the second section of Table 3.1 are resolved by Proposition 4.5, the case with $\ell(1\ 2)$ of type (2^3) is resolved by Proposition 4.6, and the last three cases listed in Table 3.1 are resolved by Proposition 4.7. The results which show each case to be of type D are also listed in Table 3.1.

This leaves only the racks listed in Table 3.2 not known to be of type D. Therefore, any twisted homogeneous rack over a simple alternating group that is not of type D is in the list given in the statement of Theorem 1.1. □

Acknowledgments

The author would like to thank Dr. Julia Plavnik and Dr. Héctor Peña Pollastri for their mentorship, and the MIT PRIMES research program for making this work possible. The author would also like to thank Dr. Leandro Vendramin for his valuable advice on an earlier version of this manuscript, and the Reviewer for their thoughtful comments and suggestions.

References

[ACG] Nicolás Andruskiewitsch, Giovanna Carnovale, and Gastón Andrés García, *Finite-dimensional pointed Hopf algebras over finite simple groups of Lie type I. Non-semisimple classes in* $\mathbf{PSL}_n(q)$, J. Algebra **442** (2015), 36–65, DOI 10.1016/j.jalgebra.2014.06.019. MR3395052

[AFGV1] N. Andruskiewitsch, F. Fantino, G. A. García, and L. Vendramin, *On twisted homogeneous racks of type D*, Rev. Un. Mat. Argentina **51** (2010), no. 2, 1–16. MR2840161

[AFGV2] N. Andruskiewitsch, F. Fantino, G. A. García, and L. Vendramin, *On Nichols algebras associated to simple racks*, Groups, algebras and applications, Contemp. Math., vol. 537, Amer. Math. Soc., Providence, RI, 2011, pp. 31–56, DOI 10.1090/conm/537/10565. MR2799090

[AFGV3] N. Andruskiewitsch, F. Fantino, M. Graña, and L. Vendramin, *Finite-dimensional pointed Hopf algebras with alternating groups are trivial*, Ann. Mat. Pura Appl. (4) **190** (2011), no. 2, 225–245, DOI 10.1007/s10231-010-0147-0. MR2786171

[AFGV4] Nicolás Andruskiewitsch, Fernando Fantino, Matías Graña, and Leandro Vendramin, *Pointed Hopf algebras over the sporadic simple groups*, J. Algebra **325** (2011), no. 1, 305–320.

[AG] Nicolás Andruskiewitsch and Matías Graña, *From racks to pointed Hopf algebras*, Adv. Math. **178** (2003), no. 2, 177–243.

[AS] Nicolás Andruskiewitsch and Hans-Jürgen Schneider, *Pointed Hopf algebras*, New directions in Hopf algebras, Math. Sci. Res. Inst. Publ., vol. 43, Cambridge Univ. Press, Cambridge, 2002, pp. 1–68, DOI 10.2977/prims/1199403805. MR1913436

[CGI] Giovanna Carnovale and Agustín García Iglesias, *θ-semisimple classes of type D in* $\mathrm{PSL}_n(q)$, J. Lie Theory **26** (2016), no. 1, 193–218. MR3384988

[FV] F. Fantino and L. Vendramin, *On twisted conjugacy classes of type D in sporadic simple groups*, Hopf algebras and tensor categories, Contemp. Math., vol. 585, Amer. Math. Soc., Providence, RI, 2013, pp. 247–259, DOI 10.1090/conm/585/11616. MR3077241

[MS] Alexander Milinski and Hans-Jürgen Schneider, *Pointed indecomposable Hopf algebras over Coxeter groups*, New trends in Hopf algebra theory (La Falda, 1999), Contemp. Math., vol. 267, Amer. Math. Soc., Providence, RI, 2000, pp. 215–236, DOI 10.1090/conm/267/04272. MR1800714

[S] Michio Suzuki, *Group theory. I*, Grundlehren der Mathematischen Wissenschaften, vol. 247, Springer-Verlag, Berlin-New York, 1982. Translated from the Japanese by the author. MR648772

Math, Science, and Technology Center, Paul Laurence Dunbar High School, Lexington, Kentucky

Email address: `jvulakh@mit.edu`

PUBLISHED TITLES IN THIS SUBSERIES

813 **Thomas Creutzig, Julia Plavnik, and David Ridout, Editors,** Quantum Symmetries, 2025

761 **Ibrahim Assem, Christof Geiß, and Sonia Trepode, Editors,** Advances in Representation Theory of Algebras, 2021

760 **Olivier Collin, Stefan Friedl, Cameron Gordon, Stephan Tillmann, and Liam Watson, Editors,** Characters in Low-Dimensional Topology, 2020

750 **A. Bourhim, J. Mashreghi, L. Oubbi, and Z. Abdelali, Editors,** Linear and Multilinear Algebra and Function Spaces, 2020

743 **H. Garth Dales, Dmitry Khavinson, and Javad Mashreghi, Editors,** Complex Analysis and Spectral Theory, 2020

680 **Sergei Gukov, Mikhail Khovanov, and Johannes Walcher, Editors,** Physics and Mathematics of Link Homology, 2016

655 **A. C. Cojocaru, C. David, and F. Pappalardi, Editors,** SCHOLAR—a Scientific Celebration Highlighting Open Lines of Arithmetic Research, 2015

654 **Carlo Gasbarri, Steven Lu, Mike Roth, and Yuri Tschinkel, Editors,** Rational Points, Rational Curves, and Entire Holomorphic Curves on Projective Varieties, 2015

638 **Javad Mashreghi, Emmanuel Fricain, and William Ross, Editors,** Invariant Subspaces of the Shift Operator, 2015

630 **Pierre Albin, Dmitry Jakobson, and Frédéric Rochon, Editors,** Geometric and Spectral Analysis, 2014

622 **S. Ejaz Ahmed, Editor,** Perspectives on Big Data Analysis, 2014

606 **Chantal David, Matilde Lalín, and Michelle Manes, Editors,** Women in Numbers 2, 2013

605 **Omid Amini, Matthew Baker, and Xander Faber, Editors,** Tropical and Non-Archimedean Geometry, 2013